T0231447

# THE DYNAMICS OF VEHICLES

## on roads and on tracks

# THE DYNAMICS OF VEHICLES
## on roads and on tracks

Edited by
H.-P. WILLUMEIT

*Proceedings of 6th IAVSD Symposium held at the Technical University Berlin, September 3-7, 1979*

*International Association for Vehicle System Dynamics affiliated to the International Union of Theoretical and Applied Mechanics*

Taylor & Francis
Taylor & Francis Group
LONDON AND NEW YORK

Published by Taylor & Francis
2 Park Square, Milton Park, Abingdon, Oxon, OX14 4RN
270 Madison Ave, New York NY 10016

Transferred to Digital Printing 2007

ISBN 90 265 0327 X

© Copyright 1980 by Taylor & Francis

No part of this book may be reproduced in any form, by print, photoprint, microfilm or any other means without written permission from the publishers.

**Publisher's Note**
The publisher has gone to great lengths to ensure the quality of this reprint but points out that some imperfections in the original may be apparent

# CONTENTS

# INTRODUCTION

H.-P. Willumeit

Institute of Automotive Engineering
Technische Universität Berlin, F.R.G.

This IAVSD Symposium on Dynamics of Vehicles on Roads and Tracks was held at the Technical University Berlin as host and organized by the Institute of Automotive Engineering.

The following scientists were appointed by the board of the IAVSD to serve as members of the Scientific Committee: M. Apetaur (CSSR), N.K. Cooperrider (USA), K. Knothe(Fed. Rep. Germany), P. Michelberger (Hungary), M. Mitschke (Fed. Rep. Germany), H.P. Pacejka (The Netherlands), A.D. de Pater (The Netherlands), L. Rus (CSSR), L. Segel (USA), R.S. Sharp (United Kingdom), S. Slibar (Austria), A. Watari (Japan), A.H. Wickens (United Kingdom), H.-P. Willumeit (Fed. Rep. Germany) chairman.

A total number of 37 papers was presented orally; these papers are printed in this publication. Five state-of-the-art papers were made available in printed form only in the Journal Vehicle System Dynamics, Volume 8, Number 4, September 1979.

Out of the 37 papers presented orally and the five state-of-the-art papers, one author came from Australia, one from Canada, four from Czechoslovakia, fourteen from Germany, four from Japan, four from the Netherlands, one from Poland, one from Sweden, four from the United Kingdom and seven from the United States of America. The 80 participants invited on behalf of the Scientific Committee came from 12 different countries all over the world.

The dominant theme of the symposium was the dynamics of systems of vehicles on roads and tracks. Specific attention was given to the presentation of new methods and techniques to be used by dealing with the problems of system dynamics.

The topics horizontal dynamics, stability, control analysis, vehicle ride on rigid and elastic support were covered during seven scientific sessions, three of which were devoted to the mechanics of road vehicles, two to the mechanics of track vehicles and two to both types of vehicles.

To give the participants the opportunity to also get acquainted with the city of Berlin two scientific sessions were grouped according to the different types of vehicles, ranging from motorcycle to truck-trailer systems and magnetically levitated trains. Technical excursions to the Berlin Transport Company (BVG), Waggon Union (Manufactury of Railway Cars), and the Berlin Traffic Guide Centre completed the scientific sessions.

The sessions were chaired by M. Apetaur (Prague), M. Mitschke (Brunswick), H.B. Pacejka (Delft), A.D. de Pater (Delft), L. Segel (Ann Arbor), A. Slibar (Vienna), A.H. Wickens (Derby).

A local organizing committee (H.-P. Willumeit, T. Clemens, U. Kramer, W. Neubert) was responsible for the day-to-day planning of the meeting and for the social activities. Due to the good work of this committee all participants - accompanying ladies included - really enjoyed the symposium.

We very much appreciate the contributions made by the following organisations, institutions, and companies:

Senator für Wirtschaft und Verkehr, Berlin
Bayerische Motoren Werke AG, München
Berliner Verkehrs-Betriebe (BVG)
Daimler-Benz AG, Stuttgart
Deutsche Forschungs- und Versuchsanstalt für Luft- und Raumfahrt, Köln
Ford of Europe, Essex
Gesellschaft von Freunden der Technischen Universität Berlin e.V.
Maschinenfabrik Augsburg-Nürnberg AG (MAN), München
Adam Opel AG, Rüsselsheim
Schwäbische Hüttenwerke GmbH, Aalen-Wasseralfingen
Volkswagenwerk AG, Wolfsburg
Wegmann & Co., Waggonfabrik und Fahrzeugbau, Kassel

Berlin, October 1979 H.-P. Willumeit

State-Of-The-Art-Papers (summaries)

# OPTIMIZATION CRITERIA FOR VEHICLES TRAVELLING ON A RANDOMLY PROFILED ROAD – A SURVEY

Tore Dahlberg

Division of Solid Mechanics, Chalmers University of Technology, Gothenburg, Sweden

SUMMARY

A 5-DOF plane vehicle model is studied. A randomly profiled road is assumed to impart normally distributed stationary vertical random displacements to the front and rear wheels. Several vehicle performance criteria, based on response mean square spectral densities, are discussed. It is emphasized that these performance criteria contain more information than do the simple response standard deviations. A computer program has been developed for optimization of two or more of the system parameters to make a vehicle response (or a weighted sum of responses) a minimum. Constraints on parameters and responses can be introduced. Ride comfort, road holding, energy absorption, fatigue failure and first-passage failure are studied.

## 1. INTRODUCTION

Random responses of a linear road vehicle model due to stationary random excitation are obtained by using the technique of input-output relation for mean square spectral densities.

The response spectral densities could serve as a basis for subjective judgement of the vehicle performance. It is also possible to integrate the spectral density to obtain the variance of the output signal. The variance is of course also a measure of the vehicle performance and can be used for design purposes. When integrating, however, some of the information contained in the response mean square spectral density is lost. In this paper some proposals are made describing how more of the information in the output signal can be retained while still having only one single scalar quantity as a design variable.

## 2. VEHICLE MODEL. EXCITATIONS. RESPONSES

The vehicle model used is a plane linear 5-DOF system (Fig. 1). The system is excited at the front and rear wheels by a random road profile.

It is assumed that the random road profile can be described as in [1]. The mean square spectral density of the displacement excitation $q_k$, $k=1,2$, then is

$$S_{qk}(\omega) = S_q(\omega) = S(\omega_0)\,(\omega_0/\omega)^{2.0} \qquad |\omega| \le \omega_0$$
$$= S(\omega_0)\,(\omega_0/\omega)^{1.5} \qquad |\omega| > \omega_0 \qquad (1)$$

Here $\omega$ is the radian frequency variable and $S(\omega_0)$ ($m^2$s/rad) the mean square spectral density at a datum radian frequency $\omega_0$. The value of $S(\omega_0)$ depends on the road quality and the vehicle speed v.

When studying responses to multiple excitations, also cross-spectral densities $S_{qk,q\ell}(\omega)$, $k \neq \ell$, are needed. The excitation of a two-wheel vehicle model can be summarized in matrix form as

$$[S_{qk,q\ell}(\omega)] = S_q(\omega)\begin{bmatrix} 1 & e^{-i\omega a_1/v} \\ e^{i\omega a_1/v} & 1 \end{bmatrix} \qquad (2)$$

The multiplier $S_q(\omega)$ is given in Eq. 1, $a_1$ is the wheel base distance, and i is the imaginary unit ($i=\sqrt{-1}$).

When using the technique of input-output relation for spectral densities, harmonic transfer functions are needed. They are here denoted by $H_{.,.}(\omega)$. For example, $H_{ri,q2}(\omega)$ gives the response $r_i$ due to a unit amplitude sinusoidal excitation $q_2$, with radian frequency $\omega$.

The auto- and cross-spectral density matrix of the responses can be written as, $i,j = 1,2...$,

$$[S_{ri,rj}(\omega)] = [H_{ri,qk}(\omega)][S_{qk,q\ell}(\omega)][H_{rj,q\ell}(\omega)]^* \qquad (3)$$

Here the asterisk indicates complex conjugate and transpose.

The standard deviation $\sigma_{ri}$ of a response $r_i$ with zero mean is the square root of the integral of the auto-spectral density $S_{ri,ri}(\omega)$. The mean square spectral density of the time derivative $\dot{r}_i(t)$ of the stationary random process $r_i(t)$ is $S_{\dot{r}i,\dot{r}i}(\omega) = \omega^2 S_{ri,ri}(\omega)$.

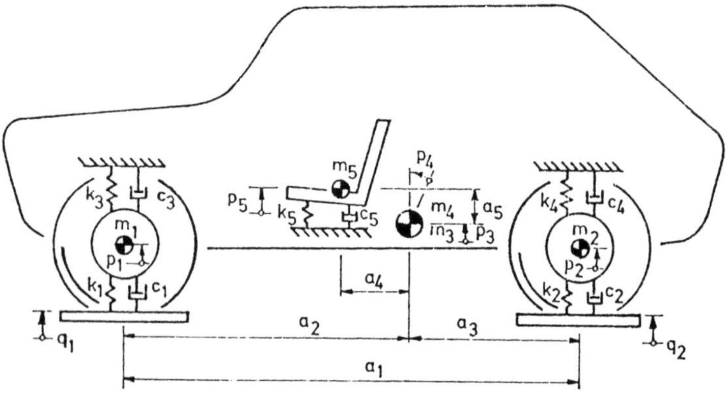


Fig. 1. Five-degree-of-freedom vehicle model. Excitations $q_1$ and $q_2$, displacements $p_1$ to $p_5$ (generalized), masses $m_1$ to $m_5$ (generalized), spring stiffnesses $k_1$ to $k_5$, damper stiffnesses $c_1$ to $c_5$, and distances $a_1$ to $a_5$.

## 3. DESIGN CRITERIA

The design criteria discussed are all based on one-figure vehicle performance measures, and they contain more of the information in the response spectral density than do the simple response standard deviation.

### 3.1 *Extreme Value*

The expected largest (or smallest) value of a response might be a better design variable than its standard deviation. The mean C of the largest value of a stationary Gaussian random process r (with zero mean value) in a time period of length T is approximately [2] (for $\ln n \gg 1$)

$$C = (\sqrt{2 \ln n} + \gamma/\sqrt{2 \ln n})\,\sigma_r \tag{4}$$

Here $\gamma$ is Euler's constant ($\gamma = 0.5772...$). Further $n = \sigma_{\dot{r}} T / 2\pi\sigma_r$ is the mean number of positive mean value crossings of the process r in the time period T.

### 3.2 *Probability of Exceedance*

Another design variable could be the probability $P_0$ that a random process r exceed a given level $r_0$ at least once in a time period of length T. Then the probability $P_0$ is composed of the two probabilities $P_1$ and $P_2$. Here $P_1$ is the probability that the stationary random process lies above $r_0$ at the starting time t=0, and $P_2$ the probability that the process r crosses the level $r_0$ at least once in the time period $0 < t \leq T$. One has approximately [3]

$$P_1 = (\sigma_r / r_0 \sqrt{2\pi})(1 - \sigma_r^2 / r_0^2) \exp(-r_0^2 / 2\sigma_r^2) \tag{5a}$$

$$P_2 = 1 - \exp(-[(\sigma_{\dot{r}} T / 2\pi\sigma_r) \exp(-r_0^2 / 2\sigma_r^2)]) \tag{5b}$$

The required compound probability $P_0$ is

$$P_0 = P_1 + (1 - P_1) P_2 \tag{5c}$$

The equations (5a,b) require that $r_0 \gg \sigma_r$, and that the level crossings can be considered independent.

The probability $P_0'$ that a process with zero mean will either exceed $r_0$ or be less than $-r_0$ can be obtained in a similar way [2].

### 3.3 *Energy Absorption*

It might be desirable to make a damper absorb as much (or as little) energy as possible. Let the system response r studied be the displacement difference between the two end points of a damper. The mean $E[\dot{W}]$ of the rate $\dot{W}$ of the energy absorption W in the damper is [4]

$$E[\dot{W}] = E[c_r \dot{r}^2] = c_r \int_{-\infty}^{\infty} \omega^2 S_r(\omega)\, d\omega = c_r \sigma_{\dot{r}}^2 \tag{6}$$

Here $c_r$ is the damper stiffness. Note that the process $\dot{r}$ has zero mean. The optimization problem then is to make $c_r\sigma^2_{\dot{r}}$ as large (or as small) as possible.

### 3.4 *Fatigue Failure. First-Passage Failure*

The mean time to fatigue failure or first-passage failure can also be estimated. The standard deviations of a stress random process and its time derivative are then needed. For formulas and restrictions, the reader is referred to the full version of this paper [5].

## 4. COMPUTER PROGRAM

A computer program for vehicle optimization has been developed. The program optimizes two or more of the components in the linear vibrating system with respect to any of the design variables described, or with respect to a weighted sum of these variables. Constraints on the design variables and on the system parameters can be introduced.

## 5. CONCLUDING REMARKS

It is concluded that the mean square spectral densities of vehicle responses can be used to formulate a variety of response functions to be used in design criteria. Extrema of these functions will give optimal system parameters.

It was found that a large amount of work can be saved in optimization procedures by using standard subroutines both for the solution of a complex linear system of equations and for the integrations and the minimization.

## 6. REFERENCES


1. ISO/TC108/WG9 (BSI Panel MEE/158/3/1), Generalized terrain-dynamic inputs to vehicles, June 1972.
2. Dahlberg, T., Parametric optimization of a 1-DOF vehicle travelling on a randomly profiled road. Journal of Sound and Vibration 55 (1977), pp 245-253.
3. Dahlberg, T., Ride comfort and road holding of a 2-DOF vehicle travelling on a randomly profiled road. Journal of Sound and Vibration 58 (1978), pp 179-187.
4. Åkesson, B.Å., Dynamic damping of wind-induced stochastic vibrations. ASME Design Engineering Technical Conference, Washington D.C. 1975, Paper 75-DET-10.
5. Dahlberg, T., Optimization criteria for vehicles travelling on a randomly profiled road - a survey. Vehicle System Dynamics 8 (1979), 239-252.

# RIDE DYNAMICS OF ARTICULATED VEHICLES – A LITERATURE SURVEY

M. A. Dokainish[1] and M. M. ElMadany[2]

[1] Department of Mechanical Engineering, McMaster University,
Hamilton, Ont.
[2] Wyle Laboratories, Colorado Springs, CO., U.S.A.

## INTRODUCTION

Over the past two decades, there has been an increased interest in the betterment of ride quality in the large articulated vehicles presently in use on our roads. The study of ride quality requires an assessment of the response of the vehicle to the inputs at the road surface. This mathematical assessment is normally termed the study of "ride dynamics" which is defined as the response in terms of mechanical vibration in the various components of the vehicle, to the random roughness that may be found in any given section of road. The mechanical vibration may lead to driver discomfort, wear and damage to vehicle components, damage or destruction of the cargo, and possibly to traction loss, and it is for these reasons that the betterment of ride quality is sought.

Two factors have contributed to the widening of interest and discussion of the ride quality in the particular form of vehicle. First, the advent of computer systems capable of dealing with complex mathematical problems has simplified the solution of the equations describing the vehicle motions.

Secondly, a lage body of relevant work is available describing the ride characteristics of simpler vehicles, such as automobiles and light trucks and these provide a ready background to an understanding of the more complex motion of the articulated vehicles.

It is the purpose of this paper to draw together, in summary form, the technical literature bearing on articulated vehicle analysis and ride comfort and safety. Thus, all phases of the problem, as described in the various papers, are included, with emphasis being placed on the means of describing the road input, the dynamics of articulated vehicles, and the methods of solution to obtain the vibrational characteristics of the vehicles. Finally, the literature presenting the indices of ride comfort and safety is reviewed and the relation of these indices to the results of the analysis is presented.

## GENERAL PROBLEM DESCRIPTION

A tractor-semitrailer vehicle moving on a road represents a complex vibratory system. It consists of the tractor and semitrailer masses coupled at the fifth wheel, supported on springing and damping elements, resting on axles and tires, and moving on the road which represents an excitation input to the tires. This complex system can be analyzed by dealing wiith the road and the vehicle elements separately and then considering the combined dynamic response. This dynamic response together with the characteristics of the vehicle are used to evaluate the ride quality, and, combined with the performance criteria, may be used to assess the ride comfort and safety.

## ROAD DESCRIPTION

A given section of road represents a specific input to a vibrating system. It is however impractical to attempt to determine the input waveform. Equally it becomes impossible to assess the profile of all roads likely to be traversed by a particular vehicle. When it is considered as well that seasonal changes in profile and surface degeneration might change the character of the road, it becomes logical to treat the road input to the vehicle system as a random phenomenon.

Three analytical models of increasing complexity have been used to describe statistically the road surface undulations. The first model describes the road as a cylindrical surface which can be defined by means of a single longitudinal track. The second model considers the surface consisting of random fluctuation in the longitudinal direction with a constant lateral slope which varies randomly for each longitudinal increment. In the third model, the road surface is considered as part of a two dimensional completely homogeneous and isotropic random process.

## ARTICULATED VEHICLE DYNAMIC ANALYSIS

Articulated vehicles have a substantially different ride performance when compared to passenger cars. The differences are due to many factors, but the prime difference is the result of the vast increase in mass and inertia. In general, suspension systems are simpler, and the power unit is designed for the heavy vehicle, with operation at much reduced power to weight ratios. Finally the ride is affected by the vehicle weight difference that will be encountered in laden and unladen condition.

Ride improvement for the cargo should not be overlooked as it is important as the ride improvement of the tractor. Both the tractor and the semitrailer are judged by their riding quality. The knowledge of the interaction of tractor and semitrailer is important. Results based on investigations of either tractor or semitrailer separately can easily lead to wrong conclusions. For example, as the semitrailer has four or five times the sprung weight of the tractor in the laden case, the semitrailer dictates the tractor ride.

The backbone of the vehicle is the frame and it has a definite influence on the vibratory motion of the vehicle. The beam vibration in the main chassis frame may result in an unpleasant motion of the cab, known as "cab-nod".

The cab is subjected to induced vibration from the beaming and pitching. It is necessary, therefore, to isolate the cab from frame vibrations while avoiding amplication of the lower frequency pitch motion of the chassis on its suspension. To accomplish these requirements it is necessary that the isolator should have low spring rates and optimum damping elements.

There are basically two types of cabs; cab-over-engine and conventional cabs and the ride characteristics may be influenced by the way the cab is mounted to the chassis.

Articulated vehicle suspension systems may vary considerably in detail, but normally comprise a spring, damper, links and rubber bushes. The main functions of the suspension are: (a) to provide vehicle support and directional control; (b) to provide effective vibration isolation from road disturbances.

As well, the suspension must be insensitive to externally applied loads. For example, cross winds and cargo weight variations should result in a minimum of vehicle vibratory motion.

The requirements necessary to fulfill these various functions have always conflicted to various degrees. Guidance requires a suspension that is neither very stiff nor very soft in order to maintain good road holding. Insensitivity to external loads requires a stiff suspension, whereas vibration isolation demands a soft suspension.

The appreciable increases in vehicle speeds and also in the laden weights have led to increase in the energy to be dissipated by vehicle suspension systems, and consequently further damping is required. This damping is essential to control vehicle motion on the suspension.

The unsprung mass (axles with their wheels and tires) constitutes an independent vibrating system in which the unsprung mass bounces between the tires and suspension springs with these spring elements acting in parallel. This vertical oscillation of the tire and wheel assembly is referred to as wheel hop. This oscillation is effectively damped by the relatively high interleaf friction of the laminated springs. Axle frequencies fall in the range of frame bending frequencies and therefore axle motions produce exciting forces at or near resonance with the frame vibration.

The tires act as vertical suspension units. They provide the primary means of support and guidance for the vehicle, and determine the vibration level transmitted to the vehicle body from the road and the force level to the road.

## RIDE COMFORT AND RIDE SAFETY

Two major factors must be considered in evaluating the ride performance of a vehicle; first, the ride comfort, both of the driver and any passenger, as well as cargo, and second, the ride safety. Ride comfort is normally interpreted as the capability of the vehicle suspension, in the particular vehicle configuration, to maintain the motion within the range of human comfort and, as well, within the range necessary to assure that there is no damage to the cargo. Ride safety is defined as the ability of the suspension to assure wheel-to-road surface contact under rough road conditions.

It is possible to obtain good ride quality while making the vehicle difficult to control, or, conversely, to obtain excellent vehicle control while destroying the driver comfort. It is clear that the design of the vehicle suspension must be a practical compromise between ride comfort and ride safety. Both performance criteria have to be considered in assessing the ride quality of any vehicle.

## REFERENCES


1. Anon., "Guide for the evolution of human exposure to whole body vibration", International Standard ISO 2631-1974(E), International Organization for Standardization, New York.

2. Bekker, M.G., "Introduction to terrain-vehicle systems", Ann Arbor, The University of Michigan Press, 1969.

3. Dodds, C.J., and Robson, J.D., "The response of vehicle components to random road surface undulations", Proc. 13, F.I.S.I.T.A. Congress, Paper 17-2D, Brussels, 1970.

4. Hanes, R.M. "Human sensitivity to whole-body vibration in urban transportation systems; A literature review", Applied Physical Laboratory for DOT, No. APL/JHU-TPR 004, May 1970.

5. Mitschke, E.M., "Influence of road and vehicle dimensions on the amplitude of body motions and dynamic wheel loads (theoretical and experimental vibration investigations)", SAE, 1962, 70, pp. 434-496.

6. Potts, G.R., and Walker, H.S., "Nonlinear truck ride analysis", ASME Transaction, Journal of Engineering for Industry, May 1974, pp. 597-602.

7. Van Deusen, B.D., "Truck suspension system optimization", SAE, Paper No. 710222.

# EFFECTS OF FREE-CONTROL VARIABLES ON AUTOMOBILE HANDLING

M. C. Good

Department of Mechanical Engineering, University of Melbourne, Parkville, Vict., Australia

Considerable efforts have been devoted over the years to determine those characteristics of an automobile which primarily determine the ease and precision with which drivers may exercise directional control - in other words, its lateral handling quality. The great majority of such investigations have concentrated on the characteristics of the vehicle's response to inputs of steering wheel angle. A review of the effects of these 'fixed-control' response characteristics has been given previously [1]. However, driver control could just as well be regarded as being by way of steering force (or torque) applied to the steering wheel. As the corresponding 'free-control' dynamic response of the automobile is quite different from its fixed-control response, it is of interest to enquire as to what effect, if any, free-control characteristics have on handling quality. That free-control responses can be important in vehicular control is evidenced by the motorcycle, for which control of roll angle by way of steer torque is much more satisfactory than by steer angle.

The relative neglect of free-control handling characteristics is possibly explained by two considerations. Firstly, drivers are able to adapt to a wide range of free-control characteristics with little change in steering task performance, whereas changes in fixed-control parameters can have a dramatic effect on the success or otherwise of manoeuvres. Secondly, the study of vehicle responses to steering torque inputs is made more difficult by the additional degrees of freedom associated with steering - and front-wheel motions, and by the non-linearities introduced by Coulomb friction in the steering system.

The small number of experimental studies of the effect of free-control variables on handling are reviewed in this paper. In view of the paucity of such studies, however, an attempt is made to establish a conceptual framework for assessing the results, by first bringing together some findings from research into manual control in compensatory tracking situations. In such situations the task of the human operator is to reproduce the system input in the system output as closely as possible over the frequency bandwidth of the input, despite disturbances. The dynamics of the controlled plant are represented by the transfer function $Y_c(j\omega)$, while the controller's operations on the system error are represented by a quasi-linear describing function $Y_p(j\omega)$ and an additive, uncorrelated output, the remnant. The operator's control output may be a manipulator position or a force exerted on the manipulator. From the research reviewed in this paper it appears that:

(i) For successful tracking and disturbance regulation the form of the open-loop transfer function $Y_p\ Y_c$ is constrained in the vicinity of the unity gain crossover frequency $\omega_c$.

(ii) The adaptive equalization characteristics of the human operator are ordinarily limited to gain changes and the generation of a single lead and a single lag.

(iii) Operators can choose between proprioceptive feedbacks of limb force and position to control their output force (a free-control strategy) or output displacement (a fixed-control strategy).

(iv) The form of the controlled plant dynamics $Y_c$ and the limb/manipulator dynamics $G_L$, together with the constraints just outlined, determine whether a fixed- or free-control strategy will be employed.

(v) For complex plants in which control forces are applied by servo actuators, with input signals derived from manipulator force or position, improved tracking performance and reduced workload can be achieved by presenting the operator with effective fixed-control plant dynamics of $\hat{Y}_c = K_c$.

(vi) There is no evidence that, when the operator is employing a fixed-control strategy, the provision of control 'feel' proportional to the control effort on the plant improves performance over that which would be obtained with a spring-restraint manipulator. Nor are data available as to the subjective response of operators to this type of control 'feel'. Such kinaesthetic feedback may be useful however in improving overall mission performance. For example, it may inform the operator of non-linearities and changes in plant dynamics and disturbances, and allow the formulation of control strategies which take actuator effort into account.

(vii) Objective measures of system performance are relatively insensitive to manipulator force characteristics when the operator adopts a fixed-control strategy, especially for simple plant dynamics. However, large manipulator inertias can cause performance losses and, for more difficult plant dynamics, there appear to be optimal manipulator spring rates.

(viii) There is a lack of information on the subjective response of human operators to manipulator force characteristics.

Application of these findings to the case of automobile steering control suggests that, where precise control of the vehicle path is required, drivers will adopt a fixed-control strategy in which they apply steering wheel position inputs to the vehicle, regardless of steering forces. Tight proprioceptive feedback of limb position then enables drivers to maintain levels of steering task performance relatively constant over wide ranges of steering force characteristics. There is some evidence that in less demanding, 'relaxed' driving scenarios drivers may adopt a free-control strategy in which steering torque is controlled. The frequency bandwidth of steering control is considerably reduced by such a strategy.

Vehicle design parameters which distinguish automobile free-control from fixed-control behaviour include the inertia of the steering wheel and front wheels, compliance and damping (both viscous and Coulomb friction) in the steering system, caster, and the efficiencies and gear ratios of the steering gear box and linkages. As with vehicle fixed-control characteristics, however, generalizable results are best obtained if the vehicle is characterized by measures of the steady-state and transient response to steering inputs. The experimental studies of driver/vehicle performance and driver opinion which are reviewed in this paper indicates that steering task performance is relatively insensitive to the free-control characteristics of the vehicle; fixed-control parameters (in particular the yaw response time) have a much stronger influence.

Despite drivers' abilities to adapt to a wide range of free-control characteristics, they show a distinct preference for certain regions of steering torque gradient and for rapid responses of the steering wheel to torque inputs (consistent with low steering wheel inertia). Recent experiments show that, within fairly wide ranges, damping of the free-control oscillatory mode has little effect on handling.

Vehicle parameters are highly interactive in their effects on handling quality, so that specification of optimum free-control characteristics must take account of the fixed-control vehicle properties also. Ranges of the primary fixed- and free-control variables which will lead to descriptions of overall handling quality as 'good' and 'fair' have been reported as follows:

| | Rated 'good' | Rated 'fair' |
|---|---|---|
| Torque gradient | : 22 - 28 N m/G | 17 - 33 N m/G |
| Steering wheel inertia | : less than 0.08 kg $m^2$ | less than 0.13 kg $m^2$ |
| Yaw rate gain | : 0.28 - 0.37 $s^{-1}$ | 0.21 - 0.44 $s^{-1}$ |
| Yaw response time | : 0.12 - 0.17 s | 0.09 - 0.21 s |

From pure tracking considerations an ideal steering force characteristic would correspond to a spring restraint (the stiffness of which would have to vary with the fixed-control response sensitivity of the vehicle). However, a steering system which provides reliable feedback of tyre aligning torques through steering feel should lead to best overall driving performance because of the information it yields about disturbances and non-linearities in the vehicle dynamic response.

REFERENCE

1. GOOD, M.C. (1977). Sensitivity of driver-vehicle performance to vehicle characteristics revealed in open-loop tests. Veh. Syst. Dynamics 6, 245-277.

# SURVEY OF WHEEL-RAIL ROLLING CONTACT THEORY

J. J. Kalker

Department of Mathematics, Delft University of Technology,
Delft,
The Netherlands

SUMMARY

This paper describes the theory of frictional rolling contact as far as it is significant for the wheel-rail system. It is divided into two parts.
The first part, mostly non-mathematical, contains a historical survey from the times of Carter and Fromm (1926) to the present day, in which all aspects of rolling contact theory are discussed. Included are a quantitative account of the results of Hertz theory (sec. 3), and a table of the creepage and spin coefficients.
The second part gives a present day account of the simplified theory (sec. 4), and of the exact linear and non-linear theory (sec. 5).
The paper closes with some recommendations for future research, of which the most pressing is a thorough investigation of the accuracy of simplified theory.

# SOME BASIC PROPERTIES OF BOGIES, AN ANALYTICAL APPROACH

C. P. Keizer

Vehicle Research Laboratory
Delft University of Technology, Delft, The Netherlands

Dynamic stability of bogies for railway vehicles depends to a large extent on the elastic properties in the horizontal plane of the primary suspension. It was shown by Wickens [1] that the stabilising effect is governed by the elastic properties of the inter-wheelset connections and that any form of inter-wheelset structure or mechanism can be characterised by two independent parameters: the bending stiffness and the shear stiffness. This article is intended as an effort to determine relations between these stiffness parameters and some basic properties of a four-wheeled bogie.

As a first approach the low speed response to lateral track irregularities is considered. With a proper choice of some dimensionless parameters and after some simplifications, a relatively simple expression for the (low speed) transfer function of a bogie is found. This expression is completely symmetric in the dimensionless bending and shear stiffness parameters. In the special case where either the bending stiffness or the shear stiffness is high, both stiffness parameters can be replaced by one single dimensionless overall stiffness parameter K. This parameter can, in a numerical sense, be regarded as a "series arrangement" of the dimensionless bending and shear stiffness parameters. The expression for the low speed transfer function then becomes:

$$\left|\frac{\hat{y}}{\hat{y}_t}\right|^2 = \frac{K^2\left(\frac{b^2}{a^2+b^2}\right)^2 + \nu^2}{K^2\left(\frac{b^2}{a^2+b^2} - \nu^2\right)^2 + \nu^2(1-\nu^2)^2} \qquad (1)$$

in which:

$\hat{y}$ = amplitude of centre of bogie
$\hat{y}_t$ = amplitude of track irregularities
$a$ = bogie wheelbase
$b$ = lateral distance of contact points between wheel and rail
$\nu$ = dimensionless frequency

As can be seen from (1), the system has zero damping for the extreme cases K = 0 and K = ∞. A simple expression can be found for the value of K, for which the transfer function has its lowest possible maximum. This value has been termed the (low speed) optimum.

In the situation where neither the bending nor the shear stiffness is high (so total flexibility is more evenly distributed over bending and shear), expression (1) is no longer valid, but numerical examples show that the expression for the "optimum" K still gives a good approximation. At higher speeds, where the inertia effects of unsprung masses have to be taken into account, the expression for the transfer function becomes more complicated. It can be shown however that in the frequency range of interest, the magnitude of the transfer function will always increase with increasing unsprung masses and increasing speed. For a given overall stiffness parameter, this effect is stronger for a high bending stiffness (flexibility concentrated in the shear distortions) than for a high shear stiffness (flexibility mainly in bending).

As the denominator of the expression for the transfer function consists of the characteristic equation of the system, this will be completely undamped for the speed for which this equation is satisfied by a purely imaginary root. This speed is commonly called the "critical speed". Although the significance of this critical speed as a characteristic property of a bogie is very limited, it may be useful to study the effect of different parameters on critical speed.

It was first found by de Pater [2] that if the mass of the bogie frame is omitted (so only the wheelset masses remain) there exist certain combinations of values for the bending stiffness, the shear stiffness and the wheelbase to track gauge ratio for which the critical speed will be infinite. In this article, only the cases will be considered where either the bending or the shear stiffness is infinite. With the dimensionless stiffness parameters mentioned earlier, it is found that the characteristic equations for both cases are identical and that for larger values of the wheelbase, infinite critical speed is produced by a rather narrow range of values for the shear or bending stiffness (according to the case), this range becoming wider for increasing wheelbase.

This situation changes drastically when the inertia of the bogie frame is taken into account: the characteristic equations are no longer identical and in the case of infinite bending stiffness no values for the other parameters can be found which result in infinite critical speed. The bogie with infinite shear stiffness however retains its potential for infinite critical speed after inclusion of the frame inertia. In general it is found that, compared to a high bending stiffness, a high shear stiffness allows a higher critical speed and, at lower speeds, a slower deterioration of dynamic behaviour with increasing speed.

Although the dynamic behaviour of a complete vehicle will to a considerable extent be determined by the properties as found for a single bogie, the interaction between body and bogie will also have an important effect. In order to obtain a qualitative indication of this effect, a simplified system is considered, consisting of a bogie with a laterally sprung mass and, again, for the bogie the special cases of infinite bending stiffness and infinite shear stiffness are compared. The system as assumed has two modes of vibration: a body mode, which experiences an additional damping due to lateral creep of the wheelsets, and a bogie mode on which, at the speed where the natural frequencies of both modes coincide, the inertia forces of the sprung mass have a destibilising effect. In the case of infinite shear stiffness, this last effect depends on the wheelbase of the bogie: for zero wheelbase it is the same as in the case of infinite bending stiffness, with increasing wheelbase it diminishes and in the (hypothetical) case of infinite wheelbase it is completely absent.

It is well known that there is a conflict between the requirements for "perfect steering" on curves and for dynamic stability at higher speeds. Assuming that a certain value for the "overall stiffness parameter" will be required for dynamic stability, the distribution of total flexibility over bending and shear can be varied and the effect of these variations on curving performance studied. It is found that a "worst case" is formed by a concentration of all flexibility in the shear distortions (infinite bending stiffness). The largest range of curve radii which can be negotiated without flange forces is given by an intermediate distribution over bending and shear of the total flexibility; once a flange force has been established however the rate of increase of this flange force and of the angle of attack of the leading wheelset with further decreasing curve radius is minimal if all flexibility is concentrated in bending (infinite shear stiffness).

## REFERENCES

1. Wickens, A.H. , Steering and dynamic stability of rail vehicles. Vehicle System Dynamics 5 (1975/1976), pp. 15-46.
2. Pater, A.D. de, Private communication.

# ASSESSMENT OF THE INFLUENCE OF THE DAMPING NON-LINEARITIES ON THE BACKGROUND SOUND PRESSURE LEVEL IN A PASSENGER CAR

Milan Apetaur and František Opička

Czechoslovak Automotive Research Institute (ÚVMV), Prague, CSSR

SUMMARY

The "background" noise inside a passenger car is caused by riding over an uneven road. Forces acting between the wheel suspension and the car body excite vibrations of the body panels which create sound pressure in the car interior. The spectrum of these forces - and therefore of the interior noise - depends strongly on the non-linearities of the suspension damping. Generally every damping non-linearity increases the background interior noise at frequencies over 30 Hz. A qualitative analysis of the interior background sound pressure level was carried out in this study for the case of a passenger car with McPherson suspension travelling with one front wheel over a poor Macadam road. Results of the computation showed that suspension damping non-linearities must be taken into account when analysing the causes of the background noise in real cars.

## 1. INTRODUCTION

Noise in the passenger compartment is produced mainly by vibrations of the body panels enclosing the car interior. These vibrations are excited (a) by airborne noise from the engine, etc., (b) mechanically by forces produced by vehicle ride and engine action. Example of interior noise spectrum is shown on Fig. 1.

Mechanically excited noise is important practically at low frequency noise range up to approximately 500 Hz only. Its spectrum at constant vehicle speed is composed (a) by discrete components caused mainly by engine unbalance, by wheel unbalance, tyre non-uniformity, etc., (b) by continuous component of stochastic nature caused by riding over an uneven road-profile. This latter component is called "the background" noise [1]. It is generally dominating on rough roads whereas on good roads the discrete tone noise is prevailing.

This study is dealing with the background noise only.

The complex dynamic system of a car is composed of elastic bodies (carbody, tyres, suspension elements, springs, etc.) bound together by practically massless elastic elements (rubber bushes, rubber mounts, etc.) Motions and forces in this system are excited by riding over an uneven road. Forces acting between the carbody and suspension elements produce vibrations of the body as a whole, of its structure and of its panels which cause the sound pressure variations in the passenger compartment.

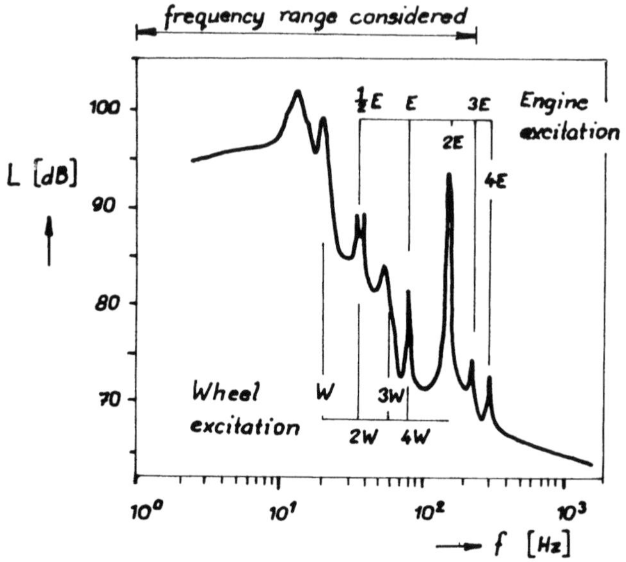


Fig. 1. Example of sound pressure level L in a car (loosely after [1]).

The influence of mainly longitudinal and side forces applied on the body by suspension links is generally small with the exception of these frequency ranges at which resonances of some subsystems (tyre, hun assembly, etc.) occur. These resonances will not be considered here.

The most dominant excitation of body vibrations originates in suspension springing and especially in its damping.

Suspension damping has a very pronounced non-linear character. Every non-linearity in a stochastically excited dynamic system causes an increase of high frequency components of its internal forces (see f.e. [3]). The composition of the background noise in a car is therefore influenced by any nonlinearity of its suspension damping.

The object of this study is to assess quantitatively the influence of damping non-linearities on the sound pressure level (SPL) of the interior background noise of a passenger car for the frequency range from 1 Hz to approximately 200 Hz. The study was conducted only theoretically and was not verified experimentally.

## 2. NON-LINEARITIES OF THE SUSPENSION DAMPING

Damping non-linearities have practically two causes, (a) non-linear relation between damping force and relative velocity of both ends of the hydraulic damper, (b) friction in the suspension and springing.

The hydraulic damping in the suspension damper is caused by the flow of the damper fluid through damper valves. The mass of the valves and of the fluid can be approximately neglected for frequencies of the relative damper deflections up to 200 Hz. The damping force $u_p$ can therefore be considered as depending on relative damper velocity $\dot{r}_p$ only, i.e.

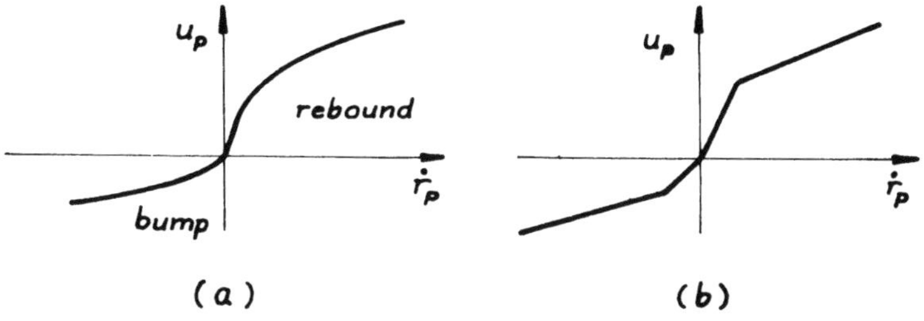


Fig. 2. Damper characteristics. (a) Real. (b) Idealised.

the damper properties can be described by its hydraulic damping characteristics $u_p(\dot{r}_p)$. This assumption does not hold for higher frequencies which however are not considered here [2].

Common dampers have degressive characteristics with different courses for bump and rebound (Fig. 2a). Non-linear characteristics were formed from straight lines in this study (Fig. 2b).

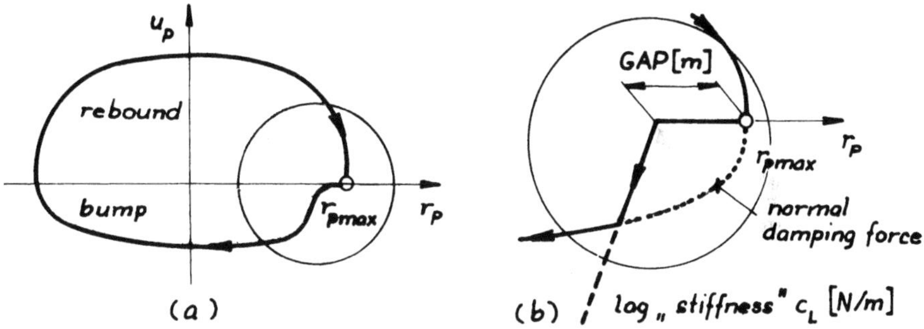


Fig. 3. Damper lag. (a) Real damper working diagram. (b) Lag modelling.

Damper misfunction occurs sometimes. Most common is the damper "lag". The damping force stays nearly zero for some deflection of the damper at the transition from rebound to bump and after overcoming this gap it reaches fastly its normal value directed by the damping characteristics and relative velocity. A typical working diagram (damping force $u_p$ - damper deflection $r_p$) of a damper with lag is shown on Fig. 3a.

Modelling of the damper lag in this study is shown on Fig. 3b. The damping force is zero for the damper deflection smaller than GAP after the transition point with maximum deflection $r_{pmax}$. After overcoming the GAP it increases linearilly (with "lag stiffness $c_l$") to reach its normal value.

It was supposed in this study that the damper lag is independent of the actual damper stroke or motion frequency, i.e. that the values of GAP and "lag stiffness" are constant in time.

Friction is present at every car suspension. It is summed up from frictional forces in the suspension damper, in ball joints of suspension and steering and eventually in the suspension spring and its attachment. In this study a McPherson suspension is discussed where a relatively high axial friction force can occur in the radially loaded telescopic strut.

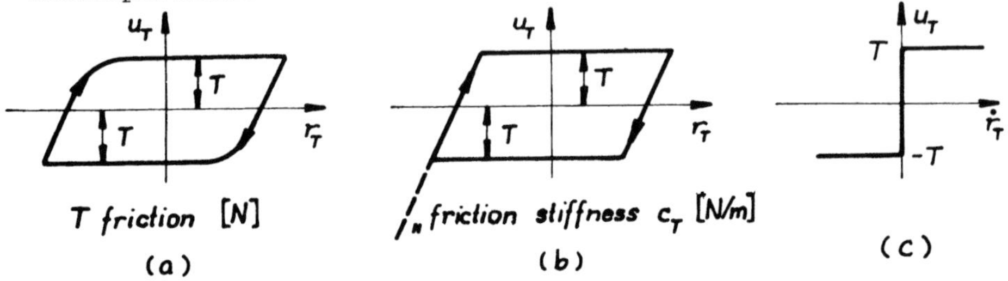


Fig. 4. Friction force. (a) Real. (b) Idealised. (c) Ideal characteristics.

A somewhat idealised real dependence of friction force $u_T$ on deflection $r_T$ of the friction element is shown on Fig. 4a. This dependence was modelled as shown on Fig. 4b. The friction is increasing linearilly (with "friction stiffness" $c_T$) after maximum deflection of the friction element was surpassed (and the sign of the relative velocity was changed) and after reaching the value of friction T it stays constant at further motion. Ideal friction $u_T$ = T sing $\dot{r}_T$ was used in the three mass model for computational reasons where apparently $c_T \to \infty$.

## 3. CAR BODY PROPERTIES

A single excitation of the car dynamic system by the road profile under one front wheel was supposed in this study. McPherson suspension was considered with helical spring concentric to the telescopic strut. Damping force of the suspension damper $u_p$, friction force $u_T$ and suspension spring force $u_S$ are acting together on the piston rod of the telescopic strut. The piston rod is mounted in the body by a rubber mount. Force $u_D$ is transmitted to the body from the piston rod through this rubber mount.

A single point excitation of the body vibrations by the force $u_D$ is resulting.

Two basic properties of the car body influence the interior noise, (a) its vibrational properties at the piston rod mount attachment, (b) its sound pressure response to force excitation at this attachment.

Vibrational properties of the body at any given point can be described in term of its driving point impedance Z [Ns/m] in function of the input force frequency f [Hz]. Results of measurements of driving point impedances in front parts of four bodies are shown on Fig. 5a. Their general course indicates that the body can be in the frequency range up to 200 Hz considered approximately as composed by parallel arrangement of a spring and a damper attached to the body mass, if smaller resonance effects are neglected (Fig. 5b).

Measured interior SPL responses to harmonic force input at front parts of three different bodies are shown on Fig. 6. If various resonances are disregarded the mean SPL response stays in the frequency range from 30 Hz to 200 Hz approximately constant. It is further known (see f.e. [4]) that interior sound pressure amplitude $p_a$ [Pa] is lineary dependent on the input force amplitude $u_{Da}$ [N].

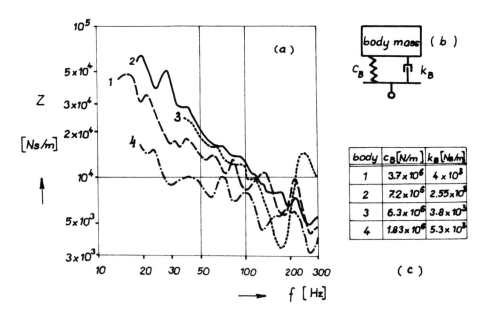

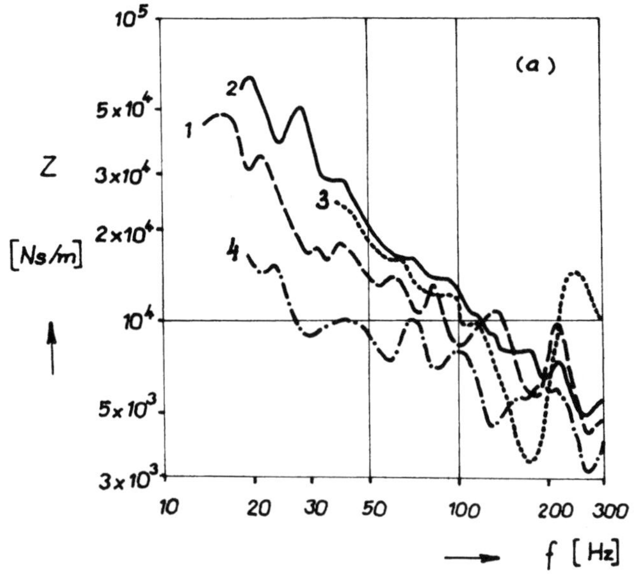


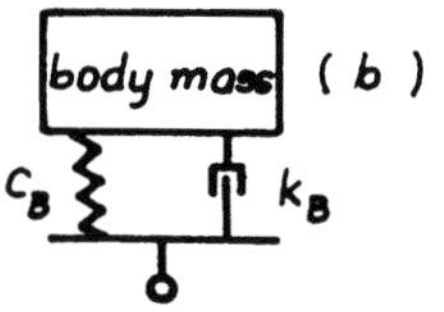


| body | $c_B$ [N/m] | $k_B$ [Ns/m] |
|---|---|---|
| 1 | $3.7 \times 10^6$ | $4 \times 10^3$ |
| 2 | $7.2 \times 10^6$ | $2.55 \times 10^3$ |
| 3 | $6.3 \times 10^6$ | $3.8 \times 10^3$ |
| 4 | $1.83 \times 10^6$ | $5.3 \times 10^3$ |

( c )

Fig. 5. (a) Measured driving point impedances Z of four car bodies: 1 - at the telescopic strut attachment point, 2, 3 - loosely after [5], 4 - loosely after [4] at the front engine mounting. (b) Idealised model. (c) Constants for the measured bodies.

As SPL

$$L = 10 \log \left(\frac{p_a}{2p_o}\right)^2 \quad \text{dB}$$

(where $p_o = 2.10^{-5}$ Pa) the transfer function of the relative sound pressure

$$H_p(f) = \frac{p_a/p_o}{u_{Da}}(f)$$

can be easily stated from the shown measurements. The course of the transfer function was approximated as shown on Fig. 7.

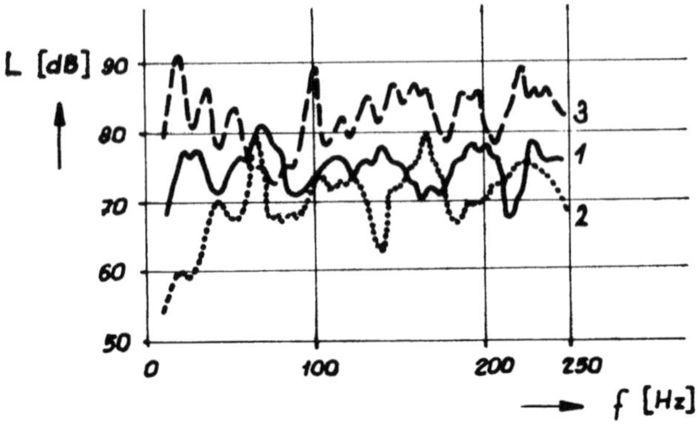


Fig. 6. Measured SPL responses L inside three different car bodies produced by constant force input with amplitude $u_{Da}$: 1 - at the telescopic strut attachment ($u_{Da}$ = 12 N), 2 - at front longeron vertically (loosely after [4], $u_{Da}$ = 6N), 3 - at front engine mounting (loosely after [1], $u_{Da}$ ; 16N).

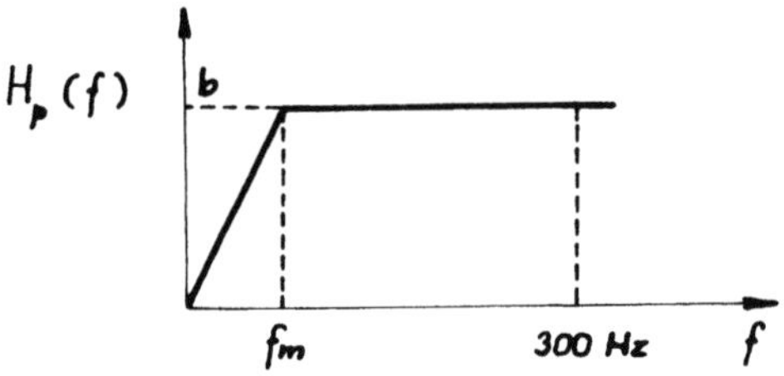


Fig. 7. Idealised relative sound pressure transfer function.

4. MATHEMATICAL MODEL AND COMPUTATIONAL PROCEDURE

The basic mathematical model used for the computations was composed by a three mass dynamic system (Fig. 8a) consisting of the body mass, wheel mass and piston rod mass (including piston rod bearing, spring support, etc.).

Road profile was taken as stochastic stationary function of distance with spectral density of common form. Its spectral density course was extrapolated to wavelengths of about 0.1 m. Vehicle speed was considered constant.

Tyre was modelled by a simple spring. Non-linearity originating in the wheel hop was included in the basic model but it did not have any practical effect under prescribed computational conditions. The property of the tyre to "swallow" very short road unevennesses (see f.e. [6]) was not taken into account.

Linear suspension spring characteristics was supposed.

The stiffness and damping of the piston rod rubber mount and of the carbody model were considered together as piston rod attachment formed by a spring with stiffness $c_D$ and a damper with damping coefficient $k_D$.

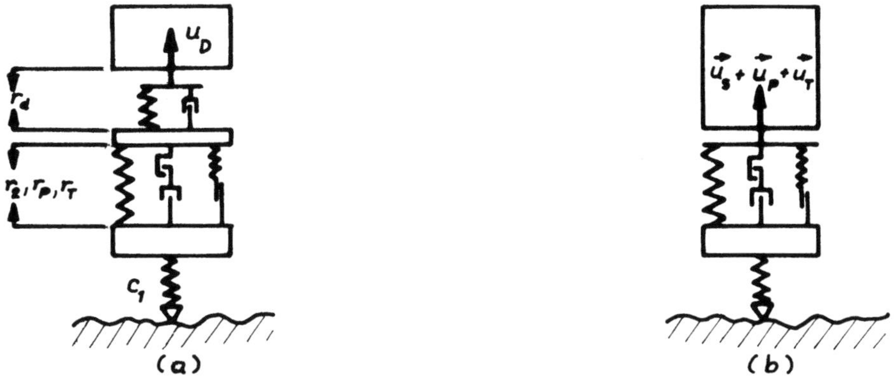


Fig. 8. Mathematical car model. (a) Three-mass model. (b) Two-mass model.

This simplification differs somewhat from reality. It is however mostly acceptable as the stiffness and especially the damping of the rubber mount are generally significantly lower than that of the body model (see Fig. 5c). The properties of the piston rod attachment are therefore mostly affected by the rubber mount parameters.

Movements and forces in this system were numerically solved in form of time-series.

The time-series of the road-profile was generated by a pseudostochastic Gaussian generator controlled for the demanded dispersion and power spectral density. The road-profile as continuous function was interpolated between the samples (with sampling frequency 600 Hz) by a 3rd grade polynom (by method of Hiroshi Akima).

The car dynamic system was described by three differential equations of second order. Their numerical integration was conducted by the Adams method with automatically varying integration step. The results in the form of time-series were given out synchronously and with the same sampling step as the time-serie of the road-profile.

The samll piston rod mass together with in practice high "friction stiffness" $c_T$ and "lag stiffness" $c_L$ form a partial vibratory system with high natural frequency (> 1000 Hz). The sampling frequency should be therefore much higher than indicated for successful integration of the differential equations of the three-mass system and the computational time would be very high. This model was therefore used with stiffness $c_T \to \infty$ and $c_L \to \infty$ only.

The computations of the influence of "friction stiffness" and "lag stiffness" were carried out in a simpler two-mass model (Fig. 8b). This model is a limiting case of the three-mass system when the piston rod attachment stiffness $c_D \to \infty$.

Time series of the force $u_D$ acting from the piston rod attachment on the body were the main results of computation in both cases.

Their autocorrelation functions were computed and from them power spectral densities $G_{uD}(f)$ of the suspension force were stated by Fourier transform.

Power spectral density $G_p(f)$ of the relative sound pressure was computed as

$$G_p(f) = H_p^2(f)\, G_{uD}(f)$$

Interior SPL is then

$$L(f) = 10 \log G_p(f) \qquad [\text{dB re } 2.10^{-5} \text{ Pa}]$$

5. NUMERICAL VALUES OF THE MODEL PARAMETERS

General parameters (Fig. 8): Body mass 350 kg. Wheel mass 50 kg. Piston rod mass 2 kg. Tyre stiffness $2{,}8.10^{-5}$ N/m. Suspension spring stiffness $1{,}4.10^4$ N/m. Piston rod attachment stiffness $c_D = 5.10^5$ N/m, ∞. Piston rod attachment damping coefficient $k_D$ = 110 Ns/m, 1100 Ns/m.

Damper lag modelling (Fig. 3): GAP = -1m (no lag), 0,0015 m, 0,003 m. Lag stiffness $c_L = 5.10^5$ N/m, $10^7$ N/m, ∞ .

Sound pressure response (Fig. 7): $b = 8.10^2$ 1/n , $f_m$ = 30 Hz.

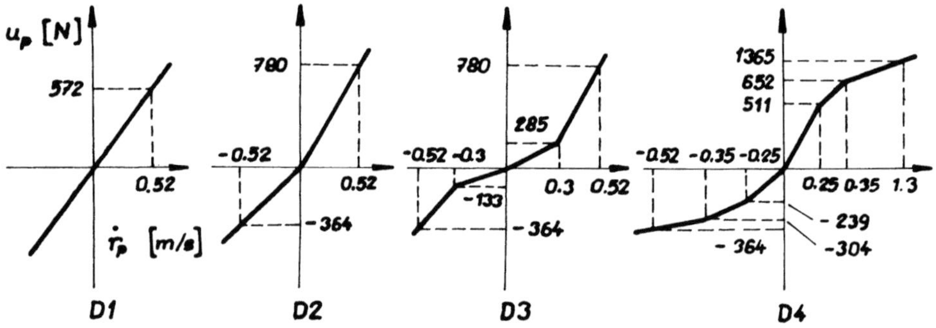


Fig. 9. Damping characteristics of the dampers considered.

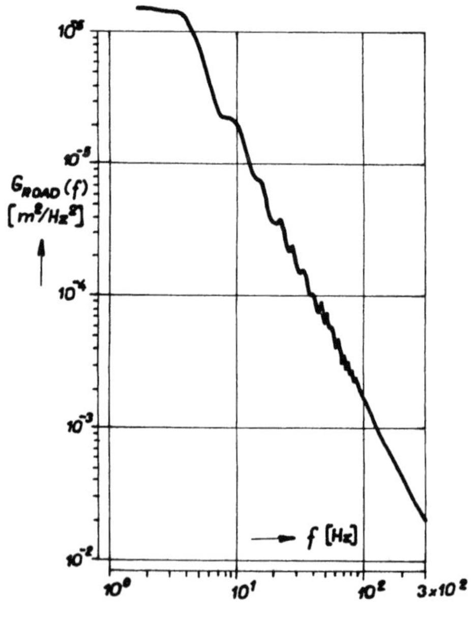


Fig. 10. Road profile spectral density.
Dispersion $1.10^{-4}$ $m^2$.

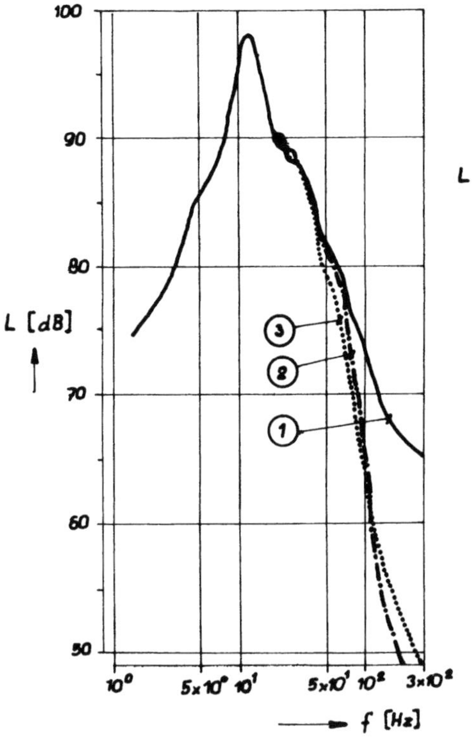


Fig. 11

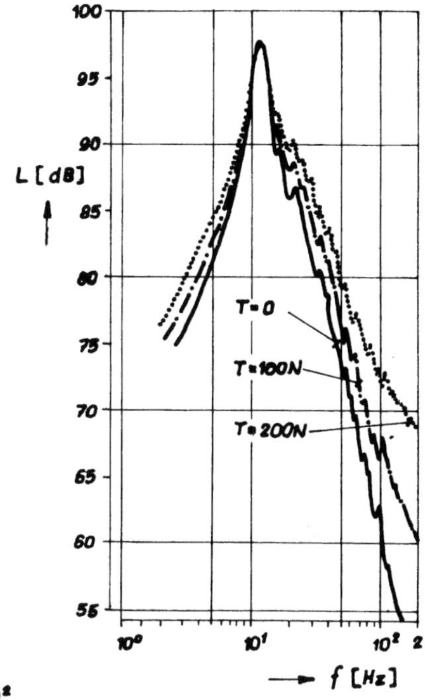


Fig. 12

Fig. 11. SPL for different piston rod attachment properties. Damper 4, GAP = -1m, T = 200 N, $c_T = \infty$ . 1 - two-mass model ($c_D = \infty$). 2 - three-mass model ($c_D = 5.10^5$ N/m, $k_D$ = 110 Ns/m). 3 - three-mass model ($c_D = 5.10^5$ N/m, $k_D$ = 1100 Ns/m).

Fig. 12. Influence of friction on SPL. Two-mass model. $c_T = 10^{10}$ N/m, damper 4, GAP = -1 m.

## 6. INFLUENCE OF DAMPING NON-LINEARITIES ON INTERIOR BACKGROUND SPL

The model in the following discussion corresponds by its parameters to a high-class passenger car travelling at 20 m/s over a poor Macadam road.

C o m p a r i s o n of the results gained with the three-mass model and with the two-mass model. The piston rod in the real car is attached to the elastic body by rubber mount of relatively small stiffness, which is taken into account by the three-mass model. It was already shown that - because of computational problems - some computations had to be carried out with the less realistic two-mass model. An objective evaluation of the results gained is therefore impossible without knowledge of the differences in SPL computed with the three-mass model and with the two-mass model.

The course of background SPL for a car system with common damping characteristics (damper 4) and normal friction (T = 200 N) is shown on Fig. 11 for the three-mass system (curves 2, 3) and for the two-mass system (curve 1).

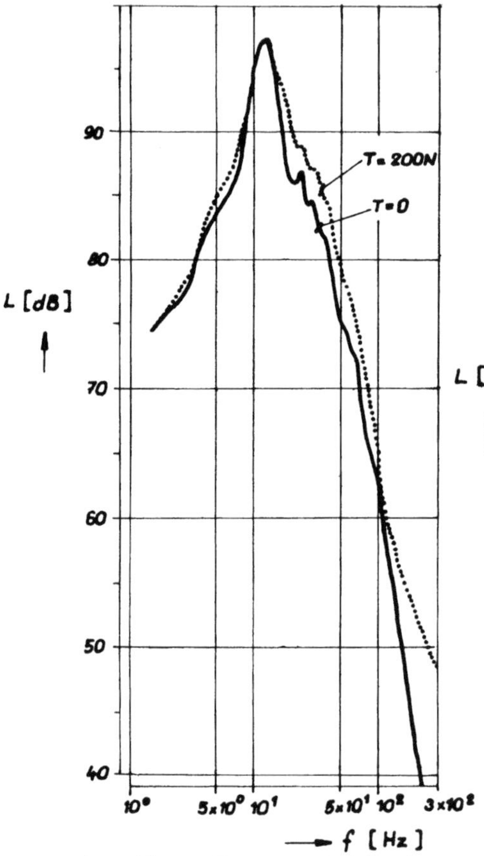


Fig. 13. Influence of friction on SPL. Three-mass model, $c_D = 5.10^5$ N/m, $k_D = 1100$ Ns/m. Damper 4, $c_T = \infty$, GAP = -1 m.

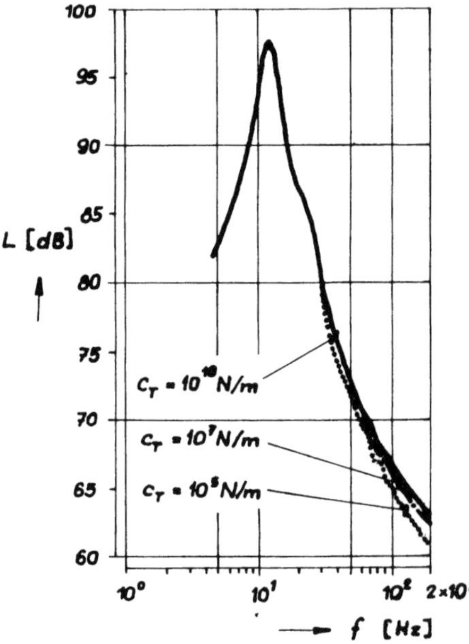


Fig. 14. Influence of "friction stiffness" $c_T$ on SPL. Two-mass model. Linear damper 1, T = 100 N, GAP = -1 m.

Infinite piston rod attachment stiffness in the two-mass system leads to higher SPL values at frequencies over 20 Hz in comparison with the values of SPL gained with the three-mass system. This result will be apparently valid for all other combinations of damping parameters.

Different damping at the piston rod attachment in the three-mass model does not change this basic trend but changes somewhat the differences of SPL in the three-mass and two-mass models. Light attachment damping (curve 2, Fig. 11) lessens the differences in the neighbourhood of the piston rod vibratory system resonant frequency (about 80 Hz) and increases the differences at high frequencies (above 110 Hz). High attachment damping (curve 3, Fig. 11) has not so pronounced resonance but increases the SPL at high frequencies over the values for light damping.

I n f l u e n c e of friction. The influence of friction on SPL in the two-mass model is shown on Fig. 12. Friction in the suspension increases the background SPL substantially. An increase of friction of 100 N causes an increase of SPL of approximately 8 dB at 200 Hz and the difference becomes bigger at higher frequencies.

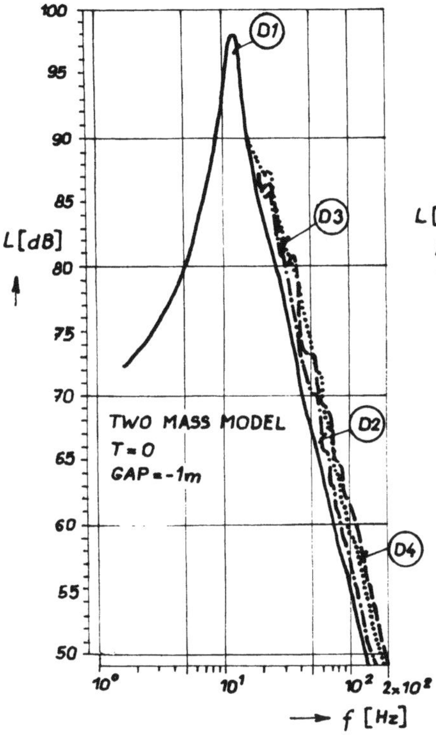


Fig. 15

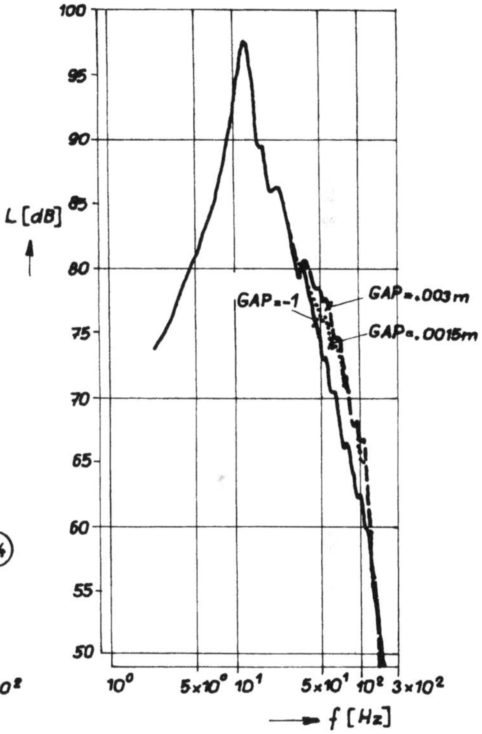


Fig. 16

Fig. 15. Influence of damping characteristics on SPL: D1 - linear, D2 - linear broken, D3 - progressive, D4 - degressive. Two-mass model. T = 0, GAP = -1 m.

Fig. 16. Influence of damper lag on SPL. Two-mass model. Damper 4, T = 0, "lag stiffness" $5.10^5$ N/m.

Similar trend can be seen in the three-mass model. The increase of SPL with increasing friction is however not so pronounced because of the filtering effect of the elastically supported piston rod mass. Even so, friction of quite normal magnitude apparently affects the background SPL of a car relatively very strongly.

I n f l u e n c e of "friction stiffness". The influence of "friction stiffness" on SPL in the two-mass system is shown on Fig. 14 for friction T = 100 N. Relatively very big "friction stiffness" changes affect little the SPL values for frequencies under 100 Hz. This result was surprising as more intensive "cushioning" of high frequency noise components with low "friction stiffness" was expected. "Friction stiffness" alone does not prevent apparently from very abrupt friction force changes which lead to high frequency components of the $u_D$ force course.

I n f l u e n c e of the damping characteristics. Non-linear dampers cause always an increase of the background SPL when compared with a linear one as can be seen on Fig. 15 for the case of the two-mass model. The damping characteristics investigated are shown on Fig. 9 (D1 is a linear, D2 linear broken, D3 progressive, D4 degressive damping characteristics). The magnitudes of their damping forces were selected in such way that the maximum SPL at wheel resonant frequency of about 12 Hz are practically the same (about 97,9 dB with zero friction, no lag), i.e. that the spectral density of the force $u_D$ at this resonant frequency is nearly the same with all damping characteristics under the prescribed excitation.

SPL produced by the commonly used degressive damper 4 is the highest among all four studied variants. The difference between SPL gained with the most favourable linear damper 1 and with the degressive damper 4 is about 5 dB at frequencies higher than 50 Hz. This difference is apparently due to steep increase of the damping force at small damper relative velocities and to rapid changes in the growth of the damping force at higher relative velocities on the degressive damper.

SPL differences between the linear damper and the non-linear ones are smaller in the more realistic three-mass model. The effect of non-linear damping characteristics can be therefore considered as nearly negligible in practice.

I n f l u e n c e of the damper lag. Influence of the damper lag on the background SPL in the two-mass model is shown on Fig. 16. Damper lag with GAP up to 3 mm accompanied with common value of "lag stiffness" affects the SPL in the frequency range between 30 Hz to 130 Hz only. High frequency noise components can be even smaller with a damper with lag than with a normal one because bump motions with very small amplitudes stay nearly undamped. Lag stiffness plays a very important role. High frequency noise components increase very intensively with increasing the "lag stiffness" in similar manner as with increasing friction. Results of the computations are not shown because very high "lag stiffness" is generally not found in practice.

## 7. CONCLUSIONS

Every non-linearity of the suspension damping increases the background SPL in a passenger car at frequencies over 20 Hz. Computations carried out with a three-mass car dynamic model have shown that the differences of the background SPL between a hypothetical linear car dynamic system and more realistic non-linear model with friction can reach values of 6 to 12 dB at 200 Hz on a poor road. The damping non-linearities present in all car suspensions must be therefore taken into account when analysing the origin of the background noise in real cars.

A linear suspension and damping would be optimal for obtaining minimal background SPL.

Non-linear damper characteristics show some influence on the background SPL but - compared to linear damper - their effect is not important.

Friction has most pronounced effect on the background SPL. Elimination of suspension friction leads therefore to lower background SPL.

Damper lag in conjuction with common "lag stiffness" increases the background SPL at frequencies around 70 Hz. Its influence is however not so pronounced as expected.

REFERENCES


1. Iha, S.K., Characteristics and Sources of Noise and Vibration and their Control in Motor cars.J. of Sound and Vibr.,47 (4) (1976).
2. Phillips, A.V., A Study of Road Noise. Symposium IME "Vibration and Noise in Motor Vehicles", C 102/71 (1972).
3. Apetaur, M., Opicka, F., Output from a Non+linear Friction Element with Stochastic Input. J. of Sound and Vibr., 62 (1)(1979).
4. Bathelt, K., Bösenberg, D., Neue Untersuchungsmethoden in der Karosserieakustik. ATZ 5/1978 (1976).
5. Rezvejakov, E.M., Tolskij, V.E., Ocenka vibroakustiticheskich charakteristik kuzova legkogo avtomobilja. Avtomobilnaya Promyshlenost, 6 (1973).
6. Yacenko, N.N., et al., Kolebanya poveski s utchetom poglashayushey sposobnosti shin pri slutchaynom vozmushenyi.Avtomobilnaya Promyshlenost, 1 (1979).

# COMFORT, VIBRATION AND STRESS OF THE BELTED TIRE

F. Böhm

Institut für Mechanik, Technische Universität Berlin, F.R.G.

SUMMARY

The object of this paper is the investigation of static and dynamic deformations and stresses occurring within car tires. The influence of vertical load, cornering force, camber and standing wave is discussed. Vibrations of the belt, lateral and radial, caused by non-uniformity, by road irregularity (cleat) and by sudden steering manoeuvres are computed. Also vibrations of the sidewall are measured and longitudinal vibrations in the acoustic range are shortly treated because they belong to the presented mechanical model.

## 1. INTRODUCTION

The state of stress and deformation of a belted tire is very complicated because of the non-isotropic behaviour of the layers.
Using simple Eulerian membrane theory (carcass and also belt) one gets for equilibrium under internal air pressure $p=p_B+p_K$ :

$$\frac{\sigma_x}{R} + \frac{\sigma_y}{\rho} = p_B \qquad \text{and} \qquad \frac{\sigma_k - \sigma_y}{\rho} = p_K \, .$$

R is the radius of the belt and $\rho$ is the radius of curvature of the belt section. $\sigma_k$ is the membran stress of the side walls (radius of curvature $r_0$ ). Because of the radial stiffened side wall it is $\sigma_k = pr_0$ and in case of a two layered belt one has $\sigma_y = \sigma_x \tan^2 \beta$ ($\pm \beta$ is the angle of the belt steel wires), so it yields for the belt stress in circumferential direction:

$$\sigma_x = R\,p\,(1-\frac{r_0}{\rho}) \qquad \text{and} \qquad \frac{p_B}{p} = (1-\frac{r_0}{\rho})\,(1+\frac{R}{\rho}\tan^2\beta)$$

These expressions are only a rough estimate and in order to get the correct shape of meridian contour one has to use numerical integration processes which are familiar to the tire designer. In practice $p_B/p$ is not constant at tread shoulder and decays to zero. Starting with this roughly sketched membrane stress system further deformations due to rolling, cornering, vibrations a.s.o. are to be treated. Side-wall deformations are mainly caused by bending in meridian direction (y) and by shear in circumferential direction (x). This shear force is reducing the belt tension force during vertical deflection of the tire. By the aid of the finite element model derived in [1], equations (14-27), nearly one hundred examples for deflections and pressure distributions in the contact patch have been computed. It turns out that the correct (nearly rectangular) shape of contact area belongs to a parabolic distribution of membrane stress $\sigma_x$ of the belt, see figure 1.

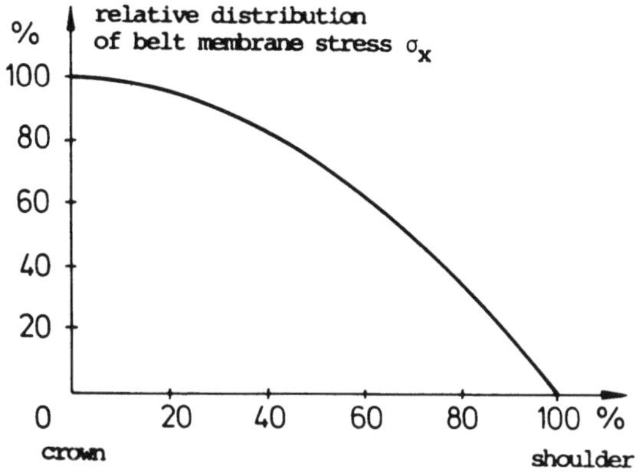


Fig. 1. **assumed membrane stress distribution to get rectangular shape of contact patch**

Until now no complete solution for the coupled belt carcass system exists, so one has to model the mechanical characteristics of the tire for the purpose of adaption to the car.

## 2. DEFORMATION AND STRESS OF THE BELT CAUSED BY INTERNAL AIR PRESSURE, VERTICAL LOAD, CORNERING FORCE AND CAMBER.

Using equations (5) derived in [1], one finds that most important influence on belt stiffness is caused by the carcass stiffness, $\frac{\partial S_k}{\partial \varepsilon_k} n_k$, given in equation (31),[1]. The circumferential belt stress is, see figure 2 :

$$\sigma_x = \frac{\partial S_k}{\partial \varepsilon_k} n_k \frac{\varepsilon_x}{\tan^4 \beta} = \frac{F' n_k}{\tan^4 \beta} \cdot \varepsilon_x \qquad (1)$$

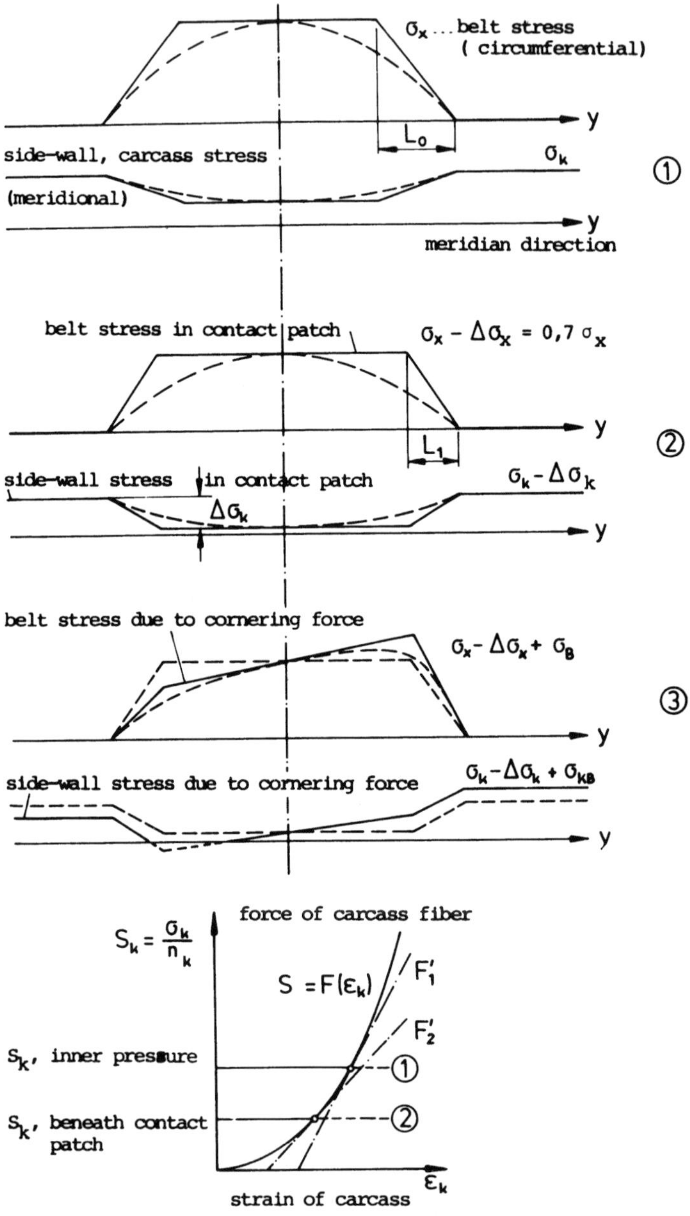


Fig.2. distribution of belt stress and carcass stress for ① inner air pressure, ② vertical load P, ③ cornering force S and stress-strain characteristic for one Rayon-fiber

where $n_k$ is the number of cords per unit length and $\pm\beta$ is the direction of the belt steel cords. For internal air pressure of p = 2 bar one gets at the equator $\sigma_x$ = 400 N/cm and for F'= 500 N (Rayon cord); tan $\beta$ = 0.325 with the aid of (1) $\varepsilon_x = 0.8 \cdot 10^{-3} = w/R$. With R = 30cm which is the radius of belt the deformation is w = 0.24 mm or 0.12 mm/bar which corresponds well with the measured deformation of 0.15 mm/bar of a 165 SR 13 steel cord passenger car tire. In a region of $L_o = \sqrt{h_k E_k / G_{PL}}$ this stress reduces at the belt boundaries to zero. The stress in meridian direction $\sigma_y$, and the shear stress between the two layers of the belt obey to the equations:

$$\left.\begin{array}{ll} \sigma_y = \sigma_x \tan^2 \beta & \tau_{xz} = \tau_{yz}/\tan\beta \\ \tau_{xz} = -\dfrac{\partial \sigma_y}{\partial y}\dfrac{1}{\tan \beta} & \tau_{yz} = -\dfrac{\partial \sigma_y}{\partial y} \end{array}\right\} \quad (2)$$

The maximum shear stress and strain are therefore

$$|\tau_{xz}|_{max} = \frac{\sigma_x \tan \beta}{L_o} \quad \text{and} \quad |\gamma_{xz}|_{max} = \frac{1}{G}|\tau_{xz}|_{max} .$$

For the same tire, with shear modulus of rubber G = 400 N/cm$^2$ it is $\tau_{xz}$ =65 N/cm$^2$ and $\gamma_{xy}$=0.16 rad = 9.2°. Because of this relatively high strain the belt has under its shoulder small cushions which reduce this strain by membrane equilibrium considerably. This means that the curvature of the carcass and the carcass membrane force carry the internal pressure force. Therefore the cushions reduce the actual belt width for internal pressure. This stress system changes to another one for vertical load.

Near contact patch the side-wall is much more convex and stress $\sigma_k$ is reduced considerably:

$$\Delta\sigma_k = \Delta S_k \, n_k = - \sigma_k \frac{r_o}{h_o} \quad H \qquad (3)$$

H is the vertical deflection, $r_o$ is the radius of the undeflected side-wall and $h_o$ is the height of the side-wall. For the 165 SR 14 tire $S_k$ reduces for normal load from 5N to approximately 2.5N. The nonlinear elasticity of the rayon cords of the carcass, shown in figure 2, reduce $L_o$ = 2.23 cm to $L_1$ = 1.56cm. But also the circumferential tension force of the belt T ($\varphi$) is reduced by side-wall shear stiffness in this direction, see figure 3.

This effect was found by Springer and Weisz [ 2 ] with the aid of the tire model drived in [ 3 ] . They found the solution for circumferential and radial deformation of the belt and the side-walls by only one nonlinear equation. Also the slip in rolling direction and the shear stiffness of the side-walls (i.e. the thickness of the side-walls) influence this solution.

For a vertical load P = 3500 N and the same tire as before there is a 30% reduction of tension within the belt. So one gets in using this solution the shear stress at the boundary $|\tau_{xz}|_{max} \doteq \dfrac{\sigma_x \cdot 0.7 \cdot \tan\beta}{L_1} =$ = 62 N/cm$^2$.

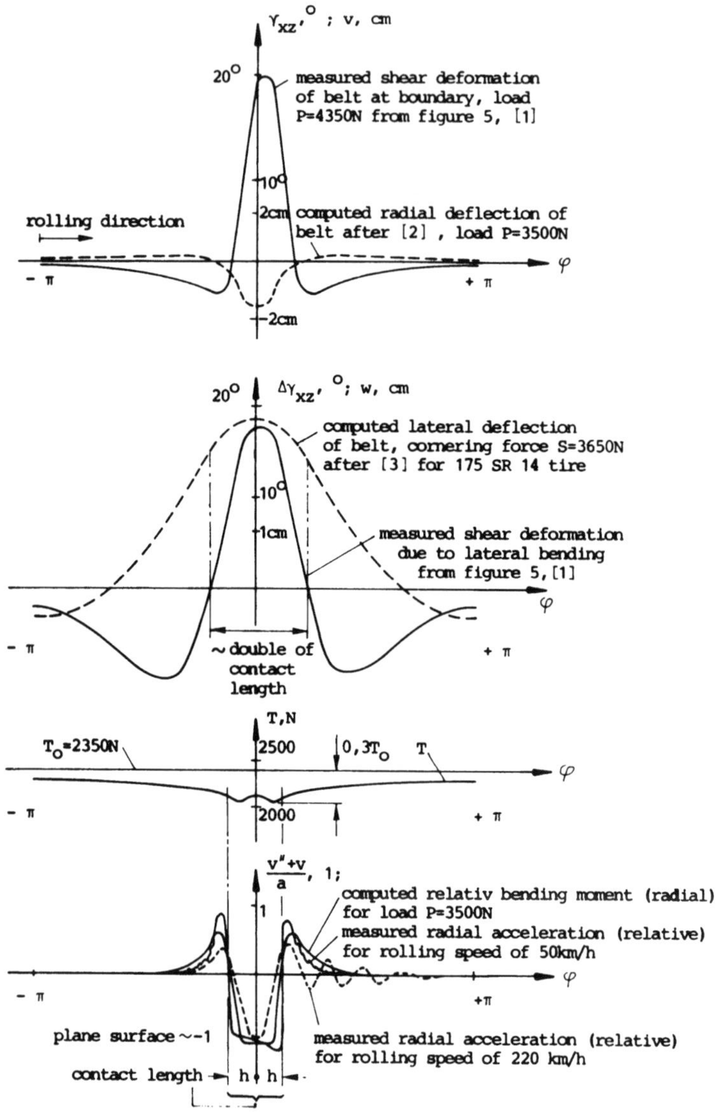


Fig. 3. $\gamma_{xz}$, w, v, $\frac{v''}{a}$ and T as function of $\varphi$

This stress is the actual one because in the contact zone there is no membrane equilibrium. So, radial deflection and shear at belt boundary are coupled. Looking at figure 3 one finds that the lines for $\gamma_{xz}$ and v are very similar. Superposition of geometrically induced shear deformations by the layers gives rise to (see [7] equations 21):

$$\gamma_{xz,\rho} = \frac{h}{R} \quad \text{and} \quad \gamma_{yz,\rho} = \frac{b}{\rho} , \tag{4}$$

here h and b are half length and width of the contact patch and $\rho$ which is the radius of curvature for a section of the belt.

The influence on the belt stress caused by a cornering force is also shown in figure 2. For a side force, for example S = 2000 N, we can compute by the belt flexibility $\gamma_B = 82\cdot10^{-6}\,(\text{Ncm})^{-1}$, see [1] figure 7, the curvature of the belt $1/\rho_B = 2\gamma_B S = 1/340\ \text{cm}^{-1}$. Assuming Euler hypothesis for bending this results in $\varepsilon_{x,max} = b/\rho_B = 4.5/340 = 1.32\cdot10^{-2}$ and $\sigma_{x,max} = 330$ N/cm. A reduction by the influence length L results in $\sigma_{x,max} = 330\cdot(b-L_1/b) = 220$ N/cm. The maximum shear stress is $|\tau_{xz}|_{max} = 46\ \text{N/cm}^2$ and $|\gamma_{xz}|_{max} = 0.112 = 6.5^o$. In figure 3 the measured shear deformation due to lateral bending is shown. Since the deformation is proportional to the strain $\varepsilon_{x,B} = b/\rho_B = w''/a^2$ the full line must be the second derivative of the dotted line near the contact patch, where the torsion angle of the belt is zero.

Springer and Weisz have also computed the radial bending of the belt. The radial bending moment is (see [3], eg. 4.6):

$$M_2 = -EI_y \frac{v''+v}{a^2} , \tag{5}$$

Here is a $\equiv$R which is the radius of the belt. The term $v/a^2$ is very small, therefore for a constant rolling speed the measured radial acceleration relative to the centripedal acceleration is $v''/a$ and this is a measure for radial bending. In case of high rolling velocity there occurs a standing wave behind the contact patch. Using a mechanical device developed by Meier-Dörnberg [10] belt deflections $w(\varphi)$ at the cornering force were measured for a tire 195 VR 14, figure 4. The width of its belt is 2b = 165 mm and therefore the flexibility is small. From figure 4 one finds for case a: $\gamma_B = 5\cdot10^{-6}\,(\text{Ncm})^{-1}$ and the curvature is $1/\rho_B = 1/2650\ \text{cm}^{-1}$. For a 165 SR 14 tire the difference is shown between a computed [1], figure 4 and a measured contact area, figure 5. The difference is due to the circumferential belt stress distribution $\sigma_x \pm\Delta\,\sigma_B$.

Lateral deflection due to a camber angle is also measured, shown in figure 4. For a wheel-load P = 3750 N the curvature of the belt is $1/\rho_C = 1/1520\ \text{cm}^{-1}$. Since the torsion angle in the contact patch is equal to the camber angle we get for a camber angle of $3^o$ by the aid of [3], eq. (4.6)

$$M_1 = -EI_x \left(\frac{w''}{a^2} + \frac{\Psi}{a}\right) = -EI_x \left(\frac{1}{\rho_C} + \frac{\gamma_C}{a}\right) \tag{6}$$

the total curvature $1/\rho_T = 1/1520 + 3/(57.3\cdot30) = 1/418\ \text{cm}^{-1}$. This deformation is equal to a deformation under the cornering force of magnitude $S_a \cdot \rho_B/\rho_T = 9900$ N. It suggests that the stress of the belt resulting from camber is very high.

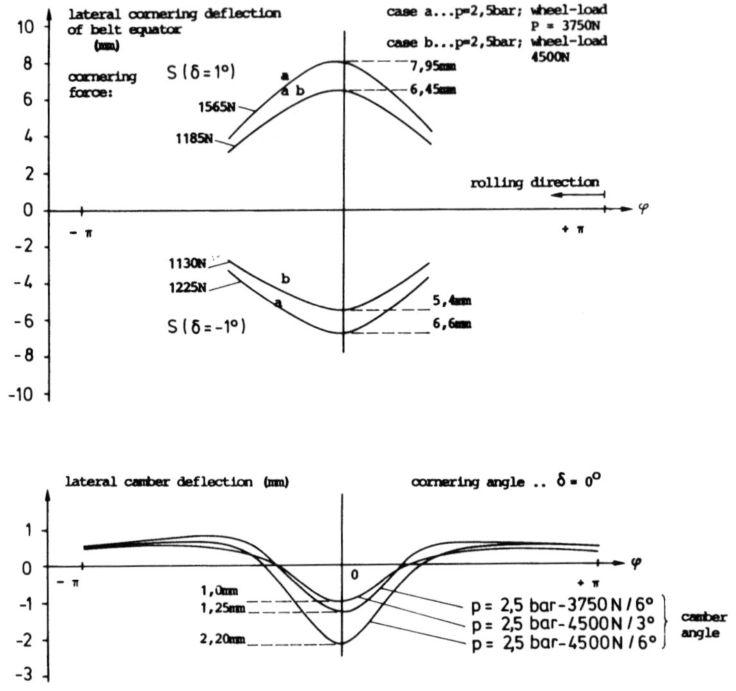


Fig. 4. w for cornering and camber as function of φ

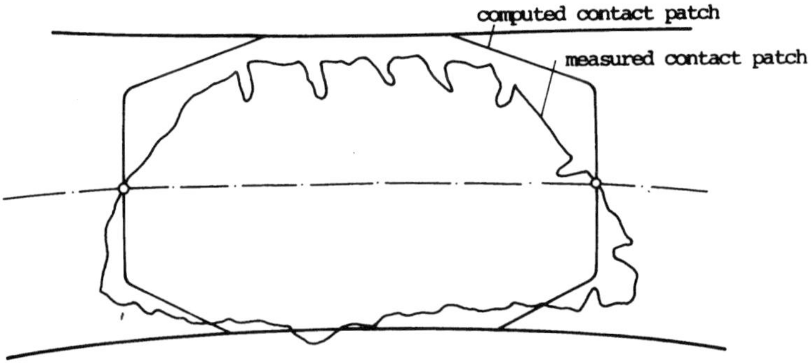


Fig. 5. Influence of $\sigma_x$-bending distribution on contact length

## 3. NON-LINEAR CHARACTERISTIC OF SIDE-WALL DEFORMATION AND DEVIATION FORCES OF THE BELT IN THE CONTACT PATCH.

In order to investigate nonlinear characteristics of the side-wall deformation the finite element model given in [1] was used to compute the load depending parameters for the elastic foundation of the belt on the carcass. A section of 1cm circumferential length with constant belt forces $\sigma_x$ was laterally deformed without and with flattening it. This was computed for several lateral loads. For instance in figure 6 one recognizes the non-flattened case for small and maximum lateral load and also the flattened section.

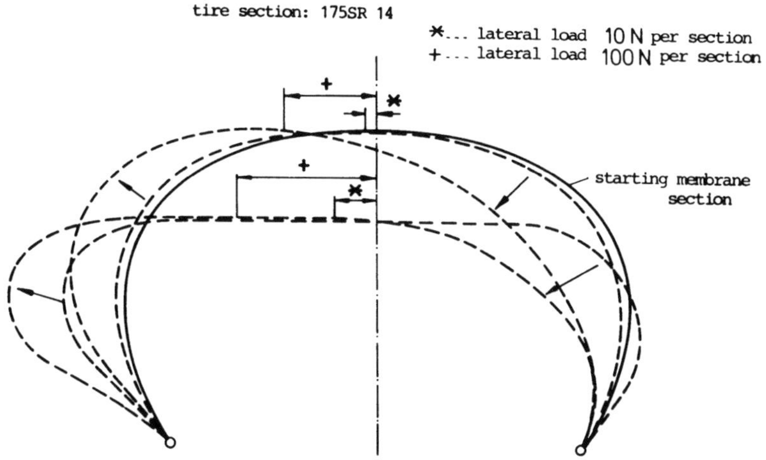


Fig. 6. deformations of tire section by vertical and horizontal loads

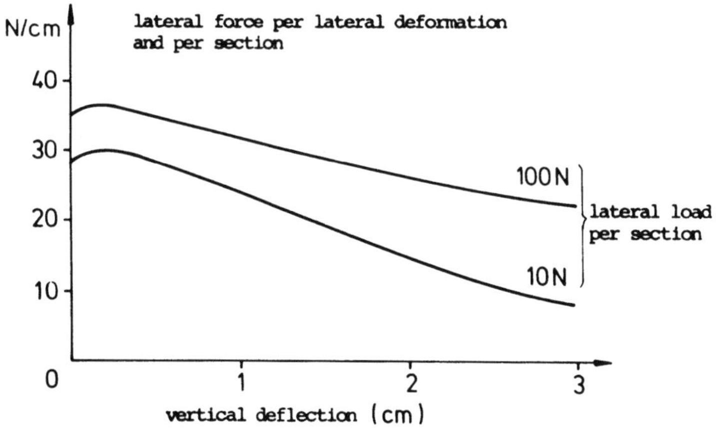


Fig. 7a. stiffness of tire section

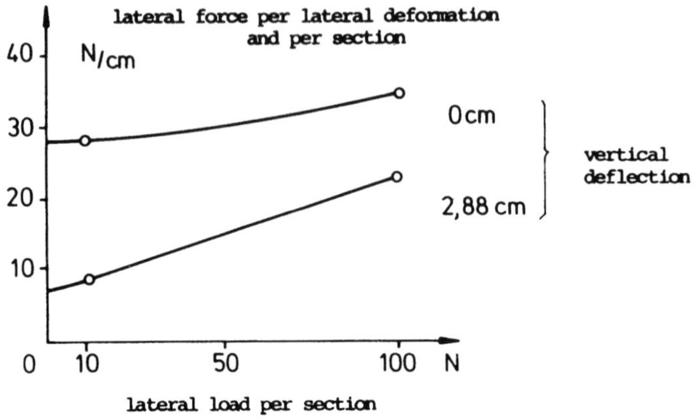


Fig. 7b .stiffness of tire section

In figure 7 it is shown that for small vertical deflections the stiffness is considerably higher than for high vertical deflections and otherwise for a small lateral force the stiffness is smaller than for a high lateral force. For small lateral non-uniformity forces the stiffness-variation with variation of vertical deflection ( i.e.Load) is one to three. This effect produces a load depending lateral stiffness of the contact patch on the whole: $c_S = c_S(P)$.

Because of the sandwich-structure of the belt changes of its curvature create a moment distribution towards the direction of the outer normal of the belt. Using [ 3 ] , equations (2.32) up to (2.34), the last one must be corrected then one gets the moment per unit length:

$$n = -\nu d \left( \frac{v'' + v}{a^2} + \frac{\tan^2 \beta}{\rho_G} \right) \tag{7}$$

$1/\rho_G$ is the change of curvature of the meridian line of the flattened belt and $\nu d$ is the stiffness of the two layers of the belt for its out of plane reactions. From figure 3 we find $v/a^2 = 1/30\ \text{cm}^{-1}$ and from the undeformed cross-section $1/\rho_G = 1/18\ \text{cm}^{-1}$ in the contact patch. From [ 3 ] eq. (2.32) it follows:

$$\nu d = (EF)\ d \tan\beta \tag{8}$$

here d is the distance of the layers and (EF) is the circumferential stiffness of the belt. In case of radial bending, shown in figure 3, one has from (7)

$$2hn = 2 \cdot 7.8 \cdot 6.7 \cdot 10^5 \cdot 0.15 \cdot 0.325 \left( \frac{1}{30} + \frac{0.1}{18} \right) = 0.17 \cdot 10^5 \text{Ncm}.$$

Using the torsional stiffness for the contact patch about the vertical axis $c_T = 1.5 \cdot 10^6$ Ncm/rad we get a cornering angle $\delta_n = 2hn/c_T = 0.17 \cdot 10^5/1.5 \cdot 10^6 = 0.0118 = 0.67^0$. There belongs a cornering force $S_n = 350$ N to this cornering angle. Now it is possible to correct this inherent cornering force by placing the belt onto the carcass unsymmetrically ( $w_0$ = lateral deviation ) with respect to the equatorline, so one shoulder has a greater rolling radius than the other shoulder. The difference is $\Delta r = w_0 \tan \alpha_G$, here $\alpha_G$ is the tangent angle at the shoulder of the belt section. Under vertical load this tire has a curved contact patch, the curvature is $1/\rho_{B,C} = w_0 \tan\alpha_G /ab$ and it produces a "camber" force of $S_C = 4/3\ (c_y bh^3)\ 1/\rho_{B,C}$, see [ 1 ], eq. (63). For $w_0 = 0.5$ cm, $\tan\alpha_G = 0.364$, a = 30cm, b = 4cm one finds $1/\rho_{B,C} = 1/660\ cm^{-1}$ and by $c_y = 55 N/cm^3$ the camber force is $S_C = 198$ N. Now there results a constant combined cornering force of $S_0 = S_n - S_C = 152N$ and there is also a remaining slip angle $\delta_0 = 0.005$ rad $= 0.26^0$.

The tire is a hand-made product. The thickness of the layers d and the symmetry of the belt $w_0$ around the belt are not constant. Therefore the figures given above vary along the equator-line. Thickness variations of the tread and of the side-wall-plates exist also. Since the half contact length is a quadratic function of the wheel-load P also $S_n$ has this dependence but on the other hand $S_C$ is of sixth order in P. Even when the cornering angle of the wheel is zero, variation of wheel-load P caused by road irregularities, produces varying non-uniformity forces, which create vibrations. The vibrations of the belt in and out of the plane of the belt are coupled. The frequency is equal to the frequency of waves on the road or on the tire circumference, effectively proportional to the rolling velocity. Vibrations caused by non-uniformity can be effectively treated by the method of Majcher [4] , who solved the tire model via the weighted residual method. He introduced as well a further degree of freedom for the nonstationary rotation of the wheel as harmonic modal functions which take into account that the contact patch acts like a boundary for the belt. This author found low frequency vibrations with a wave model of 1/2 of the belt against the wheel with a resonance in the range of $f_{1/2} = 25-30$ Hz. He measured the dynamic response of the tire after rolling over a short unevenness of the road and it turns out that the longitudinal force acting on the axle is greater than the vertical force. It follows, that short irregularities are not absorbed by the suspension.

## 4. LONGITUDINAL FORCES AND CONDITIONS FOR COMFORT

Longitudinal forces can be measured on a drum, normally used for measurements of non-uniformity, carrying a small hump or cleat. In figure 8 for a variation of internal pressure and normal load the longitudinal force for a 165 SR 14 tire is shown. At entrance the hump produces a higher rolling resistance and at outlet a smaller one and in some examples it changes to a drive of rolling. We take for the hump-profile $v_0 = H_0\ (1-x^2/l_1^2)$ where $H_0$ is the height and $2\ell_1$ the

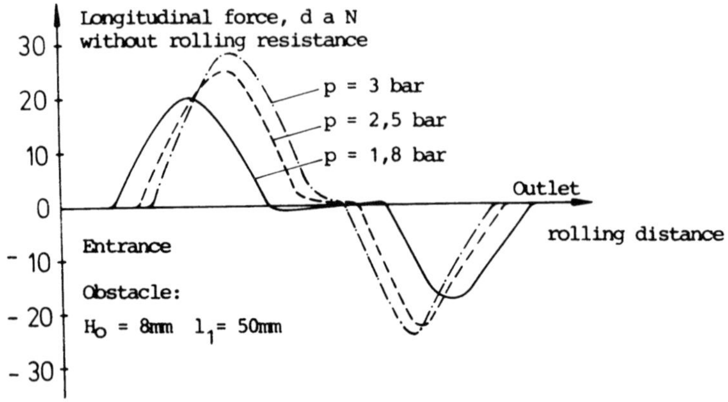


Fig.8. horizontal reaction force created by rolling over a cleat

length of the hump. For the contact patch we take the simplified equation of the belt:

$$- T_0 \frac{d^2v}{dx^2} + k_L \, v = k_L \,( h_{L_0} + v_0 ) - p \tag{9}$$

where $k_L$ is the compression stiffness of the tread and $h_{L0}$ is its thickness. In the range $-\ell_i \leqq x \leqq + \ell_i$ the solution is with $\lambda = \sqrt{\frac{k_L}{T_0}}$ :

$$v_1 = A \cosh \lambda x + h_{L_0} - \frac{p}{k_L} - \frac{2T_0H_0}{k_L\,\ell_1^2} + H_0 \,(1- \frac{x^2}{\ell_1^2}\,) \tag{10}$$

for $\ell_i < |x|$ it is

$$v_2 = B \, e^{-\lambda(x-\ell_1)} + h_{L_0} - \frac{p}{k_L} \tag{11}$$

The constants A,B are given by the boundary conditions at $x = \ell_1$ : $v_1 = v_2$ and $dv_1/dx = dv_2/dx$. This results in

$$\left.\begin{aligned} A \cosh \lambda\ell_1 - B &= \frac{2T_0H_0}{k_L\,\ell_1^2} \\ A \sinh \lambda\ell_1 + B &= \frac{2H_0}{\lambda\ell_1} \end{aligned}\right\} \tag{12}$$

For example we take $T_0$ = 4000N, $k_L = \frac{E}{h_L} \cdot 2b = \frac{750}{1} \cdot 8$ = 6000 N/cm², $H_0$ = 0.5cm, $\ell_1$ = 4cm and we get $1/\lambda$ = 1/1.222 = 0.82 cm and also A = 0.00151 and B = 0.102. Using [1], equation (35) we get the slip difference

$$s_x = h_L \frac{dv}{dx} = - h_L B \lambda e^{-\lambda x} \tag{13}$$

in the contact patch besides the hump. The longitudinal shear is $\tau_x = c_x s_x$, where $c_x$ = 280 N/cm³ is the shear stiffness of the tread, or otherwise $2bc_x = k_L/3$ because of $G = E/3$. Integrating along the contact length results in the longitudinal force:

$$U_{max} = 2b \int_0^{2h-2\ell_1} \tau_x \, dx = - \frac{k_L}{3} \cdot h_L \cdot B\lambda \int_0^{2h-2\ell_1} e^{-\lambda x} dx = \frac{k_L}{3} h_L B \left[ e^{-\lambda(2h-2\ell_1)} - 1 \right] \tag{14}$$

Within good approximation it is assumed that $B = H_0/\lambda\ell_1$ and with the above figures we get $U_{max}$ = − 435 N. This corresponds quite well with the measured maximum of the force U at entrance and at outlet with changed sign. In reality the bending stiffness $EI_y$ of the belt reduces $s_x$ by reducing dv/dx and the change of momentum of the car body is now given by

$$m_{car} \Delta v = \Delta I = \int U \, dt \doteq U_{max} \cdot \frac{h}{v} \cdot \frac{4}{3} = U_{max} \cdot t_{eff} \tag{15}$$

Here h/v is the time duration of the force. For a duration time of the shock longer than $t_{eff}$ = 33 ms one can feel the first derivative of the acceleration and it should not exceed 0.07 g/sec, this belongs to a rolling velocity which is smaller than v = 15km/h. For usual velocities the shock duration time is short and one can feel the velocity difference directly. It should not exceed $\Delta v$ = 0.03cm/sec for comfort reasons. In case of the given example it is $\Delta v = \frac{1}{m_{car}} U_{max} \frac{4h}{3v} = \frac{435 \cdot 4 \cdot 7.8}{12 \cdot 3 \cdot v} = \frac{375}{v}$ cm/sec. Therefore one does not feel the shock for rolling velocities greater than $10^4$ cm/sec or if appropriate horizontal shock absorbers are used. In general the method given in [8] for measuring vertical comfort characteristics is also used for longitudinal comfort considerations.

## 5. VIBRATIONS OF SIDE-WALL, COUPLED CIRCUMFERENTIAL AND LATERAL VIBRATIONS OF THE BELT:

Not only the belt is able to vibrate as shown in literature by a great number of measurements, also the side-wall itself creates harmonic modes which are not stationary because of the friction at the wheel rim. For a 12.00 R 20 STC truck-tire vibrations of the side-wall are shown in figure 9. In case of a) the excitation is at a

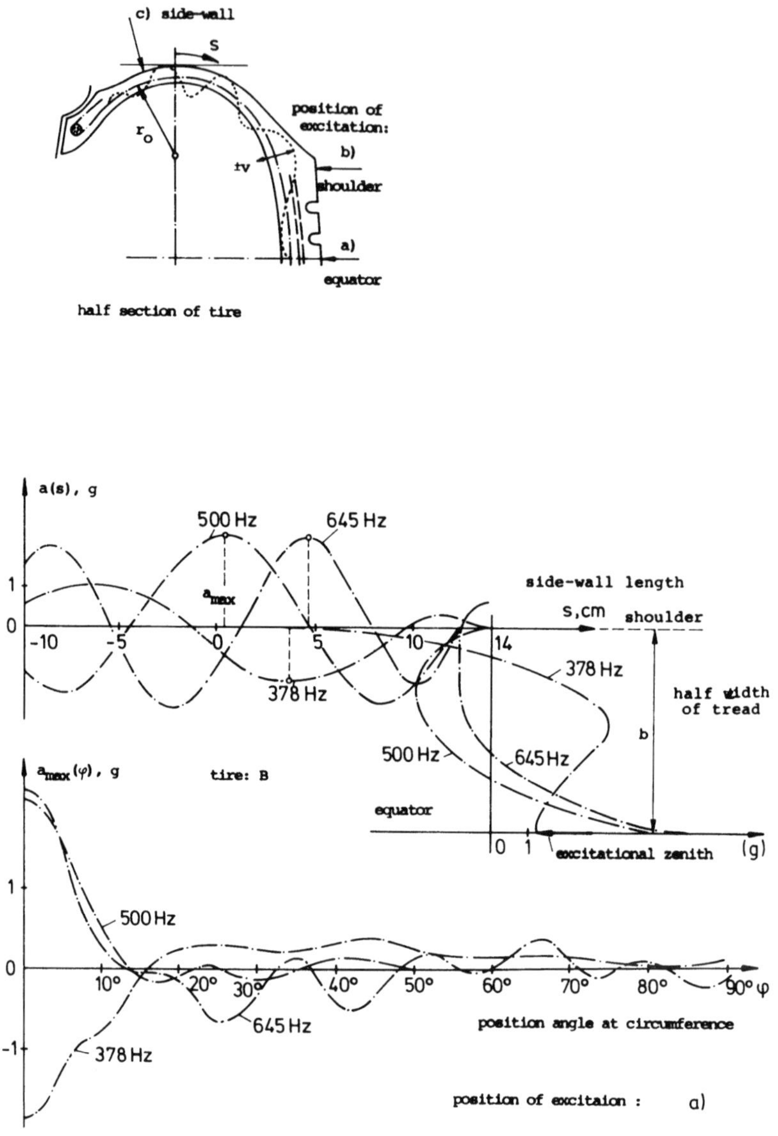


Fig. 9. vibrations normal to the surface of the side-wall of a 12.00 R20 STC-tire

point at the equator-line. Resonance frequencies exist for 378, 500, 645 Hz. In case of b) the excitation is at the shoulder and the same resonance-frequencies occur. Ignoring the bending stiffness the equation of side-wall vibration is:

$$\frac{\sigma_k}{r_0^2} ( v'' + v) = \rho_s \ddot{v} , \tag{16}$$

from which it follows $\omega^2 = \frac{\sigma_k}{\rho_s r_0^2} (n^2 - 1)$ for harmonic motion.

With $r_0$ = 9,5cm, $\rho_s = 3 \cdot 10^{-5}$N $sec^2 cm^{-3}$ , $\sigma_k = p \cdot r$ = 10bar$\cdot r_0$, n = 3 number of waves on a circle of radius $r_0$, one gets $\omega$ = 1260 $sec^{-1}$ or f = 245 Hz a.s.o. At circumference the amplitudes decrease very fast, so only the area around the contact patch is excitated by the dessin-blocks. Different tire products do not differ very much, which can be seen from figure 10, where excitation is positioned at tire

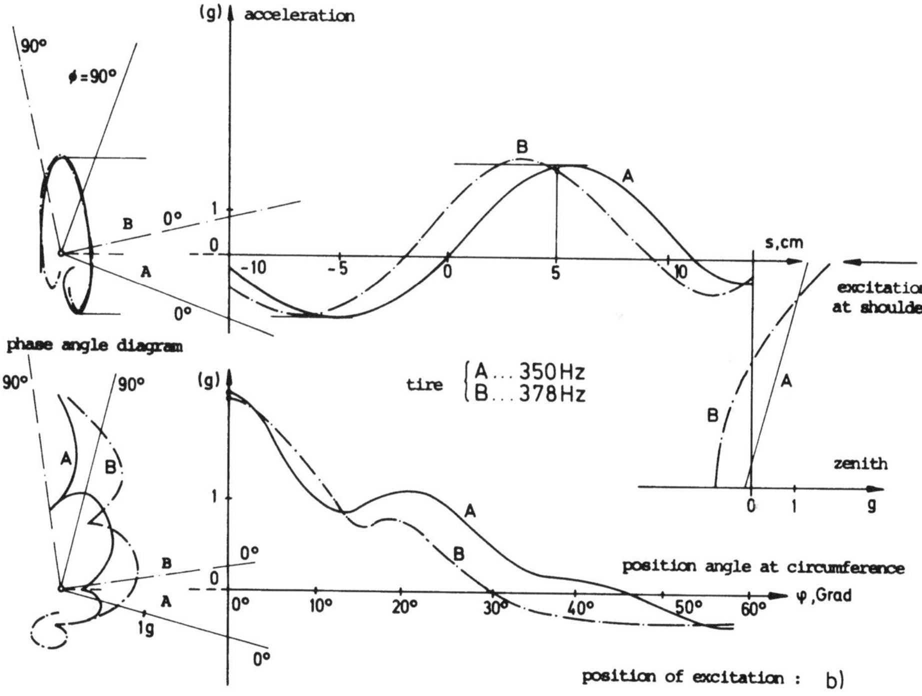


Fig. 10. vibrations of side-wall of two different tires

shoulder and also the phase angle diagram is shown.

In figure 11 the total interesting frequency range was measured.

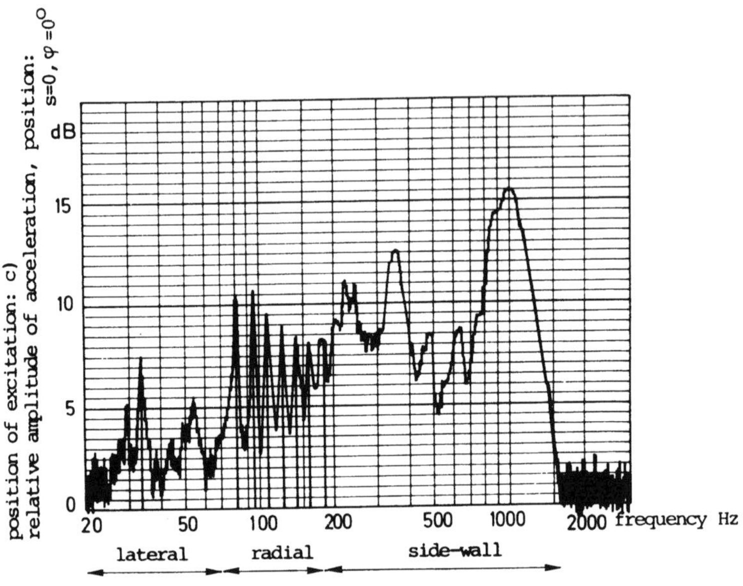


Fig.11. whole frequency range of measured side-wall vibrations

Lateral and radial nonuniformities produce sharp resonances, while geometric excitation of dessin pattern produces a hooting sound over a wide frequency range. Because mass distribution of the tread along its length is not constant $\rho_0 + \Delta\rho(\varphi)$ the rolling tire has periodic excitation. Starting with the solution $u_0$, $v_0$ of [2] the theory of small perturbations gives rise to

$$u = u_0 + \Delta u, \qquad v = v_0 + \Delta v, \qquad \rho = \rho_0 + \Delta\rho \tag{17}$$

Inserting this into equations [3] , (4.10) and (5.6) this leads to equations for the perturbed motion $\Delta u$, $\Delta v$; neglecting terms of second order:

$$\frac{EF}{a^2}(\Delta u'' + \Delta v') + \frac{EI_y}{a^4}(\Delta v''' + \Delta v') - k_\tau \Delta u - \rho_0 \Delta\ddot{u} - \rho_0 2\Omega(\Delta\dot{u}' + \Delta\dot{v})$$

$$-\rho_0(\Delta u'' + 2\Delta v' - \Delta u) = \Delta\rho_{(\varphi - \Omega t)}\Omega^2(u_0'' + 2v_0' - u_0)$$

$$\frac{T_0}{a^2}(\Delta v'' + \Delta v) - \frac{EF}{a^2}(\Delta u' + \Delta v) - \frac{EI_y}{a^4}(\Delta v'''' + \Delta v'') - k_a \Delta v - \rho_0 \Delta\ddot{v}$$

$$-\rho_0 2\Omega(\Delta\dot{v}' - \Delta\dot{u}) - \rho_0(\Delta v'' - 2\Delta u' - \Delta v) = \Delta\rho_{(\varphi - \Omega t)}\Omega^2(v_0'' - 2u_0' - v_0)$$

For high circumferential vibrations these equations can be reduced to one simple equation:

$$\frac{EF}{a^2}\,\Delta u'' - k_\tau \Delta u - \rho_0 \Delta \ddot{u} = \Delta p_{(\varphi-\Omega t)}\,\Omega^2 (u_0'' + 2v_0' - u_0) \qquad (18)$$

The most important term $v_0$ has a maximum at entrance and outlet of the contact patch. Noise generation exists at both boundaries, but is definitely greater at the outlet. After Liedl [ 5 ] longitudinal vibrations of the belt produce this noise and these are selfexcited by frictional forces. We use for the length of the sliding zone $\lambda = h - x_{\rho m}$ in the middle part of contact patch. The wave number is $n = \lambda/a$, i.e. the wave length of the belt is $2\pi\cdot(h-x_{\rho m})$, $x_{\rho m}$ is the point where sliding starts. $\omega = \frac{n}{a}\sqrt{\frac{EF}{\rho_0}} = \frac{1}{\lambda}\,v_c$, here $v_c$ is the velocity of the wave. For a nonlinear frictional oscillation the amplitude is $A = v_\rho/\omega$ , $v_\rho$ is here the sliding velocity. With the slip coefficient $s = \dfrac{U}{2\,\frac{k_L}{3} h^2}$ we get $v_\rho = v\cdot s$ and

$$A = \frac{v_\rho}{\omega} = \frac{v\cdot s}{v_c}\,\lambda \quad . \qquad (19)$$

Using piston theory of acoustics for the measurement of the noise near the contact patch this results in: $\Delta p_{acoustic}$ which is proportional to A, or with $U = \frac{1}{2}\,\rho_L\, c_W v^2$ we get $A \sim v^3/\omega$ and at least we have

$$\ln\frac{p}{p_0} = \ln\frac{\lambda}{v_c} + 3\ln v. \qquad (20)$$

This is a linear function in v for logarithmic dynamic air pressure. This linear function of the rolling velocity was also found by Denker [6] . Low frequency vibrations occur during sudden steering manoeuvres which are measured and computed by Meier-Dörnberg and Strackerjan [10] . The tire model of reference [1] predicts the instationary tire forces under a second order theory. Using equation[1], (62) and equation [10], (16) one can compare formulas which influence this cornering moment:

$$c_k \cdot \frac{\varepsilon}{v} = \frac{4}{3}\, c_x\, b^3\, h^2\, \frac{1}{\rho} \qquad (21)$$

with $c_x = 280\ \mathrm{N/cm^3}$, b = 4cm, h = 9cm one has $c_k = 198\ \mathrm{Nm^2/rad}$ which is in good agreement with the figures of [10] , table 2 $c_k = 225\text{-}240\ \mathrm{Nm^2/rad}$ found by empirical methods. To compute the influence of the mass of the belt and its gyroscopic moments we use the mode shapes of Majcher's work [ 4 ] . Outside of the contact patch $\varphi^* \leq \varphi \leq 2\pi - \varphi^*$ the lateral and torsional deformations are assumed to:

$$w(\varphi,t) = w_0(t) - w_{1/2}(t)\sin\frac{\varphi}{2} + w_1(t)\sin\varphi$$
$$\Psi(\varphi,t) = \qquad - \Psi_{1/2}(t)\sin\frac{\varphi}{2} + \Psi_1(t)\sin\varphi \; . \tag{22}$$

Since $\Psi''$ is small it can be neglected as well as $GI_p \ll EI_x$ and we find by [3], (5.2):

$$\Psi_n = \frac{n^2 w_n}{a(1+\frac{a^2 k_T}{EI_x})} = \frac{n^2 w_n}{a^*} \tag{23}$$

Now by the method of weighted residuals (Ritz-Galerkin) we have for the equation of $w(\varphi,t)$:

$$\Gamma(\varphi,t) \equiv (a^2 p(\varphi) - k_S a^2 w_0(t) - a^2\rho\ddot{w}_0(t)\cdot 1 + \left\{\left[k_T(1+\frac{T_0 a^2}{4EI_x}) + k_S a^2\right] w_{1/2}(t) + \rho a^2 \ddot{w}_{1/2}(t) - \frac{1}{4}\rho a^2\Omega^2 w_{1/2}(t)\right\}\sin\varphi/2 + \rho a^2\Omega\dot{w}_{1/2}(t)\cos\varphi/2$$
$$+ (-k_S a^2 w_1(t) - a^2\rho\ddot{w}_1(t) - a^2\rho\Omega^2 w_1(t))\sin\varphi + 2a^2\rho\Omega\,\dot{w}_1(t)\cos\varphi$$

Multiplying by 1, $\sin\varphi/2$, $\sin\varphi$ and integrating over $2\pi$ yields:

$$\rho a^2\begin{pmatrix} -2\pi & 4 & 0 \\ -4 & \pi & 0 \\ 0 & 0 & -\pi \end{pmatrix}\ddot{\underline{x}} + \rho a^2\begin{pmatrix} 0 & 0 & 0 \\ 0 & 0 & -\frac{8}{3}\Omega \\ 0 & \frac{8}{3}\Omega & 0 \end{pmatrix}\dot{\underline{x}} + \begin{pmatrix} -k_S a^2 2\pi, & k^* 4, & 0 \\ -4a^2 k_S, & k^*\pi, & 0 \\ 0, & 0, & (-k_S a^2 - \rho a^2\Omega^2)\pi \end{pmatrix}\underline{x} =$$

$$= \begin{pmatrix} -as \\ 0 \\ -M \end{pmatrix} \tag{24}$$

where $\underline{x}$ is the column vector of $w_0$, $w_{1/2}$, $w_1$ and $k^* = \left[k_T(1+\frac{T_0 a^2}{4EI_x}) + k_S a^2\right]$.

If we compare the lateral deflection w and the bending strain at the corner, which is proportional to shear the deformation $\Delta\gamma_{xz}$, we see that the assumed deformation $w(\varphi,t)$ does not hold in the region $-\varphi^* \leq \varphi \leq *$, with $\varphi^* = 2h/a$. Here, the bending of the belt produces a curvature $\frac{1}{\rho_B} = \frac{1/2\, w_{1/2}}{a^2\varphi^*}$ of the equator-line. Also we have to reduce

the maximum lateral deformation $w_0$ by $\Delta w_0 = 1/4\, w_{1/2}\varphi^*$. Static lateral deformation by a constant force S gives

$$w_{1/2} = w_o \frac{4k_S a^2}{\pi k^*} \quad \text{and} \quad w_o = \frac{S}{(2\pi - \frac{16}{\pi})k_S a} = \frac{S}{1,2k_S a} \tag{25}$$

For instance, taking figures of [10] figure 6, we have $S_{(1^0)}$=1530N and $k_S \geqq p$, assumed to $k_S$ = 28 N/cm$^2$ so from (25) it follows $w_o$ = 1530/(1.2•2.8•30) = 1.5 cm. Reducing this by $w_{1/2} \doteq w_o$ we get $\Delta w_o$ = 1/4•1.5•0.54 = 0.19 cm and $w_o - \Delta w_o$ = 1.31 cm which is a value not so far away from the measured one (0.8 cm), but perhaps $k_S$ is assumed too small. For $w_1$ we find in case of a static moment M:

$$w_1 = \frac{M}{\pi(k_S a^2 + k_T + a^2\rho\Omega^2)} \tag{26}$$

By inverting the mass-matrix one finds the equations:

$$\rho a^2 \ddot{w}_o - \frac{8}{3}\rho a^2 \Omega \frac{4}{2\pi^2 - 16}\dot{w}_1 + k_S a^2 w_o = \frac{aS}{2\pi - \frac{16}{\pi}}$$

$$\rho a^2 \ddot{w}_{1/2} - \frac{8}{3}\rho a^2 \Omega \frac{2}{2\pi - \frac{16}{\pi}}\dot{w}_1 + k^* w_{1/2} = \frac{4aS}{2\pi^2 - 16}$$

$$\rho a^2 \ddot{w}_1 - \frac{8}{3}\rho a^2 \Omega \frac{1}{\pi}\dot{w}_{1/2} + (k_S a^2 + k_T + \rho a^2 \Omega^2) w_1 = \frac{M}{\pi}$$

and the eigen-frequencies for $\Omega$ = 0:

$$\omega_o = \sqrt{\frac{k_S}{\rho}}, \qquad \omega_{1/2} = \sqrt{\frac{k^*}{\rho a^2}}, \qquad \omega_1 = \sqrt{\frac{k_S a^2 + k_T}{\rho a^2}} \tag{27}$$

These expressions $\omega_o$ and $\omega_1$ are the same as in [3]eq. (6.3) and (6.4). For $\Omega$ >0 the frequency determinant has to be solved which is of sixth order. In case of low rolling velocity one deduces form [10], figure 16, $f_{1/2}$= 46 Hz ( computed by (27) is $f_{1/2}$= 44,5 Hz ) and for high velocity $f_{1/2}$= 32 Hz. The equations (24) or the transformed one are nonlinear with respect to S and M, this excitation is given in [1]chapter 4 and 5. Looking for the total and the dissipated energy in the structure we have at first bending of the belt:

$$W_B = \frac{1}{2}\int_0^{2\pi} \frac{M_1^2}{EI_x} a\,d\varphi + \int_{-\varphi^*}^{+\varphi^*} EI_x \frac{1}{\rho^2} da\,\varphi = EI_x\left(\frac{2\pi a}{16} + 2h\right)\frac{w^2_{1/2}}{a^4} \tag{28}$$

and in the side-wall:

$$W_{SW} = 2 \cdot \frac{1}{2}\int_0^{2\pi} k_\tau w'^2 \left(\frac{b}{2a}\right)^2 a\,d\varphi = \pi a \left(\frac{b}{2a}\right)^2 k_\tau \left(\frac{w^2_{1/2}}{4} + w_1^2\right) \tag{29}$$

We take as dissipated energy 20 % of the elastic energy with $w_{1/2} = w_0 = 1.5$ cm and get

$$W_{Diss} = 0.2\ (W_B + W_{sw}) = 0.2\ (192 + 22) = 42\ \text{Ncm}$$

for one cycle of revolution. This increases the rolling resistance by $W_{Diss} / 2\pi a = \Delta U = 0.23$N. Therefore the increasing rolling resistance caused by cornering which does not stem from structural dissipation but from different circumferential slip in the contact patch.

## 6. STANDING WAVE AND STRESS OF BELT AT HIGH ROLLING VELOCITIES.

The standing wave phenomenon was measured in [ 9 ] . The wave length was measured as a function of rolling speed. The amplitude is in every case not greater than the static vertical deflection H, see figure 3 where the relative acceleration $\Omega^2 v''/a\Omega^2 = v''/a$ is shown. A formula for the amplitude is given in [ 3 ] equation (5.24) but one needs the dynamic contact length which is not known. Because of the lower longitudinal stiffness of the belt at the shoulder the amplitude of the tire tread becomes here greater which was proofed also by measurements. Buildung a smaller belt one has greater amplitudes when the mass of the tread is held constant. Damping of the oscillation becomes smaller when rolling velocity increases. The maximum strain of the belt is $\varepsilon_{x,max} \leqq H/a$. The number of waves increase with increasing velocity. For a static radial deflection H = 0.8 cm it is $\varepsilon_x \leqq 0.0266$, which is more than the strain produced by internal pressure ( for p = 2bar it is $\varepsilon_x = 0.0008$ from formula (1), chapter 2). Because of neglecting the wave deformation $u(\varphi)$ is not as high as given above.

The delamination process of the belt is caused by rupture of the chemical bonds of the macro-molecules of rubber by sulfur-chains. The density of bonds obeys the law of Arrhenius and by thermal and mechanical energy the inverse reaction to moulding and heating of the tire starts. Analogically to well known cases of technical importance the thermo-mechanical theory of molecular fluctuations leads to a proportional formula for the duration of life time (durability) under a certain stress situation:

$$L \cong \frac{1}{\sigma_{mech}}\ e^{A/RT} \tag{30}$$

Here A is the activation energy and T the absolute temperature. R is the universal gas constant. The beginning of separation of belt sheets can be measured by break down of internal pressure of the tire. For a 195/70VR10 tire temperature at tread shoulder surface, air pressure and rolling resistance were measured and in table 1 the beginning of damage of the belt corner is given.

Table 1:

| tire number | 1 | 2 | 3 | 4 | 5 | 6 | 7 | 8 | 9 |
|---|---|---|---|---|---|---|---|---|---|
| rolling velocity v,km/h | 220 | 220 | 220 | 220 | 200 | 200 | 200 | 180 | 180 |
| temperature of shoulder, °C | 80 | 75 | 78 | 75 | 105 | 100 | 100 | 82 | 100 |
| air pressure, bar | 3.15 | 3.15 | 3.15 | 3.15 | 3.15 | 3.10 | 2.80 | 2.60 | 2.50 |
| rolling resistance, N | 78 | 78 | 78 | 85 | 98 | 91 | 83 | 74 | 88 |
| camber angle, ° | 0 | 0 | 0 | 0 | 4.5 | 4.5 | 4.5 | 5 | 5 |
| running time, min | 15 | 15 | 15 | 15 | 5 | 5 | 5 | 5 | 5 |

From these results one can conclude that camber, as mentioned in chapter 2 reduces the durability very much.

## 7. CONCLUSION

Despite the fact that today no general theory for the belted tire exists it is possible by modelling the tire by the aid of mechanical methods and if one compares these solutions with measurements one can understand its behaviour and one can compute its mechanical characteristics partly. To go deeper into this field a big amount of measurement-equipment and new nonlinear computer methods are absolutely necessary.

References


1. Böhm, F., Computing and Measurement of the Handling Qualities of the Belted Tire, 2nd IUTAM Symposium on Dynamics of Vehicles on Roads and Tracks, Vienna, Sept. 19-23, 1977

2. Springer, H. and Weisz, G., Theoretische Untersuchung zum stationären Abrollvorgang des Gürtelreifens, Automobil-Industrie 3/77, Seite 87

3. Böhm, F., Mechanik des Gürtelreifens, Ingenieur-Archiv 35.Bd., Heft 2, 1966, S. 82.

4. Majcher, J.S., Simulation of radial ply tires for Ride tuning of Automobile, Diss. Univ. of Detroit, 1973

5. Liedl, W., Der Einfluß der Fahrbahn auf das Geräusch profilloser Reifen und ein Beitrag zu seiner Erklärung. Automobil-Industrie 3/77, Seite 75

6. Denker, D., Reifenprofilgeräusche und Reibbeiwerte von Reifen-Fahrbahn-Kombinationen, Teil I,II Automobil-Industrie 1/78, Seite 17 und 3/78, Seite 17

7. Böhm, F., Örtliche Einebnung eines Cord-Netzes. ZAMM 52 T 35 (1972)

8. Pilz, H. und Stumpf, H., Bewertung von Lenk- und Komfortverhalten von Reifen durch Labormessung, Kautschuk und Gummi/Kunststoffe, Heft 8, 1977, Seite 547

9. Böhm, F., Rollwulstbildung am Gürtelreifen, Sonderheft GAMM-Tagung 1966, Band 46

10. Meier-Dörnberg, K.E. und Strackerjan,B., Prüfstandsversuche und Berechnungen zur Querdynamik von Luftreifen, Automobil-Industrie 4/77, Seite 15.

# ANALYSIS OF STABILITY AND HANDLING CHARACTERISTICS OF SINGLE TRACK VEHICLES*

Pidigundla Chenchanna and Jochim Koch

Adam Opel A.G., Rüsselsheim, F.R.G.

SUMMARY

The stability and handling capabilities of the single track vehicles has been analytically and experimentally investigated. The validation of the analytical model has been examined on the basis of experimental results of straight line running stability parameters, damped natural frequency and damping ratio. Handling behaviour has been analysed using the curve negotiating capabilities, determined during test driving. Through a simple analytical model an index termed "Handling Index" (HI) has been developed, whose validity has been demonstrated using the experimental values.

## 1. INTRODUCTION

The stability and handling qualities of the single track vehicles play an important role in enhancing the accident avoidance capabilities of vehicle-rider system. Although a number of investigators [1 to 4] ** published the results of their research (majority of them concerned mainly with the theoretical analysis of straight line running stability)in recent years, the validity of the analytical models on the basis of experimental results has not been fully established. Regarding the analysis of the handling properties the concepts of UNDER- and OVERSTEER, which are well established in the field of automotive engineering, and some transient parameters were utilised [5]. But unfortunately these concepts are having certain deficiencies since the dynamic behaviour of single and two track road vehicles is basically different. In this work an attempt has been made to analyse both stability and handling behaviour of single track vehicles on the basis of analytical and experimental results.

---

* This work has been done during the employment of the authors at the "Institut für Fahrzeugtechnik, Technische Universität Berlin".

** Numbers in the brackets indicate the references.

## 2. THEORETICAL ANALYSIS OF STRAIGHT LINE RUNNING STABILITY

The motorcycle with rider in open-loop mode has been modeled as follows (fig. 1):

- Motorcycle is considered to have freedom to roll, to yaw and side slip. The steerable parts are free to rotate about the steering axis.
- Rider is assumed to have the following three subsystems:

a. Subsystem with motorcycle roll angle as input and body lean angle as output

b. Subsystem with yaw rate of the motorcycle as input and steering torque as output

c. Subsystem with steering angle as input and steering torque as output

The dynamic properties of the above three subsystems have been determined through the experiments conducted on a laboratory motorcycle simulator [6, 7].

The classical concepts of relaxation properties of tyres have been incorporated in the model. The linear cornering properties were assumed. Frame flexibilities are also included.

Based on the above model concepts the differential equations were derived, which describe the dynamic behaviour of motorcycle and rider combination at constant speed (without bracking or accelerating forces). The influence of aerodynamic forces has not been considered (see Appendix 1). The linear homogeneous equations [7] were reduced to a set of first order differential equations by introducing new variables:

$$v = v_1 \quad ; \quad \dot{v} = v_2$$

$$\gamma = \gamma_1 \quad ; \quad \dot{\gamma} = \gamma_2$$

$$\rho = \rho_1 \quad ; \quad \dot{\rho} = \rho_2$$

$$\lambda = \lambda_1 \quad ; \quad \dot{\lambda} = \lambda_2$$

$$\tau_G = \tau_{G1} \quad ; \quad \dot{\tau}_G = \tau_{G2}$$

$$\tau_H = \tau_{H1} \quad ; \quad \dot{\tau}_H = \tau_{H2}$$

$$\varkappa_B = \varkappa_1 \quad ; \quad \dot{\varkappa}_B = \varkappa_2$$

Assuming the solutions of the form

$$v_1 = v_{1o}e^{\mu t} \quad ; \quad v_2 = v_{2o}e^{\mu t} \quad ; \quad \gamma_1 = \gamma_{1o}e^{\mu t}$$

$$\gamma_2 = \gamma_{2o}e^{\mu t} \quad ; \quad \rho_1 = \rho_{1o}e^{\mu t} \quad ; \quad \rho_2 = \rho_{2o}e^{\mu t}$$

$$\lambda_1 = \lambda_{1o}e^{\mu t} \quad ; \quad \lambda_2 = \lambda_{2o}e^{\mu t} \quad ; \quad \tau_{G1} = \tau_{G1o}e^{\mu t}$$

$$\tau_{G2} = \tau_{G2o}e^{\mu t} \quad ; \quad \tau_{H1} = \tau_{H1o}e^{\mu t} \quad ; \quad \tau_{H2} = \tau_{H2o}e^{\mu t}$$

$$\varkappa_1 = \varkappa_{1o}e^{\mu t} \quad ; \quad \varkappa_2 = \varkappa_{2o}e^{\mu t}$$

and introducing in the original equations a set of algebraic equations were obtained and expressed in a matrix notation as

$$\mu \ [B] \ [X] + [A] \ [X] = 0 \qquad (1)$$

A and B are square matrices, X is a column vector of variables and $\mu$ denotes the eigenvalue. The eqn. 1 was solved using the suitable digital computer programs. The eigenvalues $\mu_i$ (i = 1,2, ...m) and the corresponding eigenvectors $[X]_i$ were determined. The real part of the eigenvalue corresponds to the decay constant whereas the imaginary part represents the damped angular natural frequency of the considered mode.

The natural frequencies and corresponding decay constants (imaginary and real parts of the eigenvalues) were calculated at different speeds and various design parameter values of a heavy type motorcycle. One set of values of damped natural frequencies $\omega_d$ (rad/s) and corresponding decay constants are shown in fig. 2. The major and important modes are categorised as:

-Monotonic roll mode termed as "Capsize mode"
-Yaw mode termed as "Weave mode"
-Steering oscillatory mode termed as "Wobble mode".

The damped natural frequency of the weave mode increases with the speed and reaches an asymptotic value with further increase of speed (fig. 2).

## 3. EXPERIMENTAL ANALYSIS

An instrumented motorcycle of heavy class was driven along a straight line and the dynamic response was measured under the excitation of nearly impulse type lateral force generated through an "Impulse generating device" (sudden expansion of compressed air through a nozzle). The test vehicle was equiped with transducers to measure the following motion variables:

| | |
|---|---|
| -Yaw rate | rate gyro |
| -Roll rate | rate gyro |
| -Steering angle | rotary precision potentiometer |
| -Steering angular rate | tachogenerator |
| -Steering torque | torque dynamometer with strain gauges |
| -Body lean angular rate | tachogenerator geared to a plastic strip attached to rider's upper body |
| -Lateral excitation force | force dynamometer with strain gauges |
| -Speed | impulse generator with necessary frequency/voltage conversion electronic circuits |

All the rate gyros and other necessary electronic circuits were integrated in the fuel tank so that the environment around the rider on the motorcycle remains unchanged compared to the production vehicle. The integrated transducer unit (steering angle, steering torque and steering angular rate) is mounted between the front fork and handle bar with insignificant dimensional changes as compared to the production motorcycle. A compact portable tape recorder with FM multiplexing unit was mounted on the luggage carrier and used to record all the dynamic variables during the test driving, which were analysed and evaluated in the laboratory. The total weight of the instrument package was kept roughly same as the weight of the petrol resulted through the reduction of the tank capacity. This means the difference in weight between the production machine and the instrumented motorcycle was kept to a minimum possible value.

## Stability Parameters

To determine the stability parameters (natural frequencies and damping constants) the vehicle was driven along a straight line at constant test speed and lateral force disturbance was applied. Simultaneously the dynamic variables such as yaw rate, roll rate, body lean angular rate, steering angle and lateral force were recorded and a sample of a record is shown in fig. 3. The measurements were made at test speeds from 20 km/h to 160 km/h. Using the yaw rate - time signal the logarithmic decrement $\delta$ at different test speeds was determined and subsequently the damping ratio $\xi$ and the damped natural frequency $f_d$ of weave mode has been calculated (The system was assumed as a mass-spring-damper system with a single degree of freedom)*

The analytical and experimental values of damping ratio $\xi$ and damped natural frequency at different speeds were shown in figs. 4 and 5. The curve describing the experimental values of damping ratio $\xi$ (fig. 5) can be divided into two zones of interest:

- Zone of nearly-zero activity of the rider;
  speed greater than 100 km/h
- Zone of activity of the rider;
  speed upto 100 km/h

The zone of activity can be further split up as follows;

- zone of intensive activity - speed 0 to 60 km/h - and
- zone of moderate activity - speed 60 to 100 km/h - .

In the zone of nearly-zero activity (system approaches to openloop operation) the validity of the analytical model is quite good, whereas in the other region certain discrepancies were observed between the theoretical and experimental values of damping ratio. The main reason for this is that the rider reacts more interactive with the vehicle (closed-loop operation) in this zone which is not considered in the model. The frequency values showed the same trend (fig. 4).

## Handling Parameters

For the analysis of handling behaviour the negotiation of a roughly 90° curve of 50 m radius at different speeds has been chosen as driving task. The rider was asked to negotiate the curve at constant test speed and during the test driving the same motion variables as the straight line running tests were recorded. A set of recorded signals are shown in fig. 6. The test speeds were varied from 30 to 70 km/h and each run was repeated twice.

For the first approximation the steering torque during the transition phase of the curve negotiation at speeds greater than 50 km/h

---

* $\xi = 1/(1+(2\pi/\delta)^2)^{1/2}$ ; $\delta$ = logarithmic decrement

could be expressed through the following equation under the assumption that the steering torque will be mainly resulted due to gyroscopic effects due to roll rate, which has been established by Ellis et al.[8] (at speeds > 50 km/h).

$$M_L = I_2 . \ddot{\lambda} - I_3 . \frac{\dot{u}}{R_v} . (\dot{\rho} + \dot{\lambda} \sin\sigma) \quad (2)$$

Since $\lambda \ll 1$

$$M_L = -I_3 . \frac{\dot{u}}{R_v} \dot{\rho} \quad (3)$$

where $M_L$ = steering torque (Nm)
$I_3$ = moment of inertia of rotating parts of the front wheel about spin axis ($kg.m^2$)
$\dot{u}$ = forward speed (m/s)
and $I_2$ = moment of inertia of steerable parts about steering axis ($kg.m^2$)

Equation 3 indicates that the steering torque $M_L$ is linearly proportional to the product of speed and roll rate (gyroscopic torque due to roll rate). For a given speed the torque $M_L$ is directly proportional to the roll rate. In fig. 7 for two test speeds the relationship between $M_L$ and $\dot{\rho}$ has been shown, from which the linearity could be seen. This justifies the above assumption to a greater extent.

If very simplified and approximated time histories of steering torque and roll rate during curve negotiation at speeds > 50 km/h are assumed as shown in fig. 8 the following relationships can be derived. From fig. 8 and eqn. 3

$$\hat{M}_L = -K_k . \dot{u} . \hat{\dot{\rho}} \quad (4)$$

where $K_k = I_3/R_v = -\hat{M}_L/(\dot{u}\hat{\dot{\rho}})$

and (^) indicates the maximum value of the variable.

The value of $K_k$ depends on the design parameters of the motorcycle and termed as "Handling Index", abbreviated as HI. In fig. 9 the experimental values of $-\hat{M}_L/(\hat{\dot{\rho}}\,\dot{u})$ for various speeds have been shown and the value of HI ($K_k$) for the test vehicle was determined as

$K_k$ = 5.40 ($N.cm.s^2/deg.m$)

From fig. 8

$$\rho_{ss} = \int_0^{T_k} \dot{\rho}(t) . dt = \int_0^{T_{k1}} \frac{\hat{\dot{\rho}}}{T_{k1}} . t . dt + \int_{T_{k1}}^{T_k} \hat{\dot{\rho}} \left(1 - \frac{1}{T_k - T_{k1}} (t - T_{k1})\right) dt$$

$$= \hat{\dot{\rho}} . \frac{T_k}{2} = \tan^{-1}(\dot{u}^2 \varkappa / g)$$

$$\hat{\dot{\rho}} = \frac{2}{T_k} . \tan^{-1}(\dot{u}^2 \varkappa / g) \quad (5)$$

Putting eqn. 5 in eqn. 4 we get

$$\hat{M}_L = - \frac{2K_k}{T_k} \cdot \dot{u} \cdot \tan^{-1}(\dot{u}^2 \varkappa / g) \tag{6}$$

where $T_k$ = transition time (sec)

$\dot{u}$ = speed (m/s)

and $\varkappa$ = curvature of the curve - 1/R - (1/m).

From the experimental data the average value of $T_k$ has been calculated and equals to 1.6 sec. This value was observed to be nearly independent of speed in the range greater than 50 km/h. Using the eqn. 6 the theoretical value of $\hat{M}_L$ at various speeds for $T_k$ = 1.6 s and radius R as parameter has been calculated and shown in fig. 10. The experimental values are also indicated in the same figure from which very good agreement could be seen at speeds $>$50 km/h (HI was kept constant). For a given maximum value of steering torque $\hat{M}_L$ , which is the limiting factor of the rider's capability, and the radius of the curve R the possible maximum curve negotiating speed could be determined. In fig. 11 the variation of $\hat{M}_L$ with speed at various values of "Handling Index" ($K_k$ - different design parameters of $I_3$ and $R_v$) for R = 50 m and $T_k$ = 1.6 s has been shown. It is observed that for a given value of $\hat{M}_L$ the maximum curve negotiating speed increases as the value of HI ($K_k$) decreases. This enhances the safety operation of the vehicle - rider system.

In fig. 12 the variation of maximum curve negotiating speed with various values of HI for $T_k$ = 1.6 s and $\hat{M}_L$ = -20 Nm at different radii has been drawn from which it could be seen that the lower values of HI ($K_k$) result in higher curve negotiating speeds for all radii.

## 4. CONCLUSIONS

Based on the results of this work following conclusions could be drawn:

The results of analytical (open-loop mode) and experimental analysis of straight line running stability exhibited fairly good validation of analytical model in the zone of nearly zero activity (speeds $>$ 100 km/h).

The analysis of handling behaviour showed that the parameter "Handling Index" (HI) in the curve negotiating task plays an important role. Based on the values of HI the curve negotiating capabilities can be estimated.

## REFERENCES


(1) Sharp, R.S., The Stability and Control of Motorcycles. J.Mech Engg. Science, vol. 13, 1971

(2) Eton, D.J., Man-Machine Dynamics in the Stabilisation of Single Track Vehicles. Ph.D. Thesis, HSRI, Univ. of Michigan,1973

(3) Weir, D.H., A Manual Control View of Motorcycle Handling. Second Int. Cong. on Automotive Safety San Francisco, 1973
(4) Singh, D.V. and Goel, V.K., Stability of Rajdoot Scooter. SAE Paper Nr. 710273
(5) Rice, R.S., Davis, J.A. and Kunkel, D.T., Accident Avoidance Capabilities of Motorcycles. Techn. Report DOT - 801810, 1976
(6) Koch, J., Chenchanna, P. and Willumeit, H.P., Lenkverhalten des Fahrers von Zweiradfahrzeugen. Entwicklungslinien in der Kraftfahrzeugtechnik, VDI, 1976
(7) Koch, J., Experimentelle und analytische Untersuchungen am Motorrad-Fahrer System, Diss. Techn. Univ. Berlin, 1979
(8) Ellis, J.R. and Hyhoe, G.F., The Steady State and Transient Handling Characteristics of a Motorcycle. Second Int. Cong. on Automotive Safety San Francisco, 1973

## APPENDIX 1

Notations

| | | |
|---|---|---|
| $\dot{v}$ | m/s | Lateral velocity |
| $\gamma$ | rad | Yaw angle |
| $\rho$ | rad | Roll angle |
| $\lambda$ | rad | Steering angle |
| $\tau_G$ | rad | Twist angle of the front fork |
| $\tau_H$ | rad | Twist angle of the main frame |
| $\varkappa_B$ | rad | Body lean angle |
| $R_V$ | m | Rolling radius of front wheel |
| $\rho_{ss}$ | rad | Steady state roll angle |
| $\sigma$ | rad | Steering head angle |

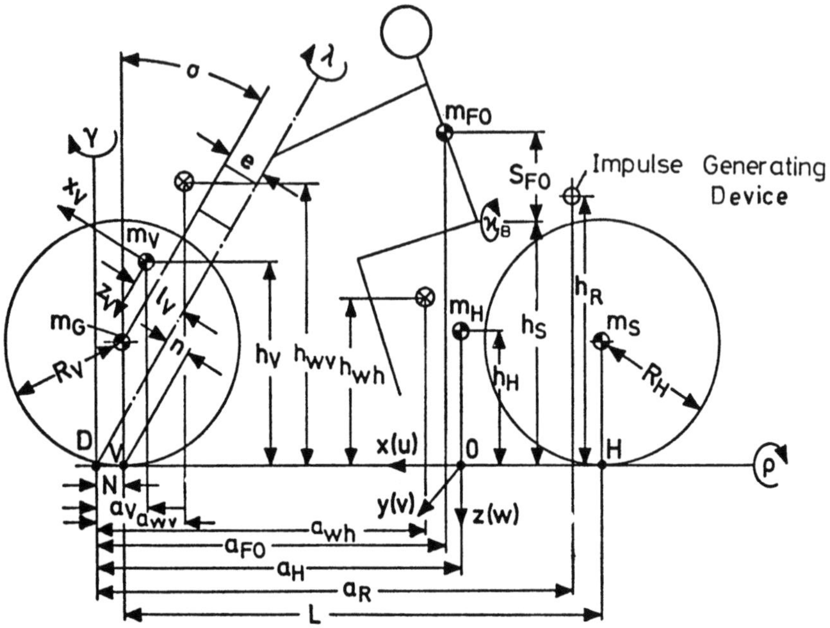


Fig. 1 Model representation of motorcycle-rider system

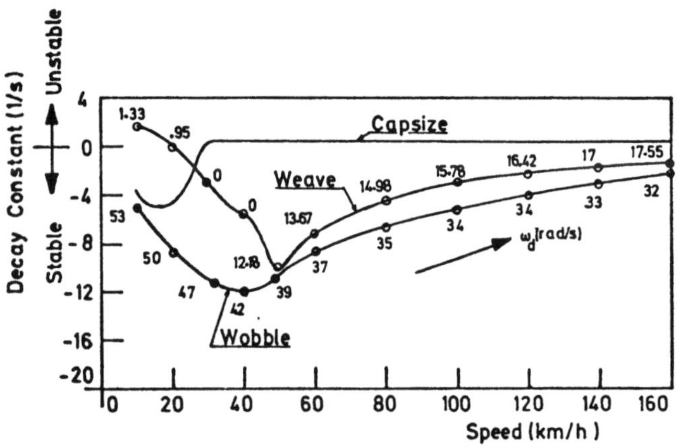


Fig. 2 Theoretical values of damped natural frequency and decay constant at various speeds

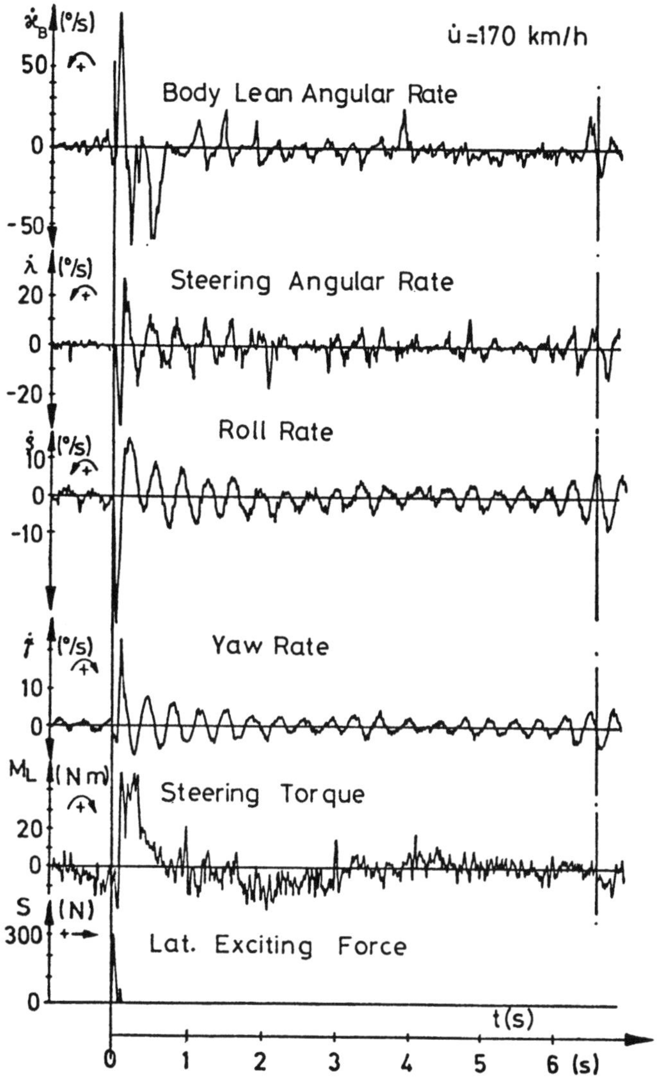


Fig. 3 Time histories of dynamic variables recorded during straight line running stability test

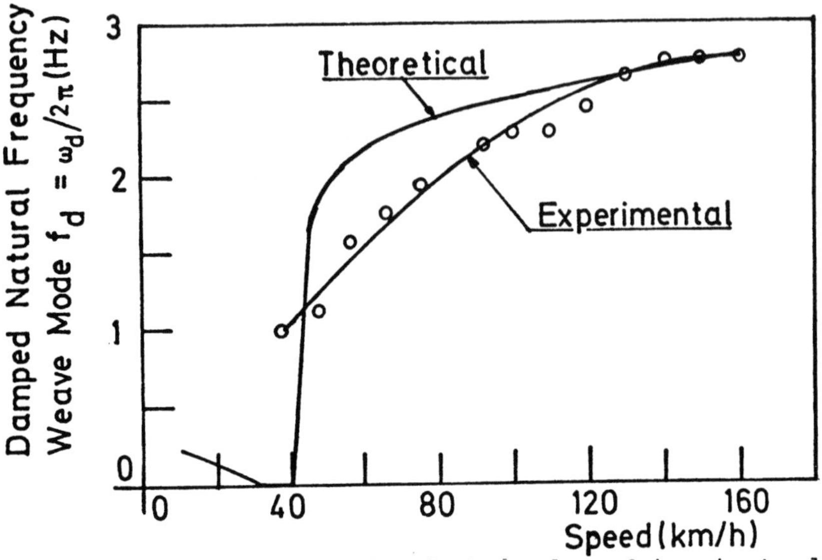


Fig. 4 Experimental and analytical values of damped natural frequency in weave mode at various speeds

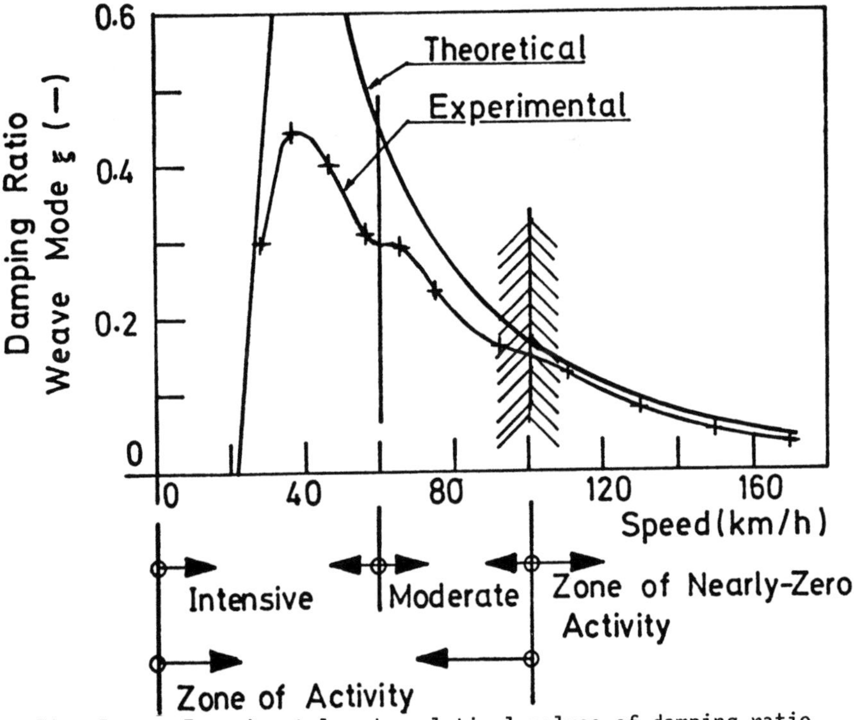


Fig. 5 Experimental and analytical values of damping ratio in weave mode at various speeds

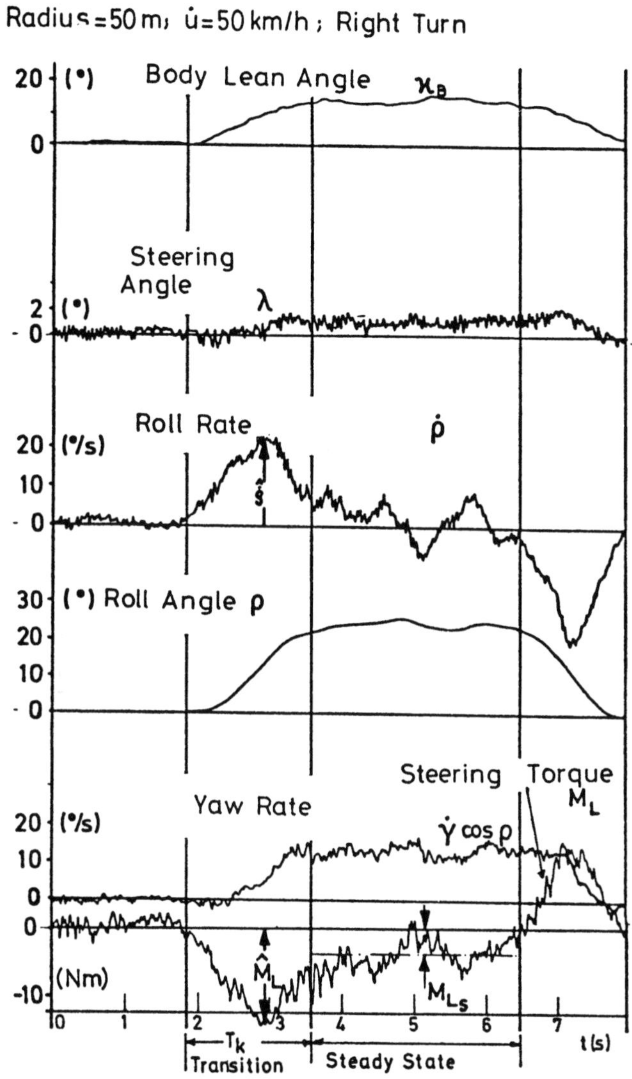


Fig. 6 Time histories of recorded variables during the curve negotiation test

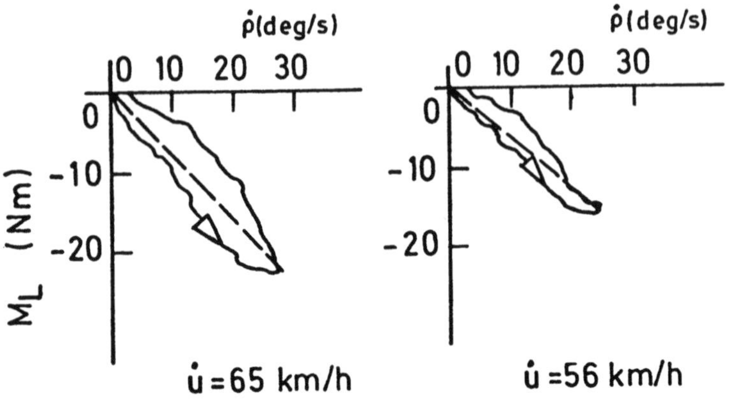


Fig. 7 Experimental relationship between roll rate and steering torque during curve negotiation test

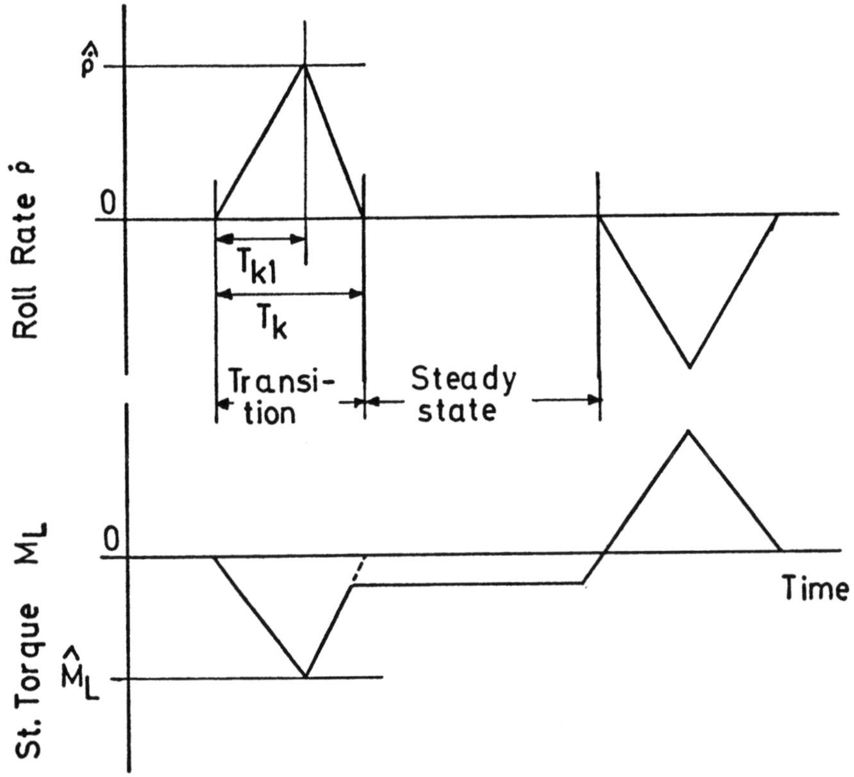


Fig. 8 Idealised time histories of roll rate and steering torque during curve negotiation

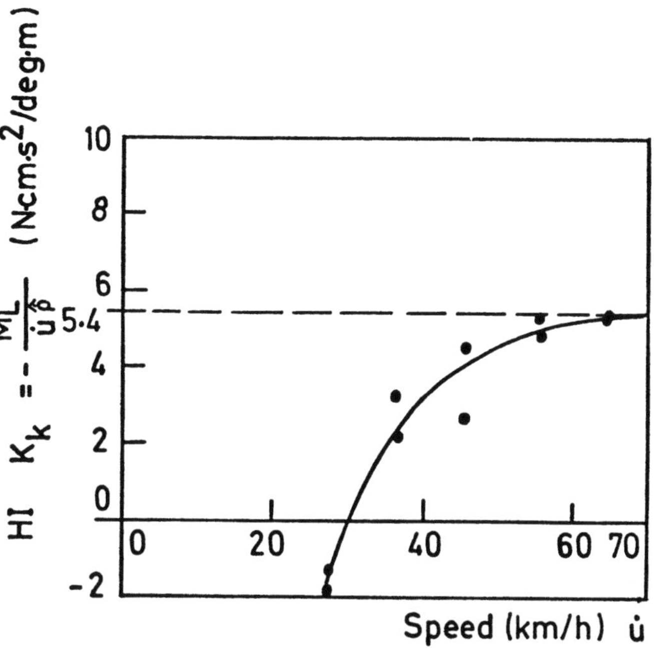


Fig. 9 Experimental values of Handling Index (HI) at various speeds

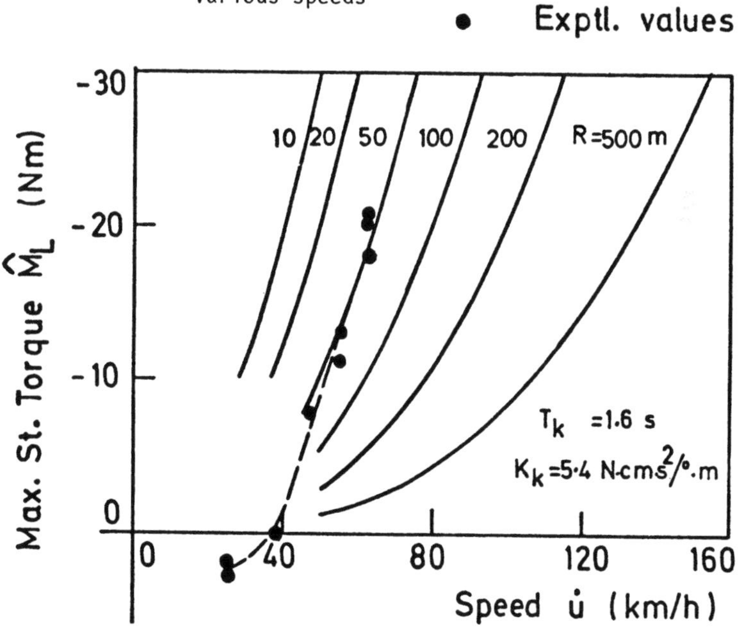


Fig. 10 Values of maximum steering torque at various speeds and radii

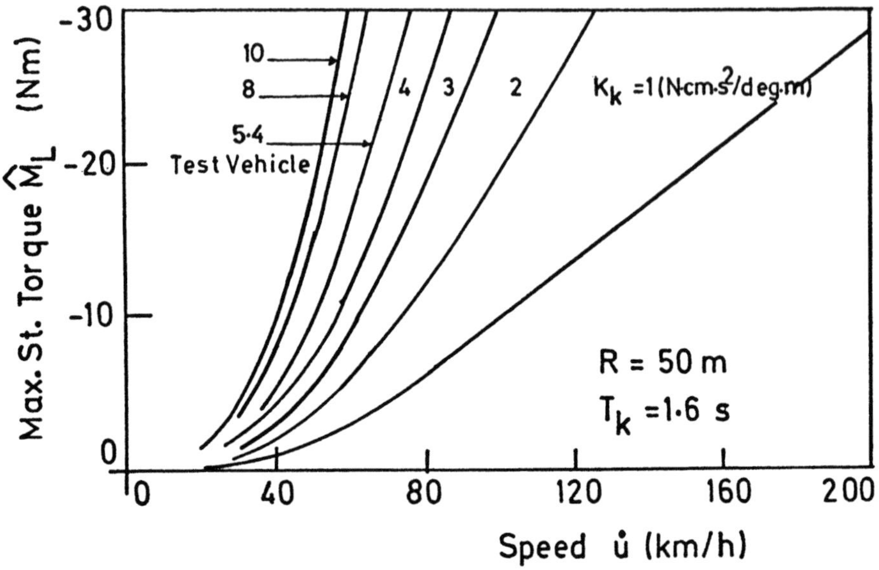


Fig. 11 Values of maximum steering torque at various speeds and Handling Index values

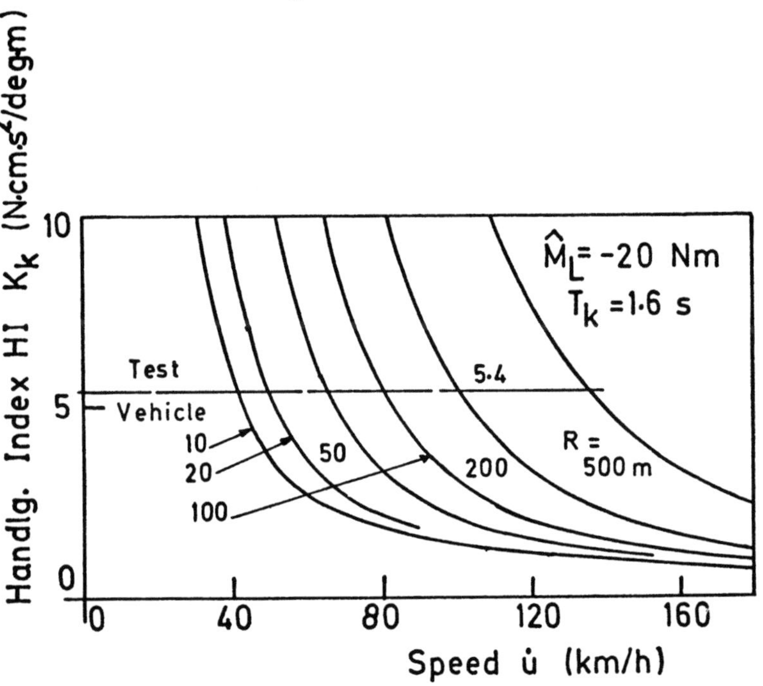


Fig. 12 Values of Handling Index at various speeds and radii

# MATHEMATICAL INVESTIGATION TO REDUCE "STEERING WHEEL SHIMMY"

Gerhard Dödlbacher

Advanced Vehicle Engineering, Ford-Werke AG, Köln, F.R.G.

SUMMARY

In the velocity range between 80 and 100 km/h 'shimmy' occurs with many passenger cars. Shimmy is a vibration of the wheel around an instantaneous axis and results in a torsional motion of the steering wheel.

As these vibrations significantly deteriorate driver comfort, the suspension design needs to be of low sensitivity. Development of such suspensions requires not only reduced amplitudes and accelerations as measured, but also confirmation with subjective ratings.

Steering wheel shimmy is caused by the vibration of the road wheel on the suspension and tire. Not only is the vibration of the wheel of importance, but also tuning other vehicle vibration systems can influence steering wheel vibrations. Due to the large number of variables a computer programme has been developed.

The mechanical model, which consists of the steering system, the suspension and vehicle vibration system, has in total 19 degrees of freedom and can be used for SLA suspensions with isolated crossmembers as well as for McPherson suspensions with rigid crossmembers.

The model also considered the geometric changes during jounce/rebound. The most important geometric parameters are:

- king pin offset
- toe change characteristic
- camber characteristic
- wheel recession.

The differential equation system has been developed using d'Alembert's law.

To investigate the phase relations both wheels are considered. The following excitations are provided:

- static unbalance at one or both wheels with/without phase relations
- dynamic unbalance
- run-out
- tyre non-uniformity
- road irregularities.

The equations have been programmed for the Hybrid Computer EAJ Pacer 600.

However, the factors, influencing shimmy also infuence other properties of the vehicle, i.e.

suspension noise transmission
brake pull
straight running
steering efforts
front end shake.

Therefore, the selection of the geometric and other variables must be a compromise between different requirements.

To validate the computed results extensive rig tests have been carried out and the measured steering wheel acceleration indicated good correlation to the computed figures.

Parameters with higher influence on shimmy are: toe-change characteristic, camber characteristic, gear box mounting stiffness, steering compliance, steering wheel moment of inertia, king pin offset, longitudinal stiffness of suspension, wheel recession, crossmember mounting.

Parameters without influence are: static toe setting, camber, wheel load, wheel rates.

The significance of the major variables is approximately as follows:

| | |
|---|---|
| geometry | 30 % |
| longitudinal stiffness of suspension | 25 % |
| longitudinal stiffness of crossmember mountings | 25 % |
| torsional stiffness of steering system | 10 % |
| moment of inertia of the steering wheel. | 10 % |

The most important parameters are discussed below:

### Wheel Recession and Longitudinal Stiffness

For each suspension design there is a wheel recession value which minimises the steering wheel acceleration. This minimum is a function of the longitudinal suspension stiffness whereby high stiffness leads generally to an improvement.

### Camber Change

During jounce the road wheel should go into positive camber. In this case the gyroscopic moments reduce vibration of the wheels.

### Steering Wheel Moment of Inertia

Increase of steering wheel moment of inertia significantly reduces the steering wheel accelerations.

### Steering System Compliance

High torsional compliance in the steering system also significantly reduces the steering wheel accelerations. Steering couplings with strong non-linear spring characteristics improve the shimmy characteristic without deterioration of the 'steering response'.

Steering Gear Mounting

The rubber mounting of the steering system should be as soft as possible in lateral direction. The limit is the influence of this mounting on vehicle handling.

Rubber Mounted Subframe

SLA front suspensions often have a rubber mounted crossmember. The longitudinal stiffness of the mounts has a significant influence on steering wheel accelerations.

New suspension systems have been built according to these theoretically determined guidelines. With these suspension systems significant improvements could be obtained and have confirmed this mathematical investigation as an effective tool for vehicle design and development.

The relation between steering wheel acceleration and subjective ratings, as well as the results of subjective evaluation of an initial suspension design and an improved version, are given.

## 1. INTRODUCTION

In the velocity range between 80 and 100 km/h 'shimmy' occurs with many passenger cars. 'Shimmy' is a vibration of the wheel around an instantaneous axis and results in a torsional motion of the steering wheel. [1]

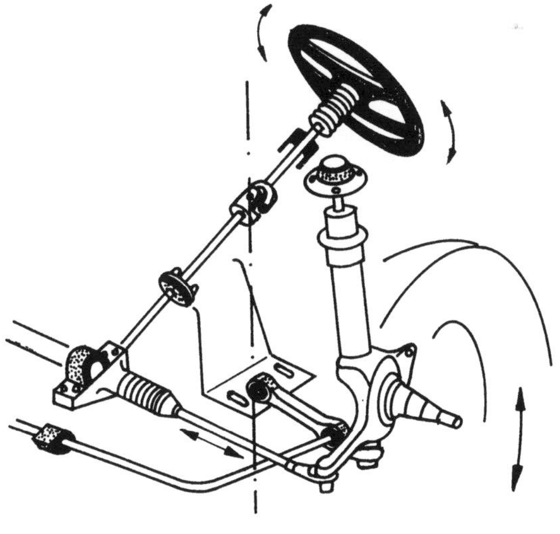

Figure 1 : Scheme of McPherson suspension.

As the resultant vibrations reduce driver comfort, the suspension design needs to be of low sensitivity. Development of such suspensions requires not only reduced amplitudes and accelerations as measured, but also confirmation with subjective ratings. A subjective evaluation method is described below.

Shimmy Rating System

| Characteristics for evaluation of steering wheel oscillations: Damping and transfer of oscillations (both hands) | Perceptibility | Visibility (simultan-ously damped) | Rating Index | Accept-ability |
|---|---|---|---|---|
| Steering wheel lightly held between finger-tips. | Not perceptible | Not visible | 10 | |
| Steering wheel lightly held between finger-tips. | Just perceptible | Not visible | 9 | |
| Steering wheel lightly held. | Just perceptible | Not visible | 8 | |
| Oscillations almost damped when steering wheel is lightly held. | Clearly perceptible | Barely visible | 7 | |
| Oscillations almost damped when steering wheel is firmly held. | Clearly perceptible | Visible with small amplitudes | 6 | |
| Oscillations will be trans-ferred to hands. Steering wheel firmly held. | Clearly perceptible Unpleasant | Visible with small amplitudes | 5 | Border-line |
| Oscillations will be trans-ferred to hands. Steering wheel firmly held. | Clearly perceptible Annoying | Distinctly visible | 4 | |
| Oscillations will be trans-ferred to hands and fore-arms. Steering wheel firmly held. | Clearly perceptible | Distinctly visible | 1-3 | |

The correlation between subjective rating and measured steering wheel acceleration is shown in figure 2.

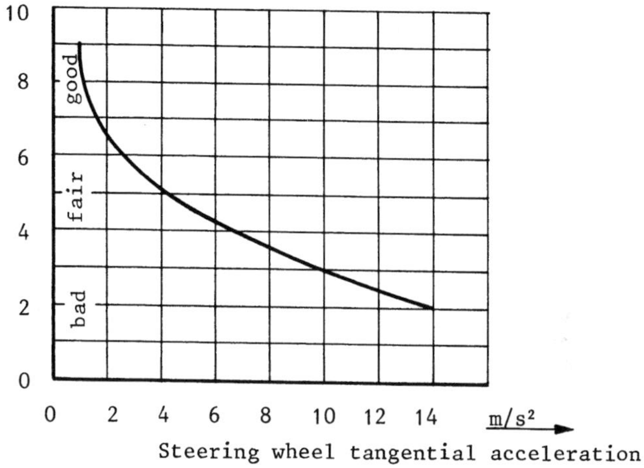


Figure 2 : Subjective rating versus acceleration

2. DEFINITION

Shimmy is caused by a vibration of the road wheel on its suspension and the tyre. The road wheel is excited by unbalance, tyre/wheel run-out, tyre non-uniformity and road irregularities. These possible kinds of excitation are shown in figures 3 and 4.

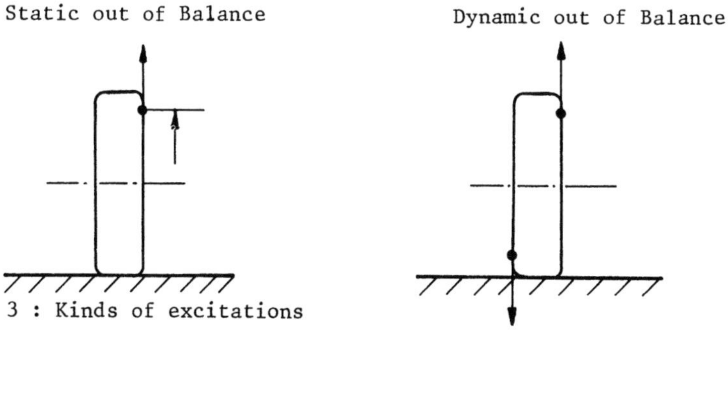


Figure 3 : Kinds of excitations

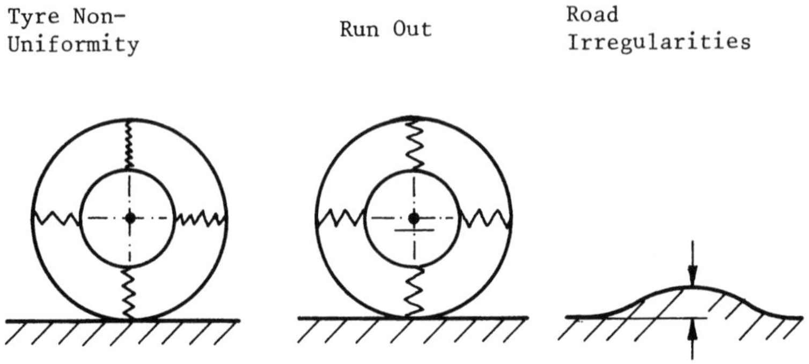


Figure 4: Kinds of excitations

Not only the torsional motion of the road wheel is important for shimmy. All the vibration systems of the front part of the vehicle influencing the steering system, e.g. engine crossmember mounting, elastically mounted engine system are significant. In addition the change of suspension geometry during jounce is of influence. Shimmy can be reduced by tuning of all coupled vibration systems. For optimisation of such a complex system a computer program has been developed.

## 3. MECHANICAL MODEL

The mechanical model is described in figures 5a + 5b.

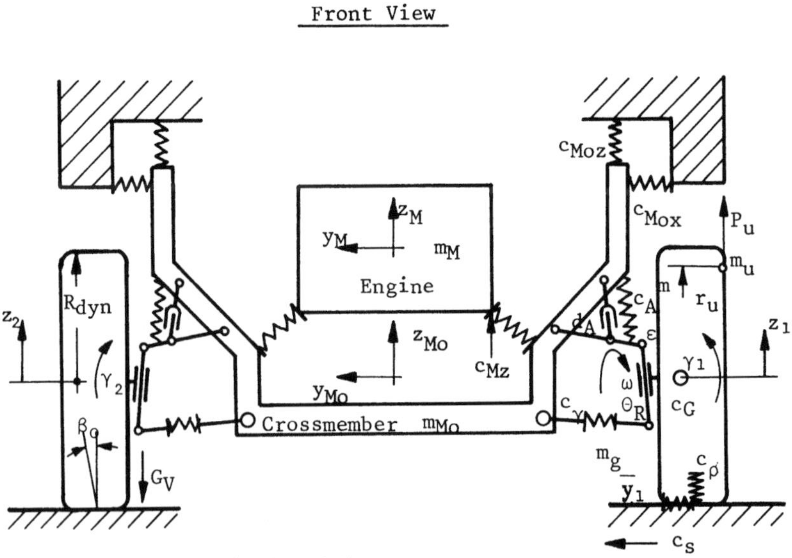


Figure 5a : Mechanical Model

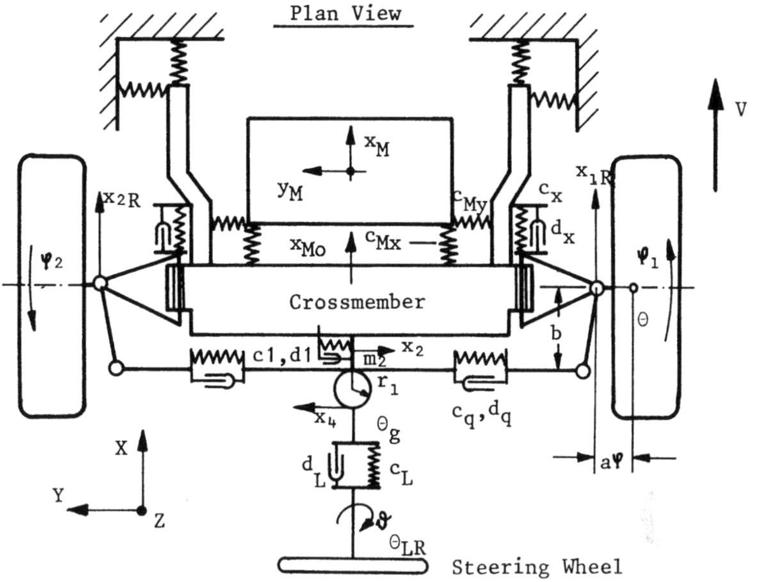


Figure 5b : Mechanical Model

This model shows a double wishbone front suspension with rubber mounted subframe as an example. Other suspension types have also been modelled.

The following main systems are considered:

Steering system with tie rods, tie rod elasticity, steering gear mounting, steering system compliance, steering wheel.

Suspension with consideration of camber, king pin inclination, castor, toe-in, wheel recession, king pin offset.

Total vehicle vibration system like rubber mounted subframe, rubber mounted engine.

In total the model has 19 degrees of freedom. [2]

## 4. DEGREES OF FREEDOM

The following degrees of freedom describe the vibration of masses of the 'shimmy' model.

Steering System

1. Rotational motion of steering wheel $\vartheta$
2. Lateral motion of elastically mounted steering gear $x_4$
3. Lateral motion of tie rod $x_2$

Suspension

| | |
|---|---|
| 4./5. Steer angle of the road wheels | $\varphi_1$, $\varphi_2$ |
| 6./7. Lateral displacement in tyre contact areas | $\bar{y}_1$, $\bar{y}_2$ |
| 8./9. Camber angle of wheels | $\gamma_1$, $\gamma_2$ |
| 10./11. Vertical displacement of wheels | $z_1$, $z_2$ |
| 12./13. Fore + aft motion of the wheel | $x_{1_R}$, $x_{2_R}$ |

Vehicle Vibration System

| | |
|---|---|
| 14. Front crossmember vertical | $z_{Mo}$ |
| 15. Front crossmember lateral | $y_{Mo}$ |
| 16. Front crossmember fore + aft | $x_{Mo}$ |
| 17. Engine vertical | $z_M$ |
| 18. Engine lateral | $y_M$ |
| 19. Engine fore + aft | $x_M$ |

## 5. DEVELOPMENT OF DIFFERENTIAL EQUATIONS

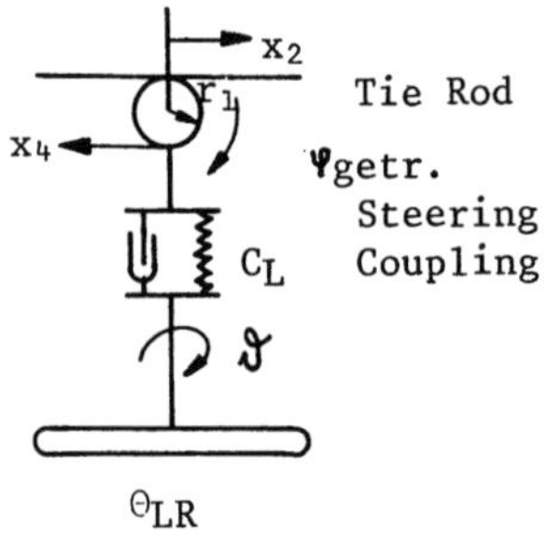


Figure 6 : Steering Wheel

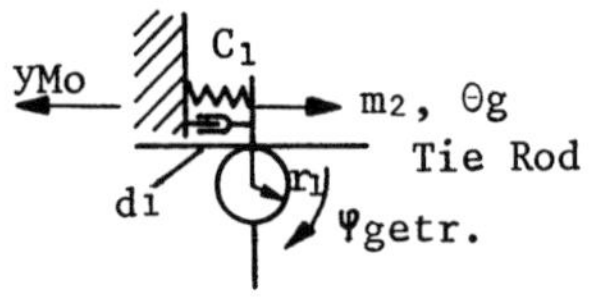


Figure 7 : Steer. Gear Mounting

$$\varphi_{getr.} \cdot r_1 - x_4 = x_2 \qquad \Big| \; \Theta_{LR} \cdot \ddot{\vartheta} + C_L \left(\vartheta - \frac{x_2 + x_4}{r_1}\right) + d_L \left(\vartheta - \frac{x_2 + x_4}{r_1}\right) = 0$$

$$m_2 \cdot x_4 + C_1 (x_4 - y_{Mo}) - \frac{C_L}{r_1} \cdot \left(\vartheta - \frac{x_2 + x_4}{r_1}\right) = 0$$

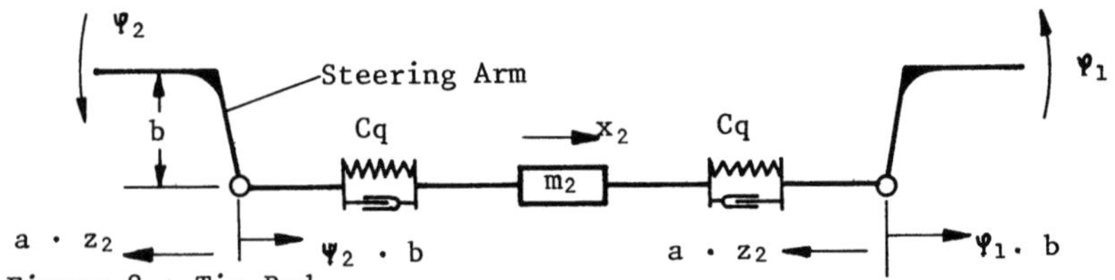


Figure 8 : Tie Rod

$$\left(\frac{\Theta g}{r_1^2} + m_2\right) \cdot \ddot{x}_2 - cq \cdot (\varphi_1 \cdot b + a \cdot z_1 - x_2) -$$

$$cq\ (\varphi_2 \cdot b - a \cdot z_2 - x_2) + \frac{c_L}{r}\left(\vartheta - \frac{x_2 + x_4}{r_1}\right) = 0$$

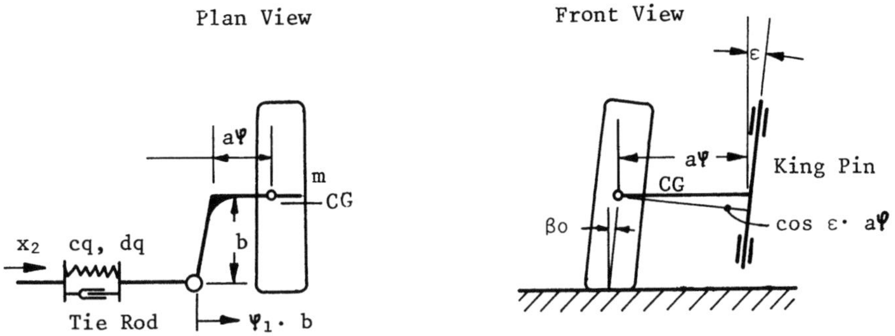


Figure 9: Moment about King Pin, Consideration of King Pin Inclination

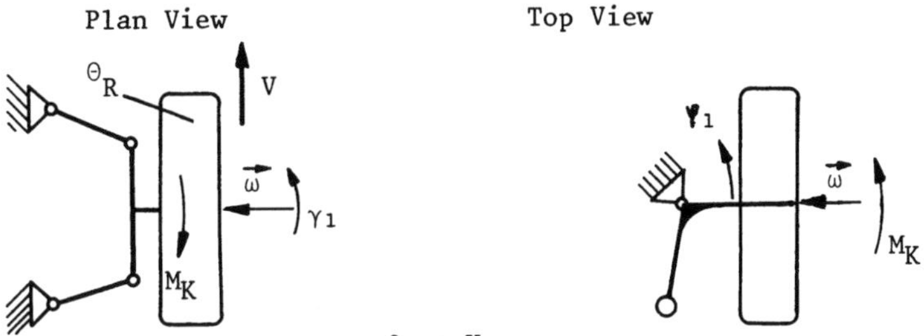


$$M_K = \frac{\Theta_R \cdot V}{Rdyn.} \cdot \dot{\gamma}_1$$

Figure 10: Gyroscopic Moments

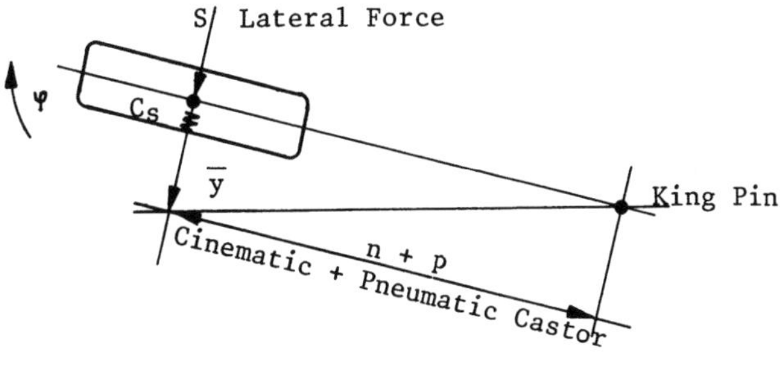


Figure 11: Simplified Tyre Model

Moments About King Pin

$$M = S\,(n + p)$$

$$S = \delta \cdot \alpha_{eff}$$

$$\alpha_{eff} = \varphi + \frac{n \cdot \dot{\varphi}}{V} - \frac{\dot{\bar{y}}}{V} \qquad [3]$$

$$M = (n + p) \cdot \vartheta \cdot \left(\varphi + \frac{n \cdot \dot{\varphi}}{V}\right) - \frac{\dot{\bar{y}} \cdot \delta}{V}\,(n + p)$$

Lateral Tyre Deformation

$$S = C_s \cdot \bar{y}_1$$

$$C_s \cdot \bar{y}_1 = \delta\left(\varphi_1 + \frac{n \cdot \varphi_1}{V} - \frac{\dot{\bar{y}}_1}{V}\right)$$

$$\delta \cdot \varphi_1 + \frac{\delta \cdot n \cdot \varphi_1}{V} - C_s \cdot \bar{y}_1 - \frac{\delta \cdot \dot{\bar{y}}_1}{V} = 0$$

More exact description of tyre behaviour is given by Pacejka [4].

Suspension Kinematics

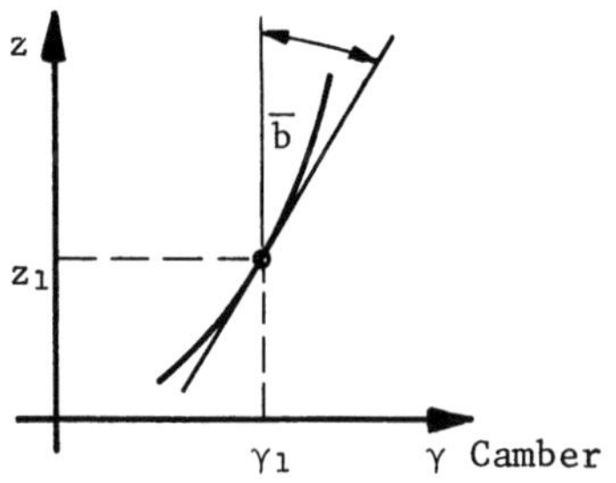


Figure 12: Camber Change

$$\tan \bar{b} = \frac{\gamma_1}{z_1}$$

$$\gamma_1 = \tan \bar{b} \cdot z_1 \approx \bar{b} \cdot z_1$$

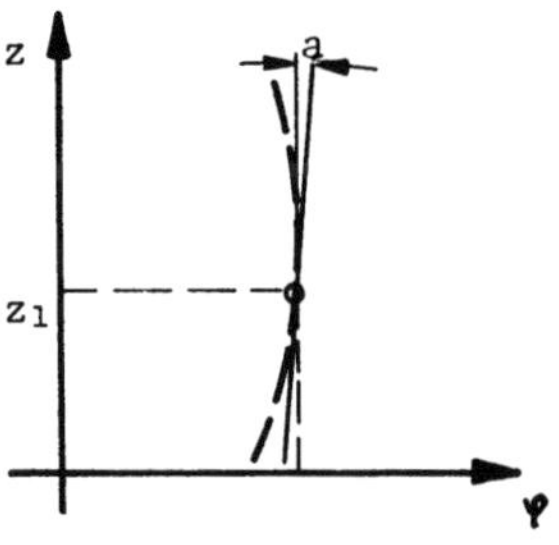


Figure 13: Toe Change, see Figure 8

$$\tan a = \frac{\varphi}{z_1}$$

$$\varphi = \tan a \cdot z_1 \approx a \cdot z_1$$

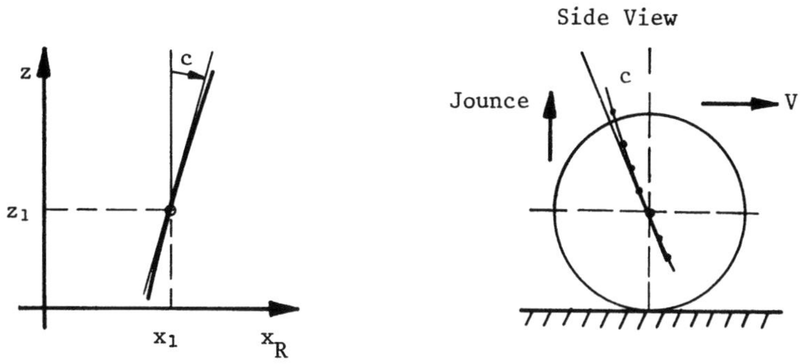


Figure 14: Recession

$\tan c = \frac{x_{1R}}{z_1}$ $\qquad x_{1R} = \tan c \cdot z_1 \approx c \cdot z_1$

## Differential Equations

$$\Theta_{LR} \cdot \ddot{\vartheta} + c_L \left(\vartheta - \frac{x_2 + x_4}{r_1}\right) + d_L \cdot \left(\vartheta - \frac{\dot{x}_2 + \dot{x}_4}{r_1}\right) = 0 \qquad (1.)$$

$$m_2 \cdot \ddot{x}_4 + c_1 \cdot (x_4 - y_{Mo}) - \frac{c_L}{r_1} \cdot \left(\vartheta - \frac{x_2 + x_4}{r_1}\right) = 0 \qquad (2.)$$

$$\left(\frac{\Theta}{r_1^2} + m_2\right) \cdot \ddot{x}_2 - cq \cdot (\varphi_1 \cdot b + a \cdot z_1 - x_2) - cq\,(\varphi_2 \cdot b - a \cdot z_2 - x_2) + \qquad (3.)$$

$$\frac{c_L}{r_1} \left(\vartheta - \frac{x_2 + x_4}{r_1}\right) = 0$$

$$(\Theta_0 + m \cdot a^2 \varphi \cos^2 \varepsilon) \cdot \varphi_1 + cq\,(\varphi_1 \cdot b + a \cdot z_1 - x_2) \cdot b + \delta\,(\varphi_1 + \qquad (4.)$$

$$\frac{n}{V} \cdot \dot{\varphi}_1) \cdot (p+n) + c_\varphi \cdot \varphi_1 + d_\varphi \cdot \dot{\varphi} - \frac{\delta}{V}\,\dot{\bar{y}}\,(p+n) - \frac{\Theta_R}{R_{dyn.}} \cdot V \cdot \dot{\gamma}_1 +$$

$$\varphi_1 \cdot \frac{G_V}{2} \cdot \varepsilon \left[R_{dyn}\,(\beta_0 + \varepsilon) + a_\varphi - \varepsilon \cdot R_{dyn}\right] = P_u \cdot a_\varphi \cdot \cos \omega t$$

$$\vartheta \cdot \varphi_1 + \frac{\delta \cdot n}{V} \cdot \dot{\varphi}_1 - c_s \cdot \bar{y}_1 - \frac{\delta}{V} \cdot \dot{\bar{y}}_1 = 0 \qquad (5.)$$

Differential Equations (continued)

$$(m+mq)\ \ddot{z}_1+z_1\cdot(c_\rho+c_A)+\dot{z}_2\cdot d_A-c_x\cdot c\cdot x_{1_R}+ \qquad (6.)$$

$$(c_\rho+c_A)\cdot\varepsilon\cdot(1-\cos\varphi_1)\cdot\left[R_{dyn}\ (\beta_o+\varepsilon)+\right.$$

$$\left.a_\varphi-\ \varepsilon\cdot R_{dyn}\right]=P_u\cdot\sin\omega t\cdot\cos\beta_o$$

$$\Theta_\gamma\cdot\ddot{\gamma}_1+c_\gamma\cdot(\gamma_1+\bar{b}\cdot z_1)+\Theta_R\cdot\ \omega\cdot\dot{\varphi}_1-R_{dyn}\cdot\delta\ (\varphi_1+\frac{n}{V}\ \dot{\varphi}_1-\frac{\dot{y}_1}{V})=0 \qquad (7.)$$

$$(m+m_q)\ddot{x}_{1_R}+c_x\ (x_{1_R}-c\cdot z_1-x_{Mo})+\ddot{\varphi}_1\cdot a_\varphi\cdot m=P_u\cdot\cos\ \omega\ t \qquad (8.)$$

$$(\Theta_o+m\cdot a_\varphi^2\cdot\cos^2\varepsilon)\ \ddot{\varphi}_2+c_q(\varphi_2\cdot b-a\cdot z_2-x_2)\ b+\delta\ (\varphi_2+\frac{n}{V}\ \dot{\varphi}_2)\ \cdot \qquad (9.)$$

$$(p+n)\ +c_\varphi\cdot\varphi_2+d_\varphi\cdot\dot{\psi}_2-\frac{\delta}{V}\cdot\dot{y}_2(p+n)\ +\frac{\Theta_R}{R_{dyn}}\cdot V\cdot\dot{\gamma}_2+\varphi_2\cdot$$

$$\frac{GV}{2}\cdot\varepsilon\left[R_{dyn}\ (\beta_o+\varepsilon\ )\ +a_\varphi-\varepsilon\ \cdot\ R_{dyn}\right]=0$$

$$\delta\cdot\varphi_2+\frac{\delta\cdot n}{V}\cdot\dot{\varphi}_2-c_s\cdot\dot{y}_2-\frac{\delta}{V}\cdot\dot{y}_2=0 \qquad (10.)$$

$$(m+m_q)\ \ddot{z}_2+z_2\cdot(c_\rho+c_A)+\dot{z}_2\cdot d_A-c_x\cdot c\cdot x_{2_R}+\ (c_\rho+c_A)\ \cdot\varepsilon(1- \qquad (11.)$$

$$\cos\varphi_2)\cdot\left[\ R(\beta_o+\varepsilon)\ +a_\varphi-\varepsilon\ R_{dyn}\right]=P_u\sin\ \omega\ t\ \cdot\ \cos\ \beta_o$$

$$\Theta_\gamma\cdot\gamma_2+c_\gamma\ \ (\gamma_2+\bar{b}\cdot z_2)\ +\ \Theta_R\ \omega\ \dot{\varphi}_2\ -\ R_{dyn}\ \cdot\ \vartheta(\varphi_2+\frac{n}{V}\dot{\varphi}_2-\frac{\dot{y}_2}{V})=0 \qquad (12.)$$

$$(m+m_q)\ \ddot{x}_{2_R}+c_x\ (x_{2_R}-c\cdot z_2-x_{M_o})\ +\ddot{\varphi}_2\cdot a_\varphi\cdot m=P_u\cdot\cos\ \omega\ t \qquad (13.)$$

$$m_{Mo}\cdot\ddot{x}_{Mo}+c_x\ (x_{Mo}-x_{1_R}+c\cdot z_1)\ +\ c_x\ (x_{Mo}-x_{2_R}+c\cdot z_2)\ + \qquad (14.)$$

$$2c_{Mox}\ \cdot\ x_{Mo}\ -\ 2c_{Mx}\ (x_M-x_{Mo})\ =\ 0$$

$$(m_{Mo}+2m)\cdot\ddot{y}_{Mo}+2c_{Moy}\cdot y_{Mo}-c_1\ (x_4-y_{Mo})\ -2c_{My}\ (y_M-y_{Mo})\ =\ 0 \qquad (15.)$$

$$m_{Mo}\cdot\ddot{z}_{Mo}+c_A\ (z_{Mo}-z_1)\ +\ c_A\ (z_{Mo}-z_2)\ +\ d_A\ (\dot{z}_{Mo}-\dot{z}_1)\ +\ d_A\ \cdot \qquad (16.)$$

$(\dot{z}_{Mo}-\dot{z}_2) - 2c_{Mo}\ (z_M-z_{Mo}) + 2c_{Moz}\cdot z_{Moz} = 0$

$$m_M \cdot \ddot{x}_M + 2c_{Mx}\ (x_M - x_{Mo}) = 0 \quad (17)$$

$$m_M \cdot \ddot{y}_M + 2c_{My}\ (y_M - y_{Mo}) = 0 \quad (18)$$

$$m_M \cdot \ddot{z}_M + 2c_{Mz}\ (z_M - z_{Mo}) = 0 \quad (19)$$

These equations have been programmed for the Hybrid Computer EAJ Pacer 600. The calculations on the computer have been carried out for all interesting parameters and for vehicles with different suspension types. In the following the maximum steering wheel acceleration is shown as function of the significant parameters. Some of these parameters reducing shimmy, however, require extensive design changes. These parameters are king pin offset and scrub radius. Other parameters reducing the torsional motion of the steering wheel, however, deteriorate other vehicle properties. Therefore, in practice the choice of the parameters must be a compromise between different requirements. The following properties of the vehicle are influenced by parameter changes: - Noise/vibration transmission of the suspension, brake pull sensitivity, steering returnability, straight running, parking effort, steering response, front end shake -.

## 6. INFLUENCE OF PARAMETERS ON WHEEL 'SHIMMY'

Parameters with great influence are: Toe change, camber change, stiffness of steering gear mounting, steering compliance, steering wheel moment of inertia, king pin offset, longitudinal suspension stiffness, wheel recession, castor, engine mounting system, subframe mounting.

Roughly the amount of influence is as follows:

| | |
|---|---|
| Geometry | 30 % |
| Longitudinal stiffness of suspension | 25 % |
| Longitudinal stiffness of subframe mounting | 25 % |
| Steering compliance | 10 % |
| Moment of inertia of steering wheel | 10 % |
| | 100 % |

In the following the most important parameters should be discussed:

### 6.1 Wheel Recession and Total Longitudinal Suspension Stiffness

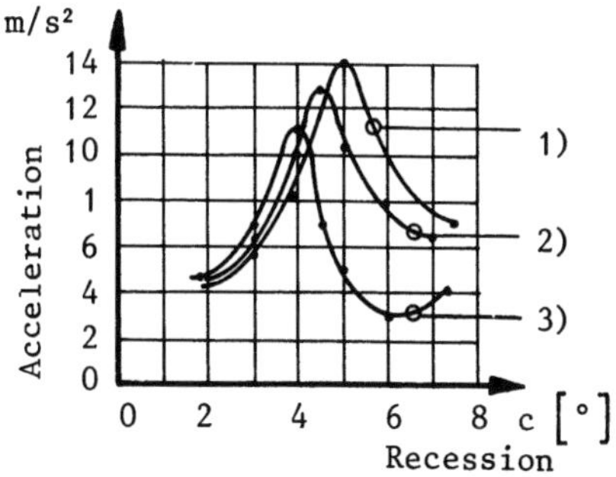

Figure 15: Influence of Wheel Recession

For the vehicle under investigation the relation between steering wheel acceleration and wheel recession, with respect to suspension stiffness is shown in figure 15. With higher longitudinal stiffness the masimum is reduced and shifted to lower wheel recession values.

1) $c_x = 0{,}9 \cdot 10^6$ N/m
2) $c_x = 1{,}5 \cdot 10^6$ N/m
3) $c_x = 2{,}2 \cdot 10^6$ N/m

## 6.2 Toe Change

The calculation indicated the significant influence of toe change on the steering wheel acceleration. Contrary to the wheel recession the toe characteristic has no influence in combination with other parameters.

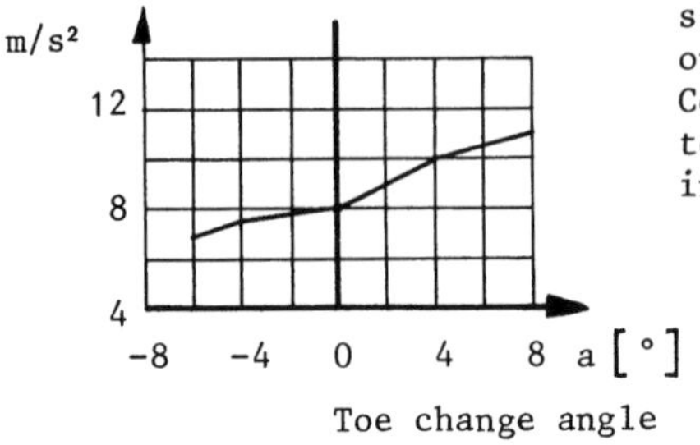


Figure 16: Influence of Toe Characteristic

## 6.3 Camber Characteristic

The wheels should move to slight positive camber during jounce. Thus the vibration on the wheel is reduced by the gyroscopic moment.

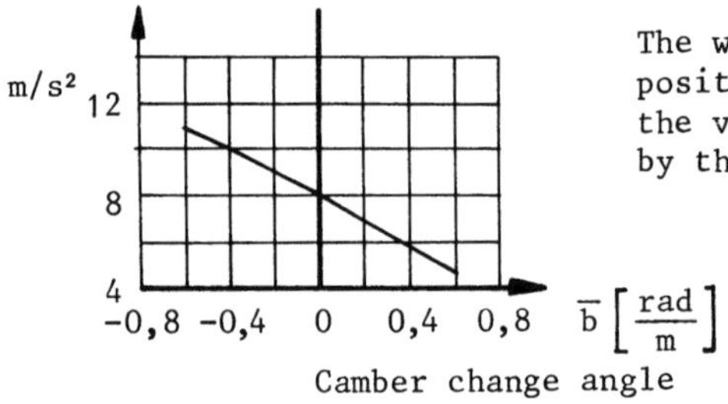


Figure 17: Influence of Camber Characteristic

## 6.4 Moment of Inertia of Steering Wheel and Steering Compliance

Increase of the steering wheel moment of inertia as well as steering compliance reduces the steering wheel acceleration. Steering couplings installed in the steering system have a strong non-linear characteristic. A characteristic which is soft around zero leads to an important improvement in shimmy.

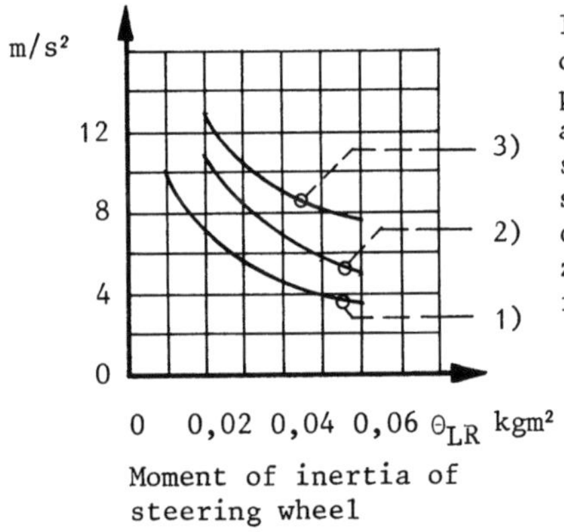


1) $c_L = 40 \frac{N \cdot m}{rad}$

2) $c_L = 100 \frac{N \cdot m}{rad}$

3) $c_L = 120 \frac{N \cdot m}{rad}$

Figure 18: Influence of Moment of Inertia

## 6.5 Elastical Mounting of Steering System

The elastically mounted steering system should be as soft as possible in lateral direction. A strong non-linear characteristic is necessary not to deteriorate handling.

### 6.6 Elastical Mounting of Subframe

Double wishbone long and short axles often are fixed on a subframe which is rubber mounted to the body. The longitudinal stiffness of these bushes has a decisive influence on shimmy. The longitudinal stiffness should be as low as possible. (Non-linear characteristic required.)

## 7. COMPARISON OF TESTS AND COMPUTER RESULTS

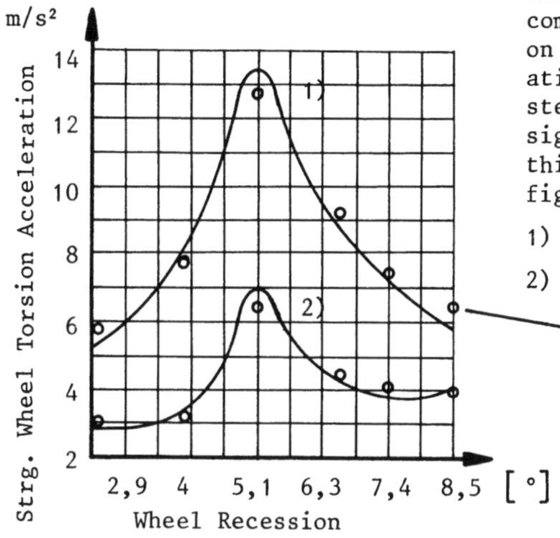


The computer results have been compared with tests carried out on a drum test rig. The accelerations have been measured on the steering wheel and other significant points. A result of this comparison is shown in figure 19.

1) Hard steering coupling

2) Soft steering coupling

Figure 19: Comparison of Tests and Computer Results

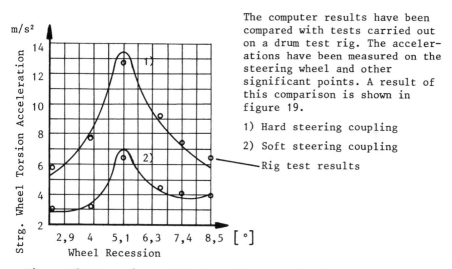

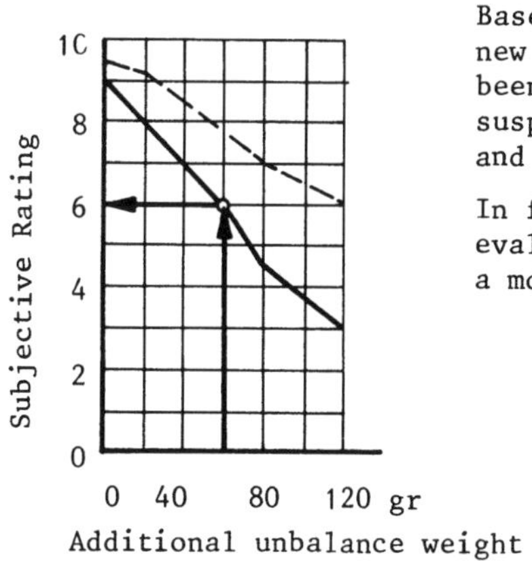


Based on these computer investigations new SLA and McPherson suspensions have been designed and built. With these suspension road tests were carried out and shimmy was subjectively evaluated.

In figure 20 the results of a subjective evaluation for the initial condition and a modified suspension design are shown.

Figure 20: Subjective Rating System

NOTATIONS

$a$ toe change coefficient

$a_\varphi$ king pin offset

$b$ length of steering arm

$\bar{b}$ camber change coefficient

$c_1$ lateral stiffness of steering gear mounting

$c$ recession coefficient

$c_A$ wheel rate

$c_L$ steering coupling sitffness

$c_{Mx}$ engine mounting fore and aft

$c_{My}$ engine mounting lateral

$c_{Mz}$ engine mounting vertical

$c_{Mox}$ crossmember mounting fore and aft stiffness

$c_{Moy}$ crossmember mounting lateral stiffness

$c_{Moz}$ crossmember mounting vertical stiffness

$c_q$ tie rod elasticity

$c_s$ lateral tyre stiffness

$c_x$ suspension stiffness fore and aft

$c_\gamma$ camber stiffness

$c_\rho$ radial tyre stiffness

$c_\varphi$ torsional stiffness

$d_A$ shock absorber damping

$d_L$ steering system damping

$d_x$ suspension damping fore and aft

$d_\varphi$ torsional damping

$G_V$ front axle load

$m$ wheel mass

$m_z$ mass of steering system

$m_M$ engine mass

$m_{Mo}$ subframe mass

$m_q$ mass of track control arm

$n$ kinematic castor

$n_p$ pneumatic castor

$P_u$ unbalance force

$r_1$ pinion gear

NOTATIONS continued

$R_{dyn}$ dynamic tyre radius

$r_u$ radius of unbalance

V vehicle velocity

$\beta_o$ camber angle

$\varepsilon$ king pin inclination

$\delta$ tyre coefficient

$\Theta_o$ moment of inertia of road wheel (vertical axis)

$\Theta_g$ moment of inertia of rotating masses of steering gear system

$\Theta_{LR}$ moment of inertia of steering wheel

$\Theta_R$ moment of inertia of road wheel (rotational axis)

$\Theta_\gamma$ moment of inertia of road wheel (camber)

$\omega$ angular velocity of road wheel

REFERENCES

1. Döhring, E. und Becker, F.: Die Lenkungsunruhe der McPherson-Achsen, ATZ 5, 1979, P. 155 - 162

2. Dödlbacher, G. und Gaffke, H.G.: Untersuchung zur Reduzierung der Lenkungsunruhe, ATZ 7/8, 1978, P. 317 - 322

3. Zomotor, A.: Beitrag zur Theorie der Lenkungsunruhe, Automobilindustrie 4, 1970, P. 101 - 109

4. Pacejka, H.: The Wheel Shimmy Phenomenon, Doctoral Thesis Delft 1966

# NONLINEAR RANDOM RESPONSE OF ARTICULATED VEHICLES

M. A. Dokainish[1] and M. M. ElMadany[2]

[1] Department of Mechanical Engineering, McMaster University,
Hamilton, Ont.
[2] Wyle Laboratories, Colorado Springs, CO., U.S.A.

SUMMARY

This paper deals with the nonlinear modelling of articulated vehicle suspensions in heave and pitch modes. The nonlinearity considered is the Coulomb friction presented in the vehicle suspension laminated springs. The main purpose of this theoretical analysis is to investigate the effect of Coulomb friction forces on the ride comfort of the articulated vehicle using the equivalent linearization technique.

## 1. INTRODUCTION

The dynamic response of the articulated vehicles to random surface undulations has been treated using simplified vehicle models by assuming linear suspension systems. The linear analyses of articulated vehicle dynamics have provided an invaluable understanding of the effect of the various vehicle parameters on the riding behaviour of the vehicle [1, 2, 3, 4, 5]. The assumption of linearity is justifiable to simplify the consideration of basic features of vehicle ride motion, but system nonlinearities should be included in the practical design of the suspension. The nonlinear characteristic that is frequently present in articulated vehicle suspension is the dry friction. The frictional force generated in the leaf springs depends on the frictional surfaces, their material and quality. This force is in the range of 7-20% of the static load on the front axles, and 15-30% of the static load on the rear axles for the semielliptical laminated springs [6]. The static frictional force will cause the vehicle to vibrate on the tires, since there is no deflection of the suspension springs until the tire deflection produces sufficient force to overcome the breakaway friction. This type of nonlinearity may be expected to influence the ride dynamics of the vehicle to an appreciable degree. Consequently, there is a need to evaluate the effect of the dry friction on the general behaviour of the vehicle.

Typically, the response of the articulated vehicle, treated as a discrete nonlinear dynamical system subjected to random excitation, is modelled by a set of second order nonlinear stochastic differential equations. The nonlinear model can be analyzed using simulation techniques by utilizing analogue and digital computers to integrate the equations of motion [7, 8, 9]. However, the practicality of such techniques is severely limited by the expense of computer time and the need to simulate the vehicle response to a wide variety of initial conditions to thoroughly examine the riding quality of the vehicle. Therefore, an analytical technique; equivalent linearization technique, is sought for the analysis of the nonlinear motion response of the articulated vehicle.

In this paper, the equivalent linearization technique is adapted to give a technique applicable to the articulated vehicle ride problem. The influence of dry friction on vehicle riding quality is reported, and the nature of the vibratory motions is described. It is the object of this study to examine this nonlinearity in order to gain a measure of its relative importance on the dynamic response of the articulated vehicle to the random road surface undulations. This will give a guide to the accuracy which must be attained in assessing its effects for a real vehicle so as to enable successful prediction of vehicle response to be carried out. The results have been obtained from an analysis of nonlinear equations of motion written for an articulated vehicle, modelled in heave and pitch modes.

## 2. ROAD DESCRIPTION

It has been recognized from a number of measurements on roadways that their roughness is normally distributed and can be accurately represented as a stationary random process [10]. As the power spectral density of the input process is a quantity suitable for the analysis of the dynamic behaviour of the vehicle on which a random process is acting, the road random disturbances are represented by their power spectral densities. References [11-12] discuss the spectral densitites that closely approximate available experimental data and propose that road roughness can be described by

$$\begin{aligned} s(n) &= s(n_o)\left(\frac{n}{n_o}\right)^{-r_1}, \qquad n \leq n_o \\ &= s(n_o)\left(\frac{n}{n_o}\right)^{-r_2}, \qquad n \geq n_o, \end{aligned} \tag{1}$$

where

$s(n)$ is the spatial spectral density of the road roughness, $(m^2/c/m)$.

$s(n_o)$ is the roughness coefficient (the value of the spectral density at the discontinuity frequency, $n_o$).

$n$ is the spatial frequency of road roughness

$n_o$ is $1/2\pi$ c/m

$r_1, r_2$ are waviness.

## 3. ARTICULATED VEHICLE MODEL

The articulated vehicle system adopted for this study is shown in Fig. 1. The vehicle is considered to be travelling over an uneven road at a constant forward velocity. The tractor and semitrailer are allowed to translate in the longitudinal and vertical directions and to pitch except as constrained by the coupling system between the two units. The unsprung mass of each wheel and axle assembly is represented by a mass having vertical translation freedom only. Each suspension is represented by a viscous damper and a linear spring in parallel with a source of dry friction of constant value, that is, a source which provides a force opposing the sliding but otherwise independent of sliding velocity. Each tire acts as a linear spring-damper system having point contact with the road surface.

By considering the constraints imposed by the fifth wheel on the motion of the tractor and semitrailer, the model includes six degrees of freedom; heaving and pitching motions of the tractor centre of gravity, $(Y_t, \theta_t)$, pitching motion of the semitrailer centre of gravity, $(\theta_s)$, and the vertical motions of the three axle suspension systems, $(Y_1, Y_2, Y_3)$.

## 4. ANALYSIS OF ARTICULATED VEHICLE MOTION USING EQUIVALENT LINEARIZATION TECHNIQUE

The vehicle system can be considered as an n-degree-of-freedom system connected by nonlinear elements. The mathematical equations which describe the response of the system may be written as:

$$M\ddot{\bar{x}} + C\dot{\bar{x}} + K\bar{x} + g(\dot{\bar{y}},\bar{y}) = \bar{f}(t), \tag{2}$$

where $\bar{x}' = [Y_t, \theta_t, \theta_s, Y_1, Y_2, Y_3]$ is the generalized displacement vector

$\bar{y}$ is the relative displacement vector across the nonlinear elements

$M, C, K$ are the mass, damping and stiffness matrices, respectively

$\bar{g}(\dot{\bar{y}},\bar{y})$ is a nonlinear vector function of the dependent variable $\bar{y}$ and its derivative $\dot{\bar{y}}$

$\bar{f}(t)$ is the excitation force vector

Because of the scarcity of exact solutions of Eq. 2 when the function $\bar{g}(\dot{\bar{y}},\bar{y})$ is nonlinear, attention has been directed toward equivalent linearization technique which gives an approximate analysis. The principle of the technique is to replace the nonlinear dynamical system (2) by an auxiliary linear system for which the exact analytical formula for solution is known. The replacement is made so as to be optimum with respect to some measure of the difference between the original and the auxiliary system. The auxiliary system is called equivalent linear system. While this equivalent linear system is effectivley linear, the system response depends on signal amplitude, a basic characteristic of nonlinear behaviour.

The equivalent linear system is defined by the linear differential equation

$$M\ddot{\bar{x}} + (C + C_e)\dot{\bar{x}} + (K + K_e)\bar{x} = \bar{f}(t), \tag{3}$$

where $C_e$ and $K_e$ are two arbitrary matrices. These matrices are to be determined so that the mean square value of the difference, $\bar{e}$, between the original system and the equivalent linear system is minimized for every stationary Gaussian random vector $\bar{x}$. The difference $\bar{e}$ is defined by

$$\bar{e} = \bar{g}(\dot{\bar{y}},\bar{y}) - C_e\dot{\bar{x}} - K_e\bar{x} \tag{4}$$

The coefficients of the matrices $C_e$ and $K_e$ can be determined analytically if the nonlinear function $\bar{g}(\dot{\bar{y}},\bar{y})$ can be decomposed into a sum of simpler nonlinear elements, each of these nonlinear elements depends solely on the relative displacement and velocity of the masses of the system. Let the nonlinear function $\bar{g}(\dot{\bar{y}},\bar{y})$ take the following form

$$g_i\ (\dot{\bar{y}},\bar{y}) = \sum_{j=1}^{m} q_{ij}\ (\dot{y}_j,y_j), \tag{5}$$

where $q_{ij}(\dot{y}_j,y_j)$ denotes one of the nonlinear frictional forces acting along the direction of the generalized coordinate $x_i$ due to the velocity and displacement $\dot{y}_j$ and $y_j$, of the nonlinear element j.

By minimizing the mean square error, and using the mathematical properties of the multi-dimensional Gaussian distribution, it has been proved in references [13-14] that each nonlinear element $q_{ij}(\dot{y}_j,y_j)$ may be replaced by a combination of linear elements according to the rule

$$q_{ij}\,(\dot{y}_j, y_j) \rightarrow \alpha^e_{ij}\,\dot{y}_j + \gamma^e_{ij}\,y_j, \tag{6.a}$$

where
$$\alpha^e_{ij} = E[q_{ij}(\dot{y}_j, y_j)\dot{y}_j]/E[\dot{y}^2_j], \tag{6.b}$$

$$\gamma^e_{ij} = E[q_{ij}(\dot{y}_j, y_j)y_j]/E[y^2_j], \tag{6.c}$$

and the elements of $c^e_{ij}$ and $k^e_{ij}$ of the matrices $C_e$ and $K_e$ are linear combinations of $\alpha^e_{ij}$ and $\gamma^e_{ij}$; $i,j = 1, \ldots, n$.

Let the friction force be given by the following expression

$$q_{ij}(\dot{y}_j, y_j) = F_{ij}\ \mathrm{sgn}\ (\dot{y}_j), \tag{7}$$

then equation (6) can be written as

$$\alpha^e_{ij} = E[F_{ij}\ \mathrm{sgn}(\dot{y}_j)\ \dot{y}_j]/E[\dot{y}^2_j] \tag{8.a}$$

$$\gamma^e_{ij} = E[F_{ij}\ \mathrm{sgn}(\dot{y}_j)\ y_j]E[y^2_j] \tag{8.b}$$

The relative displacements and velocities, $y_j$ and $\dot{y}_j$, will not be Gaussian because of nonlinearity, however, for the equivalent linearization technique they are taken to be Gaussian. Representing the probability density, $p(y_j, \dot{y}_j)$, by two-dimensional Gaussian distribution with zero mean ($E[y_j\ \dot{y}_j] = 0$),

$$P(y_j, \dot{y}_j) = \frac{1}{2\pi\sqrt{E[y^2_j]E[\dot{y}^2_j]}}\ \exp\{-\frac{1}{2}\ [\ \frac{y^2_j}{E[y^2_j]} + \frac{\dot{y}^2_j}{E[\dot{y}^2_j]}\ \} \tag{9}$$

Substituting equation (9) into equations (8.a) and (8.b), the following solution can be obtained since $p(y_j, \dot{y}_j)$ is symmetric

$$\alpha^e_{ij} = \frac{\sqrt{2}}{\sqrt{\pi}}\ F_{ij}/\sqrt{E[\dot{y}^2_j]} \tag{10.a}$$

$$\gamma^e_{ij} = 0 \tag{10.b}$$

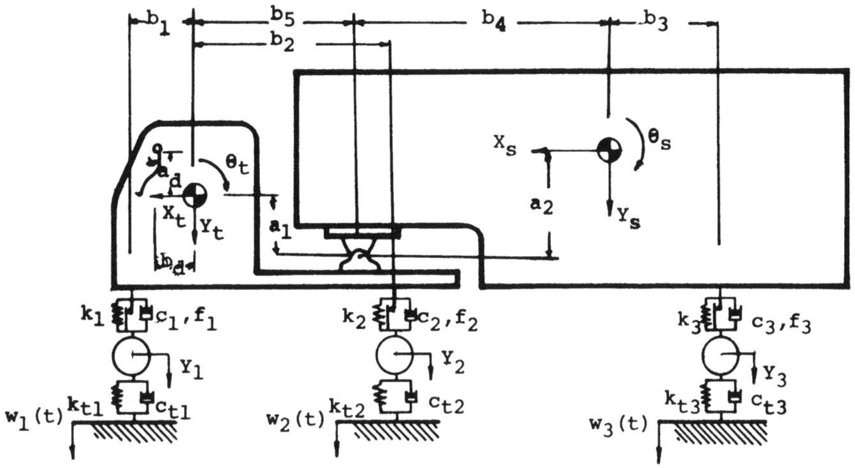


Fig. 1 Articulated vehicle model.

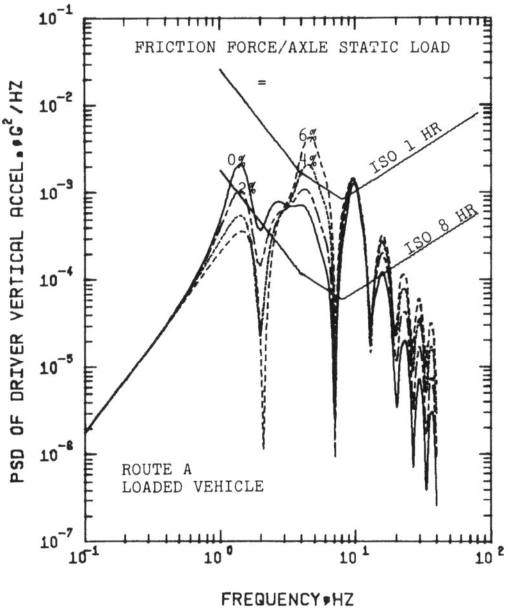


Fig. 2 Effect of dry frictional forces on driver acceleration spectra.

Table 1 Baseline values for articulated vehicle model

| | | | | | |
|---|---|---|---|---|---|
| I. | Tractor | | | | |
| | a. General | | | | |
| | sprung mass ($M_t$) | | | | 6400 kg |
| | front unsprung mass ($M_1$) | | | | 360 kg |
| | rear unsprung mass ($M_2$) | | | | 1450 kg |
| | pitch moment of inertia ($I_t$) | | | | 3390 $kgm^2$ |
| | b. Dimensions | | | | |
| | $b_1$ | 1.27 m | | $b_5$ | 1.4 m |
| | $b_2$ | 2.03 m | | $a_1$ | -0.3 m |
| | $b_d$ | 0.51 m | | $a_d$ | 0.95 m |
| II. | Semitrailer | | | | |
| | a. General | | | | |
| | sprung mass ($M_s$) | | | | 15180 kg |
| | unsprung mass ($M_3$) | | | | 1450 kg |
| | pitch moment of inertia ($I_s$) | | | | 152000 $kgm^2$ |
| | b. Dimensions | | | | |
| | $b_3$ | 3.96 m | | | |
| | $b_4$ | 4.57 m | | | |
| | $a_2$ | 1.02 m | | | |
| III. | Suspension Characteristics | | | | |
| | $k_1$ | 357 kN/m | | $c_1$ | 11.5 kNs/m |
| | $k_2$ | 630 kN/m | | $c_2$ | 29 kNs/m |
| | $k_3$ | 630 kN/m | | $c_3$ | 29 kNs/m |
| IV. | Tire Characteristics | | | | |
| | $k_{t1}$ | 1560 kN/m | | $c_{t1}$ | 0.7 kNs/m |
| | $k_{t2}$ | 5250 kN/m | | $c_{t2}$ | 1.2 kNs/m |
| | $k_{t3}$ | 5250 kN/m | | $c_{t3}$ | 1.2 kNs/m |
| V. | Vehicle Speed | | | 80 km/h | |
| VI. | Road Characteristics | | | | |
| | Roughness Coefficient | | | $s(n_o) = 13.8 \times 10^{-6}\ m^3/c$ | |
| | Waviness | | | $r_1 = 2.1,\ r_2 = 1.42$ | |

## 5. EFFECT OF DRY FRICTION

The power spectral density curves for the vertical accelerations at the driver position are shown in Fig. 2. The dry friction force in each suspension is considered to be a percent of the corresponding axle static load. The curves are plotted for different values of the ratio of the friction force to the axle static load; 0% (linear model), 2%, 4% and 6%. The computation has been carried out using the baseline values for the articulated vehicle parameters given in Table 1.

The results show that inclusion of the dry friction in the analysis appreciably affect the system response. Increasing the dry friction force diminishes the peak response in the low frequency region corresponding to the bounce frequency. However, the response increases considerably in the high frequency region by increasing the dry friction force. Specifically, the vibration comfort in the frequency range 3-6 Hz becomes worse and reaches ISO 1 hour boundary. In other words, increasing the friction will form a second resonance peak in the range of 3-6 Hz. These frequencies are in the range of human body resonances and as can be seen from the figure it is deleterious to the ride motion.

It may be seen from the figure that increasing the dry friction will stiffen the suspension and more energy will be transmitted to the sprung masses. The spectrum exceeds the 8-hour ISO guide in both the low and high frequency range for the linear system. However, increasing the friction made the spectra exceed the 1-hour ISO guide in the high frequency range. This ride is definitely rough corresponding to 80 km/h.

To show the effect of tractor front and rear suspension friction dampings on the vehicle vibrational responses, the amount of dry frictional forces are varied while all other vehicle parameters are held constant at the baseline values. Fig. 3 illustrates three-dimensional plot of the rms vertical accelerations at the tractor centre of gravity, while Fig. 4 illustrates three-dimensional plot of the rms vertical accelerations at the driver's location. The rms accelerations are plotted in terms of the ratio of the dry frictional forces of each axle to the corresponding axle static loads. From a study of the plots it may be concluded that the dry friction damping in vehicle suspension is not unidirectional. The dry friction damping improves the ride comfort expressed in rms acceleration up to a certain extent. At the tractor centre of gravity, the optimum friction dampings are different for different locations in the vehicle structure. The optimum values of dry friction are zero at the tractor front axle and 4% of the static load at the tractor rear axle for the rms acceleration. The rms driver vertical accelerations are minimum for frictional forces of zero values at the front and rear axles.

The difference in the optimum values of the frictional forces that provide minimum values of the rms accelerations at the tractor

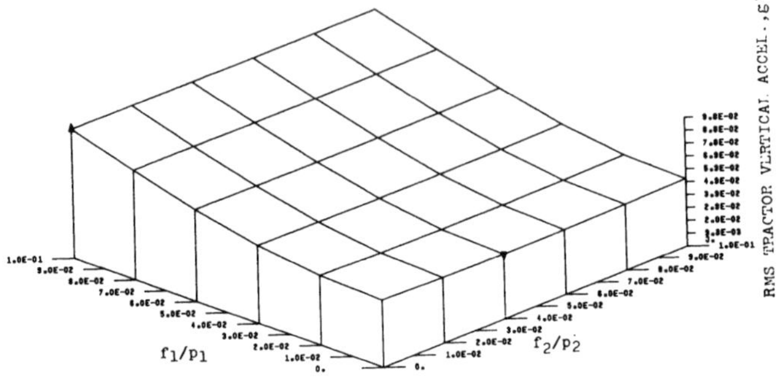


Fig. 3 Effect of dry frictional forces on rms tractor acceleration.

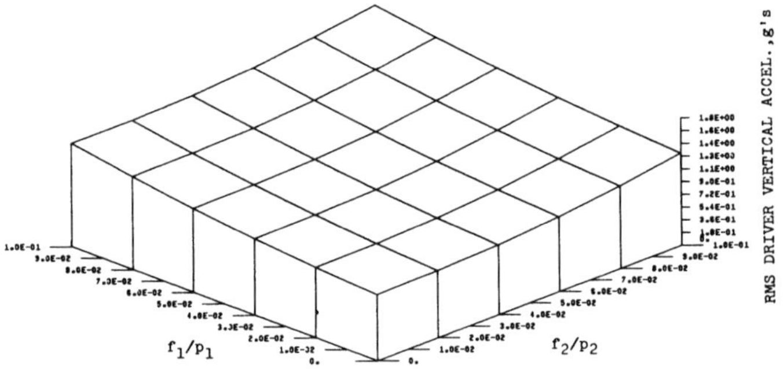


Fig. 4 Effect of dry frictional forces on rms driver acceleration.

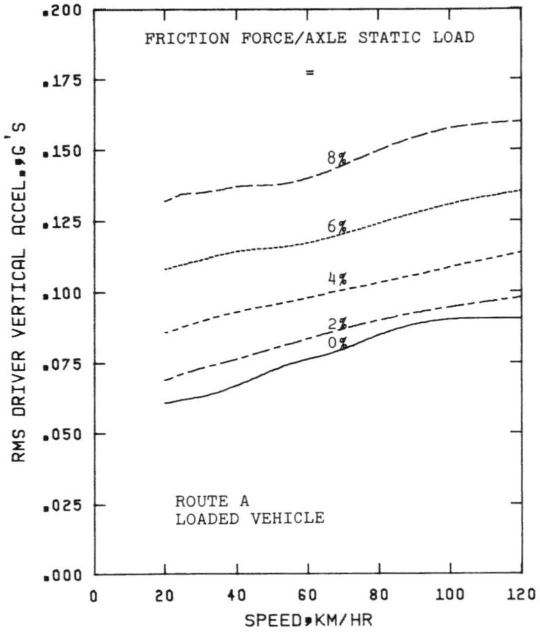


Fig. 5 Effect of speed on rms driver acceleration - varying frictional forces.

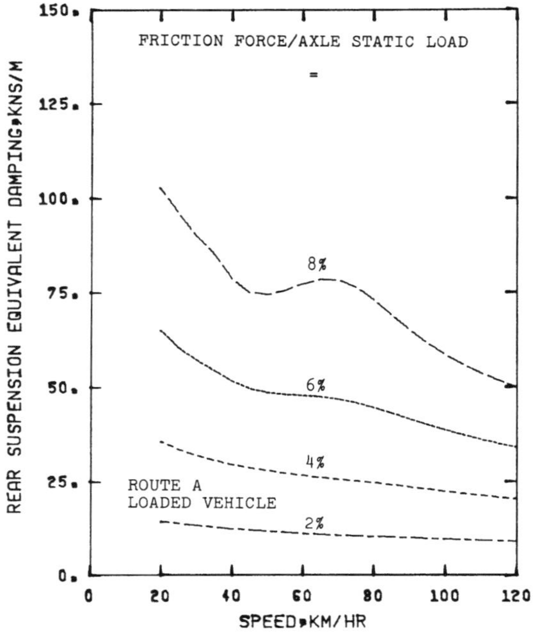


Fig. 6 Effect of speed on suspension equivalent dampings - varying frictional forces.

centre of gravity and the driver's position is due to the tractor pitching acceleration which has been found to increase with increasing the dry friction dampings. Beyond the optimum values of the frictional forces increasing frictional forces will increase the rms accelerations and cnsequently decrease the comfort level.

Figure 5 shows the variation with vehicle speed of the rms vertical accelerations at the driver position for different values of the dry frictional forces. In general, the rms accelerations are increased with increasing vehicle speed. It may be inferred from the figure that there is a possibility of resonant condition at about 100 km/h for the linear system. For a small amount of dry friction, 0-2% of the axle static load, the linear system gives reasonable results, however, for higher dry friction, the effect of firction should be considered in obtaining the response and the linear system is not adequate for vehicle analysis.

In Fig. 6, the suspension equivalent dampings at the tractor rear axle are given as function of vehicle speed for different values of dry frictional forces. It may be observed from the figure that the equivalent dampings of high dry frictional forces are very high at low vehicle speed. In this case the suspension would seem to be very stiff and the vibrations of the unsprung masses would significantly transfer to the vehicle body without considerable attenuation. However, at high speeds, the equivalent dampings are significantly reduced.

## 6. CONCLUDING REMARKS

The Random analyses show that the analytical techniques could be exceptionally effective tools in the hands of the engineering designers. Using the analytical techniques would allow the design of particular components to be checked in terms of their effect on the vibrational response of the overall vehicle structure. These techniques may also serve as aids in designing the tests and in interpreting the test results.

The dry friction dampings are detrimental to the ride comfort according to the ISO criteria. Excessive friction dampings increase the transmitted force and hence lead to an increase in shock wave transmission to the vehicle body.

REFERENCES


1. Dokainish, M.A., and ElMadany, M.M., Dynamic Response of Tractor-Semitrailer Vehicle to Random Input. Proceedings of the 5th VSD 2nd IUTAM Symposium on Dynamics of Vehicles on Roads and Tracks, September 1977, pp. 237-255.

2. Dokainish, M.A., and ElMadany, M.M., Random Response of Tractor-Semitrailer System. VSD 9 (1980), pp. 87-112.

3. Crosby, M.J., and Allen, R.E., Cab Isolation and Ride Quality. SAE, Paper No. 740294, 1974.

4. Foster, A.W., A Heavy Truck Cab Suspension for Improved Ride. SAE, Paper No. 780408, 1978.

5. Ribarits, J.I., Aurell, J., and Andersers, E., Ride Comfort Aspects of Heavy Truck Design. SAE, Paper No. 781067, 1978.

6. Janeway, R.N., A Better Ride for Driver and Cargo-Problems and Practical Solutions. SAE, Special Publication, No. 154.

7. Metcalf, W.W., The Ride Behaviour of Multi-Element Vehicle Traversing Cross-Country Terrain. Cornell Aeronautical Laboratory, Cornell University, Buffalo, New York, 1961.

8. Van Deusen, B.D., Truck Suspension System Optimization. Journal of Terramechanics, 1973, Vol. 9, No. 2, pp. 83-100.

9. Potts, G.R., and Walker, H.S., Nonlinear Truck Ride Analysis. ASME Transactions, Journal of Engineering for Industry, May 1974, pp. 597-602.

10. Robson, J.D., and Dodds, C.J., Stochastic Road Inputs and Vehicle Response. VSD 5 (1975/1976), pp. 1-13.

11. LaBarre, R.P., Forbes, R.T., and Andrew, S., The Measurement and Analysis of Road Surface Roughness. MIRA Report No. 1970/5, 1969.

12. Robson, J.D., and Dodds, C.J., The Responses of Vehicle Components to Random Surface Undulations. Proc. 13, FISITA Congress, Paper 17-2D, Brussels, 1970.

13. Yang, I., Stationary Random Response of Multidegree-of-Freedom Systems. Ph.D. Thesis, California Institute of Technology, 1970.

14. ElMadany, M.M., Random Response of Articulated Vehicles. Ph.D. Thesis (in progress), McMaster University, Hamilton, Ontario, Canada, 1979.

# THE USE OF SPECTRAL ANALYSIS TECHNIQUE TO INVESTIGATE NONLINEAR STEERING VIBRATION

M. A. Dorgham

Faculty of Technology, the Open University, Milton Keynes, United Kingdom

SUMMARY

Test runs were conducted using an instrumented test vehicle, designed and constructed to investigate the effects of certain parameters on the steering oscillation of a road vehicle. These parameters included changes in suspension geometry, vehicle speed, tyre inflation pressure and wheel imbalance.

The time histories of the recorded signals of steering oscillations and body motion, obtained at different excitation conditions, were recorded and analysed. Power spectral density and coherence functions of the recorded signals were estimated to yield a statistical description of the oscillations in the frequency domain and to establish the coupling between the steering oscillations and body motion.

Inspection of the frequency spectrum of the steering oscillations showed that certain frequencies were dominant. These frequencies were found to be functions of the excitation input to the system. The non-uniform pattern of the forces and moments caused by tyre-wheel irregulatiries excited other harmonics which are linearly related in magnitude to the dominant frequencies. Coupling between the harmonics of the body motion and the steering oscillations was investigated.

## 1. INTRODUCTION

The liability of self-sustained oscillations is inherent in the vehicle system when the vehicle is in motion. This is partially due to coupling effects between the rotational motion about the steering axis and different modes of motion of the vehicle components whilst a flow of energy is fed to the system by the motion of the vehicle. In addition, factors such as clearances, dry friction and the behaviour of visco-elastic elements, which exist in the tyre-suspension-steering system, increase the susceptibility of steering mechanisms to self-sustained oscillations. The phenomenon may be triggered in such a dynamic system by factors such as road surface conditions or wheel irregularities.

This paper reports on experimental work which was conducted as a part of an analytical and experimental study (1) to investigate the mechanism which generates steering oscillations. It also describes the advantages of using spectral analysis to yield a statistical description of the effects of certain design parameters on steering vibration in the frequency domain. These parameters include changes in suspension geometry, vehicle speed, tyre inflation pressures and wheel imbalance. Test runs were conducted using an instrumented test vehicle designed and constructed for this purpose.

## 2. THE EXPERIMENTAL VEHICLE

The design objective of the test vehicle was to produce a generalised test device which allowed for the attachment of different types of suspension and steering systems. The test permitted controllable changes in the geometrical and inertial properties of the vehicle as well as alterations to the suspension and steering parameters; enabling the device to be used for studying the characteristics of a particular road vehicle, newly designed or otherwise, in a variety of simulated dynamic configurations. The device also has the capability of displaying the effects of these configurations on the steering vibration phenomenon. The constructed chassis, is towed by a motor vehicle through a tow bar having a ball joint at each end, allowing the chassis to move freely whilst the driving motor vehicle is towing it. A constant torque device, Figure (1), is attached to the steering hand-wheel to simulate torque applied by the driver.

Measurement of the resilient properties of the suspension linkages was achieved by mounting the test vehicle rigidly on a specially made frame while the suspension units were displaced by 4 position servo controlled hydraulic jacks. Sinusoidal and constant velocity inputs were applied to the suspension units at different frequencies. The forces produced through the system, and the displacement and velocity across the units, were measured and recorded (as shown in Figure 2). The resilient propoerties of the suspension system are shown to be amplitude and frequency dependent.

The test vehicle was instrumented to measure the steering hand-wheel angular movement and body motion. Various parameters recorded are listed in Table (1) together with the transducers used for their measurement, their location, nominal range and accuracy.

The main runway of Cranfield airfield was used for road testing. This track is about 2Km in length and its surface is adequately smooth and flat. The friction number of the track surface, as measured with an M.L. Aviation Mu-Meter, was in the region of 70-80.

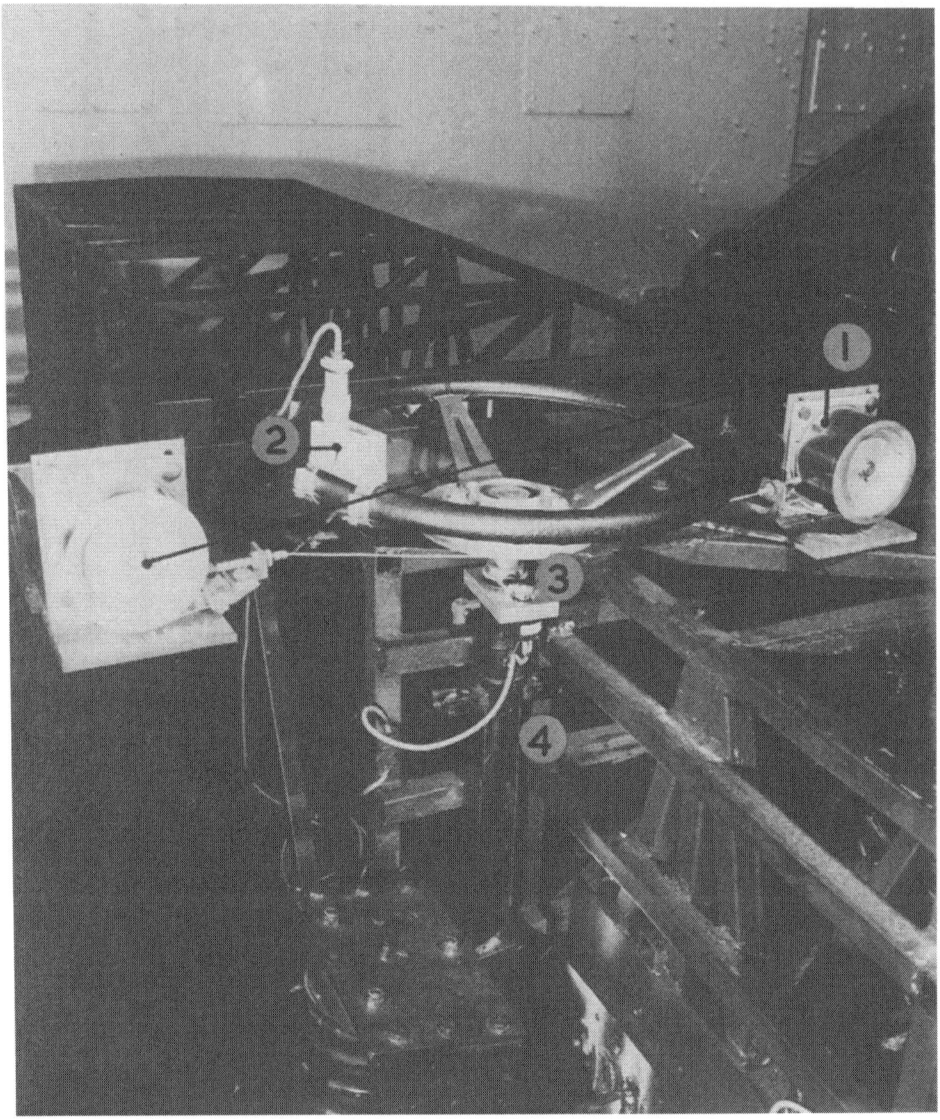


1. Constant steering torque mechanism
2. Force - balance accelerometer
3. Rotary potentiometer
4. Steering column

FIG (1) STEERING SYSTEM ARRANGEMENT

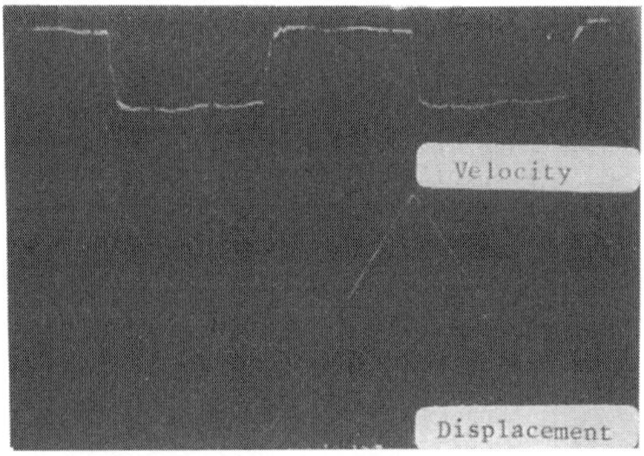


(a) Triangular Input

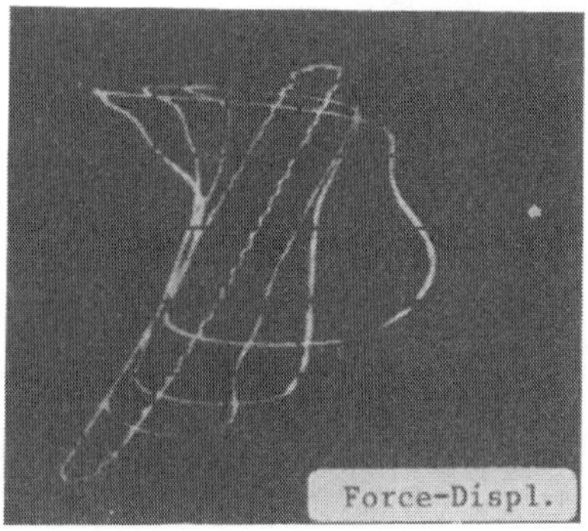


(Frequencies 1-5-7-9)

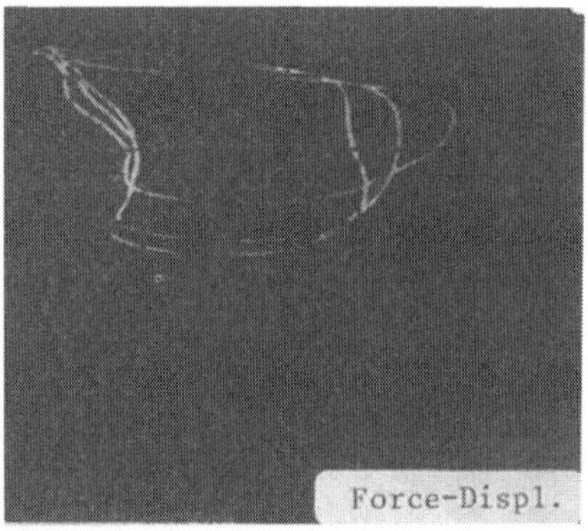


Frequencies 10-12-15)

(b) Force/displacement relationship for a typical front suspension

FIG ( 2 ) RESILIENT CHARACTERISTICS OF SUSPENSION AT DIFFERENT FREQUENCIES

| TAPE CHANNEL | SIGNAL RECORDED | | TRANSDUCER DEVICE | LOCATION | TRANSDUCER RANGE | ACCURACY |
|---|---|---|---|---|---|---|
| | PARAMETER | SYMBOL | | | | |
| 2 | STEERING HAND WHEEL ANGLE | $\delta_{sw}$ | ROTARY POTENTIOMETER | GEARED TO HUB OF STEERING HAND WHEEL | ± 20 Deg | 0.5% |
| 3 | ROLL/YAW RATE | P/r | RATE GYRO (SMITH MINIATURE) | INSTRUMENTATION BACK MOUNTED TO THE CHASSIS | ± 40 Deg per sec | 1% |
| 4 | PITCH RATE | q | RATE GYRO (SMITH MINIATURE) | INSTRUMENTATION BACK MOUNTED TO THE CHASSIS | ± 20 Deg per sec | 1% |
| 6 | STEERING HAND WHEEL ACCELERATION | $\ddot{\delta}_{sw}$ | SCHAEVITZ FORCE BALANCE ACCELEROMETER | FIXED TO THE RIM OF THE STEERING HAND WHEEL | ± 5 g | 0.05% |
| 7 | FORWARD SPEED | U | SPERRY TACHO-GENERATOR | ARTICULATED FIFTH WHEEL TRAILED BEHIND THE VEHICLE | 0-70 mph | 0.5% |

TABLE ( 1 ) TRANSDUCER DESCRIPTION

The tyre wheel assemblies, fitted to the experimental vehicle, were statically and dynamically balanced, individually and as a part of the vehicle, before initiating the tests. Unbalanced weights were then, attached to the front wheels at a fixed radius to investigate the effect of wheel imbalance on the steering oscillations.

## 3. MEASURED PARAMETERS

150 test runs were conducted using the free rolling experimental vehicle for variations in the following parameters:-

i) Suspension geometry
ii) Different tyre construction
iii) Inflation pressure of the tyre
iv) Wheel imbalance

The time histories of the following variables were recorded simultaneously:-

i) Steering hand-wheel angle
ii) Steering hand-wheel angular acceleration
iii) Roll, pitch and yaw rates of the test chassis
iv) Forward speed of the test chassis

The recorded signals, obtained at different excitation conditions, were recorded and analysed. Spectral analysis of the experimentally measured data was conducted by manipulating the recorded data on an Interdata Model 70 mini-computer based system.

## 4. RESULTS AND DISCUSSIONS

Power spectral density (PSD) was estimated, Figure (3), to establish the characteristic frequency and the amplitude levels of the recorded signals and to describe how the total energy of the signal was distributed between all its constituent frequency components. A plot of the PSD of the steer angle oscillations on a linear amplitude/linear frequency scale showed the single dominant peak, whilst a logarithmic amplitude/linear frequency scale showed a number of other significant resonancies. Inspection of the frequency spectrum of the steering oscillations showed that three frequencies $f_1$, $f_2$ and $f_3$ ($f_1$ = 1 3HZ, $f_2$ = 10-13 HZ, $f_3$ = 31-34 HZ) were dominant.

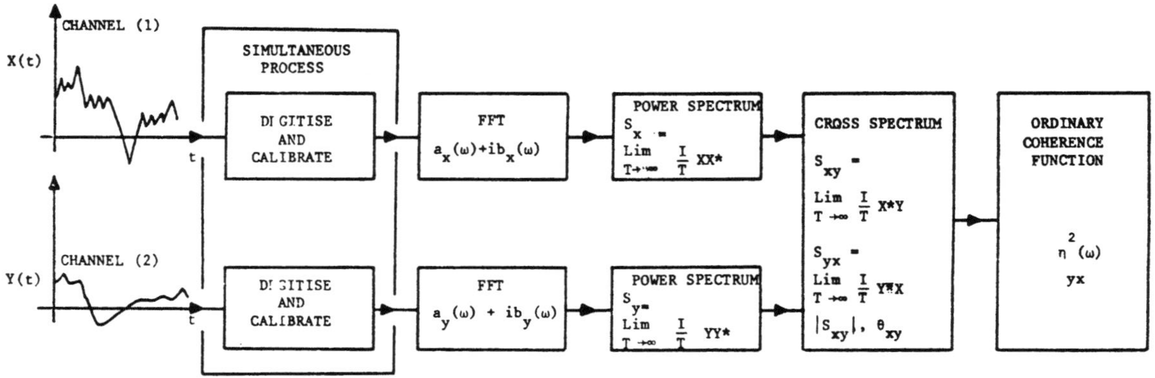


FIGURE (3) SPECTRAL ESTIMATES

These frequencies are thought to be the characteristic frequencies of the three-rotor steering system. The frequencies, $f_1$, $f_2$ and $f_3$, are found to be dependent on the inertial and resilient characteristics of the tyre-suspension system, excitation input to the system and vehicle acceleration. Excitation inputs are attributed to disturbances caused by stochastic inputs transmitted from the road surface, irregularities and changes in configuration of tyre-wheel assembly, and by inputs transferred from the vehicle body via the suspension units. Examples of the recorded results are shown in Figures (4-13).

The non-uniform pattern of the forces and moments caused by tyre-wheel irregularities excited other harmonics which are linearly related in magnitude to the three dominant frequencies $f_1$, $f_2$ and $f_3$.

Cross spectral density and coherancy functions of both steering hand-wheel angular displacements and acceleration and the rates of body displacements in roll, pitch and yaw were calculated. The analysis established a coupling effect between harmonics of the body motion in the roll, pitch and yaw modes and steering oscillations, and indicated how this coupling was more pronounced with an increase of wheel imbalance, a decrease in inflation pressure and changes in suspension geometry.

Effects of changes in different parametres on the harmonics $f_1$, $f_2$ and $f_3$ of the oscillations may be summarized as follows:-

I: Effect of Vehicle speed:

An increase of vehicle accel increased the frequencies and PSD values of the harmonics. The effects were more pronounced for the frequency of harmonic $f_2$ and its multiples $f_2$, $f''_2$ and for the PSD values of harmonic $f_3$.
($f'_2 = 2f_2$, $f''_2 = 3f_2$)
The speed at which the oscillation started was increased by increasing the vehicle acceleration. Acceleration and deceleration of the vehicle also affected the formation of the wave packets (or beats) observed in the oscillations and the frequencies and PDS values of their harmonics.

II: Effects of inflation pressure of the front tyres on the Steering Oscillations:-

i) An increase in the inflation pressure of the front tyres produced the following effects:

a) An increase in the frequencies of the three main harmonics, $f_1$, $f_2$, $f_3$.

b) A slight increase in the PSD values of the harmonics $f_1$ and $f_2$.

c) A considerable increase in the PSD value of the harmonic $f_2$.

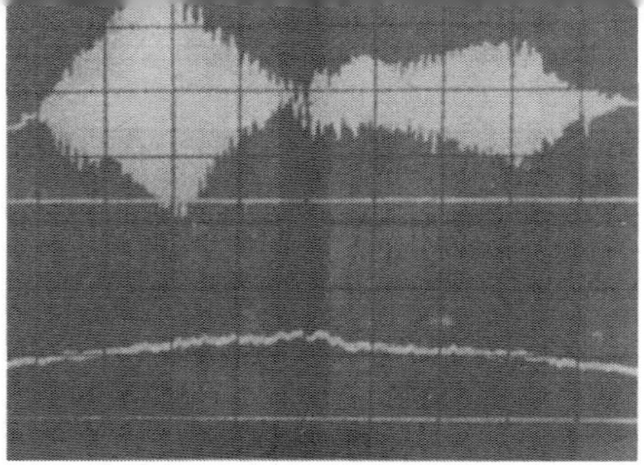

a. **Inflation pressure = 10 psi**

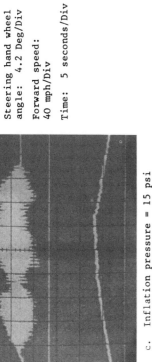

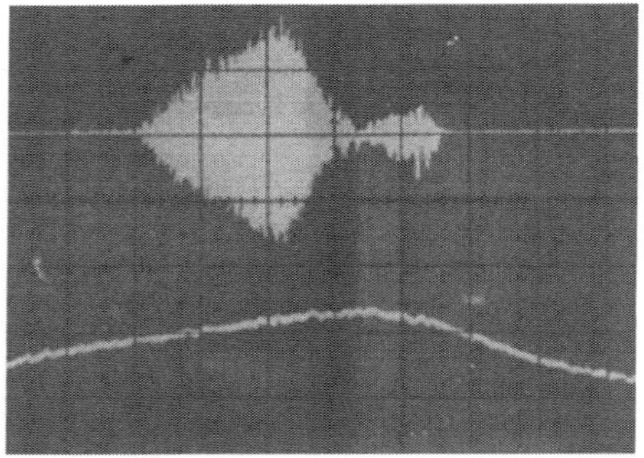

b. Inflation pressure = 12 psi

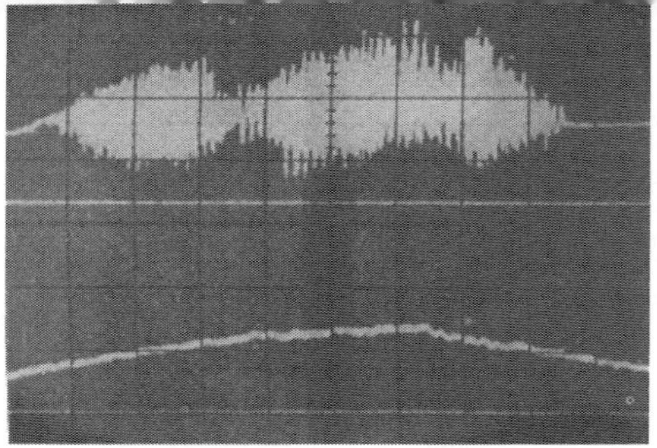

c. Inflation pressure = 15 psi

Scale (a, b, c)

Steering hand wheel angle: 4.2 Deg/Div

Forward speed: 40 mph/Div

Time: 5 seconds/Div

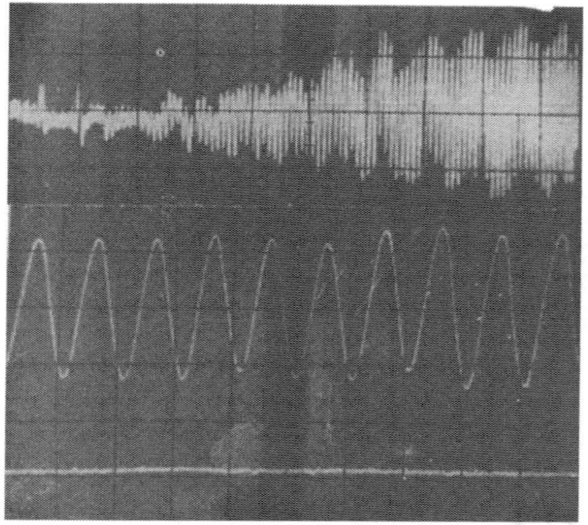

Scale:

Steering hand wheel angle: 2.1 Deg/Div

Time: 1 Sec/Div (Pressure: 15 psi)

Scale:

Steering hand wheel angle: 2.1 Deg/Div

Time: 0.1 Sec/Div (Pressure 15 psi)

Test Condition: unbalance weight = 950 gms on both front wheels at radii equal to 155 mm. Constant torque applied at the steering hand wheel (cross ply bald tyres).

FIG ( 4 ) STEERING HAND WHEEL OSCILLATION **MEASURED AT DIFFERENT INFLATION PRESSURES**

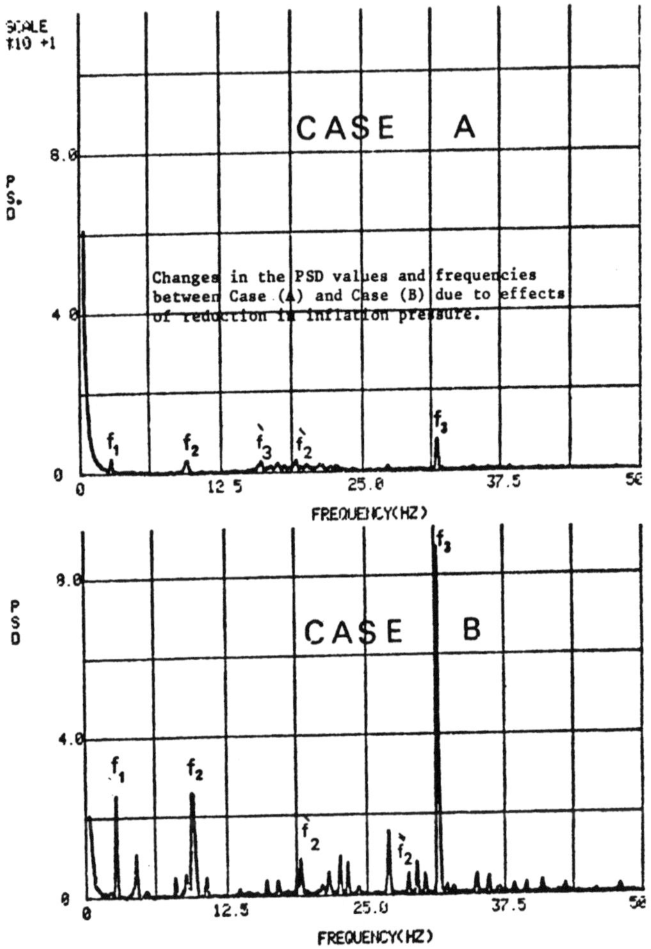


CASE A: Inflation pressure of the front tyres are 24 psi and 15 psi
CASE B: Inflation pressure of the front tyre, are 20 psi and 10 psi

FIG ( 5 ) EFFECTS OF PARTIAL INFLATION PRESSURE AT FRONT WHEELS
(NO UNBALANCED WEIGHTS, RADIAL PLY TYRE).

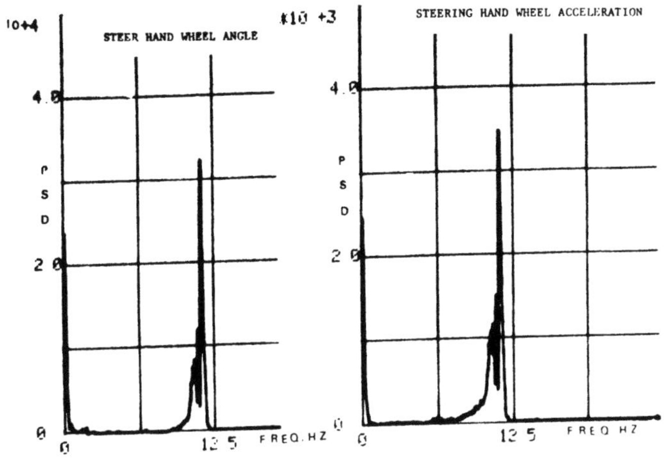


Fig (6) Power Spectral Density (PSD) of Steering Hand Wheel Oscillations (angular displacement and acceleration, linear PSD Scale)

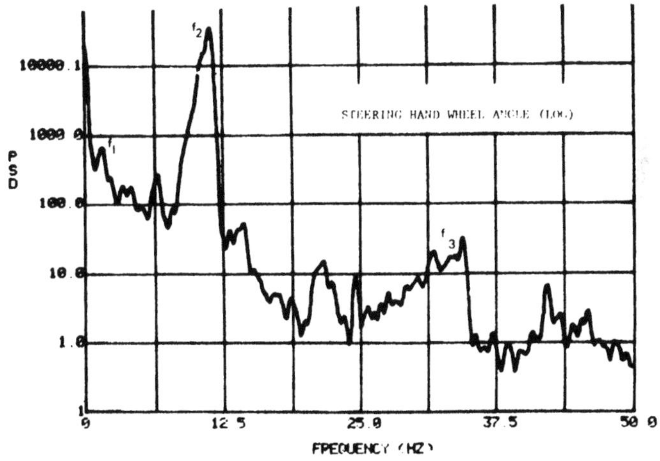


Fig (7) Power Spectral Density (PSD) of Steering Hand Wheel Oscillations (angular displacement, logarithmic PSD scale)

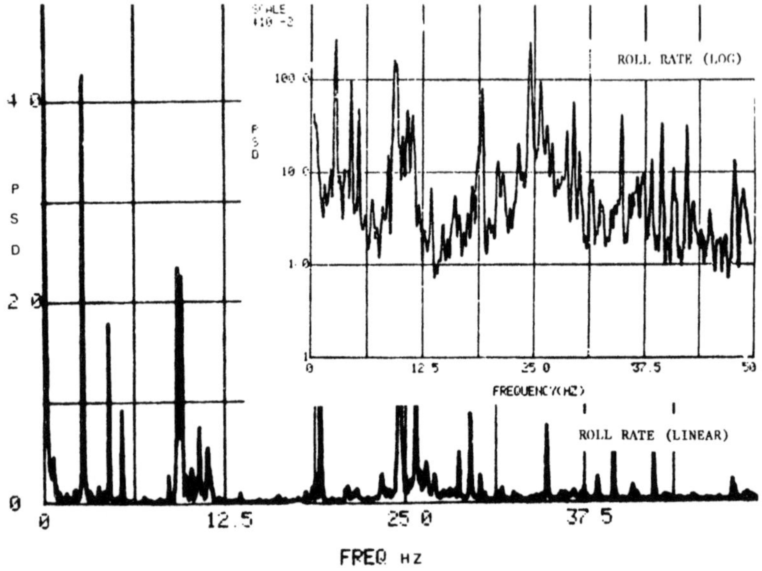


Fig. (8) Power Spectral Density (PSD) of Vehicle Body Motion in Roll (Roll Velocity linear and logarithmic PSD scale).

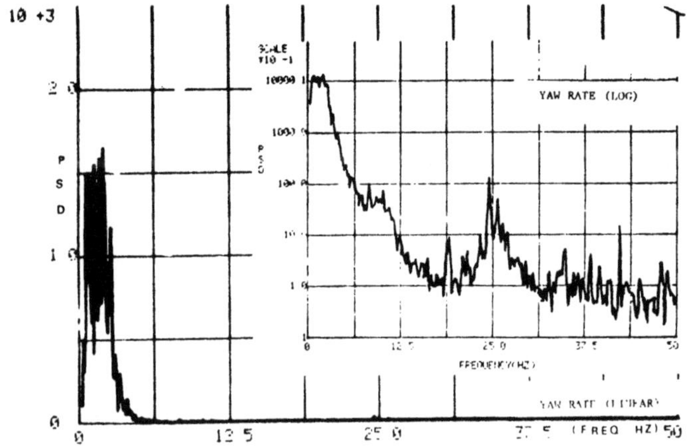


Fi.g (9) Power Spectral Density (PSD) of Vehicle Body Motion in Yaw (Yaw velocity, linear and logarithmic PSD scale)

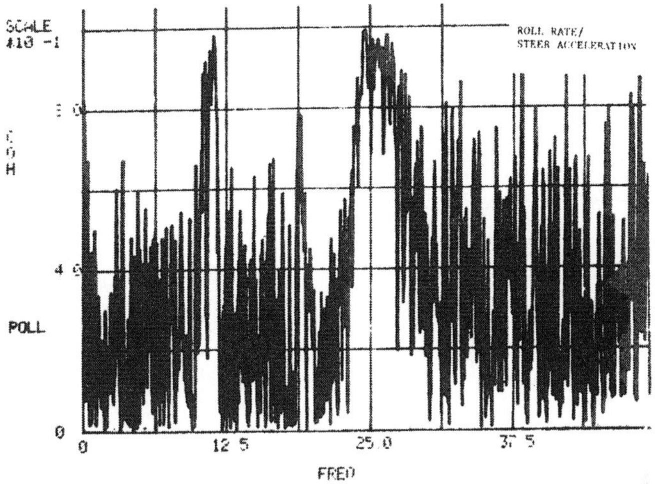


Fig. (10) Coherence Estimates of Signals of Vehicle Body 'Roll Rate' and Steering Hand Wheel Angular.

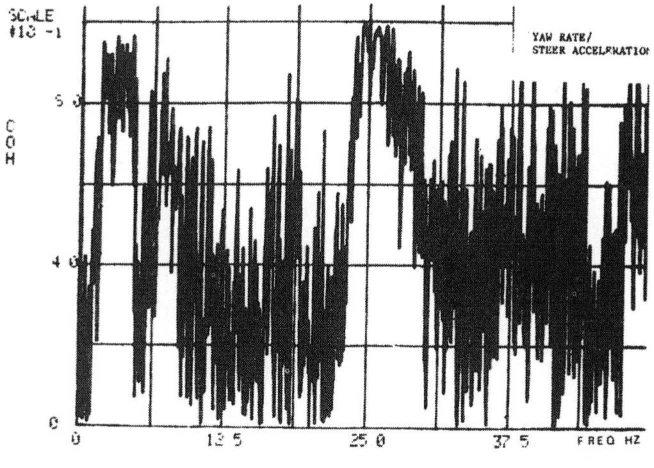


Fig. (11) Coherence Estimates of Signals of Vehicle Body 'Yaw rate' and Steering Hand Wheel Angular Oscillations

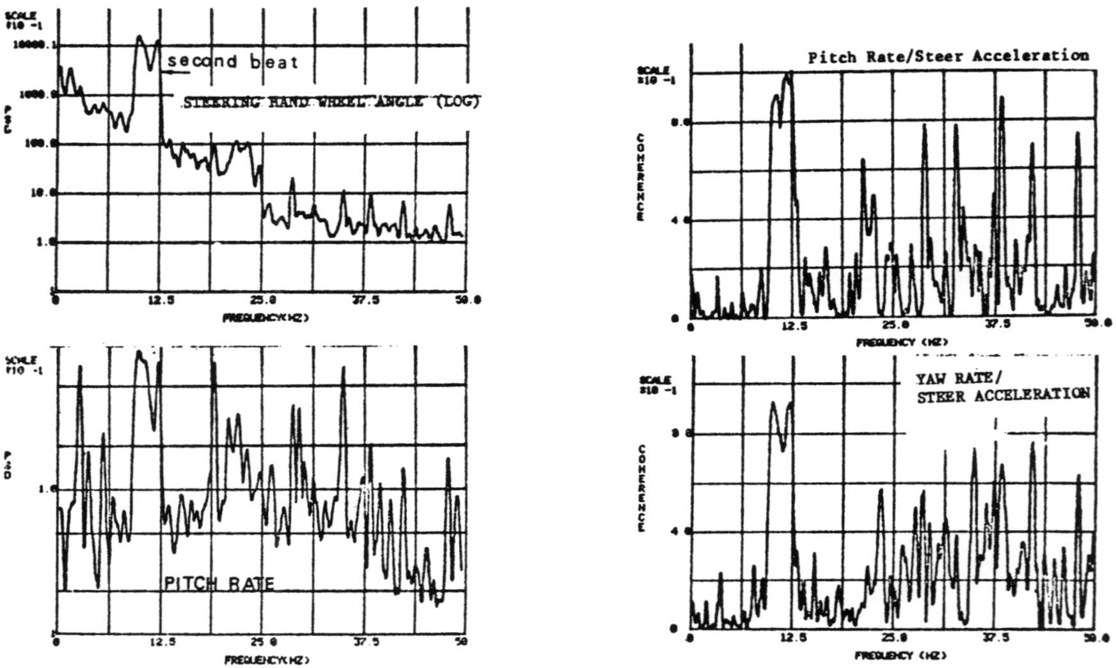


Fig. (12) Spectral Analysis of a pitch rate, yaw rate and steering acceleration signal.

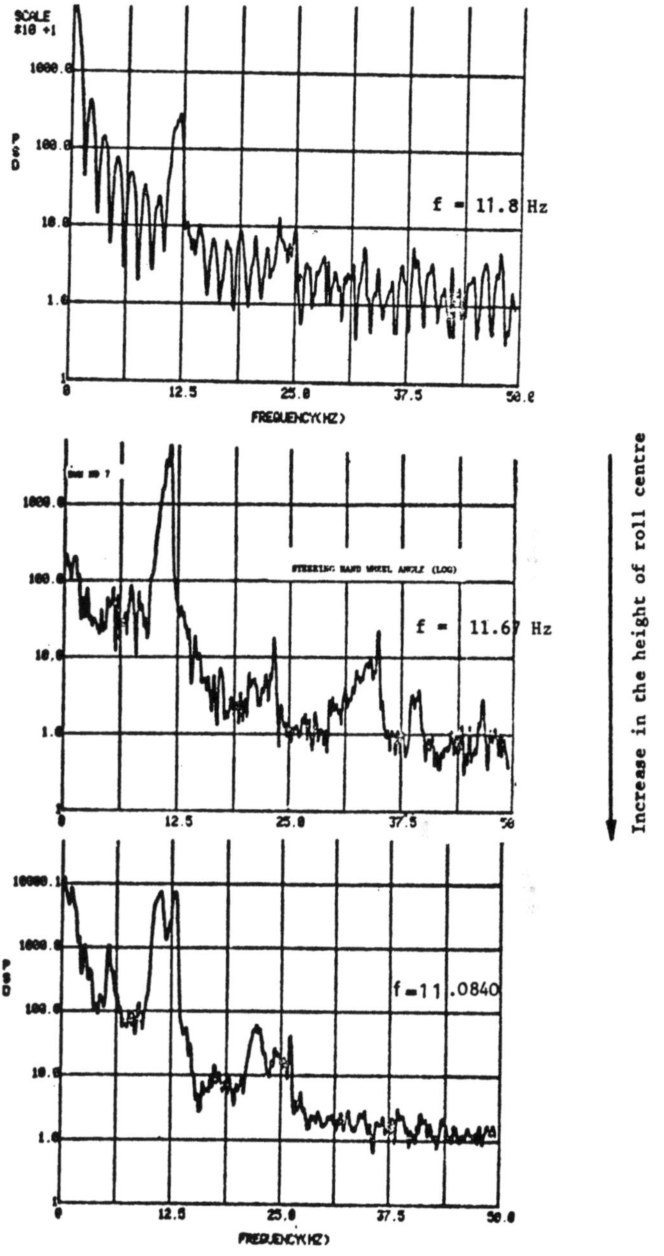


FIG ( 13 ) EFFECTS OF CHANGES OF THE HEIGHT OF THE ROLL CENTRE ON STEERING OSCILLATIONS (IMBALANCE WEIGHT = 600 gms ATTACHED

d) An increase in the energy of the oscillation of the steering angular displacement and acceleration represented by the r.m.s value obtained from the PSD - frequency plot in each case.

ii) Partial inflation pressure at the front tyres has the following effects:-

a) An increase in the PSD of the third harmonic ($f_3$)

b) A decrease in the PSD of the first harmonic ($f_1$) and a significant decrease in the PSD value of the second harmonic ($f_2$).

c) An increase in the energy of the system represented by r.m.s. obtained from the steering angle and acceleration PSD/frequency plots.

III: Effects of unbalanced weights attached to front wheels on the steering oscilations:

The following effects of wheel imbalance have been observed:

i) Effects of decreasing the tyre inflation pressure while constant imbalance weights attached to front wheels:

a) Unbalanced weight was attached to one front wheel:

i) An increase in the frequencies of the harmonics of the oscillations

ii) An increase in the r.m.s. values of the harmonics of the oscillation (for both signals of the steering hand wheel angle and acceleration)

iii) PSD values of the harmonics are increased

These effects were more significant on the frequency $f_2$ and its multiples $f_2'$ and $f_2''$. The effects were less pronounced on the frequencies $f_3$ and $f_3'$ ($f_3' = \frac{1}{2} f_3$).

When partial pressure at the front wheel was considered the frequencies and r.m.s. values of the harmonics were sensitive to changes in the inflation pressure. The rate of increase of frequency and r.m.s. values with decrease in inflation pressure was higher than in the case of no partial inflation pressure.

b) Unbalance weights were attached to both front wheels:

Similar effects to those explained in case (a) were observed except the values of PSD and r.m.s. of the harmonics were higher in this case. The rate of increase of frequency and energy of the vibration system with respect to decreases in inflation pressures was higher than in case (b).

The wave packets or beats that appeared in the oscillation pattern have been also affected. The duration time of each beat has increased and the time interval between the consecutive beats was shorter in this case than that occuring in case (a).

ii) Effects of increasing the unbalance weights attached to front wheels while the inflation pressure of the front tyre was kept constant:

(a) Unbalance weight was attached to one front wheel:

1) A decrease in the frequency of the harmonics
2) The PSD values of the harmonics increased
3) A decrease in the r.m.s. values of the harmonics (for both signals of the steering hand wheel angle and acceleration)

The rate of decrease in both frequency and r.m.s. values was higher at high inflation pressure. These effects were more significant on the frequency $f_2$ and its multiples $f_2'$ and $f_2''$. The effects were less pronounced on the frequencies $f_3$ and $f_3'$. The rate of change of these effects was then the rate of change of these effects in case (i).

(b) Unbalance weights were attached to both front wheels:

1) A decrease in the frequency of the harmonics
2) The PSD values of the harmonics increased
3) An increase in the r.m.s. values of the harmonic.

The rate of decrease in frequency and of increase in r.m.s. values are higher at high inflation pressure.

Partial imbalance results in an increase of the frequency and a decrease in the r.m.s. values of the harmonics (with respect to the values of these parameters when each of the unbalanced weights at both front wheels was equal to the largest unbalanced weight used in the partial imbalance case).

Effects of bald cross-ply tyres fitted to the front suspension resulted in lower frequencies and higher PSD values at high pressures and low unbalanced weight than the equivalent cases when radial tyre applied. At lower pressure and high unbalanced weights the high changes in rolling radius (due to the lower stiffnesses of the bald cross-ply tyre) reduced the inertial and resilient characteristics of the system. These cases resulted in higher frequencies than those when the radial tyres were subjected to similar operating conditions.

IV: Effects of changes in Suspension Geometry on the Steering Oscillations:

i) An increase in the unbalance weight while the inflation pressure of the tyre was kept constant

Unbalance weight was attached to one front wheel only:

An increase in the height of the roll centre resulted in these effects:

1) A decrease in the frequencies of the harmonics of the oscillations
2) A decrease in the PSD values of the oscillations
3) A decrease in the energy of the vibration represented by r.m.s. value in the PSD-frequency plot.

These effects were more significant on the frequency $f_2$ $f_2'$, $f_2''$.

ii) The above mentioned effects were increased by reducing the inflation pressure of the tyre. These effects were also increased in the cases where unbalance weight was attached to one of the front tyres than in the cases where unbalance weights were attached to both front wheels.

Changes in the suspension geometry changed the energy in different modes of body motion and affected the coupling between these modes and the steering oscillations.

V Effects of Coupling between the motion of the Vehicle Body in Roll, Pitch and Yaw on the Steering Oscillations:

i) Coupling effects between the vehicle motion in roll mode and the steering oscillations:

Coupling took place between different harmonics of the signals as observed from coherence estimates. The effect was significant around the frequencies $f_2$ and its multiples $f_2'$ and $f_2''$ ($f_2'$ in particular.) These effects were increased by increasing the magnitude of wheel imbalance. A reduction in inflation pressure of the tyre also increased the coupling effects. These effects were more significant around the frequency $f_2$. Coupling was more pronounced in the cases where two unbalanced weights were attached to the front wheels than in the cases where unbalanced weight was attached only to one front wheel. Partial wheel imbalance and inflation pressure at the front tyres increased coupling effects. Increase of the height of roll centre decreased the coupling effects.

ii) Coupling effects between the vehicle motion in pitch mode and the steering oscill ations:-

The harmonics of the pitch motion of the vehicle were shown to have a strong coupling with the harmonics of the steering oscillations,

| Parameter | Vehicle Speed | | Inflation press | | Increase in Front Wheel imbalance | | | | | Increase in the height of roll cent centre |
|---|---|---|---|---|---|---|---|---|---|---|
| | | | | | Const inflation press | | | | | |
| | Accel | Decel | Reduction | Partial | one wheel | two wheels | Partial imbalance | Reduction in tyre pressure | | |
| Frequency | Slight increase | Slight increase | Decrease especially $f_3$ | more pronounced effects | Decrease especially $f_2, f'_2, f''_2$ | more pronounced effects on all parameters | more pronounced effects on all parameters | decrease in general (depends on rolling radius) | | decrease (depends on imbalance and tyre pressure) |
| PSD | increase | decrease | Increase especially $f_3$ | | Increase especially $f_2, f'_2, f''_2$ | | | increase | | slightly decreased |
| Energy (rims) | increase | slight decrease | increase | | increase especially $f_2, f'_2, f''_2$ | | | increase | | decrease |
| Starting Speed | increase | - | decrease | | decrease | | | decrease | | decrease |
| Speed at which oscillation vanishes | - | increase | decrease | | decrease | | | decrease | | decrease |
| Beat formation | decrease time interval between beats | - | increase in the time at which beating takes place | | decrease time interval between consecutive beats - increase in the time at which beating takes place | | | More pronounced effect | | decrease |

TABLE ( 2 ) EFFECTS OF CHARGES OF DIFFERENT PARAMETERS ON THE STEERING OSCILLATIONS PHENOMENON AS OBTAINED FROM EXPERIMENTAL RESULTS

especially at low inflation pressure, over most of the frequency range from 0 to 50 Hz. Significant coupling occurred around the frequency $f_2$ and its multiple frequencies $f_2$ and $f_2$. The effects were increased with an increase in wheel imbalance and were slightly decreased with an increase in the height of the roll centre. Partial inflation pressure and partial wheel imbalance of the front tyres also resulted in an increase of the coupling effects. An increase in the height of the roll centre slightly decreased the coupling effects. Effective pitch motion at front wheels may be expressed in terms of equivalent bounce movements and it would appear that coupling effects between the motion of the vehicle in bounce mode and the steering oscillations takes place in a pattern similar to the pitch case.

iii) Coupling effects between the vehicle motion in yaw mode and the steering oscillations.

Coupling effects increased by increase in unbalance weights and were more significant when unbalance weights were attached to both front wheels more than when unbalance weight was only attached to one wheel. A decrease in the inflation pressure of the tyres enhanced the coupling effects over the whole range from 0-50 Hz. Coupling effects were even more accentuated in the case of bald cross-ply tyres than in the case of radial tyres.

Partial inflation pressure and imbalance of front wheels also increased the coupling effects. An increase in the height of the roll centre reduced the coupling effects.

A summary of the experimental results is shown in Table (2).

## 5. CONCLUSIONS

The steering oscillation phenomenon is shown to be a complex problem and is a function of many inter-related variables. The following are the conclusions obtained from experimental results.

It has been concluded that excitation of the steering oscillations is a function of the internal resistance of the suspension-steering system, changes in the magnitude and direction of the moments about the steering axes and the motion of the body in different modes as well as the vehicle forward speed and acceleration.

Geometric and structural non-uniformities of the tyre result in variation of stiffness and non-uniform distribution of mass in the radial, lateral and longitudinal directions of the tyre. The effects of these tyre non-uniformities are to vary the forces and moments developed in these directions and to change the rolling radius.

Similar changes arise due to the vertical oscillations of the tyre-wheel assembly and the re-distribution of vehicle loads in the vertical, lateral and longitudinal directions. These changes result in variations in the direction and magnitude of the forces applied to the tyres and lead to dynamic imbalance effect, and magnifies the dynamic forces generated by initial tyre irregularities. The excitation force due to tyre non-uniformities is a function of the square of vehicle speed. Increases in the longitudinal forces of the tyre usually rise at a higher rate than the square of the vehicle speed and the excitation forces are high when the wheel rotational frequency is near to one of the frequencies, especially in the longitudinal direction, of the tyre-suspension-steering system.

Changes in the rolling radius and in the contact forces of the tyre assume an oscillatory pattern and the moment of these cyclic forces about the steering axis also assumes an oscillatory character. In the meantime, a continuous fluctuating energy, contributed by the road spectrum and the continuous deformation process of the rolling tyre, flows into the vibrating system. When the energy of the vibrating system is enough to bring a synchronism between the harmonics of the oscillations and the frequencies of the system, self-excited oscillations start. Additionally, coupling between harmonics of the vehicle motion in different modes and the oscillatory motion about the steering axis in the presence of energy flow to the system initiates self-excited oscillations. These oscillations have in their make-up prominent frequencies which have been identified as the characteristic frequencies of the three-rotor steering system. The excitation itself affects the internal resistance of the system and changes its resilient properties which are amplitude and frequency dependent (as shown in Figure (3)). These changes result in variations in the frequency of the system and subsequently a 'shift' occurs between the frequencies of oscillation and the frequencies of the system, i.e. synchronism vanishes. (1)

The oscillation may appear again if the shift between the effects of the oscillatory motion about the steering axis and the internal characteristics of the system vanishes. The process continues[(2)] in such a way which may be described as a 'feed-back cause and effect manner' and may assume a hunting fashion until the 'shift' between frequencies is wide enough to stop coupling action.

The speed at which the oscillations are initiated depends on the state of the inertial and resilient properties of the steering system, degree of coupling between the oscillations about the steering axis and body motion and the amount of energy flowing into the system.

---

[(1)] In the meantime, changes in the inertial and resilient properties of the suspension-steering system, due to changes in geometrical configuration and in excitation inputs, also result in variations of the forces generated in the system and in their moments about the steering axes.

[(2)] Hunting also occurs in the system due to other factors such as: backlash, difference in front tyre rolling radii.. etc. while resilient elements in the system give rise to limit cycle effects.

## 6. ACKNOWLEDGEMENTS

This paper is based on the work conducted by the author through Science Research Council Grants No B/RG/3.27.6 and B/RG/87068 under the direction of Professor J.R. Ellis at Cranfield Institute of Technology.

## REFERENCES

(1) DORGHAM, M.A. 'The Non Linear Vibration of the Steered Wheels of Road Vehicles' PhD Thesis, Cranfield Institute of Technology (1975/6)

# MEASUREMENT OF THE DRIVER'S TRANSFER AND COHERENCE FUNCTION DURING A FATIGUETEST IN A SIMULATOR

Kambiz Garavy

Institute of Automotive Engineering, Technische Universität Berlin, F.R.G.

SUMMARY

Transfer functions and coherence functions of drivers are measured in a closed loop man-machine-system. Input and output signals are random data; by use of Fast-Fourier analyses transfer function and coherence function can be computed. The temporal variations of this function during an eight hour test should give information about the physiological condition of the drivers. The change of transfer function and coherence function, caused by driver's fatigue, is shown exemplary for one subject.

## INTRODUCTION

Most of the accidents in street traffic are caused by driver's failure. In many cases the physiological condition of the driver is the reason for fatal accidents. This report deals with the measurement of driver's fatigue, especially its effects on the driver's behaviour in a closed loop man-machine-system.

Input and output signals in this system are random data, they are signals in a low frequency range between 0.1 and 2.0 Hz. By use of Fast-Fourier analyses transfer function and coherence function can be computed. This is the way to define the unknown element in this closed loop system, the human operator. His behaviour in the operator-vehicle-system is strongly influenced by his decreasing physical condition, due to prolonged driving.

## SIMULATOR

Since the test persons were stressed by a twenty hour sleep deprivation we considered driving in real traffic as too dangerous. Therefore, we decided to perform the experiment on a driving simulator.

We used a bandsimulator (Fig. 1) which means that the landscape - including streets, houses, trees, etc. - are fixed on a band. This band is running over two rolls, driven by an electric engine.

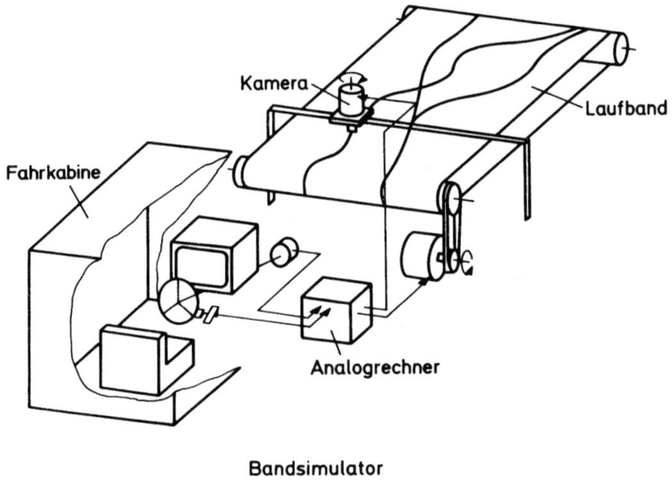


Fig. 1: Bandsimulator

The driver sits in a cabin, facing a TV-monitor and controlls with steering wheel and driving pedals the speed of the band and the movement of the camera. An analog-computer simulates the lateral and longitudinal dynamics of a symplified vehicle as well as the driving noises.

## EXPERIMENTAL DESIGN

The subjects were eleven male students in the age of 23 to 27 years. The experiment was devided into three sections:

- Section 1: Adaption phase to avoid the unwanted effects of learning behaviour.

- Section 2: On the testday the subjects were supervised from 8.00 a.m. till 8.00 p.m.
- Section 3: The test started at 8.00 p.m. till next morning at 4.00 a.m. with hourly measurements.

TRANSFER AND COHERENCE FUNCTION

For a linear system with random input and output data the transfer function is defined:

$$H(f) = \frac{G_{xy}(f)}{G_x(f)} \quad (1)$$

$G_x(f)$ : input power spectrum

$G_{xy}(f)$ : cross power spectrum

The degree to which the system output is the result of the input is defined by the coherence function:

$$\gamma^2_{xy}(f) = \frac{G_{xy}(f)^2}{G_x(f)\ G_y(f)} \quad (2)$$

$G_y(f)$ : output power spectrum

The coherence is a value ranging from 0 to 1.0; it gives a perfect indication of the causal relation between input and output.

If extraneous noise is present in the input or output data (Fig. 2), it is necessary to separate the unwanted noises from the signal.

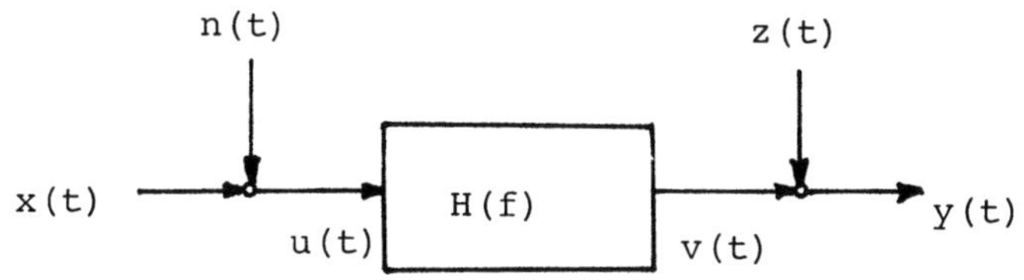


Fig. 2: Diagram of the generalized measurement problem

This problem is solved by Eq. 3 - 6:

- Incoherent output noise only

$\gamma^2_{xy}\ G_y = G_u$ Spectral output at y due to u (3)

$(1 - \gamma^2_{xy})\, G_y = G_z$ Spectral output at y due to z (4)

- Incoherent input noise only

$$\gamma^2_{xy}\, G_x = G_u \quad \text{Spectral input at x due to u} \qquad (5)$$

$$(1 - \gamma^2_{xy})\, G_x = G_n \quad \text{Spectral input at x due to n} \qquad (6)$$

The coherence function, like the transfer function, is defined for a linear, stationary system. However, if the system is nonlinear, the products of the nonlinearity are treated as system noise. So it will happen that the value of the coherence will be reduced from 1.0. If the coherence is zero, there is no input-output-relation, and any measurement of the transfer function is invalid.

OPERATOR-VEHICLE CONTROL SYSTEM

Fig. 3 shows the closed loop of the man-machine-control system. The driver gets his input from the monitor as a visual command signal.

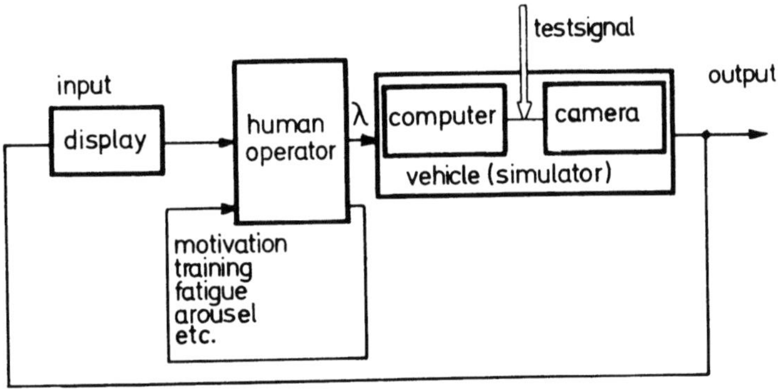


Fig. 3: Operator-Vehicle Control System

His output is the steerangle $\lambda$ as input for the computer for controlling the position of the camera.

The problem is that only output $\lambda$ can be measured but the input remains unknown. Therefore, the transfer function can't be found because input and output are needed. For that reason it is necessary to take an artificial input, here called "testsignal" (Fig. 3). This is a random stationary signal with 10 Hz bandwidth which works as a disturbance noise to the camera. This noise produces an additional motion of the camera and forces the driver to compensate by steering.

This is the way to get a closed loop system with measurable input and output (Fig. 4).

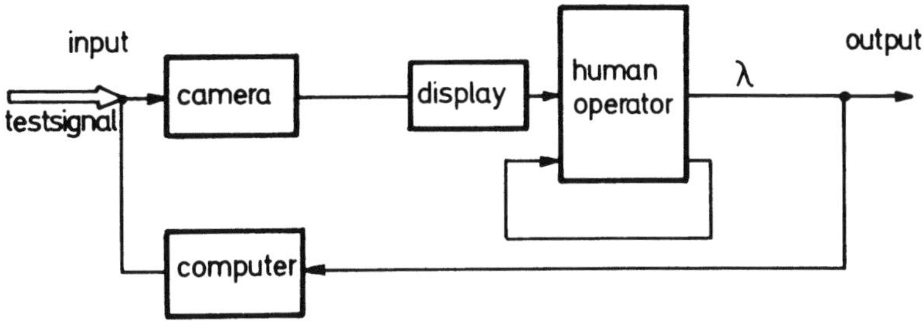


Fig. 4: Operator-Vehicle Control System

During the measurement of the driver's transfer function the real street course is now an extraneous disturbance noise. Of course, the coherence function can never reach the value 1.0.

RESULTS

Fig. 5 and Fig. 6 show two measurements of transfer and coherence function of one of the test drivers. The first measurement (Fig. 5) is made at 8.00 p.m., the second one (Fig. 6) at 4.00 a.m.

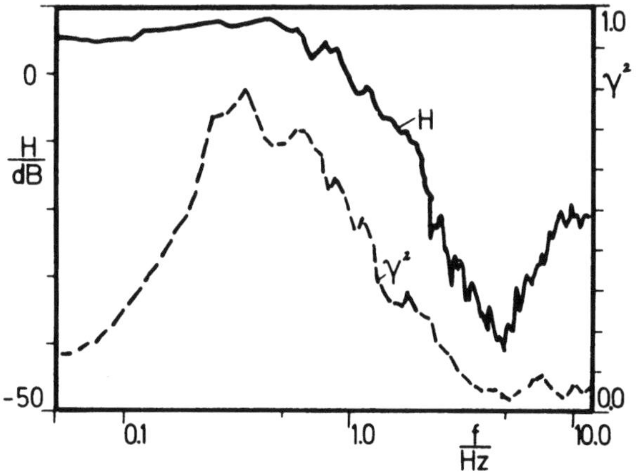


Fig. 5: First Measurement of Transfer (H) and Coherence ( ) at 8.00 p.m. (Subject No. 7)

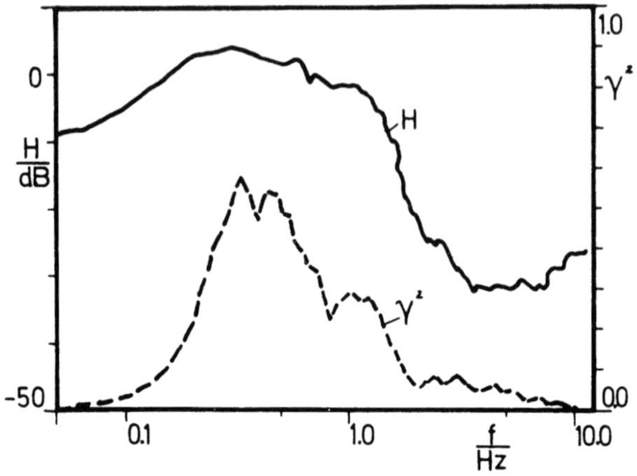


Fig. 6: Last Measurement of Transfer (H) and Coherence ( )
at 4.00 a.m. (Subject No. 7)

Very interesting is the great difference between the two coherence functions. The decrease of the coherence is not continuos. After each break the coherence increases for a short period again but than decreases very rapidly.

In most cases, the lowest level was reached at 4.00 o'clock in the morning after the eight hour test. The bandwidth and the absolute level have decreased by 50 to 100 %.

The comparison of the transfer functions shows a significant decrease in the low-frequency range; between 0.2 and 1.2 Hz the decrease is less significant.

During the last measuring period some of the test persons were almost unable to control the "vehicle" which resulted in several serious accidents. Of course, without any consequences for the driver in the simulator, but the errors could have led to fatal accidents in real traffic.

REFERENCES

1. Bendat, J.S., Piersol, A.G., Random Data: Analyses and Measurement Procedures. Wiley Interscience, N.Y.
2. Schlitt, H., System Theorie für regellose Vorgänge. Springer Verlag, Berlin.
3. Roth, P.R., Effective Measurements Using Digital Signal Analyses. IEEE Spectrum, April 1971.
4. Etschberger, K., Leistungsfähigkeit und Regelungsverhalten des Menschen bei der Nachführung stochastischer Signale. Springer Verlag, Berlin, 1975.
5. McRuer, Graham, Krandall, Reisener, Human Pilot Dynamics in Compensatory Systems. AFFDL-TR-65-15, July 1965

# PASSENGER CARS WITH SINGLE-AXLE TRAILER – INFLUENCE OF THE PARAMETERS OF THE MOTOR-CAR-TRAILER UNIT ON DRIVING STABILITY

R. Gnadler and M. K. Zabadneh

Institut für Maschinenkonstruktionslehre, Abt.
Kraftfahrzeugbau, Universität Karlsruhe, F.R.G.

SUMMARY

The driving stability of individually operated motor-cars has reached a very high level today, whereas the driving stability of the motor-car-trailer units still requires extensive development. Thus, it was necessary to develop a conception to improve it.

Up to now, no theoretical investigations are known the results of which fully agree with practical experience. The reason for this fact was mainly based on the circumstance that only the stability of the velocity in the lateral direction was investigated in these papers, but not the lateral motion itself. Furthermore, the participation of the driver has usually been disregarded.

To solve this task a three-dimensional calculation model was developed with reference to Gnadler's model. According to Ljapunow's methods the disturbance equations could be determined from the equations of motion of the undisturbed and the disturbed state. Thus it was possible to investigate the influence of the main motor-car-trailer unit parameters on stability. The result was that the driving stability could be considerably improved by optimizing the most essential parameters with relatively low expenditure.

## 1. INTRODUCTION

Unexpected reactions in the road-holding ability for example of a caravan, can be attributed again and again to the fact that, up to now, this problem has not sufficiently been investigated, and this was done in a few papers only. Two-dimensional calculation models were utilized; in this connection the actual three-dimensional condition was not explained realistically enough nor could it only be determined by a few parameters. Accordingly, the results could not satisfactorily answer the still unsolved questions.

The three-dimensional calculation model of Gnadler [1], which divides the vehicle into superstructure and wheel masses where each mass can move in its six possibilities of motion, was extended to the calculation of a towing vehicle with a trailer. Due to this fact the most important parameters of the motor-car-trailer unit can be varied over a large spectrum, as is shown in examples, so that any caravan or trailer

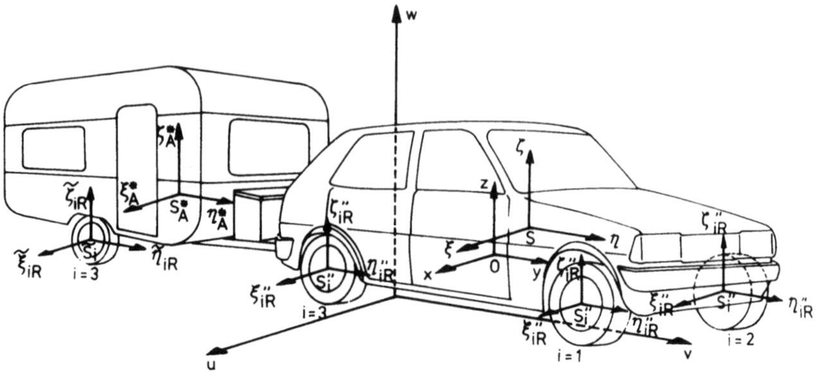

Fig. 1. Motor-car-trailer unit consisting of towing vehicle and single-axle trailer with axes of coordinates for the individual wheel and superstructure masses.

type which may be used can be incorporated in this calculation.

## 2. EXPLANATION OF THE EQUATIONS OF MOTION

The operating condition of the motor-car-trailer unit (Fig. 1) is illustrated by the motion of its 2 superstructures and 6 wheel masses in relation to the inertial coordinate system u,v,w, which is absolutely stationary. The equations of motion can be most appropriately set up by the Lagrange equations. In order to achieve this the energies or non-potential forces (relative to the inertial system) occurring in various forms, must be determined. To calculate the kinetic energy of any mass of the motor-car-trailer unit, any point of this mass is connected with the origin of the inertial system by means of a radius vector and then this vector is derived with respect to time. Thus the absolute velocity of this point or the kinetic energy of this particle of mass is obtained.

By integration via the entire mass, its kinetic energy is calculated. In a similar way the calculation of the other forms of energy and the non-potentional forces are obtained, that is, for the whole motor-car-trailer unit. In order to achieve this, the radius vectors must be divided into their individual components, which must be calculated in the corresponding coordinates. Those coordinates which are subjected to the compulsory design conditions are expressed by the remaining coordinates (degrees of freedom). Thus the equations of motion [4] were set up.

## 3. EXPLANATION OF THE STABILITY INVESTIGATION

The stability can be investigated according to Ljapunow's method (Malkin [3], Leipholz [2]). By disturbing the motor-car-trailer unit (mathematically) and then subtracting the equations of motion of the undisturbed state from the equations of the disturbed condition, one thus obtains the so-called disturbance equations. The solutions of the disturbance equations set up are then investigated for a stable road-holding ability. With progressing time the disturbance must decrease until it disappears completely. If, however, the disturbance increases with progressing time the motion will become unstable. In these two cases the linear terms of the disturbance equations can be decisive, no matter how the terms of the higher order are constituted. The terms of the higher order have only an influence when a critical case in the mathematical sense occurs (i.e. if a root of the so-called characteristic equation vanishes).

A critical case of this kind could e.g. occur, if one regards the motion of the towing vehicle transverse to the driving direction without considering the participation of the driver. This indicates that the towing vehicle has no self-acting resetting forces transverse to the driving direction.

When driving in circles, it is true, "apparent" resetting forces occur in contrast to the state of driving in a straight line. They act, however, only on the radius on the road and must not be treated as resetting forces.

Broadly speaking, this fact can be read directly from the equations of motion. If one neglects this important fact one has only investigated the stability of the lateral velocity, but not that of the lateral motion itself. Those who apply this method argue that a transverse low-speed deviation can be controlled by the driver. This, however, would only be true, if the participating of the driver were taken into account as already mentioned. A resetting force can only be achieved by the driver.

In the following the driver is taken into account. In this connection he is requested to keep the course without any lateral deviation. This driver is well-known as the so-called "ideal" driver.

## 4. CARRYING OUT THE STABILITY CALCULATION

As criteria for driving in circles with a constant speed on a plane road as mentioned, the motions transverse to the driving direction as well as vibrations around the vertical and the longitudinal axes are determined, both for the towing vehicle and the trailer. In this connection the lateral motion of the towing vehicle is determined by the degree of freedom of the steering wheel manoeuvres of an "ideal" driver. The lateral motion of the trailer itself does not include any degree of freedom, since it necessarily results from the other degrees of freedom.

Thus, with the degrees of freedom determined, five in all, the most important criteria for driving safety are taken into account with regard to the investigated driving manoeuvres since vertical and pitch movements as well as a deviation in longitudinal direction theoretically are not expected, and, in case they would, this can be regarded as temporary phase.

The Hurwitz-criterium enables the determination of the stability without being obliged to solve the disturbance equations completely. This is favourable for the objective of these investigations since they do not deal with the stability behaviour of a predetermined type of trailer, but quite generally, with influencing the driving stability of any single-axle trailer.

For example calculated parameters of the trailer were chosen in such a way that they correspond to an average lorry trailer or caravan. The individual parameters were then varied over a wide range so that they practically correspond to all types of trailers.

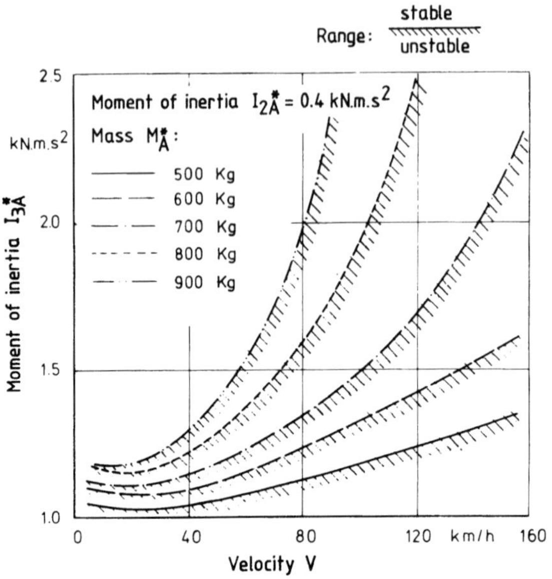


Fig. 2. Limiting velocity V depending on the moment of inertia $I^*_{3A}$ (vertical axis - trailer) for a constant moment of inertia $I^*_{2A}$ (longitudinal axis - trailer). One curve is attributed to each value of the trailer mass $M^*_A$.

From the theoretical results of this investigation a series of measures arises to improve stability. The extensive observation of vehicles with trailers being in operation confirms that these measures are not completely new. Some may already have been applied by this or that manufacturer or driver, basing solely on their experience.

## 5. RESULTS FOR INFLUENCING THE MOTOR-CAR-TRAILER UNIT PARAMETERS ON DRIVING STABILITY

In the following figures the effects of a variation of most dissimilar kinds of parameters on stability are shown. The curves shown separate the stable from the unstable area. These curves explain the course of the limiting velocity with which the motor-car-trailer unit just becomes unstable.

### 5.1. Principal moments of inertia and masses of the motor-car-trailer unit

It has been demonstrated by the fact that an increase of the moment of inertia $I^*_{3A}$ around the vertical axis of the trailer is improving the driving stability, which increases to such a degree the smaller the trailer mass is (Fig. 2). In contrast to this, the influence of the moment of inertia $I^*_{2A}$ around the longitudinal axis of the trailer turned out to be relatively small.

Furthermore, it is quite obvious that an increase of the moment of inertia around the longitudinal axis of the towing vehicle has a very

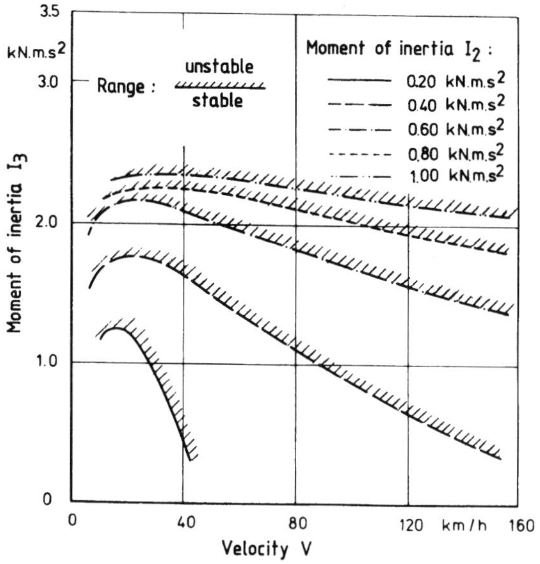


Fig. 3. Influence of the moments of inertia $I_2$ (longitudinal axis - towing vehicle) and $I_3$ (vertical axis - towing vehicle) on the stability. One curve is attributed to each $I_2$-value.

stabilizing effect, whereas an increase of the moment of inertia $I_3$ around the vertical axis is harmful with regard to the stability limit (Fig. 3).

These results can be explained as follows:
If the members of a motor-car-trailer unit come into yawing vibrations caused by any disturbance, the greater portion of the disturbance energy will be absorbed by the member with the smaller moment of inertia. Of course, it is much easier for the driver to control the towing vehicle than the trailer. Thus the motor-car-trailer unit becomes more stable if one either increases $I_{3A}^*$ (trailer) or reduces $I_3$ (towing vehicle). In contrast to this, a part of the disturbance energy is stored in the spring system as potential energy during the rolling motions. The influence of the moment of inertia around its longitudinal axis is reduced by the hard spring of the trailer. The opposite applies to the towing vehicle, whose springs are rather soft.

Considering the driver's participation, the direct effect of the mass of the towing vehicle is eliminated to a greater degree - and with an ideal driver, it will be eliminated completely. The idea that the mass of the towing vehicle should remain within a certain relation to the one of the trailer is not correct, since the mass can only indirectly influence stability by the moments of inertia and by the lateral forces of the towing vehicle.

## 5.2. Tyres and elasticities in the wheel suspension of the trailer

In connection with the tyre forces the term of the cornering stiffness coefficient with regard to the tyres of the towing vehicle is used in

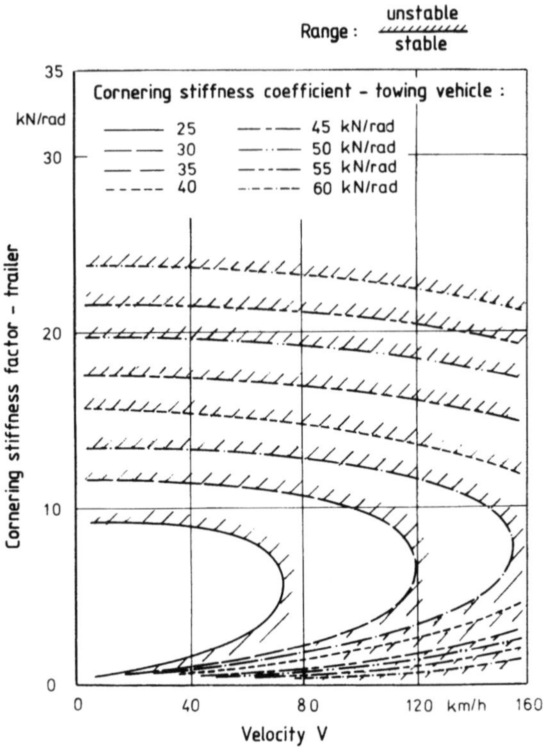


Fig. 4. Stability limit depending on the cornering stiffness factor (product: cornering stiffness coefficient - transfer factor of the slip angle considering the lateral elasticity of the trailer axle). The curves only differ in so far as the tyres of the towing vehicle were changed.

this investigation [4] and, in addition, also the term of the cornering stiffness factor is used with regard to the tyres of the trailer. This cornering stiffness factor consists of the product of the cornering stiffness coefficient of the trailer tyre and a transfer factor in order to take into account the lateral elasticity of the trailer axle.

Thus, stability always improves (Fig. 4) with steadily increasing cornering stiffness coefficient (at the towing vehicle). The cornering stiffness factor of the trailer is a different matter. First the limiting velocity increases with a rising cornering stiffness factor to reach a maximum value and then it decreases again until it reaches the value zero. The stability curves here resemble a parabola and have no common axis, which means, that there is no optimum cornering stiffness factor or no optimum tyres on the trailer which would match to every towing vehicle. Even if all the wheels of the motor-car-trailer unit are equipped with the same tyres the cornering stiffness factor remains much smaller than the cornering stiffness coefficient since the transfer factor is smaller than 1.

The lower branches of the stability curves of Fig. 4 are hardly

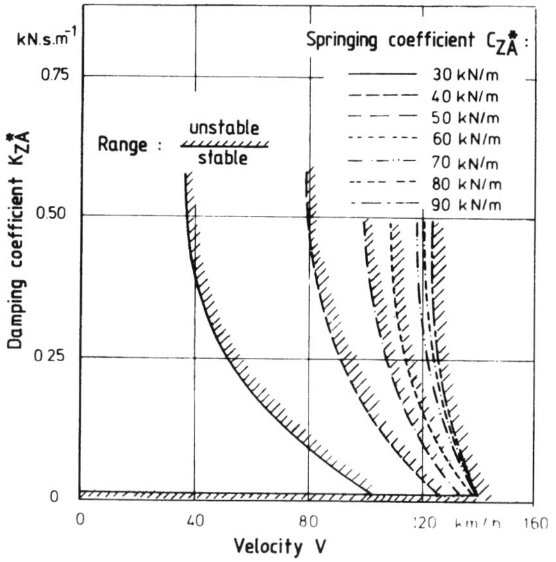


Fig. 5. Influence of the damping coefficients of the shock-absorbers $K_{Z\mathring{A}}^*$ and the spring rates $C_{Z\mathring{A}}^*$ of the trailer (direction z-axis in Fig. 1) on stability.

applicable to normal caravans and lorry trailers. These curve ranges apply to trailers only with very small wheels, e.g. trailers with folding tents. In this connection the cornering stiffness factor must be increased in order to approximate the optimum condition, that is, by increasing the transfer factor, e.g. by utilizing a more rigid connection between wheel and superstructure in lateral direction.

If one considers the tyres of normal caravan and lorry trailers, which as a rule, are of similar size as those of the towing vehicle itself, one comes to the upper areas of the stability curves in the illustration mentioned. With these tyre sizes one attempts to keep the cornering stiffness factor small in order to approximate the optimum condition. This can be achieved by using either diagonal instead of radial tyres on the trailer or a smaller transfer factor when choosing the wheel suspension. The manufacturers of the larger trailers have always made use of the second measure. Due to practical experience it became evident that a certain elasticity between the wheels and the superstructure in lateral direction is indispensable for stability. This measure which is only a remedy against small vibrations, is limited, since otherwise the trailer would vibrate in lateral direction.

### 5.3. Tyres, springs and damping elements of the motor-car-trailer unit

If one considers the influence of the vertical spring and damping action of the motor-car-trailer unit on its motion with regard to the degrees of freedom investigated here, it seems to be appropriate also to include the influence of the tyres into this consideration.

The interrelation between spring and damping action of the trailer

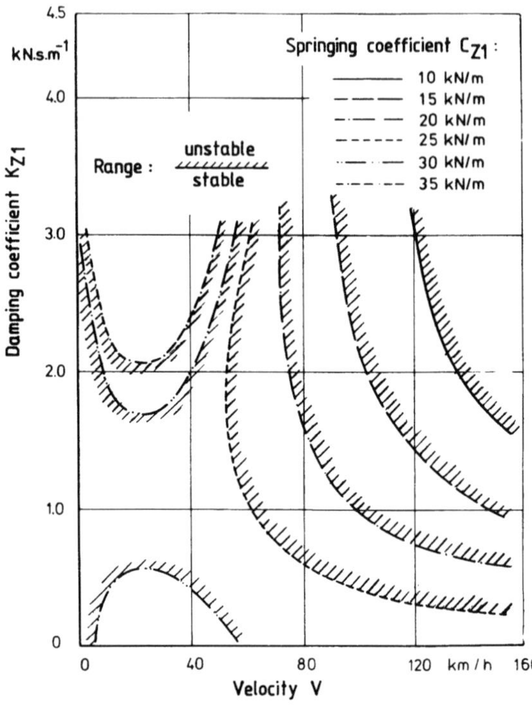


Fig. 6. Influence of the limiting velocity V by the damping coefficient or spring rate, $K_{Z1}$ or $C_{Z1}$ (wheel 1 of the towing vehicle, s. Fig. 1). Here is $K_{Z1} = K_{Z2} = K_{Z3} = K_{Z4}$ or $C_{Z1} = C_{Z2} = C_{Z3} = C_{Z4}$, Curve parameter is $C_{Z1}$.

is shown in Fig. 5. A minimum degree of damping is indispensable for stability. The optimum damping required is, however, so small that many trailers can drive in a stable way even without shock absorbers. Their damping effect is achieved by friction in the spring elements including the tyres. Too soft springs of a trailer act extremely negative on stability with an increasing damping coefficient. From a certain hardness of the springs the worsening effect of the damping remains negligibly small.

A further increase of cornering stiffness of the tyres of the motor-car-trailer unit improves this effect with harder springs of the trailer and gets worse with softer ones. This also explains the necessity of using much harder springs on trailers than on passenger cars.

The investigation carried out here showed that the springs of the trailer can be optimized. If the hardness reaches a certain value, an additional increase of the spring rate cannot provide any stability improvement. This limit can be called the optimum with regard to safety and comfort.

The springs of the towing vehicle influence the stability just contrary to those of the trailer. That means that soft springs of the towing vehicle are of advantage both in solo and in trailer operation. If the spring rate of the towing vehicle increases considerably (Fig.

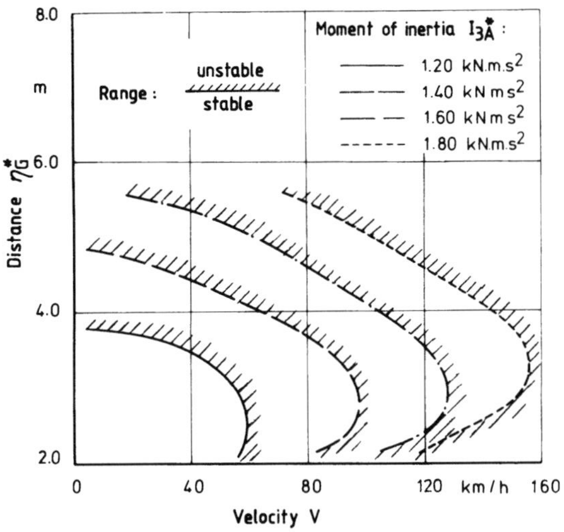


Fig. 7. Interrelation between limiting velocity V and drawbar length $\eta_G^*$ (distance of coupling point to trailer axle) including the moment of inertia $I_{3A}^*$ (vertical axis - trailer). It was assumed that the centre of gravity of the trailer is situated on the trailer axle. Curve parameter is $I_{3A}^*$.

6), the stability curves will show a hyperbola-like form. The stability area thus covered by them is practically useless.

The influence of the damping coefficient of the towing vehicle here acts negatively on the stability (the shock-absorbing of the trailer shows a similar effect). With higher damping coefficients the increase of the damping effect on the stability of the motor-car-trailer unit is reduced to inefficiency.

From the analytical point of view the following can be stated about the spring and damping of the motor-car-trailer unit: Both the spring system of the towing vehicle and its moment of inertia and cornering stiffness coefficients influence the stability contrary to the corresponding parameters of the trailer. The damping forces in the vertical direction, however, are nothing but disturbance forces, with respect to the remaining axes, i.e. no matter whether they act on the towing vehicle or on the trailer. The calculation results show that all shock absorbers to a certain degree act negatively on the stability.

## 5.4. Drawbar length and position of the centre of gravity of the trailer

Usually the distance from the coupling point to the axle of the trailer is defined as drawbar length. Since the influence of the drawbar length very strongly depends on the position of the centre of gravity of the trailer, first of the centre of gravity of the trailer was fixed on the axle of the trailer for Fig. 7.

As a disturbance is transferred from the towing vehicle to the trailer, long drawbars (analytically) are of advantage, whereas short drawbars are better for stability when the disturbance is transferred into the opposite direction. As a result, there must be an optimum for the drawbar length. This optimum length can be seen in Fig. 7 as already mentioned. Therefore, no generally valid evidence can be given, whether longer or shorter drawbars are of advantage with respect to stability.

### 5.5. Coupling point and centre of gravity of the towing vehicle

If the centre of gravity of the towing vehicle is transferred to the rear on the longitudinal axis, this measure will have a deteriorating effect on the stability because the towing vehicle will be subjected to a steadily increasing oversteering effect. This behaviour is also well-known when driving the car without trailer.

Reducing the distance from the centre of gravity of the towing vehicle to the trailer coupling, which occurs as one transfers the centre of gravity as mentioned, affords, when solely being considered, substantial advantages for stability. Thus, both effects must be considered together.

Accordingly, towing vehicles with a shorter rear end are of advantage. It can be explained thus that the driver while trying to keep his course is developing steering motions and consequently yawing motions around the vertical axis of the towing vehicle. The transfer of these motions decreases the smaller the distance is from the coupling to the centre of gravity of the towing vehicle.

### 5.6. Height of coupling point and centres of gravity of the motor-car-trailer unit

The stability limit reacts less sensitive to an increase in height of the centre of gravity of the trailer than an increase in height of the towing vehicle (soft springs). This can be explained by the fact that trailers with different loads resulting in a low or high centre of gravity can be operated without very extensive differences with regard to their road holding ability.

## REFERENCES

1. Gnadler, R., Das Fahrverhalten von Kraftfahrzeugen bei instationärer Kurvenfahrt mit verschiedener Anordnung der Hauptträgheitsachsen und der Rollachse. Dissertation, Universität Karlsruhe (1971).
2. Leipholz, H., Stabilitätstheorie. B. G. Teubner, Stuttgart (1968).
3. Malkin, J. G., Theorie der Stabilität einer Bewegung. München (1959).
4. Zabadneh, M. K., Personenkraftwagen mit einachsigem Anhänger - Einfluß der Gespannparameter auf die Fahrstabilität. Dissertation, Universität Karlsruhe (1978).

# THE THEORY OF STABILITY AND CONTROLLABILITY IN CONSIDERATION OF SUSPENSION AND STEERING COMPLIANCES

H. Harada[1], T. Hashimoto[1] and A. Watari[2]

[1] Toyota Motor Co.
[2] Japan Automobile Research Institute, Tokyo, Japan

SUMMARY

Recently, the compliances of suspension and steering system become very important in order to meet various demands in the market, in the areas of stability, controllability, ride comfort, noise and vibrations.

In this report, the effects of these compliances on directional control dynamics, which have only been considered qualitatively, are analized theoretically on a quantitative basis.

The values of stability factor K, yawing natural frequency $f_y$ and its damping ratio $\zeta$, and the crosswind sensitivity coefficient of a vehicle with or without a trailer, etc., can be estimated by means of the equivalent cornering stiffness of tires, in consideration of lateral force compliance steer, lateral force compliance camber and aligning torque compliance steer of the suspension and the torsional rigidity of the steering system.

Furthermore, the optimum zone of $f_y$-K characteristics is obtained by experiments and compared with the optimum characteristics of a simple vehicle-driver system in a theoretical analysis.

## 1. INTRODUCTION

Recently, for reasons of fuel economy, vehicles have been reduced in weight and radial tires are more installed, and so the demands for a higher quality of ride comfort and a greater reduction in noise and vibrations have become more severe. In order to solve these problems, this paper dealt with the effects of steering and suspension compliances on directional control dynamics theoretically and quantitatively, as these compliances play a great role in optimizing consumer demands by improving stability and controllability.

## 2. EQUIVALENT CORNERING STIFFNESS

The vehicle model of this paper is simplified to be a linear system for the analysis of ordinary handling where drivers should not require greater acceleration than 0.4 G. Side forces, aligning torques and camber forces of tires are all assumed to have linear characteristics.

The positive directions of variables are shown in Fig.1 and 2 in the case of left turn. The steer angles of the front wheels consist of the rotational angle of the steering system and of the additional steer angles caused by the rolling of the sprung mass (roll steer) and by the lateral force and self aligning torque on the front tires (compliance steer). The camber angles of front wheels are also the sum of a geometrical change caused by the rolling of the sprung mass (roll camber) and an elastic change caused by the lateral force on front tires (compliance camber). They are, therefore, expressed by the linear combinations of each variable as follows:

$$\beta_f=\beta+(\frac{\partial\beta}{\partial\phi})_f\phi+(\frac{\partial\beta}{\partial S})_f S_f-(\frac{\partial\beta}{\partial T})_f T_f \qquad (1)$$

$$\gamma_f= (\frac{\partial\gamma}{\partial\phi})_f\phi+(\frac{\partial\gamma}{\partial S})_f S_f \qquad (2)$$

and similarly on rear wheels,

$$\beta_r=(\frac{\partial\beta}{\partial\phi})_r\phi+(\frac{\partial\beta}{\partial S})_r S_r-(\frac{\partial\beta}{\partial T})_r T_r \qquad (3)$$

$$\gamma_r=(\frac{\partial\gamma}{\partial\phi})_r +(\frac{\partial\gamma}{\partial S})_r S_r \qquad (4)$$

Here, initial steer and camber angles are omitted because their effects on dynamics are cancelled out by the right and left wheels.

Tire side forces and self aligning torques are expressed as mean values of right and left wheels as follows:

$$S_f=C_f(\beta_f-\frac{\nu+\ell_f\dot\psi}{V})-Ct_f\gamma_f \qquad (5)$$

$$T_f=n_f S_f \qquad (6)$$

$$S_r=C_r(\beta_r-\frac{\nu-\ell_r\dot\psi}{V})-Ct_r\gamma_r \qquad (7)$$

$$T_r=n_r S_r \qquad (8)$$

Putting Eqs.1,2,3 and 4 into Eqs.5 and 7, respectively, side forces are derived as follows:

$$S_f=C_f{}^*(\beta+\varepsilon_f\phi-\frac{\nu+\ell_f\dot\psi}{V}) = C_f{}^*\alpha_f \qquad (9)$$

$$S_r=C_r{}^*(\varepsilon_r\phi-\frac{\nu-\ell_r\dot\psi}{V}) = C_r{}^*\alpha_r \qquad (10)$$

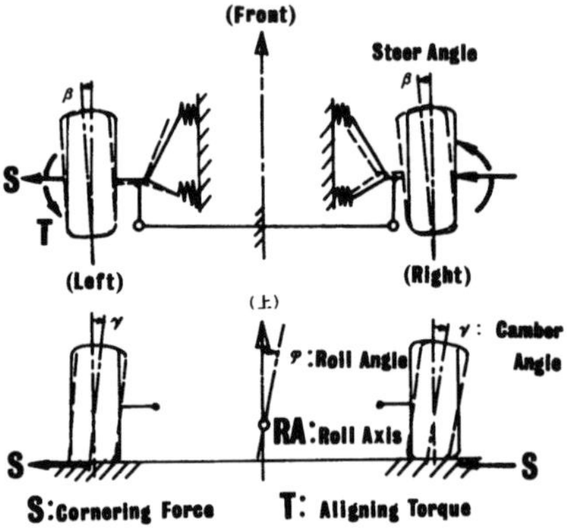


Fig.1 Positive Directions of Variables

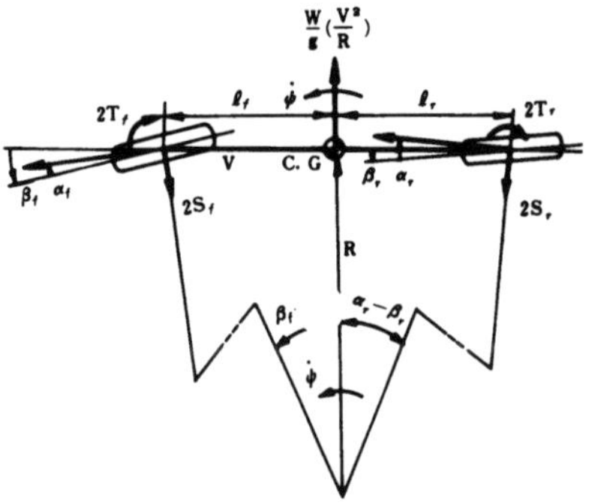


Fig.2 Vehicle Model

$$\text{where } \varepsilon_{f,r}=(\frac{\partial\beta}{\partial\phi})_{f,r}-(\frac{\partial\gamma}{\partial\phi})_{f,r}Ct_{f,r}/C_{f,r} \qquad (11)$$

$$C_{f,r}{}^*=C_{f,r}/[1-C_{f,r}\{(\frac{\partial\beta}{\partial S})_{f,r}-n_{f,r}(\frac{\partial\beta}{\partial T})_{f,r}-(\frac{\partial\gamma}{\partial S})_{f,r}Ct_{f,r}/C_{f,r}\}] \qquad (12)$$

The equilibrium of the moments about the king pin axis is given by the next relation:

$$K_s(\frac{\delta}{N}-\beta) = 2(n_0S_f+T_f) = 2(n_0+n_f)S_f \qquad (13)$$

Putting Eq.9 into Eq.13, steer angle is derived as follows:

$$\beta=[K_s(\frac{\delta}{N})-2(n_0+n_f)C_f^*(\epsilon_f\phi-\frac{\nu+\ell_f\dot{\psi}}{V})]\ /\ [K_s+2(n_0+n_f)C_f^*] \tag{14}$$

Putting Eq.14 into Eq.9, side forces of front tires can be expressed as follows:

$$S_f = C_f'(\frac{\delta}{N}+\epsilon_f\phi-\frac{\nu+\ell_f\dot{\psi}}{V}) = C_f'\alpha_f \tag{15}$$

where $C_f'=C_f^*/[1+2(n_0+n_f)C_f^*/K_s]$ (16)

In Eqs.10 and 15, if all compliances $(\frac{\partial\beta}{\partial S}),(\frac{\partial\beta}{\partial T}),(\frac{\partial\gamma}{\partial S})$ and $1/K_s$ are zero, then $C_f'$ and $C_f^*$ are equal to $C_f$, and $C_r^*$ is equal to $C_r$, and so Eqs.15 and 10 are equal to Eqs.5 and 7, respectively.

From these relations, a dynamic analysis with regard to the compliances can be easily made by changing cornering stiffness $C_f$ to $C_f^*$ or $C_f'$ and $C_r$ to $C_r^*$ in the analysis without compliances. From Eqs.12 and 16, the following relations can be derived.

$$\frac{1}{C_{f,r}^*}=\frac{1}{C_{f,r}}-(\frac{\partial\beta}{\partial S})_{f,r}+n_{f,r}(\frac{\partial\beta}{\partial T})_{f,r}+(\frac{\partial\gamma}{\partial S})_{f,r}\cdot\frac{Ct_{f,r}}{C_{f,r}} \tag{17}$$

$$\frac{1}{C_f'}=\frac{1}{C_f^*}+\frac{2(n_0+n_f)}{K_s} \tag{18}$$

These relations mean that each spring corresponding to each compliance is assumed to be connected in series. We defined $C_f^*$, $C_r^*$ and $C_f'$ as equivalent cornering stiffness.

In the case of ordinary passenger cars, the evaluated values of $C_f'/C_f$ is about 0.6-0.8 for recirculating ball type steering system and about 0.8-0.9 for rack and pinion type steering system, and $C_r^*/C_r$ is about 0.7-0.9 for independent suspension and about 0.9-1.0 for rigid axle suspension. Therefore, the effects of compliances should not be neglected, as the the effects of compliances are very important in vehicle dynamics.

## 3. DIRECTIONAL CONTROL CHARACTERISTICS

### 3.1. Stability Factor and Compliances

The effects of compliances on understeer/oversteer characteristic, which represents typical stationary property on controllability, are discussed in this chapter.

Assuming the lateral forces of front and rear axles to be : $F_{yf}\doteqdot 2S_f\cdot\cos\beta_f=2S_f$ and $F_{yr}\doteqdot 2S_r\cos\beta_r=2S_r$ , the equations of equilibrium of side forces, self aligning torques and centrifugal force on C.G. are derived as follows:

Lateral forces : $2S_f+2S_r = \frac{W}{g}(\frac{V^2}{R})$ (19)

yawing moments : $2S_f(\ell_f+n_0-n_f) -2S_r(\ell_r+n_r) = 0$ (20)

Therefore,

$$2S_f = \frac{W_f}{g}\left(\frac{V^2}{R}\right) \qquad \text{where } W_f = W\ell_r^*/(\ell_f^* + \ell_r^*) \tag{21}$$

$$2S_r = \frac{W_r}{g}\left(\frac{V^2}{R}\right) \qquad \text{where } W_r = W\ell_f^*/(\ell_f^* + \ell_r^*) \tag{22}$$

As shown in Fig.2, the relation between steer and slip angles of front and rear tires are given by

$$R\ (\ \beta_f - \beta_r \quad -\alpha_f + \alpha_r\ ) = \ell \tag{23}$$

or in another form, combining Eqs.10 and 15,

$$\frac{\delta}{N} + (\varepsilon_f - \varepsilon_r)\phi - (\alpha_f - \alpha_r) = \frac{\ell\dot{\psi}}{V} = \frac{\ell}{R} \tag{24}$$

$$\text{where } \phi = \frac{W_o r}{K_\phi - W_o r}\left(\frac{V^2}{R}\right) = \frac{\Phi}{g}\left(\frac{V^2}{R}\right) \tag{25}$$

$$\alpha_f = \frac{S_f}{C_f'} = \frac{W_f}{2gC_f'}\left(\frac{V^2}{R}\right) \tag{26}$$

$$\alpha_r = \frac{S_r}{C_r^*} = \frac{W_r}{2gC_r^*}\left(\frac{V^2}{R}\right) \tag{27}$$

Especially at low speed, the following relation is given by the Ackermann steering geometry,

$$\frac{\delta_0}{N} = \frac{\ell}{R_0} \tag{28}$$

Combining Eqs.24 and 28,

a) Circular Turn with Constant Radius : $R=R_0$=Const.

$$\frac{\delta}{N} + (\varepsilon_f - \varepsilon_r)\phi - (\alpha_f - \alpha_r) = \frac{\ell}{R_0} = \frac{\delta_0}{N} \tag{29}$$

Putting Eqs.25,26 and 27 into Eq.29,

$$\frac{\delta}{\delta_0} = 1 + K.V^2 \tag{30}$$

$$\text{where } K = \frac{1}{\ell g}\left[\ \frac{W_f}{2C_f'} - \frac{W_r}{2C_r^*} \quad -\Phi(\varepsilon_f - \varepsilon_r)\ \right] \tag{31}$$

b) Circular Turn with Constant Steering Wheel Angle : $\delta=\delta_0$=Const.

$$\frac{\delta_0}{N} + (\varepsilon_f - \varepsilon_r)\phi - (\alpha_f - \alpha_r) = \frac{\ell}{R} \tag{32}$$

Similarly,

$$\frac{R}{R_0} = 1 + K.V^2 \tag{33}$$

In Eqs.30 and 33, K is a stability factor defined at the steering wheel, not at the front wheel, and is a typical characteristic of understeer/oversteer property. And from Eqs.11,17 and 18, the stability factor K is differently expressed by the design parameters as follows:

$$K = \frac{1}{\ell g}\left[\frac{W_f}{2}\left\{\frac{1}{C_f}-\left(\frac{\partial\beta}{\partial S}\right)_f+n_f\left(\frac{\partial\beta}{\partial T}\right)_f+\frac{Ct_f}{C_f}\left(\frac{\partial\gamma}{\partial S}\right)_f\right\} - \frac{W_o r}{K_\phi - W_o r}\left\{\left(\frac{\partial\beta}{\partial\phi}\right)_f - \frac{Ct_f}{C_f}\left(\frac{\partial\gamma}{\partial\phi}\right)_f\right\}\right]$$

$$-\frac{1}{\ell g}\left[\frac{W_r}{2}\left\{\frac{1}{C_r}-\left(\frac{\partial\beta}{\partial S}\right)_r+n_r\left(\frac{\partial\beta}{\partial T}\right)_r+\frac{Ct_r}{C_r}\left(\frac{\partial\gamma}{\partial S}\right)_r\right\} - \frac{W_o r}{K_\phi - W_o r}\left\{\left(\frac{\partial\beta}{\partial\phi}\right)_r - \frac{Ct_r}{C_r}\left(\frac{\partial\gamma}{\partial\phi}\right)_r\right\}\right]$$

$$+\frac{1}{\ell g}\cdot\frac{(n_0+n_f)W_f}{K_s} \quad (34)$$

In order to adjust the contributions of each compliance factor to the value of K, the following variational equation is useful.

$$\Delta K = \frac{1}{2\ell g}\left\{\frac{W_r}{C_r}\left(\frac{\Delta C_r}{C_r}\right) - \frac{W_f}{C_f}\left(\frac{\Delta C_f}{C_f}\right) + (W_f\Delta CS_f - W_r\Delta CS_r)\right\}$$

$$-\frac{\phi}{\ell g}(\Delta\varepsilon_f - \Delta\varepsilon_r) - \frac{W_f}{g}\cdot\frac{(n+n_f)}{\ell}\cdot\frac{1}{K_s}\left(\frac{\Delta K_s}{K_s}\right) \quad (35)$$

where $\Delta CS_{f,r} = -\Delta\left(\frac{\partial\beta}{\partial S}\right)_{f,r} + n_{f,r}\Delta\left(\frac{\partial\beta}{\partial T}\right)_{f,r} + \Delta\left(\frac{\partial\gamma}{\partial S}\right)_{f,r} Ct_{f,r}/C_{f,r}$ (36)

$$\Delta\varepsilon_{f,r} = \Delta\left(\frac{\partial\beta}{\partial\phi}\right)_{f,r} - \Delta\left(\frac{\partial\gamma}{\partial\phi}\right)_{f,r} Ct_{f,r}/C_{f,r} \quad (37)$$

## 3.2. Steering Effort on Steady State Turning

Steering torque $T_H$ on steering wheel in steady state turning can be calculated from Eqs.13 and 21, and given as follows:

$$T_H = \frac{K_s}{\eta N}\left(\frac{\delta}{N}-\beta\right) = \frac{(n_0+n_f)}{\eta N}\cdot\frac{W_f}{g}\left(\frac{V^2}{R}\right) \quad (38)$$

where $\eta$ is the total efficiency of the steering system.
Combining Eqs.28,33 and 38, steering sensitivity regarding torque can be derived as follows:

$$\frac{T_H}{\delta} = \frac{1}{\eta N^2}\left(\frac{n_0+n_f}{\ell}\right)\frac{W_f}{g}\left(\frac{V^2}{1+K.V^2}\right) \quad (39)$$

This means that the effect of compliances can be estimated through the value of stability factor K.

## 3.3. Yawing Response and Compliance

Vehicle response to quick steering input and its convergency are in general evaluated by the frequency response of yaw velocity. In this chapter, the characteristics of yawing **response** are dealt with in the relation to the compliance factors.

The equations of lateral and yawing motion are derived, using equivalent cornering stiffness previously defined, as follows:

$$\frac{W}{g}(\dot{v}+V\dot{\psi}) = 2C_f'\left(\frac{\delta}{N}+\varepsilon_f\phi - \frac{v+\ell_f\dot{\psi}}{V}\right) + 2C_r^*\left(\varepsilon_r\phi - \frac{v-\ell_r\dot{\psi}}{V}\right) \quad (40)$$

$$I\,\ddot{\psi} = 2C_f'\ell_f^*\left(\frac{\delta}{N} + \varepsilon_f\phi - \frac{\nu + \ell_f\dot{\psi}}{V}\right) - 2C_r^*\ell_r^*\left(\varepsilon_r\phi - \frac{\nu - \ell_r\dot{\psi}}{V}\right) \quad (41)$$

where $\phi$ is assumed to be proportional to centrifugal acceleration and given by

$$\phi = W_o r/(K_\phi - W_o r).V\dot{\psi} \quad = \Phi.V\dot{\psi}/g \quad (42)$$

Using Laplace transformation of Eqs.40 and 41,

$$\begin{bmatrix} \frac{W}{2g}s + \frac{C_f'+C_r^*}{V} & \frac{W}{2g}V + \frac{C_f'\ell_f - C_r^*\ell_r}{V} - (C_f'\varepsilon_f + C_r^*\varepsilon_r)\frac{\Phi}{g}V \\ \frac{C_f'\ell_f^* - C_r^*\ell_r^*}{V} & \frac{I}{2}s + \frac{C_f'\ell_f\ell_f^* + C_r^*\ell_r\ell_r^*}{V} - (C_f'\ell_f^*\varepsilon_f - C_r^*\ell_r^*\varepsilon_r)\cdot V\Phi \end{bmatrix} \begin{bmatrix} L(\nu) \\ L(\dot{\psi}) \end{bmatrix}$$

$$= \begin{bmatrix} C_f' \\ C_f'\ell_f^* \end{bmatrix} \frac{L(\delta)}{N} \quad (43)$$

From this equation, the frequency response of yaw velocity to steering wheel angle is given as follows:

$$G(s) = \frac{L(\dot{\psi})}{L(\delta)} = \frac{2C_f'\ell_f^*}{N\ I}\left[\frac{s + (g/V)(2C_r^*/W_r)}{s^2 + 2\zeta\omega_y s + \omega_y^2}\right] \quad (44)$$

where yawing natural frequency $\omega_y = 2\pi f_y$ and its damping ratio $\zeta$ are given as follows:

$$f_y = \frac{\ell}{\pi}\sqrt{\frac{gC_f'C_r^*}{I\ W}\left(\frac{\ell_f^* + \ell_r^*}{\ell}\right).\left(\frac{1}{V^2} + K\right)} \quad (45)$$

$$\zeta \doteqdot \frac{1}{2\ f_y V}\left[\frac{g(C_f'+C_r^*)}{W} + \frac{C_f'\ell_f\ell_f^* + C_r^*\ell_r\ell_r^*}{I} - \frac{C_f'\ell_f\varepsilon_f - C_r^*\ell_r\varepsilon_r}{I}\cdot\frac{\Phi}{g}\cdot V^2\right] \quad (46)$$

Putting s=0 in Eq.44, the gain of steady state yaw velocity G(0) is given by the following equation and this corresponds to $G_\delta^\gamma|_0$ mentioned in the reference [5,6],

$$G(0) = \frac{1}{N\ell}\cdot\frac{V}{1+K.V^2} = G_\delta^\gamma|_0 \quad (47)$$

The variation of $f_y$ value according to the variations of each compliance can be calculated as follows:

$$\frac{\Delta f_y}{f_y} \doteqdot \frac{1}{2}\left(\frac{\Delta C_f'}{C_f'} + \frac{\Delta C_r^*}{C_r^*} + \frac{\Delta K}{1/_{V^2} + K}\right) \quad (48)$$

From Eqs.45 and 48, it can **easily be** seen how to adjust each compliance factor to obtain the desired values of $f_y$ for better handling characteristic of the vehicle.

### 3.4. An Example of Control and Response parameters

An example of the contributions of directional control parameters to response characteristics such as $f_y$ and K is shown in Table 1 and 2. Comparing cornering compliances, the effects of the tires are predominant. However, it can be said that each compliance factor has almost the same effect on K because the differences between front and rear for each compliance factor are nearly equal.

Table.1. Directional Control Parameters (TOYOTA STARLET)

| | | Front | Rear |
|---|---|---|---|
| Weight (Sprung Weight) | kg | $W=970$ ($W_0=870$) | |
| Weight Distribution | kg (%) | $W_1=466$ (48%) | $W_2=504$ (52%) |
| Wheelbase | m | $l=2.300$ | |
| Distance from C.G. to Axle | m | $l_f=1.196$ | $l_r=1.104$ |
| Tread | m | $t_f=1.300$ | $t_r=1.277$ |
| C.G. Height | m | $h=0.505$ | |
| Yawing Moment of Inertia | $kgms^2$ | $I=122.3$ | |
| Size (Infration Pressure) | ($kg/cm^2$) | 145SR13(1.7) | 145SR13(1.7) |
| Cornering Stiffness | kg/deg | $C_f=47.5$ | $C_r=55.0$ |
| Pneumatic Trail | m | $n_f=.03$ | $n_r=.03$ |
| Camber Stiffness | kg/deg | $C_{tf}=1.5$ | $C_{tr}=1.5$ |
| Lateral Force Compliance Steer | deg/kg | $(\frac{\partial\beta}{\partial S})_f=.0009$ | $(\frac{\partial\beta}{\partial S})_r=.0$ |
| Aligning Torque Compliance Steer | deg/kgm | $(\frac{\partial\beta}{\partial T})_f=.0$ * | $(\frac{\partial\beta}{\partial T})_r=.0$ |
| Roll Steer | deg/deg | $(\frac{\partial\beta}{\partial\phi})_f=-.20$ | $(\frac{\partial\beta}{\partial\phi})_r=.026$ |
| Roll Camber | deg/deg | $(\frac{\partial\gamma}{\partial\phi})_f=.75$ | $(\frac{\partial\gamma}{\partial\phi})_r=.0$ |
| Roll Stiffness | kgm/deg | $K_f=41.2$ | $K_r=11.8$ |
| Roll Center Height | m | $h_f=.020$ | $h_r=.280$ |
| Roll Arm Length | m | $r=.356$ | |
| Steering Rigidity | kgm/deg | $K_s=14.4$ * | |

* ---- $(\frac{\partial\beta}{\partial T})_f$ is included in $K_s$.

Table 2. Response Parameters

| | |
|---|---|
| Dynamic Load, kg | |
| $W_f = W\frac{l_r^*}{l_f^*+l_r^*} = 478$ | $W_r = W\frac{l_f^*}{l_f^*+l_r^*} = 492$ |
| where $l_f^*=l_f-n_f+n_0=1.166$ | $l_r^*=l_r+n_r=1.134$ |
| Roll Gradient, deg/g | |
| $\phi = W_0r/(K_f+K_r-W_0r.\pi/180) = 6.5$ | |
| Cornering Compliance, deg/g | |
| $0.5W_f/C_f = 5.03$ | $0.5W_r/C_r = 4.46$ |
| $-0.5W_f(\frac{\partial\beta}{\partial S})_f = -.22$ | $-0.5W_r(\frac{\partial\beta}{\partial S})_r = .0$ |
| $0.5W_f(\frac{\partial\beta}{\partial T})_f n_f = 0$ * | $0.5W_r(\frac{\partial\beta}{\partial T})_r n_r = .0$ |
| $0.5W_f(\frac{\partial\beta}{\partial S})_f C_{tf}/C_f = .03$ | $0.5W_r(\frac{\partial\beta}{\partial S})_r C_{tr}/C_r = .06$ |
| $-\phi(\frac{\partial\beta}{\partial\phi})_f = 1.59$ | $-\phi(\frac{\partial\beta}{\partial\phi})_r = -.20$ |
| $\phi(\frac{\partial\gamma}{\partial\phi})_f C_{tf}/C_f = .15$ | $\phi(\frac{\partial\gamma}{\partial\phi})_r C_{tr}/C_r = .0$ |
| $(n_0+n_f)W_f/K_s = 1.23$ | ---------- |
| Total $\Sigma_f = 7.81$ | $\Sigma_r = 4.32$ |
| Equivalent Cornering Stiffness, kg/deg ($C_{f,r}^*/C_{f,r}$ %) | |
| $C_f'=0.5W_f/\Sigma_f=30.6$ (64%) | $C_r^*=0.5W_r/\Sigma_r=56.4$ (100.2%) |
| Stability Factor, $s^2/m^2$ | |
| $K = \pi(\Sigma_f-\Sigma_r)/(l.g.180) = 0.00267$ | |
| Yawing Natural Frequency, Hz. (at V=100km/h) | |
| $f_y = \frac{180}{\pi^2}\sqrt{\frac{gC_f'C_r^*\,l.(l_f^*+l_r)}{W\,I}.(\frac{1}{V^2}+K)} = 1.00$ | |

## 4. CROSSWIND SENSITIVITY COEFFICIENT AND COMPLIANCE

Crosswind sensitivity, which is very important for highway driving, primarily depends on the amount of aerodynamic force and moment, namely on body style. However small refinements can be made by suspension and steering compliances.

### 4.1. Crosswind Sensitivity Coefficient Sw

Sw is defined as the lateral acceleration per unit aerodynamic side force in steady state with fixed control [3]. While a vehicle is blown by aerodynamic side force F on its A.C. and the steering wheel is fixed , the lateral acceleration $\alpha$ of the vehicle has the following relation in steady state.

$$\frac{\alpha}{V^2} = \frac{1}{R} \tag{49}$$

From the equilibrium of lateral forces and roll and yaw moments as shown in Fig.3, the following relations can be derived,

$$2S_f = \frac{W_f}{g}\alpha - F\cdot\frac{\ell_r^*+d}{\ell_f^*+\ell_r^*} = 2C_f'\alpha_f \tag{50}$$

$$2S_r = \frac{W_r}{g}\alpha - F\frac{\ell_f^{*}-d}{\ell_f^{*}+\ell_r^{*}} = 2C_r^{*}\alpha_r \qquad (51)$$

$$\phi = \frac{(W_o/g)\alpha r - Fe}{K_\phi - W_o r} = \frac{\Phi}{g}\alpha - \frac{F\,e}{K_\phi - W_o r} \qquad (52)$$

where $d=\ell.Cyaw/Cy-(\ell_r-\ell_f)/2$ (53)

$e=\ell.Croll/Cy-h_0$ (54)

aerodynamic coefficients:
- Cyaw : yawing moment
- Croll : rolling moment
- Cy : side force

Putting Eqs.49 and 52 into Eq.24, gives

$$\frac{\ell\alpha}{V^2} = \frac{\delta}{N} + (\varepsilon_f-\varepsilon_r)\left(\frac{\Phi}{g}\alpha - \frac{F\,e}{K_\phi - W_o r}\right) - (\alpha_f-\alpha_r) \qquad (55)$$

Putting $\alpha_f$ and $\alpha_r$ obtained from Eqs.50 and 51 into Eq.55,

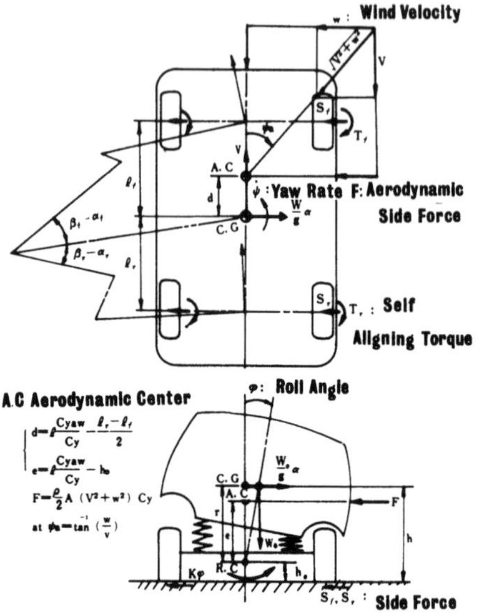


Fig.3 Vehicle Model

$$\frac{\ell\alpha}{V^2} = \frac{\delta}{N} - \frac{1}{2g}\left[\frac{W_f}{C_f'} - \frac{W_r}{C_r^{*}} - 2(\varepsilon_f-\varepsilon_r)\Phi\right]\alpha + \frac{1}{2\ell^{*}}\left[\frac{\ell_r^{*}+d}{C_f'} - \frac{\ell_f^{*}-d}{C_r^{*}} - 2\ell^{*}(\varepsilon_f-\varepsilon_r)\frac{e}{K_\phi - W_o r}\right].F \qquad (56)$$

Putting $\delta=0$ for the case of fixed control of steering wheel at the straight ahead position, crosswind sensitivity coefficient Sw is derived as follows:

$$Sw = \frac{\alpha}{F} = V^2.\frac{\frac{1}{2\ell\ell^{*}}\left[\frac{\ell_r^{*}}{C_f'} - \frac{\ell_f^{*}}{C_r^{*}} + d\left(\frac{1}{C_f'}+\frac{1}{C_r^{*}}\right)\right] - \frac{e(\varepsilon_f-\varepsilon_r)}{\ell(K - W_o r)}}{1 + \frac{1}{2\ell g}\left[\frac{W_f}{C_f'} - \frac{W_r}{C_r^{*}} - 2(\varepsilon_f-\varepsilon_r)\frac{W_o r}{K_\phi - W_o r}\right].V^2} \qquad (57)$$

Namely, Sw is decided by design parameters and the position of the A.C. , and the effects of each compliance factor can be estimated quantitatively from this equation. To assess the effect of understeer/oversteer characteristic, Sw is also expressed in another form as follows:

$$Sw = \frac{V^2}{1+K.V^2}\left[\frac{g}{W}K + \frac{d}{2\ell\ell^{*}}\left(\frac{1}{C_f'}+\frac{1}{C_r^{*}}\right) + \frac{(\varepsilon_f-\varepsilon_r)}{\ell}\cdot\frac{(r-e)}{(K_\phi - W_o r)}\right] \qquad (58)$$

The 1st, 2nd and 3rd terms mean the effect of understeer/oversteer property, equivalent cornering stiffness and roll steer, respectively. Of course, the 2nd and 3rd terms depend on the position of A.C., and if A.C. coincides with C.G., they disappear and Sw depends only on the stability factor K.

## 4.2. Steering Correction Angle for Keeping Straight Running

Putting $\alpha=0$ and $\delta=\delta_c$ in Eq.56, the steering wheel angle $\delta_c$ to keep the vehicle on a straight line can be derived as follows:

$$\delta_c = \frac{N}{2\ell^*}\left[\frac{\ell_f^*-d}{C_r^*} - \frac{\ell_r^*+d}{C_f'} - \frac{2\ell^* e(\varepsilon_f-\varepsilon_r)}{K_\phi - W_o r}\right].F \tag{59}$$

or in another form for normalizing,

$$\frac{1}{F\ell}\left(\frac{\delta_c}{N}\right) = -Sw.\left(\frac{1+K.V^2}{V^2}\right) \tag{60}$$

Therefore the effects of compliances on $\delta_c$ can be estimated by the values of K and Sw [2].

## 4.3. Crosswind Sensitivity Coefficient of A Vehicle with Trailer $Sw_T$

Fig.4 defines parameters for trailers and these parameters are marked with (').

Sw given by Eq.58 is assumed to be a function of A.C. position and expressed as Sw(d,e). Two lateral forces (F : aerodynamic side force on A.C. and $F_h$ : lateral force on the hitch point of the towing car) are acting on the towing car in Fig.4. Therefore, the lateral acceleration $\alpha$ of the towing car can be given by superposing the effects of the two forces as follows:

$$\alpha = F.Sw(d,e)+F_h.Sw(-\ell_h,e_h) \tag{61}$$

where

$\ell_h$ is the distance from C.G. to the hitch point.

$e_h$ is the distance from the hitch point to the roll axis of towing car.

From the equilibrium of lateral forces on the trailer, $F_h$ is calculated by the following equation.

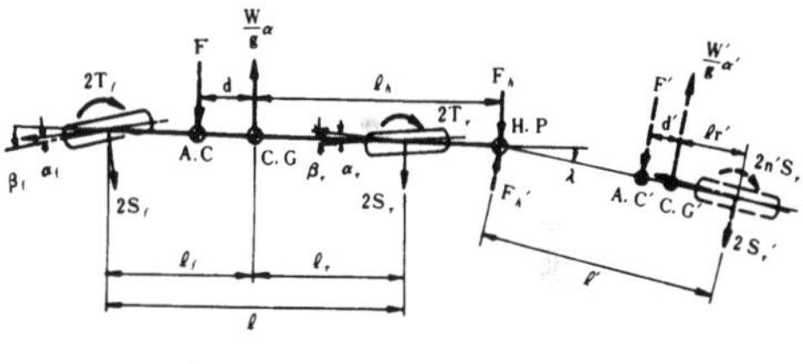


Fig.4 Vehicle - Trailer Model

$$F_h = F_h' = F'\frac{\ell_r'+d'}{\ell+n'} - \frac{W_h}{g}\alpha' \tag{62}$$

$$\text{where } W_h=W'(\ell_r'+n')/(\ell'+n') \tag{63}$$

Assuming that the relative yaw angle $\lambda$ between the towing car and the trailer is small, and that $F_h'$ and $\alpha'$ are nearly equal to $F_h$ and $\alpha$, respectively, then $\alpha$ can be obtained from Eqs.61 and 62 as follows:

$$\alpha = \frac{F.Sw(d,e) + F'(\ell_r'+d')/(\ell'+d').Sw(-\ell_h,e_h)}{1+(W_h/g).Sw(-\ell_h,e_h)} \tag{64}$$

Omitting rollsteer effects, $Sw_T$ defined as $\alpha/F$ is derived as follows:

$$Sw_T = \frac{V^2}{1+K_T.V^2}\left[\frac{g}{W}K + \frac{d}{2\ell\ell^*}\left(\frac{1}{C_f'}+\frac{1}{C_r^*}\right)\right.$$

$$-\left(\frac{F'}{F}\right)\left(\frac{\ell_r'+d'+n'}{\ell'+n'}\right).\frac{1}{2\ell\ell^*}\left(\frac{\ell_h-\ell_r^*}{C_f'}+\frac{\ell_h+\ell_f^*}{C_r^*}\right)\,] \tag{65}$$

where $$K_T=K-\frac{W_h}{g}\cdot\frac{1}{2\ell\ell^*}\left(\frac{\ell_h-\ell_r^*}{C_f'}+\frac{\ell_h+\ell_f^*}{C_r^*}\right) \tag{66}$$

$K_T$ is the stability factor of a vehicle with trailer
In these way, the effects of compliances on crosswind sensitivity are also estimated by the equivalent cornering stiffness.

## 5. EXPERIMENTAL AND THEORETICAL ANALYSIS FOR OPTIMUMCHARACTERISTICS

As chapters 3 and 4 show, the effects of each compliance factor on the characteristics of stability and controllability can be estimated by means of the equivalent cornering stiffness as an intermediate parameter, and designers can realize the optimum characteristics by reducing equivalent cornering stiffness to suitable compliance factors for suspension, steering and tires. And then, it becomes necessary to decide the optimum value or preferable zone of the directional control characteristics.

### 5.1. Optimum Characteristics of $f_y$ and K

As Eq.45 shows, $f_y$ and K do not have an independent relation. The $f_y$-K diagram, which is calculated for a vehicle with dimensions in Fig.5, shows that $f_y$ mainly depends on $C_r^*$, and K mainly on $C_f'$, and that $f_y$ increases slightly as K increases. From this net diagram, it is quite easy to find the desired combination of tire characteristics to realize the optimum values of $f_y$ and K.

The relations between the measured values of $f_y$ and K regarding many European and Japanese passenger cars are shown in Fig.6. European cars have a little weaker understeer characteristics, especially by sporty cars. However, they have good drive feeling generally because they have such high $f_y$ values. Generally speaking, control feeling is good in higher $f_y$, within a reasonable range of K. If K is too small, steering is too sensitive, and if K is too large, steering response is too dull. From Fig.6, the preferable zone of $f_y$ and K can be clearly recognized. The optimum characteristics for passenger cars are recommended empirically to be inside the area surrounded by dotted lines, namely $f_y \geq 0.8$ Hz, $0.002 \leq K \leq 0.005$ $s^2/m^2$, at vehicle speed of 100 km/h.

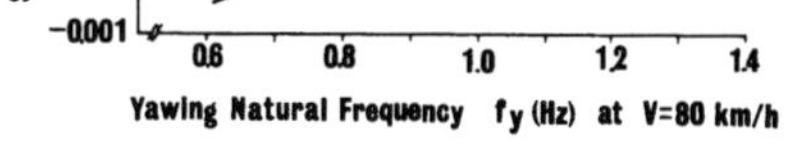

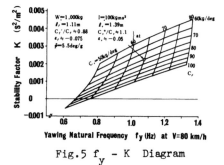

Fig.5 $f_y$ - K Diagram

### 5.2. The Optimum Zone Recommended by System Technology Inc.

System Technology Inc. recommended the optimum zone of yawing response for American full-size and intermediate cars as shown in Fig.7 [5,6].

In this figure, the vertical axis shows steady state yaw velocity gain $G^{\gamma}_{\delta}|_0$, which has a relation with stability factor K as in Eq.47, and the horizontal axis represents effective time constant Te, which is estimated by determining the frequency at which yaw velocity to steering wheel angle has 45 degrees of phase lag.

From Eq.47, K can be calculated as follows:

$$K=\frac{1}{N\ell}\cdot\frac{1}{VG^{\gamma}_{\delta}|_0 \text{ at 50mph}}-\frac{1}{V^2} \quad (67)$$

whrere V=22.2 m/s.

Te nearly coincides with the inverse of natural frequency $\omega_y$, and $\omega_y$ is roughly proportional to 1/V, therefore $f_y$ is obtained as follows:

$$f_y \doteq \frac{\sqrt{1/22.2^2+K}}{\sqrt{1/27.8^2+K}}\left(\frac{1}{2\pi Te}\right)_{\text{at 50mph}} \quad (68)$$

Putting wheelbase $\ell$=9.25ft=2.82m and overall gear ratio N=20 for the test vehicle of S.T.I., and using Eqs.67 and 68, the optimum zone recommended by S.T.I. is transformed to the zone surrounded by the solid lines and dashed lines in the $f_y$-K diagram of Fig.6. They partly coincide with our recommendation which is shown by the hatched area.

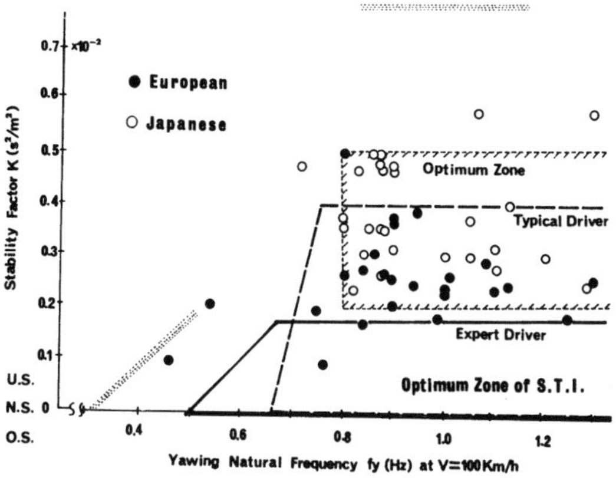


Fig.6 Measured Values of $f_y$ and K

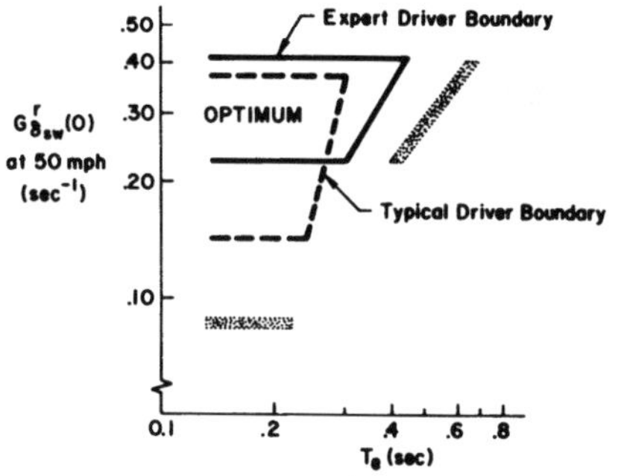


Fig.7 The Optimum Zone of S.T.I.

5.3. Theoretical Analysis for Optimum Directional Control Dynamics

In order to explain the preferable zone mentioned above, the following theoretical analysis was made. Using a simple vehicle-driver system as in Fig.8, drivers steer the steering wheel so as to run on a desired course x causing vehicle response y. The mean square value J of course deviation is calculated in order to estimate the optimum control.

Using the transfer function of driver $F_1(s)$ and vehicle $F_2(s)$, mean square value J is calculated as follows:

$$J=\frac{1}{2\pi}\int\left|X(s)-Y(s)\right|^2_{s=j\omega}d\omega=\frac{1}{2\pi}\int\left|\frac{X(s)}{1+F_1(s)\,F_2(s)}\right|^2_{s=j\omega}d\omega \quad (69)$$

where X(s) and Y(s) are the Laplace transformation of x and y.

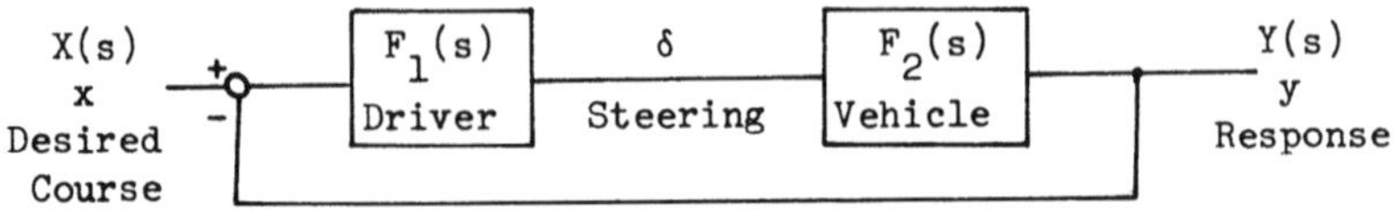


Fig.8 Vehicle - Driver Model

Regarding the typical case of step steering input, namely the vehicle changes its trajectory from a straight line to a circle, and the Laplace transformation of the desired course x is given by

$$X(s) = \frac{x_0}{s} \tag{70}$$

where $x_0$ is the carvature of the desired course.
Assuming that driver steers to correct the difference in angle between the desired course and the vehicle direction, and chosing the carvature of the vehicle trajectory as a vehicle response, the equation of steering wheel angle is expressed by [4],

$$T_n\dot{\delta} + \delta = G \int V(y-x_0)dt \tag{71}$$

where $T_n$ is a time constant of steering motion.
$G$ is a gain of steering correction.
Therefore, the transfer function of driver's steering motion is given by,

$$F_1(s) = \frac{L(\delta)}{L(y-x_0)} = G.\frac{V}{s}\left(\frac{1}{1+T_n s}\right) \tag{72}$$

From the relations of $Y(s)=(L(\nu)+V.L(\dot{\psi}))/V^2$ and Eq.43,

$$F_2(s) = \frac{L(y)}{L(\delta)} = \left(\frac{2C_f'}{M\ I\ V^2}\right).\left[\frac{Is^2+2C_r{}^*\ell_r\ell^*s/V+2C_r{}^*\ell^*}{s^2+2\zeta\omega_y s+\omega_y{}^2}\right] \tag{73}$$

Putting Eqs.70, 72 and 73 into Eq.69, J can be calculated by the following formula:

$$J=\frac{1}{2\pi}\int\left|\frac{B_0s^3+B_1s^2+B_2s+B_3}{A_0s^4+A_1s^3+A_2s^2+A_3s+A_4}\right|^2_{s=j\omega}d\omega \tag{74}$$

$$=\frac{\pi}{2}\cdot\frac{B_3{}^2(A_2A_1-A_3A_0)/A_4+A_1(B_2{}^2-2B_3B_1)+A_3(B_1{}^2-2B_2B_0)+B_0{}^2(A_3A_2-A_4A_1)/A_0}{A_3(A_2A_1-A_3A_0)-A_4A_2}$$

where $B_0=T_n$, $B_1=1+2\zeta\omega_yT_n$, $B_2=2\zeta\omega_y+\omega_y{}^2T_n$, $B_3=\omega_y{}^2$,
$A_0=B_0$, $A_1=B_1$, $A_2=B_2+2GC_f'/(M.V)$, $A_3=B_3+4GC_f'C_r{}^*\ell_r\ell^*/(M.I.V)$,
$A_4=4GC_f'C_r{}^*\ell^*/(M.I.V)$

Fig.9 shows the relation between J and K according to the variations of $C_f'$, in the case of a vehicle with the dimensions shown in Fig.9. In this figure, J has the minimum value $J_{min}$ at a weak understeer at high speed, and the minimum point moves to oversteer side as the vehicle's speed decreases. These results coincide qualitatively with the facts that proper understeer is better at higher speeds, and sensitive or sharp steering is better at lower speeds [2].

Fig.10 shows the calculated results of J at the combinations of $f_y$ and K, which are decided by the combinations of front and rear cornering stiffness $C_f'$ and $C_r^*$. In this figure, the values of J are the radiuses of circles and their centers correspond to the values of $f_y$ and K, and $J_{min}$ of every $C_r^*$ values are shown by black circles. The range of black circles is the preferable zone of K and $f_y$. It is obvious that the black circles become slightly smaller as $f_y$ increases.

From this analysis of a simple model, we can recognize the validity of the recommended optimum zone in Fig.6, although this result does not coincide quantitatively with the experimetal results. However, this will be solved by further researches of vehicle-driver model as a closed loop system, for example, by refining the driver model to predictive or programed steering, or by the establishment of the proper evaluation function included in driver's tasks.

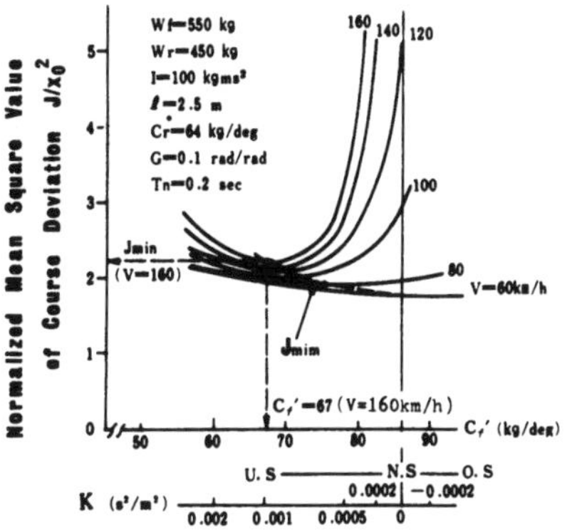


Fig.9 An Example of Optimum Control

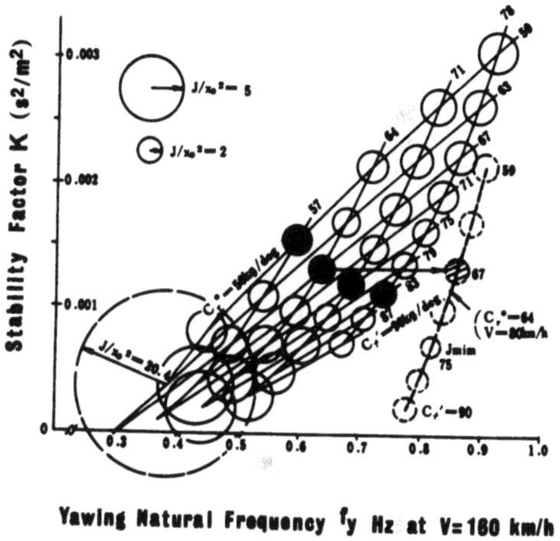


Fig.10 An Example of Optimum Values of $f_y$ and K

## 6. CONCLUSION

The typical characteristics of stability and controllability are derived from a simple vehicle model, and the concept of equivalent cornering stiffness is introduced in order to discuss those problems quantitatively with regard to suspension and steering compliances.

By using the equivalent cornering stiffness, the contributions of compliances to the characteristics such as stability factor, yawing natural frequency, crosswind sensitivity coefficient, etc., can be readily established, and the roles of each compliance factor can be determined and assigned during the design process to optimize those characteristics.

## 7. Nomenclature

| | |
|---|---|
| C.G. | : center of gravity of vehicle. |
| A.C. | : aerodynamic center of vehicle. |
| f,r | : subscript designating front and rear axle. |
| $V$, $\nu$ | : forward and lateral velocity of vehicle at C.G. |
| $\dot{\psi}$ | : yaw velocity of vehicle. |
| $\phi$ | : roll angle of sprung mass. |
| $\delta$, $\delta_0$ | : steering wheel angle. where $\delta_0 = \delta\ (V \to 0)$. |
| $\alpha_f$, $\alpha_r$ | : slip angle of front and rear tires. |
| $\beta_f$, $\beta_r$ | : steer angle of front and rear wheels. |
| $\gamma_f$, $\gamma_r$ | : camber angle of front and rear wheels. |
| $R$, $R_0$ | : radius of turning. where $R_0 = R\ (V \to 0)$. |
| $\alpha$ | : lateral acceleration of vehicle at C.G. |
| $g$ | : acceleration of gravity. where $g = 9.80\ m/s^2$ |
| $W_f$, $W_r$ | : equivalent axle load. |
| $W_0$ | : sprung weight of vehicle. |
| $I$ | : yawing moment of inertia of vehicle. |
| $\ell$ | : wheelbase. |
| $\ell_f$, $\ell_r$ | : horizontal distance from C.G. to front and rear axle. |
| $n_0$ | : caster trail. |
| $\ell_f^*$, $\ell_r^*$ | : effective distance from C.G. to front and rear axle. where $\ell_f^* = \ell_f + n_0 - n_f$, $\ell_r^* = \ell_r + n_r$. |
| $\ell^*$ | : equivalent wheelbase. where $\ell^* = \ell_f^* + \ell_r^*$. |
| $h$ | : C.G. height. |
| $r$ | : distance from C.G. to roll axis. |
| $K_\phi$ | : total roll stiffness about roll axis. |
| $K_s$ | : torsional rigidity of steering system about king pin axis. |
| $N$ | : overall gear ratio of steering system. |
| $S_f$, $S_r$ | : side force of front and rear tires. |
| $T_f$, $T_r$ | : self aligning torque of front and rear tires. |
| $C_f$, $C_r$ | : cornering stiffness of front and rear tires. |
| $n_f$, $n_r$ | : pneumatic trail of front and rear tires. |
| $Ct_f$, $Ct_r$ | : camber stiffness of front and rear tires. |
| $C_f^*$, $C_r^*$ | : equivalent cornering stiffness modified by compliances of front and rear suspensions. |
| $C_f'$ | : equivalent cornering stiffness of front tires in consideration of front suspension and steering compliances. |
| $(\frac{\partial\beta}{\partial\phi})_f$, $(\frac{\partial\beta}{\partial\phi})_r$ | : roll steer coefficient. |
| $(\frac{\partial\gamma}{\partial\phi})_f$, $(\frac{\partial\gamma}{\partial\phi})_r$ | : roll camber coefficient. |
| $\varepsilon_f$, $\varepsilon_r$ | : equivalent roll steer coefficient. |
| $(\frac{\partial\beta}{\partial S})_f$, $(\frac{\partial\beta}{\partial S})_r$ | : lateral force compliance steer coefficient. |
| $(\frac{\partial\beta}{\partial T})_f$, $(\frac{\partial\beta}{\partial T})_r$ | : aligning torque compliance steer coefficient. |
| $(\frac{\partial\gamma}{\partial S})_f$, $(\frac{\partial\gamma}{\partial S})_r$ | : lateral force compliance camber coefficient. |
| $w$ | : crosswind velocity. |
| $d$, $e$ | : horizontal and vertical distance from C.G. to roll axis. |

K : stability factor.

$\Phi$ : roll angle at lateral acceleration of 1.0 G. where $\Phi = W_O r/(K_\phi - W_O r)$

$f_y$ : yawing natural frequency.

Sw : crosswind sensitivity coefficient of vehicle.

$Sw_T$ : crosswind sensitivity coefficient of vehicle with trailer.

$\delta_c$ : steering wheel angle required for straight running against crosswind.

## 8. REFERENCES


1. Mitschke :
   Dynamic der Kraftfahrzeuge. Springer Verlag. Aug.1971.
2. Tsuchiya, Harada and Watari :
   Some Experimental and Theoretical Analyses on The Disturbed Motion of A Vehicle as A Closed Loop System. 16th FISITA Congress-Tokyo, 1976.
3. Watari, Tsuchiya and Iwase :
   On The Crosswind Sensitivity of An Automobile. 15th FISITA Congress-Paris, 1975.
4. Braess :
   Zur theoretischen Optimierung der Fahrverhaltens von Kraftfahrzeugen. ATZ.Vol.74 (1972), Nr.5.
5. McRuer, et al.:
   Automobile Controllability : Driver/Vehicle Response for Steering Control. Vol.1. Summary Report. PB-2402080, Feb. 1975.
6. Weir and Dimarco :
   Correlation and Evaluation of Driver/Vehicle Directional Handling Data. SAE Paper 780010.

# THE APPLICATION OF QUASILINEARIZATION TO THE LIMIT CYCLE BEHAVIOUR OF THE NONLINEAR WHEEL-RAIL SYSTEM

Wolfgang Hauschild

Institut für Mechanik, Technische Universität Berlin, F.R.G.

SUMMARY

The spatial kinematic constraints of arbitrary wheel and rail profiles can exactly be analysed by solving a system of nonlinear equations. The nonlinear equations for longitudinal and lateral creepages and spin are derived. The nonlinear equations of motion of a restrained wheelset on straight and curved track are set up and reduced. The external loads exerted on the wheelset are expanded. Nonlinear creep loads are evaluated, using different methods of saturation.

The nonlinear differential equations of motion are transformed into quasilinear differential equations and then into nonlinear algebraic equations, in order to calculate periodic solutions or limit cycles. The quasilinear equations are set up in form of 2 wheelset models. In the simple model only 3 nonlinear functions of kinematics are replaced by their describing functions. In the complex model describing functions for the external loads are computed.

Numerical results of limit cycles are presented for a wheelset with 2 or 3 degrees of freedom on straight and curved track.

## 1. INTRODUCTION

As most elements of railvehicles are nonlinear, e.g. kinematics,creep loads, primary and secondary suspension, the motions are examined more exactly, using nonlinear differential equations instead of linearized versions. Motion on straight track is described by an autonomous system whereas rail irregularities and curved track create a forced input. Besides the analogue and digital simulation there are analytical methods, which are able to directly evaluate periodic solutions.

The approximate calculation of periodic solutions of the nonlinear differential equations of the wheel-rail system was carried out by Cooperrider, Hedrick et al. [1,2,3,4] by means of quasilinearizing the simplified nonlinear kinematics and by applying a linear theory of creep loads. This method is improved by considering not only geometric nonlinearities but also the changing contact areas and the nonlinear creep loads.

The deficiencies of the linear curving theory of Boocock [5] are eliminated in the quasistatic curving theory of Elkins, Gostling [6] by allowing for nonlinearities of geometry, contact areas and creep loads. This method is expanded into a dynamic curving theory by adding the curving terms to the equations of motion used for limit cycle studies.

By quasilinearization [7,8] the assumed periodic solution is put into the system of nonlinear differential equations as a fourier polynomial of single period. After each nonlinear function is replaced by its fourier coefficient, i.e. describing function, quasilinear differential equations and a system of nonlinear algebraic equations are set up. Frequency, amplitudes and phases of possible limit cycles are solutions of this system of equations. By perturbing the amplitudes, the stability of the limit cycle is determined.

The variational method of Galerkin, which uses fourier polynomials of higher periods, may clear up, whether higher harmonics are significant and whether several simplifications are justified [9].

## 2. SPATIAL KINEMATICS OF A WHEELSET

The spatial kinematics of the wheel-rail system [10] can exactly be analysed by solving a system of nonlinear algebraic equations. Rail irregularities as well as elasticities of wheelset and rail can be taken into account. The profiles enter as a set of polynomial coefficients or as a set of coordinates. The recent version of the programme KINEMA does not search for multipoint contact. Some coordinate systems and notations are shown in Fig. 1.

### 2.1 Contact point coordinates

The nonlinear equations result from 3 conditions:

1. Wheel and rail contact points are the same points in space.
2. Vectors normal to the surface in the wheel and rail contact points are parallel.
3. Wheel and rail cannot penetrate each other.

The first 2 conditions yield 4 vector equations or 10 scalar equations

$$\begin{aligned} \mathbf{r}_{00} + \mathbf{T}_R \cdot [\, \mathbf{p}_{00} + \mathbf{T}_{P1,2}\, \mathbf{p}_{1,2} \,] - \mathbf{g}_{00} - \mathbf{T}_G \cdot [\, \mathbf{s}_{00} + \mathbf{T}_{S1,2}\, \mathbf{s}_{1,2} \,] &= 0 \\ [\, \mathbf{T}_R \cdot \mathbf{T}_{P1,2}\, \mathbf{n}_{P1,2} \,] \times [\, \mathbf{T}_G \cdot \mathbf{T}_{S1,2}\, \mathbf{n}_{S1,2} \,] &= 0 \end{aligned} \tag{2.1}$$

The third condition allows the approximate computation of the coordinates of the contact points by means of one-dimensional nonlinear minimization. This pre-iteration is necessary, in order to assure convergence even under situations of jumping contact points.

### 2.2 Velocity and angular velocity

The mass element in the momentary contact point on the wheel moves with an absolute velocity displayed in the O-coordinate system

$$\begin{aligned} \mathbf{v}_{OP1,2} &= \mathbf{T}_K^T\, \mathbf{v}_{OOP1,2} = \mathbf{T}_K^T \frac{d}{dt} [\, \mathbf{R}_{000} + \mathbf{r}_{00} + \mathbf{p}_{00} + \mathbf{p}_{10,20} \,] = \\ &= \mathbf{T}_K^T\, \mathbf{R}'_{000} + \frac{\delta}{\delta t} \mathbf{r}_{00} + (\mathbf{T}_K^T\, \boldsymbol{\omega}_{000}) \times \mathbf{r}_{00} + (\mathbf{T}_K^T\, \boldsymbol{\omega}_{00P}) \times [\, \mathbf{r}_{P1,2} - \mathbf{r}_{00} \,] - \\ &- \mathbf{T}_R [\, \boldsymbol{\omega}_{RP} \times \mathbf{p}_{00} \,] + \mathbf{T}_R \frac{\delta}{\delta t} \mathbf{p}_{1,2} \end{aligned} \tag{2.2}$$

The absolute velocity of the mass element in the rail contact point is expanded in a similar manner, noting that a particular mass element

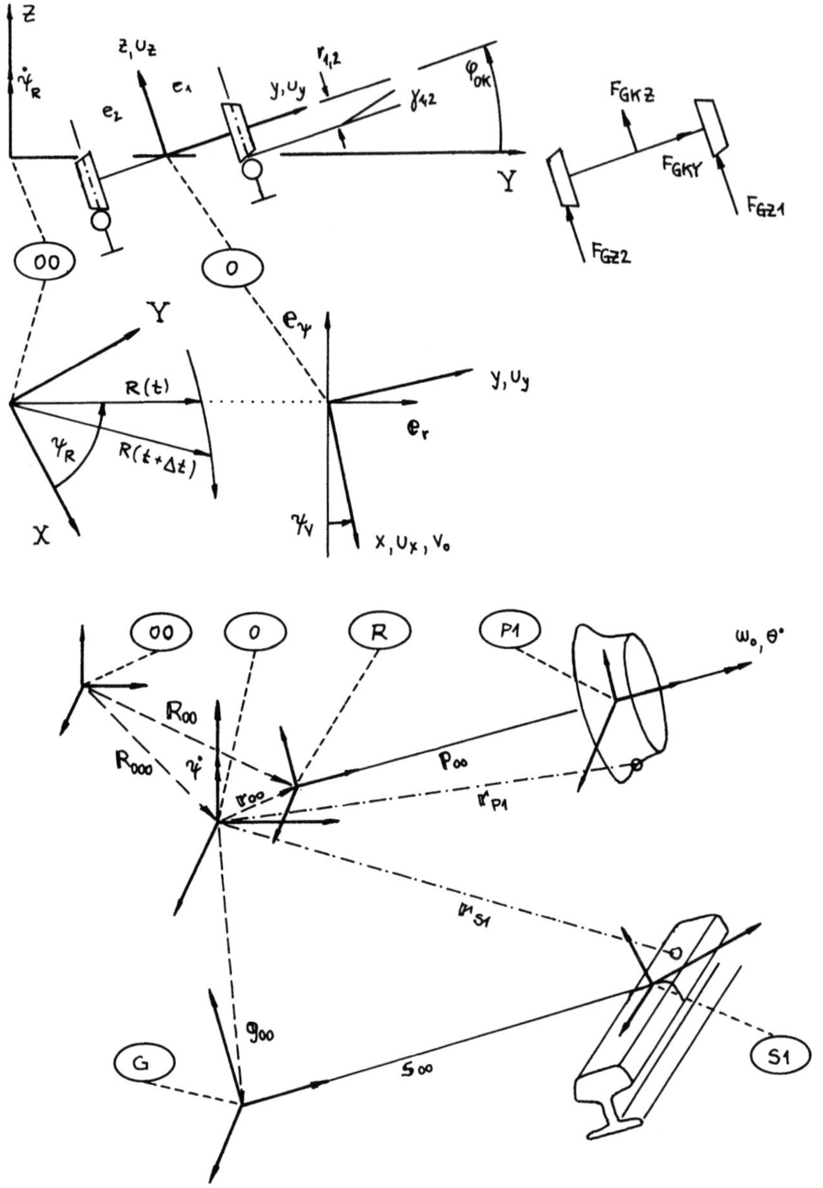

$\omega_{ij}$ angular velocity of system j relative to system i, shown in i

$T_K, T_R, T_G, T_{P1,2}, T_{S1,2}$ matrices to rotate between coordinate systems OO-O, O-R, O-G, R-P1,2, R-S1,2

Fig. 1. Coordinate systems and some notations

is not moving due to rail irregularities but only due to elasticities. Thus a rigid rail yields $\mathbf{v}_{OS\,1,2} = 0$.

The absolute angular velocity of mass elements in the contact points of the wheel and a rigid rail read

$$\begin{aligned} \boldsymbol{\omega}_{OP\,1,2} &= \mathbf{T}_K^T\, \boldsymbol{\omega}_{OOP\,1,2} = \mathbf{T}_K^T\, \boldsymbol{\omega}_{OOO} + \boldsymbol{\omega}_{OR} + \mathbf{T}_R\, \boldsymbol{\omega}_{RP\,10,20} \\ \boldsymbol{\omega}_{OS\,1,2} &= \mathbf{T}_K^T\, \boldsymbol{\omega}_{OOS\,1,2} = \mathbf{T}_K^T\, \boldsymbol{\omega}_{OOO} + \boldsymbol{\omega}_{OG} + \mathbf{T}_G\, \boldsymbol{\omega}_{GS\,1,2} = 0 \end{aligned} \tag{2.3}$$

The rolling velocity is defined [11] as the mean of the relative velocity between the mass element and the contact point on the wheel and the relative velocity between the mass element and the contact point on the rail

$$\begin{aligned} \dot{\mathbf{R}}_{P1,2} &= \dot{\mathbf{R}}_{S1,2} = \frac{d}{dt}[\mathbf{T}_K^T\, \mathbf{R}_{OOO} + \mathbf{r}_{P1,2}] \\ \mathbf{v}_{OR\,1,2} &= -\frac{1}{2}[(\mathbf{v}_{OP1,2} - \dot{\mathbf{R}}_{P1,2}) + (\mathbf{v}_{OS\,1,2} - \dot{\mathbf{R}}_{S1,2})] \end{aligned} \tag{2.4}$$

## 2.3 Longitudinal and lateral creepages and spin

The relative velocities and relative angular velocities have to be transformed into coordinates centered at the rail contact points and normalized with the rolling velocity.

$$\begin{aligned} \mathbf{v}_{BD\,1,2} &= [\mathbf{T}_R\, \mathbf{T}_{P1,2}\, \mathbf{T}_{B1,2}]^T\, (\mathbf{v}_{OP\,1,2} - \mathbf{v}_{OS\,1,2}) \\ \boldsymbol{\omega}_{BD\,1,2} &= [\mathbf{T}_R\, \mathbf{T}_{P1,2}\, \mathbf{T}_{B1,2}]^T\, (\boldsymbol{\omega}_{OP\,1,2} - \boldsymbol{\omega}_{OS\,1,2}) \end{aligned} \tag{2.5}$$

$$\begin{bmatrix} \nu_{\xi\,1,2} \\ \nu_{\eta\,1,2} \\ \nu_{\zeta\,1,2} \end{bmatrix} = \frac{1}{|\mathbf{v}_{OR\,1,2}|} \begin{bmatrix} v_{BD\,1,2\;x} \\ v_{BD\,1,2\;y} \\ \omega_{BD\,1,2\;z} \end{bmatrix} \tag{2.6}$$

In general these transformations have to be done numerically. In case of wheel-rail kinematics without rail irregularities and elasticities the creepages may be approximated by

$$\begin{bmatrix} \nu_{\xi 1,2} \\ \nu_{\eta 1,2} \\ \nu_{\zeta 1,2} \end{bmatrix} = \frac{1}{v_{R1,2}} \begin{bmatrix} v_0 c\psi_V + \dot{u}_x - [\omega_0 + \dot{\Theta}]\, r_{1,2} \mp [\dot{\psi} - v_0/R\, c\psi_V]\, e_{1,2} + v_0 s\psi_V \psi_V \\ -v_0 c\psi_V \psi\, c\gamma^*_{1,2} + [\dot{u}_y + r_{1,2}\dot{\varphi}]\, c\gamma^*_{1,2} + [\dot{u}_z \pm e_{1,2}\dot{\varphi}]\, s\gamma^*_{1,2} + v_0 \varkappa\, c\gamma^*_{1,2} \\ [\dot{\psi} - v_0/R\, c\psi_V]\, c\gamma^*_{1,2} - [\omega_0 + \dot{\Theta}]\, s\gamma_{1,2} \end{bmatrix} \tag{2.7}$$

$$\gamma^*_{1,2} = \gamma_{1,2} + \varphi \;;\quad s = \sin \;;\quad c = \cos \;;\quad \varkappa = -c\psi_V \psi_V + s\psi_V$$

$$v_{R1,2} = 1/2\, [v_0 \cos\psi_V + \dot{u}_x + (\omega_0 + \dot{\Theta})\, r_{1,2} \mp (\dot{\psi} - v_0/R \cos\psi_V)\, e_{1,2}] \tag{2.8}$$

# 3. NONLINEAR EQUATIONS OF MOTION OF A RESTRAINED WHEELSET

The complete equations of motion in 6 degrees of freedom, set up by the Newton-Euler or the Lagrangian method,are reduced to 6 simplified equations. The principle of virtual displacements is applied, the constraints of the system are introduced and 2 superfluous coordinates are eliminated. As a basis for further treatment 4 simplified equations of motion remain. Coordinate systems are shown in Fig. 1.

## 3.1 Equations of motion by Newton-Euler treatment

The equations of motion of a rigid body without constraints are set up by the methods of Newton-Euler or Lagrange. The wheelset has been treated with both methods, until the results were identical.

The translational equations are displayed in the O-coordinate system. Therefore the external forces $K^{(a)}$ have to be set up in the same system.

$$m\frac{d^2}{dt^2}[T_K^T R_{000} + r_{00}] = m\,[T_K^T \ddot{R}_{000} + \frac{\delta}{\delta t}(T_K^T \omega_{000}) \times r_{00} + (T_K^T \omega_{000}) \times$$
$$\times [(T_K^T \omega_{000}) \times r_{00}] + 2\,(T_K^T \omega_{000}) \times \frac{\delta}{\delta t} r_{00} + \frac{\delta^2}{\delta t^2} r_{00}]$$
$$= \Sigma\, K^{(a)} \qquad (3.1)$$

The rotational equations however are displayed in the R-coordinate system, which is a non-rotating system of principal axes through the center of mass. The external moments $M^{(a)}$ related to the center of mass have to be expressed in that system.

$$\omega = T_R^T T_K^T \omega_{00R} + \omega_{RP} \qquad (3.2)$$
$$D_S^T = \{\Theta_X \omega_X\ ;\ \Theta_Y \omega_Y\ ;\ \Theta_Z \omega_Z\} \qquad \Theta_X, \Theta_Y, \Theta_Z = \text{const}$$
$$\frac{d}{dt} D_S = \frac{\delta}{\delta t} D_S + (T_R^T T_K^T \omega_{00R}) \times D_S = \Sigma\, M^{(a)} \qquad (3.3)$$

## 3.2 Simplified equations of motion

As the complete equations are difficult to manage, all higher order terms are cancelled, except curving terms and gyro terms.

The 3 translational equations in the O-coordinate system read

$$m\ddot{u}_X + d_X \dot{u}_X + c_X u_X - F_{GXO} - F_{SXO} = F_A - m\alpha\,[s\psi_V + \beta c\psi_V]$$
$$m\ddot{u}_Y + d_Y \dot{u}_Y + c_Y u_Y - F_{GYO} - F_{SYO} = -Q_Y - mg s\varphi_K - m\alpha\,[c\psi_V - \beta s\psi_V]\,c\varphi_K \qquad (3.4)$$
$$m\ddot{u}_Z + d_Z \dot{u}_Z + c_Z u_Z - F_{GZO} - F_{SZO} = -Q_Z - mg c\varphi_K - m\alpha\,[-c\psi_V + \beta s\psi_V]\,s\varphi_K$$
$$\alpha = R^{\bullet\bullet} - R\dot{\psi}_R^2 \qquad \beta = R^{\bullet}/R\dot{\psi}_R$$

The 3 rotational equations in the R-coordinate system read

$$\Theta_X \ddot{\varphi} + d_Z e_{dZ}^2 \dot{\varphi} + c_Z e_{cZ}^2 \varphi - \Theta_Y \omega_0 \dot{\psi} - M_{GXR} - M_{SXR} = 0$$
$$\Theta_Y \ddot{\theta} \qquad - M_{GYR} - M_{SYR} = M_A \qquad (3.5)$$
$$\Theta_X \ddot{\psi} + d_X e_{dX}^2 \dot{\psi} + c_X e_{cX}^2 \psi + \Theta_Y \omega_0 \dot{\varphi} - M_{GZR} - M_{SZR} = 0$$

The external loads are subdivided in

- applied loads suspension, creep, weight, curving
- reaction loads normal forces in the contact areas

## 3.3 Introduction of constraints

The virtual work of the system is determined by means of the principle of virtual displacements. For ease of computation it should be remembered, that the virtual work of reaction loads vanishes. The system is premultiplied by the virtual displacements $\delta r^T$ or $\delta\varphi^T$, the kinematics have to be considered and all terms have to be rearranged.

$$\delta r^T = \{\delta u_x\ ; \quad \delta u_y \quad ; \quad \delta u_z \ \} \qquad \delta u_z = f_{11}\,\delta u_y + f_{12}\,\delta\psi$$
$$\delta\varphi^T = \{\delta\varphi\ ;\ \delta\theta + \delta\psi\sin\varphi\ ;\ \delta\psi\cos\psi\} \qquad \delta\varphi = f_{21}\,\delta u_y + f_{22}\,\delta\psi \tag{3.6}$$

Superfluous coordinates are eliminated and 4 equations of motion remain, in order to describe the 4 degrees of freedom of a wheelset.

## 3.4 Simplified equations of motion including constraints

These equations will again be reduced by cancelling almost all kinematic couplings. Later on the gyro term and the coupling in the mass matrix will be neglected as well.

$$\begin{aligned}
&m\ddot{u}_x && + d_x\dot{u}_x && && + c_x u_x && - F_{SX} && && = F_A + F_{GKX} \\
&(m+\Theta_x f_{21}^2)\ddot{u}_y && + d_y\dot{u}_y && - \Theta_y\omega_0 f_{21}\dot{\psi} && + c_y u_y && - F_{SY} && - F_{GY} && = F_{GKY} \\
&\Theta_y\ddot{\theta} && && && && - M_{SY} && && = M_A \\
&\Theta_x\ddot{\psi} && + d_x e_{dx}^2\dot{\psi} && + \Theta_y\omega_0 f_{21}\dot{u}_y && + c_x e_{cx}^2\psi && - M_{SZ} && - M_{GZ} && = 0
\end{aligned} \tag{3.7}$$

| | |
|---|---|
| curve and gravitation | $F_{GKX}$ , $F_{GKY}$ |
| gravitation | $F_{GY}$ , $M_{GZ}$ |
| creep | $F_{SX}$ , $F_{SY}$ , $M_{SY}$ , $M_{SZ}$ |
| traction | $F_A$ , $M_A$ |

In the limit cycle studies the wheelset is assumed to move with constant velocity $v_0$ . Therefore the longitudinal degree of freedom $u_x$ vanishes and 3 coupled equations for the lateral, yaw and rotational degrees of freedom $u_y, \psi, \theta^\circ$ remain. The fourth equation is used to deter- the driving force or moment, which is necessary to sustain the motion.

# 4. CURVE-, GRAVITATION- AND CREEP-LOADS

Aside from traction and suspension other external loads exist. Curve and gravitational forces are the resultant from centrifugal forces and weight. By elimination of the vertical displacement the gravitational stiffness builds up. It is in fact a measure for the change in potential energy due to lateral $u_y$ and yaw $\psi$ movement. If the external load due to curve and gravitational forces is substituted by the reaction forces in the contact areas, an approximation for the gravitational stiffness and for the normal forces is gained. Creep loads are gathered by summing up the individual creep loads in the contact areas.

## 4.1 Curve and gravitational forces

Centrifugal forces are exerted on the wheelset and the coupled mass. The lateral components cancel with gravitational forces for a specific cant angle $\varphi_{0K}$ of the curved track. For arbitrary cant angles $\varphi_K = \varphi_{0K} + \varphi_{1K}$ the resulting forces to the free wheelset become

$$\begin{aligned}
F_{KX} &= \qquad\qquad - m\alpha\,[s\psi_V + \beta c\psi_V] && = 0 \\
F_{KY} &= -mg\,s\varphi_K - m\alpha\,[c\psi_V - \beta s\psi_V]\,c\varphi_K && = -mg\,[s\varphi_K - \tan\varphi_{0K}\,c\varphi_K] = -mg\,\varphi_{1K} \\
F_{KZ} &= -mg\,c\varphi_K - m\alpha\,[-c\psi_V + \beta s\psi_V]\,s\varphi_K && = -mg\,[c\varphi_K + \tan\varphi_{0K}\,s\varphi_K] = -mg\,(1+\varphi_{0K}^2)
\end{aligned} \tag{4.1}$$

and respectively to the coupled mass

$$-Q_X = 0 \quad ; \quad -Q_Y = -Q\varphi_{1k} \quad ; \quad -Q_Z = -Q\,(1+\varphi_{0k}^2) \tag{4.2}$$

Summing up both parts yield the resulting forces to the wheelset

$$\begin{aligned} F_{GKX} &= -Q_X + F_{KX} \equiv 0 \\ F_{GKY} &= -Q_Y + F_{KY} = -[Q+mg]\,\varphi_{1k} \\ F_{GKZ} &= -Q_Z + F_{KZ} = -[Q+mg]\,[1+\varphi_{0k}^2] \end{aligned} \tag{4.3}$$

$$\tan\varphi_{0k} = \frac{1}{g}\frac{v_0^2}{R}\cos\psi_V \tag{4.4}$$

4.2 Gravitational stiffness and creep loads

Treating the simplified equations of motion with the principle of virtual displacements, the gravitational stiffness follows

$$\begin{aligned} F_{GY} &= F_{GKZ}\,f_{11} = -[Q+mg]\,[1+\varphi_{0k}^2]\,f_{11} \\ M_{GZ} &= F_{GKZ}\,f_{12} = -[Q+mg]\,[1+\varphi_{0k}^2]\,f_{12} \end{aligned} \tag{4.5}$$

and the creep loads rearrange

$$\begin{aligned} F_{SX} &= F_{SX0} &&= F_{\xi 1} + F_{\xi 2} - \psi\,[F_{\eta 1} c\gamma_1^* + F_{\eta 2} c\gamma_2^*] \\ F_{SY} &= F_{SY0} + F_{SZ0} f_{11} + M_{SXR} f_{21} &&= F_{\eta 1}[c\gamma_1^* + f_{21} r_1 c\gamma_1] + F_{\eta 2}[c\gamma_2^* + f_{21} r_2 c\gamma_2] \\ & &&\quad + \psi\,[F_{\xi 1} + F_{\xi 2}] \\ M_{SY} &= M_{SYR} &&= -F_{\xi 1} r_1 - F_{\xi 2} r_2 - M_{\zeta 1} s\gamma_1 - M_{\zeta 2} s\gamma_2 \\ M_{SZ} &= M_{SZR} + F_{SZ0} f_{12} + M_{SXR} f_{22} &&= -F_{\xi 1} e_1 + F_{\xi 2} e_2 + M_{\zeta 1} c\gamma_1 + M_{\zeta 2} c\gamma_2 \end{aligned} \tag{4.6}$$

4.3 Gravitational forces and normal forces

Another method is capable of evaluating the gravitational forces and at the same time the normal forces.

On a wheelset subjected only to the external force $F_{GKZ}$ (cf. Fig.1), the vertical reactions $F_{GZ\,1,2}$ in the contact are

$$F_{GZ\,1,2} = -\frac{e_{2,1}}{e_1 + e_2} F_{GKZ} = \frac{e_{2,1}}{e_1 + e_2}\,[Q+mg][1+\varphi_{0k}^2] \tag{4.7}$$

Resolving for horizontal components $F_{GY\,1,2}$ and resultant $N_{1,2}^*$ yield

$$\begin{aligned} F_{GY1,2} &= -F_{GZ\,1,2}\tan(\gamma_{1,2}+\varphi) \\ N_{1,2}^* &= \frac{F_{GZ\,1,2}}{\cos(\gamma_{1,2}+\varphi)} \end{aligned} \tag{4.8}$$

and the gravitational stiffness becomes

$$\begin{aligned} F_{GY} &= F_{GY1} + F_{GY2} &&= -N_1^* s\gamma_1^* - N_2^* s\gamma_2^* \\ M_{GZ} &= [-e_1 F_{GY1} + e_2 F_{GY2}]\,\psi &&= [e_1 N_1^* s\gamma_1^* - e_2 N_2^* s\gamma_2^*]\,\psi \end{aligned} \tag{4.9}$$

On straight track the resultant $N_{1,2}^*$ is an approximation for the normal forces $N_{1,2}$. In case of curved track this has to be corrected to gain the normal forces

$$N_{1,2} = N_{1,2}^* + \frac{1}{2} F_{GKY} \sin(\gamma_{1,2} + \varphi) \tag{4.10}$$

This approximation is rather useful, because the influence of cant angle deficiency on the normal forces can be examined even for the simple wheelset model (cf. Sec.6.1)

$$N_{1,2} = \frac{1}{2} [Q + mg][1 \mp \gamma_0 \varphi_{1k}] = N_0 [1 \mp \Delta N/N_0] \tag{4.11}$$

These quasistatic approximations are questionable in case of large wheelset accelerations, i.e. flange contact. An exact computation of normal forces is possible, if the system with constraints Eq.(3.7) is extended by 2 equations from the system without constraints Eq.(3.4). These additional equations are solved for the normal forces.

## 5. CREEP LOADS

The creep loads in the contact area are computed at minimal expense of time according to the linear theory and at significant expense of time according to the simplified nonlinear theory of Kalker [11,12]. The method of creep load saturation corrects the results of the linear theory without additional computation time. A method to generate quasilinear Kalker coefficients is presented.

### 5.1 Creep load saturation

The coefficient of friction limits the tangential force in the contact area $F_R \leq \mu N$. The maximum spin moment is assessed $M_\varsigma \leq 3/16 \pi c \mu N$. This assessment is given, when the normal stress of Hertz's theory is integrated across a circular contact area of radius $c = \sqrt{ab}$.

Three methods of creep load saturation are presented in Fig.5

$F_R = \sqrt{F_\xi^2 + F_\eta^2}$ ; $M_\varsigma$ — tangential force, spin moment

$f_R = F_R/(\mu N)$ ; $f_R = |M_\varsigma|/(3/16 \pi c \mu N)$ — creep load coefficient

<u>Linear saturation</u> a) $0 \leq f_R \leq 1$ $\quad f_R^* = f_R$ (5.1)

b) $1 \leq f_R$ $\quad f_R^* = 1$

<u>Exponential saturation</u> $f_R^* = 1 - \exp(-f_R)$ (5.2)

<u>Cubic saturation</u> a) $f_R^* = f_R + [\frac{1}{c^2}(3-2c)] f_R^2 + [\frac{1}{c^3}(c-2)] f_R^3$

$c \leq f_R$ $\quad f_R^* = 1$ (5.3)

The saturated creep forces and moments read

$$F_{\xi,\eta}^* = f_R^*/f_R \cdot F_{\xi,\eta} \quad ; \quad M_\varsigma = f_R^*/f_R \cdot M_\varsigma \tag{5.4}$$

### 5.2 Quasilinear creep coefficients

The 5 individual terms of the linear creep theory may be saturated independently. If these nonlinear functions are replaced by their descibing functions, quasilinear Kalker coefficients $c_{ikN}$ develop, which depend on the coefficient of friction.

$$
\begin{aligned}
F^{*}_{\xi 11}(\nu_\xi) &= -\mu N f_R^{*} \operatorname{sign}(\nu_\xi) \qquad\qquad f_R = |F_{\xi 11}|/(\mu N) \\
c_{11N} &= -\frac{1}{Gc^2}\,\frac{4}{\pi \bar{\nu}_\xi} \int_0^{\pi/2} F^{*}_{\xi 11}(\bar{\nu}_\xi \sin\phi)\,\sin\phi\; d\phi \\
F^{**}_{\xi 11}(\nu_\xi) &= -Gc^2\, c_{11N}\, \nu_\xi
\end{aligned}
\tag{5.5}
$$

## 6. QUASILINEAR EQUATIONS OF MOTION

A simple and a complex wheelset model is presented. Both models are suited for straight and curved track and for 2 or 3 degrees of freedom. The driving force or moment on the wheelset, which is necessary to sustain the movement, can be determined with the complex model.

In the simple model the only nonlinearities are 3 functions of geometry, which are replaced by their describing functions. The conditions in the contact area remain unchanged. This model is almost identical to studies of Cooperrider and Hedrick [1,2,3,4]. On curved track the rotational degree of freedom $\Theta^{\bullet}$ is eliminated in advance, similar to Boocock, Elkins and Gostling [5,6], but using the describing functions of geometry.

In the complex model all nonlinear creep-, gravitational- and curving loads are computed, expanded into a fourier series and approximated by the first harmonic. Thus the changing conditions in the contact area are taken into account.

### 6.1 Wheelset model 1

The nonlinear wheel-rail kinematics are replaced by 3 describing functions (Fig.2).

$$
\begin{aligned}
u_y &= u_{y0} + u_y^{*} = u_{y0} + \bar{u}_y \sin\omega t \\
\varphi\,(u_y) &= \phi_0(u_{y0},\bar{u}_y) + \phi_1(u_{y0},\bar{u}_y)\, u_y^{*} = \phi_0 + \phi_1 u_y^{*} \\
\frac{r_1 - r_2}{2e_0}(u_y) &= \Lambda_0(u_{y0},\bar{u}_y) + \Lambda_1(u_{y0},\bar{u}_y)\, u_y^{*} = \Lambda_0 + \Lambda_1 u_y^{*} \\
\frac{\gamma_1 + \gamma_2}{2}(u_y) &= \Gamma_0(u_{y0},\bar{u}_y) + \Gamma_1(u_{y0},\bar{u}_y)\, u_y^{*} = \Gamma_0 + \Gamma_1 u_y^{*}
\end{aligned}
\tag{6.1}
$$

The creep loads are computed by means of the linear theory. The conditions in the contact patch, especially the area of the ellipse, are constant. In case of curved track a correction term has to be added. Due to cant deficiency there is a resulting lateral force, which in turn creates different normal forces (cf.Sec.4.3) and different contact areas on the left and right wheel.

$$
\begin{aligned}
c &= \sqrt{ab} = p\, N^{1/3} \\
c_{1,2}^{m} &= p^m N^{m/3} = p^m N_0^{m/3}\left[1 \mp m\,\tfrac{1}{3}\,\tfrac{\Delta N}{N_0}\right] = p^m N_0^{m/3}\left[1 \mp m n\right]
\end{aligned}
\tag{6.2}
$$

A nontrivial solution for the rotational velocity is now possible. Neglecting the inertial term, the creep moment vanishes and $\Theta^{\bullet}$ expands.

$$
M_{SY} = -F_{\xi 1} r_1 - F_{\xi 2} r_2 = 0 \tag{6.3}
$$

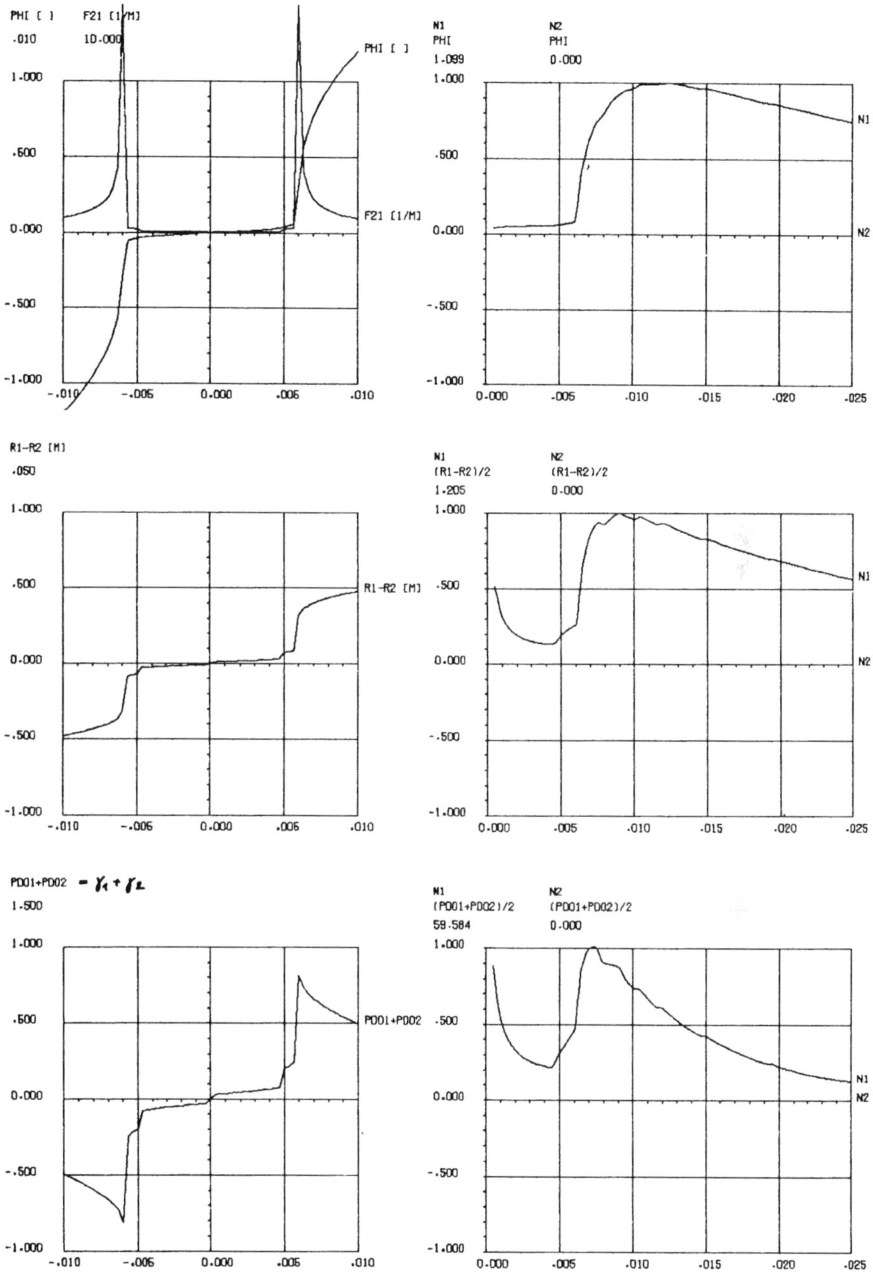


Fig. 2. Geometry functions and corresponding describing functions
Wheel profile UIC ORE S1002
Rail profile UIC 60 cant 1/40 gauge 1435 mm

$$\Theta^{\cdot} = 2n \left[ \frac{v_0}{r_0^2} e_0 \frac{r_1 - r_2}{2e_0} + \frac{e_0}{r_0}\psi^{\cdot} - \frac{e_0}{r_0}\frac{v_0}{R} \right]$$

$$= 2n \left[ \frac{v_0}{r_0^2} e_0 \left( \Delta_0 + \Delta_1 u_y^* \right) - \frac{e_0}{r_0}\frac{v_0}{R} + \frac{e_0}{r_0}\psi^{\cdot} \right] \qquad (6.4)$$

This result is put into the equations of motion (3.7) considering the expressions for the creepages and for curve-, gravitation- and creep loads. A quasilinear system of differential equations is set up, describing the motion of a wheelset with 2 degrees of freedom on straight and curved track. Kalker coefficients $C_{ik}$ may be replaced by their quasilinear counterparts $C_{ikN}$ .

$$\mathbb{M}\ddot{\mathbf{x}}^* + \mathbb{D}\dot{\mathbf{x}}^* + \mathbb{S}\mathbf{x}^* = \mathbb{R} \qquad (6.5)$$

$$\mathbb{M} = \begin{bmatrix} m & 0 \\ 0 & \Theta_z \end{bmatrix} \qquad \mathbf{x}^* = \begin{bmatrix} u_y^* \\ \psi^* \end{bmatrix}$$

$$\mathbb{D} = \begin{bmatrix} d_y & 0 \\ 0 & d_x e_{dx}^2 \end{bmatrix} + 2Gc^2 \frac{1}{V_R} \begin{bmatrix} C_{22}\left[1 + r_0\phi_1\right]^2 & cC_{23}\left[1 + r_0\phi_1\right] + X_{n6}\frac{e_0}{r_0} \\ cC_{32}\left[1 + r_0\phi_1\right] & e_0^2 C_{11} + c^2 C_{33} + X_{n4}\frac{e_0}{r_0} \end{bmatrix}$$

$$\mathbb{S} = \begin{bmatrix} c_y & 0 \\ 0 & c_x e_{cx}^2 \end{bmatrix} + 2Gc^2 \frac{v_0}{V_R} \begin{bmatrix} -cC_{23}\left[1 + r_0\phi_1\right]\frac{1}{r_0}\Gamma_1 + X_{n6}\frac{e_0}{r_0^2}\Delta_1 & -C_{22}\left[1 + r_0\phi_1\right] \\ C_{11}\frac{e_0^2}{r_0}\Delta_1 - c^2 C_{33}\frac{1}{r_0}\Gamma_1 + X_{n4}\frac{e_0}{r_0^2}\Delta_1 & -cC_{32} \end{bmatrix}$$

$$+ \left[Q + mg\right] \begin{bmatrix} \Gamma_1 + \phi_1 & 0 \\ 0 & -e_0\gamma_0 \end{bmatrix}$$

$$\mathbb{R} = -2Gc^2 \frac{v_0}{V_R} \begin{bmatrix} -cC_{23}\frac{1}{r_0}\Gamma_0\left[1 + r_0\phi_1\right] + X_{n6}\frac{e_0}{r_0^2}\Delta_0 - cC_{23}\frac{1}{R}\left[1 + r_0\phi_1\right] - X_{n6}\frac{e_0}{r_0}\frac{1}{R} \\ C_{11}\frac{e_0^2}{r_0}\Delta_0 - c^2 C_{33}\frac{1}{r_0}\Gamma_0 + X_{n4}\frac{e_0}{r_0^2}\Delta_0 - C_{11}e_0^2\frac{1}{R} - c^2 C_{33}\frac{1}{R} - X_{n4}\frac{e_0}{r_0}\frac{1}{R} \end{bmatrix}$$

$$- \left[Q + mg\right] \begin{bmatrix} \Gamma_0 + \phi_0 + \varphi_{1k} \\ 0 \end{bmatrix} - \left\{ \begin{bmatrix} c_y & 0 \\ 0 & c_x e_{cx}^2 \end{bmatrix} + 2G_c^2 \frac{v_0}{V_R} \begin{bmatrix} 0 & -C_{22}\left[1 + r_0\phi_1\right] \\ 0 & -cC_{32} \end{bmatrix} + \left[Q + mg\right] \begin{bmatrix} 0 & 0 \\ 0 & -e_0\gamma_0 \end{bmatrix} \right\} \begin{bmatrix} u_{y0} \\ \psi_0 \end{bmatrix}$$

$$X_{n6} = 6n^2 \left( cC_{23}\gamma_0\left[1 + r_0\phi_1\right] \right)$$

$$X_{n4} = 4n^2 \left( -C_{11}e_0 r_0 + 2c^2 C_{33}\gamma_0 \right)$$

$$V_R = v_0 + \frac{1}{2}\Theta^{\cdot} = v_0 \left[ 1 - e_0 n \left( 1/R - \Delta_0/r_0 \right) \right]$$

$$n = 1/3\,\gamma_0\varphi_{1k}$$

## 6.2 Wheelset model 2

The simplified equations of motion (3.7) have to be subdivided for a pulled or an axle-driven wheelset.

$$m\ddot{u}_y + d_y\dot{u}_y + c_y u_y - F_{SY} - F_{GY} = F_{GKY}$$

$$\Theta_z\ddot{\psi} + d_x e_{dx}^2\dot{\psi} + c_x e_{cx}^2\psi - M_{SZ} - M_{GZ} = 0 \qquad (6.6)$$

pulled $\Theta_y\,\Theta^{\cdot\cdot} - M_{SY} = 0$     axle-driven $-F_{SX} - F_{GKX} = 0$ (6.7)

$F_A = -F_{SX} - F_{GKX}$     $M_A = \Theta_y\ddot{\Theta} - M_{SY}$

This is a system of 3 differential equations containing linear parts nonlinear parts $\mathbb{F}(\dot{\mathbf{x}},\mathbf{x})$ and a right hand side $\mathbb{P}$ , which is constant or slowly varying [7,8]

$$\mathbb{M}\ddot{\mathbf{x}} + \mathbb{D}_L\dot{\mathbf{x}} + \mathbb{S}_L\mathbf{x} + \mathbb{F}(\dot{\mathbf{x}},\mathbf{x}) = \mathbb{P} \qquad (6.8)$$

The nonlinear part $\mathbb{F}$ is expanded by means of the assumed periodic

$$x = x_0 + x^* = x_0 + A \sin \phi \tag{6.9}$$

solution into a fourier series, approximated by the first harmonic and expressed in terms of describing functions

$$\mathbb{F}(\dot{x}, x) = \mathbb{N}_0 + \mathbb{N}_1 x^* + \frac{1}{\omega} \mathbb{N}_2 \dot{x}^* \tag{6.10}$$

$$\begin{Bmatrix} \mathbb{N}_0 \\ \mathbb{N}_1 \\ \mathbb{N}_2 \end{Bmatrix} (x_0, A, \phi) = \frac{1}{\pi A} \int_0^{2\pi} \mathbb{F}(\dot{x}, x) \begin{Bmatrix} A/2 \\ \sin\phi \\ \cos\phi \end{Bmatrix} d\phi$$

Entered into Eq.(6.8), a quasilinear system of differential equations develops

$$\begin{aligned} &M \ddot{x}^* + [\mathbb{D}_L + \tfrac{1}{\omega} \mathbb{N}_2] \dot{x}^* + [\mathbb{S}_L + \mathbb{N}_1] x^* = 0 \\ &- \mathbb{S}_L x_0 - \mathbb{N}_0 + \mathbb{P} = \mathbb{R} = 0 \end{aligned} \tag{6.11}$$

The damping and stiffness matrices of the periodic part are functions of the amplitudes, phases and frequency of the assumed periodic solution. The constant part is a nonlinear algebraic system of equations, describing the static equilibrium.

The changing conditions in the contact area are taken into account, because the kinematics and creep loads are computed for every position of the wheelset.

## 7. PERIODIC SOLUTIONS

The search for periodic solutions of quasilinear differential equations can be done with different methods [13], e.g. by means of minimization or search of least damped eigenvalues. A nonlinear algebraic system of equations seems to be the most general method, which may also be applied to solve for forced and statistical response. A combination of eigenvalue search and nonlinear equations appears adviseable, in order to cope with systems of many degrees of freedom.

The quasilinearization is an approximate method with important restrictions: Periodic solutions of the nonlinear equations do not necessarily exist in the vicinity of periodic solutions of the quasilinear equations and vice versa. Stability calculations have the same restrictions. The variational method of Galerkin however, which uses fourier polynomials of arbitrary order, may answer these questions.

### 7.1 Nonlinear algebraic system of equations

Both wheelset models give a quasilinear system of differential equations for the periodic solutions and a nonlinear algebraic system of equations for the static equilibrium.

$$M \ddot{x}^* + \mathbb{D}(x_0, A, \phi) \dot{x}^* + \mathbb{S}(x_0, A, \phi) x^* = 0 \tag{7.1}$$

$$\mathbb{R}(x_0, A, \phi) = 0 \tag{7.2}$$

If the assumed periodic solution $x^* = X e^{j\omega t}$ is inserted into Eq.(7.1), n sufficient conditions Eq.(7.3) and one necessary conditions Eq.(7.4) follow. Together with Eq.(7.2) they form a nonlinear algebraic system of equations, governing the existence of the periodic solutions.

$$[(S - \omega^2 M) + j\omega D] X = 0 \tag{7.3}$$

$$\det \{(S - \omega^2 M) + j\omega D\} = 0 \tag{7.4}$$

## 7.2 Periodic solutions and stability

The nonlinear equations are solved, using the hybrid method due to Powell [14]. The reference lateral amplitude $\bar{U}_Y$ is chosen constant and its phase angle is set to zero. The velocity $V_0$ and the limit cycle frequency $\omega$ take their positions as variables.

The stability of the limit cycle is analysed by means of the characteristic equation (7.4). The amplitude vector is slightly disturbed, and if the limit cycle is stable, the system will return to the former periodic solution.

The programme ZYKEL computes limit cycles of a restrained wheelset on straight and curved track, using both wheelset models with 2 or 3 degrees of freedom.

# 8. FIRST RESULTS

A wheelset with profile UIC ORE S1002 on UIC 60 rails with cant of 1/40 and gauge of 1435 mm has been analysed. The wheelset is suspended softly without damping. Limit cycles on straight track and on curved track are presented. A constant radius of curvature of 1000 m and a cant deficiency of $-\varphi_{1K}=0$ and $-\varphi_{1K}=-.1$ is chosen. The lateral amplitude $\bar{U}_Y$ is varied from 0 to 10 mm in 50 steps. The integration over a period is done in 60 steps for model 1 and 30 steps for model 2. Calculation times refer to a CDC 6500 computer.

The evaluation of stability and necessary traction give until now unsatisfactory results.

## 8.1 Straight track

Limit cycle calculations with model 1 and linear or quasilinear Kalker coefficients differ above lateral amplitudes of 4.8 mm, i.e. flange contact (Figs.3,4). For wheelset displacements below flange contact model 1 is qualified and should be preferred to the complex model 2 because of its short computation time of 10 sec.

The Kalker coefficients C32 and C33 are not signifant, whereas C23 must not be neglected. Quasilinear Kalker coefficients allow a quick assessment of the influence of the coefficient of friction (Fig.4).

Limit cycle calculations with model 2 differ significantly from model 1 for wheelset displacements above flange contact (Figs.3,6), where conditions in the contact area change drastically.

All methods of creep load saturation, including ROLCON (Fig.3), produce results, which are slightly different below flange contact and almost identical above flange contact. Whereas model 2 with creep load saturation takes 600 sec, the respective computation time with ROLCON multiplies by a factor of 20.

The rotational velocity $\Theta^{\cdot}$ does not yield a significant result.

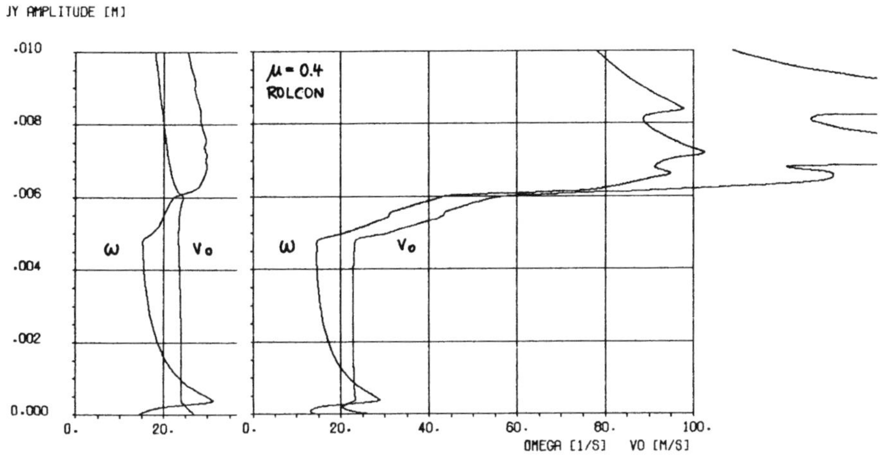


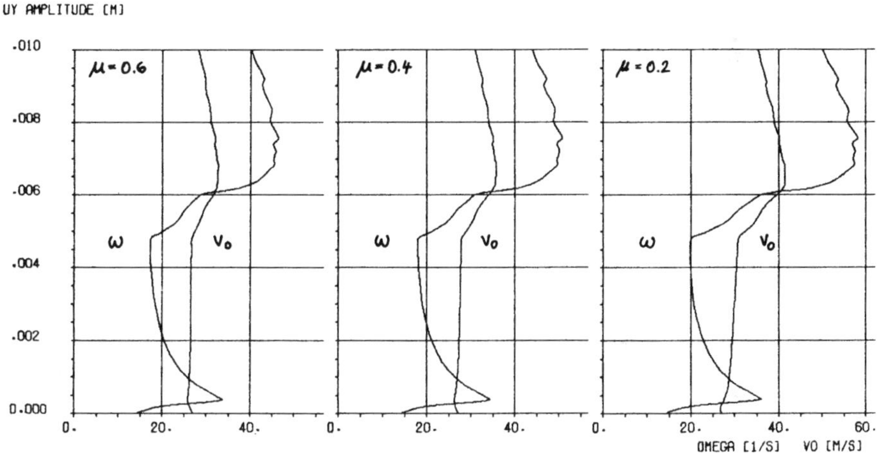


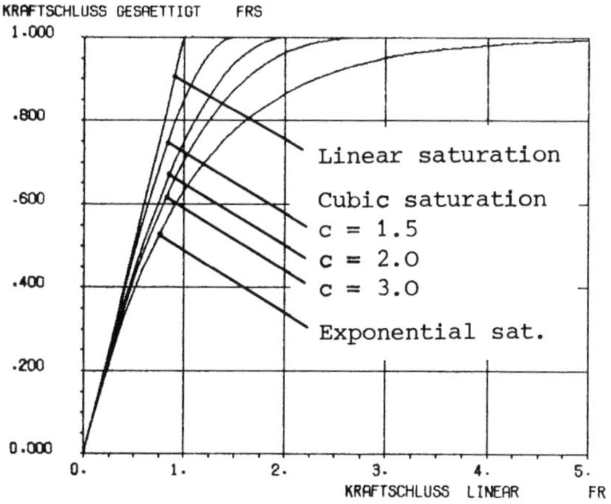


Fig. 3. above
Straight track
left Model 1 with linear creep
right Model 2 with ROLCON

Fig. 4. center
Straight track
Model 1 with quasilinear Kalker coefficients

Fig. 5. left
Characteristics of creep load saturation

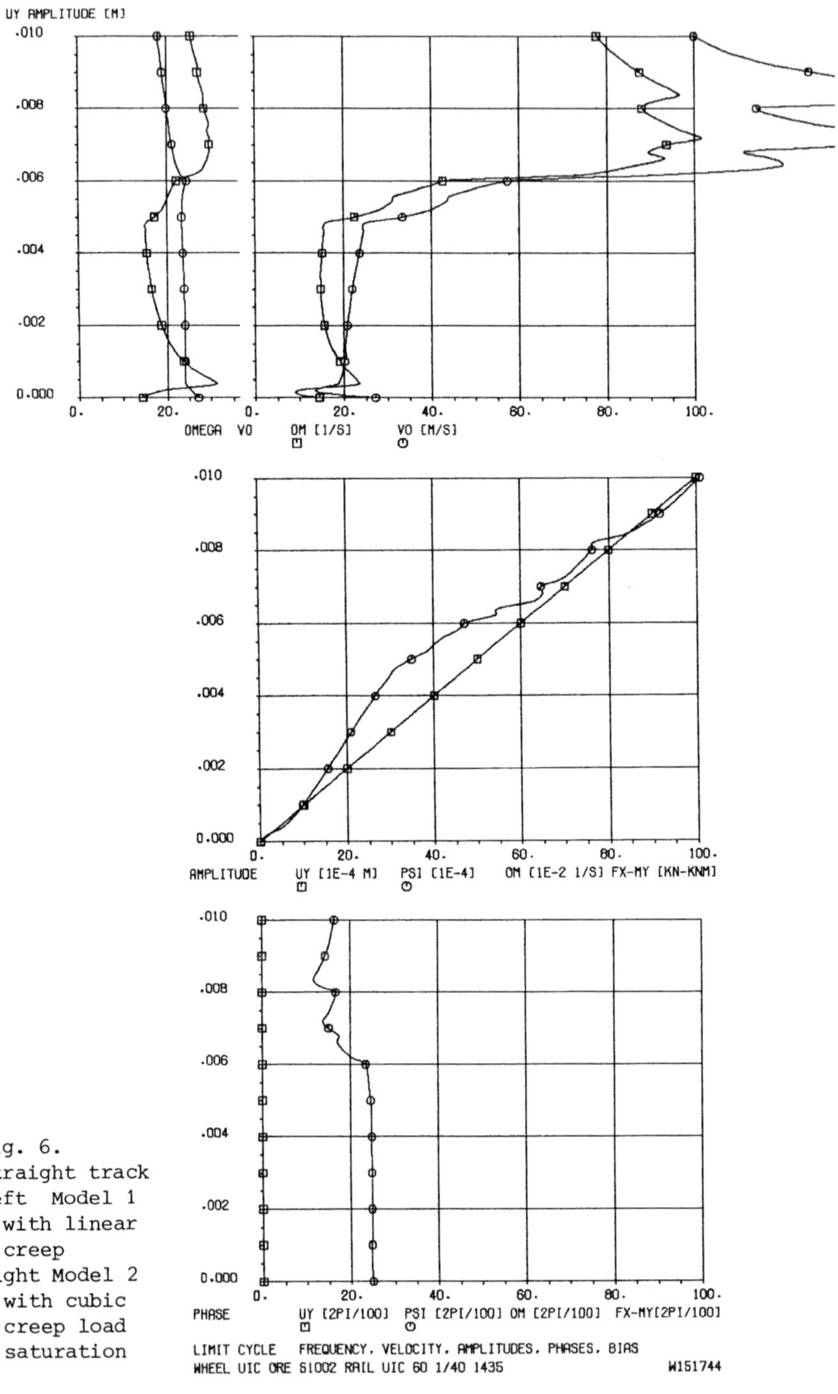


Fig. 6.
Straight track
left Model 1
with linear
creep
right Model 2
with cubic
creep load
saturation

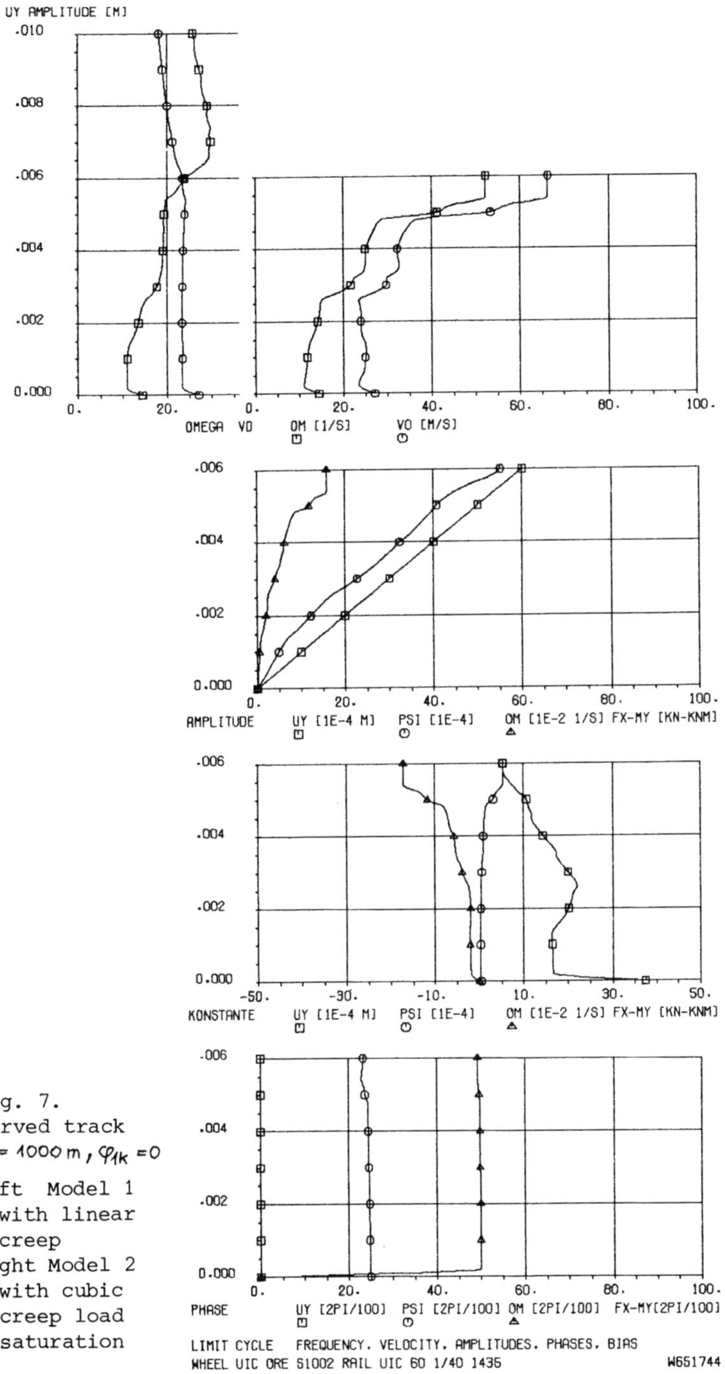


Fig. 7.
Curved track
$R = 1000\,m,\ \varphi_{1k} = 0$
left Model 1 with linear creep
right Model 2 with cubic creep load saturation

## 8.2 Curved track

Limit cycle calculations with models 1 and 2 differ above flange contact, which starts with lateral displacements of 2.2 mm bias and 2.6 mm amplitude (Fig.7). Above displacements of 0.5 mm bias and 5.4 mm amplitude the method does not converge due to flange contact at a displacment of 5.9 mm (cf.Fig.2)

Calculations with cant deficiency support the fact, that the wheelset yaws inwards but stays at the equilibrium lateral displacement [5].

Calculations with 3 instead of 2 degrees of freedom converge faster, while the results are almost identical.

The curved track takes 20 % more computation time than the straight track.

## 9. CONCLUSIONS

Quick parameter variations of the limit cycle behaviour of a wheelset on straight and curved track are carried out with the simple model 1, as long as there is no severe flange contact. The significant nonlinearities of the system are well reproduced by 3 describing functions of geometry. Nonlinearities in the contact area are not significant.

For limit cycle studies above flange contact, the complex wheelset model has to be used. The significant nonlinearities of the system are the nonlinear kinematics and the changing area of the contact ellipse. The method of creep load saturation is not significant.

In contrast to the sophisticated Galerkin method, the describing function technique can be introduced into existing programmes of linear railway dynamics without much complication. Thus a nonlinear response of complete vehicles can approximately be predicted, and the design can be optimized.

ACKNOWLEDGEMENT This report is result of my research in the railway dynamics group at the Institut für Luft- und Raumfahrt. Plot routines and subroutines for the compressed tabulation of intermediate results are developed by Jörg Maurer, a student tutor in our research group.

## REFERENCES

1. Cooperrider,N.K.,Hedrick,J.K.,Law,E.H.,Malstrom,C.W., The Application of Quasilinearization Techniques to the Prediction of Nonlinear Railway Vehicle Response. Vehicle Systems Dynamics, Vol. 4, No. 2-3.
2. Hull,R., Cooperrider,N.K., Influence of Nonlinear Wheel-Rail Contact Geometry on Stability of Rail Vehicles. Journal of Engineering for Industry, TRANS ASME 76-WA / RT-2.
3. Hannebrink,D.N.,Lee,H.S.H.,Weinstock,H.,Hedrick,J.K., Influence of Axle Load, Track Gauge and Wheel Profile on Rail Vehicle Hunting Journal of Engineering for Industry, TRANS ASME 76-WA / RT-3.
4. Hedrick,J.K., Cooperrider,N.K., Law,E.H., The Application of Quasilinearization Techniques to Rail Vehicle Dynamic Analyses. Final Report FRA-OR 1977.
5. Boocock,D., The Steady State Motion of Railway Vehicles on Curved Track. Journal of Mechanical Engineering Science Vol. II, No. 6 1969.

6. Elkins,J.A., Gostling,R.J., A General Quasi-static Curving Theory for Railway Vehicles. Proc. 5th. VSD-IUTAM Symposium on Dynamics of Vehicles on Roads and Tracks, Vienna 1977.
7. Silyak,D., Nonlinear Systems. Wiley, New York 1969.
8. Gelb.A., Vander Velde,W.E., Multiple Input Describing Functions and Nonlinear Systems Design. Mc Graw Hill, New York, 1968.
9. Moelle,D., Steinborn,H., Gasch,R., Computation of Limit Cycles of a Wheelset using a Galerkin Method. 6th. IAVSD-IUTAM Symposium on Dynamics of Vehicles on Roads and Tracks, Berlin 1979.
10. Hauschild,W., Die Kinematik des Rad-Schiene Systems. 2.Institut für Mechanik TU Berlin, 1977.
11. Kalker,J.J., On the Rolling Contact of Two Elastic Bodies in the Presence of Dry Friction. Doctoral Thesis, Delft 1967.
12. Knothe,K., Moelle,D., Steinborn,H., ROLCON - Ein schnelles, vielseitiges Digitalprogramm zum rollenden Kontakt. ILR Mitt. 55, Institut für Luft- und Raumfahrt TU Berlin, 1978.
13. Gasch,R., Hauschild,W., Moelle,D., Zur Berechnung periodischer Lösungen der nichtlinearen Schwingungsdifferentialgleichungen des Rad-Schiene Systems. ILR Mitt. 52, Institut für Luft- und Raumfahrt TU Berlin, 1978.
14. Rabinowitz,Ph., Numerical Methods for Nonlinear Algebraic Equations Gordon and Breach, London, 1970.

# STATISTICAL LINEARIZATION OF THE NONLINEAR RAIL VEHICLE WHEELSET

J. Karl Hedrick and I. A. Castelazo

Department of Mechanical Engineering, Massachusetts Institute of Technology, Cambridge, MA, U.S.A.

SUMMARY

The method of statistical linearization is applied to predict the stationary statistical response of the nonlinear two-degree-of-freedom wheelset subjected to random alignment track irregularities. The lateral wheelset model includes large contact angle wheel/rail geometric nonlinearities as well as nonlinear creep force saturation effects.

The paper presents a frequency domain algorithm to compute the equivalent linear system and then investigates the accuracy of the method by comparing the results with digital simulations of the nonlinear equations. A simplified approximation method is presented to include the effects of creep force saturation.

The results show that the method predicts the rms response of the system quite accurately and also predicts the onset of hunting with surprising accuracy.

## 1. INTRODUCTION

The concept of equivalently linearizing a nonlinear system about an assumed input signal form is not a new one [1,2]. These methods have, to a limited extent, been applied to predict the nonlinear response of rail vehicles. The method of equivalent linearization has been predominantly applied for the case of the input signal being either a sinusoid or a Gaussian random variable. In rail vehicle analysis the single sinusoid is a good approximation to describe the system in a hunting condition [3,5] and a Gaussian probability density function is a good approximation for the system below its critical speed [5,7]. More general signal forms have been used to describe both hunting and forced statistical response [4,6] for those applications when increased accuracy is desired.

In this paper the method of statistical equivalent linearization assuming a Gaussian probability density function will be applied to a simple two degree-of-freedom wheelset model and the accuracy of this method will be investigated.

## 2. STATISTICAL LINEARIZATION

The theoretical foundations of statistical linearization are presented in several textbooks, e.g. [8] and will not be repeated here. In this paper it will be assumed that all signals are zero mean, Gaussian random variables and that all nonlinearities are single input and memoryless. Thus, all nonlinearities will be replaced by an equivalent linear function whose gain, k, is chosen to minimize the mean squared error between the nonlinear output and its linear approximator, it is straightforward to show that this leads to:

$$f(x) \simeq kx \tag{1}$$

$$k \triangleq \frac{E[xf(x)]}{E[x^2]} = \frac{1}{\sigma_x^2}\int_{-\infty}^{\infty} xf(x)p(x)dx \tag{2}$$

$$p(x) = \frac{e^{-\left(\frac{x}{\sigma_x}\right)^2 \cdot \frac{1}{2}}}{\sqrt{2\pi}\,\sigma_x} \tag{3}$$

For a given nonlinearity, f(x), Eq. 2 is computed for various values of $\sigma_x$ and stored in a table, i.e., $k(\sigma_x)$ is a constant for a specified value of the input standard deviation, $\sigma_x$. k is often called a describing function gain.

## 3. NONLINEAR WHEELSET EQUATIONS

In the Extensive Summary [9] of this paper the full nonlinear two degree-of-freedom wheelset equations were presented. They are repeated* here for completeness. Figures 1a and 1b illustrate the constrained wheelset. In the equations the following assumptions have been made:

1. The c.g. of the wheelset is traveling down tangent track with a constant forward velocity, V.
2. The wheelset remains in contact with the rail at all times, i.e., the wheelset vertical and roll motion with respect to the track is constrained.
3. The wheelset roll, $\phi$, and yaw, $\theta_w$, angles are small.

---

*The equations presented in [9] were more general in that they included cross-level inputs and rail flexibility.

4. The vertical and roll inertial forces have a negligible effect on the lateral dynamics.

5. The rails are rigid.

6. The predominant lateral input is lateral alignment variations, i.e., gage and cross-level inputs have been neglected.

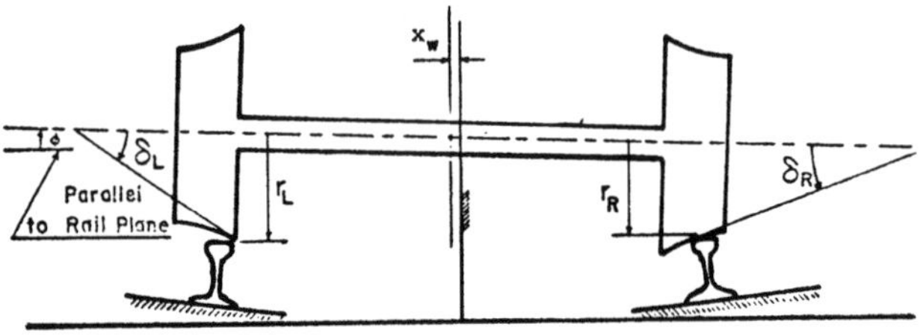


Figure 1a. Rail Vehicle Wheelset (rear view).

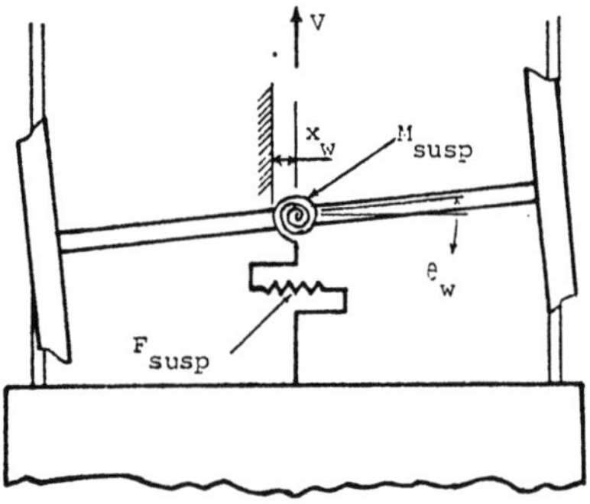


Figure 1b. Suspended rail vehicle wheelset (top view).

$$M_w\ddot{x}_w - \frac{VI_{wx}\delta_o}{r_o a}\dot{\theta}_w - F_{Lx}\cos(\delta_L+\phi) - F_{Rx}\cos(\delta_R-\phi) + L_A\Delta_1(x_w-x_a) = F_{susp}(x_w) \quad (4)$$

$$I_{wy}\ddot{\theta}_w + \frac{VI_{wx}\dot{\phi}}{r_o} - aL_A\delta_o\theta_w - a(F_{Rz}-F_{Lz}) - M_L\cos(\delta_L+\phi) - M_R\cos(\delta_r-\phi) = M_{susp}(\theta_w) \quad (5)$$

where:*

$x_w$ = lateral displacement of wheelset c.g.

$\theta_w$ = yaw angular displacement

$\frac{VI_{wx}\delta_o\dot{\theta}_w}{r_o a}$, $\frac{VI_{wx}\dot{\phi}}{r_o}$ = Gyro forces

$F_{Lx}, F_{Rx}$ = lateral components of creep forces on left and right wheels (in the contact plane)

*In Eq. (5) the yaw gravitational force has been approximated by a linear expression.

$F_{Lz}, F_{Rz}$ = longitudinal components of creep forces (in the contact plane)

$M_L, M_R$ = spin creep moments about the normal axis (in the contact plane)

$L_A \Delta_1$ = gravitational stiffness force in the wheelset lateral direction where $L_A$ = axle load and $\Delta_1$ is a nonlinear function of the wheel/rail geometry.

$a$ = half track gage

$r_o$ = centered rolling radius

$r_L, r_R$ = left and right rolling radius

$\delta_L, \delta_R$ = left and right contact angle (between wheelset axle and tangent to contact plane)

$\delta_o$ = contact angle in centered position

$\phi$ = wheelset roll angle with respect to rail

$x_a$ = lateral displacement of rail centerline due to alignment irregularities

$F_{susp}$ = lateral suspension connecting wheelset to truck frame (inertial space)

$M_{susp}$ = yaw suspension connecting wheelset to truck frame (inertial space)

*Creep Forces*

The terms, $F_{Lx}$, $F_{Rx}$, $F_{Lz}$, $F_{Rz}$, $M_L$, $M_R$ in equations (4) and (5) represent the left and right lateral, longitudinal and spin moment forces between the wheel and rail. In most previous papers on the analytical prediction of wheelset stability and forced response a linear creep force assumption has been made, in this paper creep force saturation will be taken into account by an approximation method similar to that proposed by Vermeulen and Johnson, [10] i.e.,

$$F'_{Lx} = -f_{11}\xi_{xL} - f_{12}\xi_{SPL} \qquad F'_{Rx} = -f_{11}\xi_{xR} - f_{12}\xi_{SPR} \tag{6}$$

$$F'_{Lz} = -f_{33}\xi_{ZL} \qquad F'_{Rz} = -f_{33}\xi_{ZR} \tag{7}$$

$$F'_{rL} = \sqrt{F'^{2}_{Lx} + F'^{2}_{Lz}} \qquad F'_{rR} = \sqrt{F'^{2}_{Rx} + F'^{2}_{Rz}} \tag{8}$$

$$F_{rL} \stackrel{\Delta}{=} g(F'_{rL}),\ F_{rR} \stackrel{\Delta}{=} g(F'_{rR}) \tag{9}$$

where $g(F'_r)$ is defined by:

$$g(F'_r) = \begin{cases} \mu N\left[\dfrac{F'_r}{\mu N} - \dfrac{1}{3}\left(\dfrac{F'_r}{\mu N}\right)^2 + \dfrac{1}{27}(F'_r/\mu N)^3\right], & F'_r < 3\mu N \\ \mu N & , F'_r \geq 3\mu N \end{cases} \tag{10}$$

Finally, the creep forces in the contact plane are defined by:

$$F_{Lx} = \frac{F'_{Lx} \cdot F_{rL}}{F'_{rL}} \qquad F_{Rx} = \frac{F_{Rx} \quad F_{rR}}{F'_{rR}} \tag{11}$$

$$F_{Lz} = \frac{F'_{Lz} \cdot F_{rL}}{F'_{rL}} \qquad F_{Rz} = \frac{F'_{Rz} \cdot F_{rR}}{F'_{rR}} \tag{12}$$

The linear spin moment terms are:

$$M_L = -f_{22}\xi_{SPL} + f_{12}\xi_{XL}, \quad M_R = -f_{22}\xi_{SPR} + f_{12}\xi_{XR} \tag{13}$$

The creepages in the contact plane at the left and right contact points are:

$$\xi_{XL} = \left(\frac{\dot{x}_w + r_L\dot{\phi}}{V} - \theta_w\right)\cos(\delta_L + \phi) + \frac{a\dot{\phi}}{V}\sin(\delta_L + \phi) \tag{14}$$

$$\xi_{XR} = \left(\frac{x_w + r_R\dot{\phi}}{V} - \theta_w\right)\cos(\delta_R - \phi) + \frac{a\dot{\phi}}{V}\sin(\delta_R - \phi) \tag{15}$$

$$\xi_{ZL} = \left(1 - \frac{r_L}{r_o}\right) - \frac{a\dot{\theta}_w}{V}, \quad \xi_{ZR} = \left(1 - \frac{r_R}{r_o}\right) + \frac{a\dot{\theta}_w}{V} \tag{16}$$

$$\xi_{SPL} = \frac{\dot{\theta}_w}{V}\cos(\delta_L + \phi) - \frac{\sin(\delta_L)}{r_o} \tag{17}$$

$$\xi_{SPR} = \frac{\dot{\theta}_w}{V}\cos(\delta_R - \phi) + \frac{\sin(\delta_R)}{r_o} \tag{18}$$

*Suspension Forces*

Reference [9] presented a general nonlinear suspension form that can be used to describe most rail vehicle suspensions. In this paper a simplified linear form will be used:

$$F_{susp}(x_w) = K_w x_w + C_x \dot{x}_w \tag{19}$$

$$M_{susp}(\theta_w) = K_\theta \theta_w \tag{20}$$

*Wheel/Rail Profile Geometry*

(i) lateral gravitation "stiffness" force - the net lateral component of the normal contact force is approximately given by $L_A\Delta_1(x_w - x_a)$ where $L_A$ is the axle load and $\Delta_1$ is defined by:

$$\Delta_1 \triangleq \frac{\tan(\delta_L + \phi) - \tan(\delta_r - \phi)}{2 - \frac{1}{a}[r_L \tan(\delta_L+\phi) + r_R \tan(\delta_r - \phi)]} \qquad (21)$$

This expression includes the effects of large contact angles and assumes the wheelset is in vertical and roll equilibrium.

If the assumption of linear creep forces is made, the following wheel profile nonlinearities result after Eqs. (6)-(18) are substituted into eqs. (4) and (5).

(ii) lateral force due to spin creepage - the term, $\frac{2f_{12}}{r_o}\Delta_2(x_w-x_a)$, results in the lateral equation where $\Delta_2$ is defined by:

$$\Delta_2 \triangleq \frac{1}{2}\ [\sin(\delta_L)\cos(\delta_L + \phi) - \sin(\delta_R)\cos(\delta_R - \phi)] \qquad (22)$$

(iii) yaw moment due to spin creepage - the term $\frac{f_{22}}{r_o}\Delta_2(x_w-x_a)$ results in the yaw equation where $\Delta_2$ is defined by eq. (22).

(iv) yaw moment due to rolling radii difference - the term, $\frac{2a^2 f_{33}}{r_o}[r_L-r_R/2a]$ results in the yaw equation where $r_L-r_R$ is the rolling radii difference.

(v) Gyro moment due to $\dot{\phi}$ - the term $\frac{VI_{wx}\dot{\phi}}{r_o}$ appears in the yaw equation, where $\phi(x_w-x_a)$ is the wheelset roll angle.

*Simulation Equations for the Linear Creep Case*

After Eqs. (6)-(20) are substituted into Eqs. (4) and (5) and some approximations* are made, the following equations result:

$$M_w\ddot{x}_w - \frac{VI_{wx}\delta_o}{r_o a}\dot{\theta}_w + L_A\Delta_1(x_w - x_a) + \frac{2f_{11}}{V}(\dot{x}_w - V\theta_w)$$

$$+ \frac{2f_{12}\dot{\theta}_w}{V} - \frac{2f_{12}}{r_o}\Delta_2(x_w - x_a) + K_x x_w + C_x\dot{x}_w = 0 \qquad (23)$$

---

*In order to simplify the resulting equations it was assumed that:

$$\cos^2(\delta_L + \phi) + \cos^2(\delta_r - \phi) \simeq 2, \text{ and}$$

$$\sin^2(\delta_L + \phi) + \sin^2(\delta_r - \phi) << \cos^2(\delta_L + \phi)+\cos^2(\delta_R - \phi)$$

These approximations are reasonably valid for contact angles less than 30°.

$$I_{wy}\ddot{\theta}_w + \frac{VI_{wx}\dot{\phi}}{r_o} - aL_A\delta_o\theta_w + \frac{2a^2 f_{33}\dot{\theta}_w}{V}$$

$$+ \frac{2a^2 f_{33}}{r_o}\left[\frac{r_L - r_R}{2a}\right] + \frac{2f_{22}\dot{\theta}_w}{V} - \frac{2f_{22}\Delta_2(x_w - x_a)}{r_o}$$

$$- \frac{2f_{12}}{V}(\dot{x}_w - V\theta_w) + K_\theta\theta_w = 0 \qquad (24)$$

*Equivalent Linearization Equations for the Linear Creep Case*

The wheel/rail geometry nonlinearities are replaced by the following linear approximations:

$$\Delta_1 \approx \delta_1 \cdot \frac{(x_w - x_a)}{a} \qquad (25)$$

$$\Delta_2 \approx \delta_2 \cdot \frac{(x_w - x_a)}{a} \qquad (26)$$

$$\frac{r_L - r_R}{2a} \approx \lambda \cdot \frac{(x_w - x_a)}{a} \qquad (27)$$

$$\phi \approx a_1 \cdot \frac{(x_w - x_a)}{a} \qquad (28)$$

where $\delta_1, \delta_2$, $\lambda$, and $a_1$ are describing functions gains and are computed numerically for a given wheel/rail profile by applying Eq. (2) and storing the results in a table for various values of the standard deviation of $x_w - x_a$. Once expressions (25) - (28) are substituted into (23) and (24) the resulting equations are in equivalent linear form.

*Simulation Equations for the Nonlinear Creep Case*

Several simplifying approximations were made in Eqs. (23) and (24) in order to arrive at a convenient form of the nonlinear equations for equivalent linearization. This will also be useful when evaluating the results since any difference between the nonlinear digital simulation and equivalent linearization results will be due to the linearization approximation. For the case of nonlinear creep no approximations for the digital simulation will be made in Eqs. (4) - (20).

*Equivalent Linearization Equations for the Nonlinear Creep Case*

Theoretically the equivalent linearization of a multiple-input nonlinearity is a difficult procedure since the equivalent gains become functions of all the input variables and their correlation with each other. In order to avoid this difficulty a very heuristic, ad hoc procedure is proposed to account for the effects of creep force saturation.

The proposed procedure is simply to ignore the multiple input, cascaded features of Eqs. (6) - (12) and to replace the Vermeulen-Johnson saturation function Eq. (10) with an equivalent linear function*, i.e.,

$$g(F'_r) \simeq K_c F'_r \tag{29}$$

If $K_c$ is chosen to minimize the mean square error between g and $K_c F'_r$, then the result (Eq. (2)) is:

$$K_c \triangleq \frac{E[F'_R g(F'_R)]}{E[F'^2_R]} \tag{30}$$

In order to evaluate the expectation implied by Eq. (30), the probability density function of $F'_R$ must be known, it is clear from Eq. (8) that $F'_R$ will not be Gaussian, in fact, it will not even have a zero mean. It can be shown that if $F'_x$ and $F'_y$ are Gaussian then $F'_R \triangleq \sqrt{(F'_x)^2 + (F'_y)^2}$ will be a Rayleigh process. Eq. (30) was evaluated for both Gaussian and Rayleigh distributions and quite similar results we obtained as shown in Figure 2.

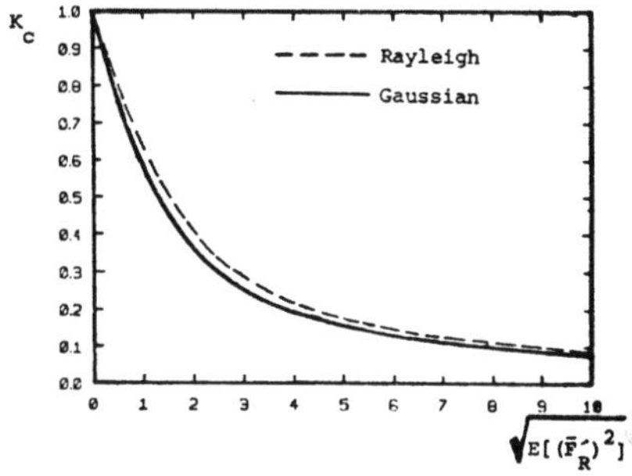


Figure 2. Creep force describing function gain.

In Figure 2, $\bar{F}'_R$ is the normalized resultant linear creep force, i.e., $\bar{F}'_R \triangleq F'_R/\mu N$.

Substituting Eq. (29) into Eqs. (10) and (11) results in a nice simplification, i.e.:

$$F_{Lx} \simeq K_{cL} F'_{Lx} \quad , \quad F_{Rx} \simeq K_{cR} F'_{Rx} \tag{31}$$

$$F_{Lz} \simeq K_{cL} F'_{Lz} \quad , \quad F_{Rz} \simeq K_{cR} F'_{Rz} \tag{32}$$

---

* $g(F_r)$ should be replaced by $K_o + K_c F'_r$ since $F'_r$ does not have a zero mean, in this treatment the term $K_o$ will be neglected.

A further simplification can be made if the mean square values of the left and right resultant linear creep forces ($F'_{rL}$, $F'_{rR}$) are assumed to be equal, then $K_{cL} = K_{cR} \triangleq K_c$.

In order to evaluate $K_c$ it is necessary to compute $E[(\bar{F}'_R)^2]$ for either the left or right creep force. Using the equivalently linearized expressions for the creep forces results in:

$$F'_x \approx -f_{11}\left(\frac{\dot{x}_w}{V} - \theta_w\right) - \frac{f_{12}\dot{\theta}_w}{V} + \frac{f_{12}}{r_o}\,\frac{\delta_2(x_w - x_a)}{a} \tag{33}$$

$$F'_z \approx f_{33}\left(\frac{\lambda(x_w - x_a)}{r_o} + \frac{a\dot{\theta}_w}{V}\right) \tag{34}$$

$$E[(\bar{F}'_R)^2] = \frac{1}{(\mu N)^2}\left[E[(F'_x)^2] + E[(F'_z)^2]\right] \tag{35}$$

Substituting (25)-(28), (31) and (32) into (4) and (5) results in the equivalent linear wheelset equations:

$$M_w\ddot{x}_w - \frac{VI_{wx}\delta_o}{r_o a}\dot{\theta}_w + L_A\frac{\delta_1(x_w - x_a)}{a}$$
$$+ 2K_c\left\{\frac{f_{11}}{V}(\dot{x}_w - V\theta_w) + \frac{f_{12}}{V}\dot{\theta}_w - \frac{f_{12}}{r_o}\delta_2\frac{(x_w - x_a)}{a}\right\}$$
$$+ k_x x_w + C_x\dot{x}_w = 0 \tag{36}$$

$$I_{wy}\ddot{\theta}_w + \frac{I_{wx}V a_1}{r_o a}\dot{x}_w - aL_A\delta_o\theta_w$$
$$+ 2a^2K_c f_{33}\left(\frac{\dot{\theta}_w}{V} + \frac{\lambda(x_w - x_a)}{r_o a}\right) + \frac{2f_{22}\dot{\theta}_w}{V}$$
$$- \frac{2f_{22}}{r_o}\delta_2\frac{(x_w - x_a)}{a} - \frac{2f_{12}}{V}(\dot{x}_w - V\theta_w)$$
$$+ K_\theta\theta_w = 0 \tag{37}$$

## 4. ANALYSIS AND SIMULATION METHODS

In this section the results of the statistical linearization algorithm are compared with the results of nonlinear digital simulation.

### *Statistical Linearization Algorithm*

A frequency domain algorithm was used to solve for the stationary response of the equivalent linear system defined by Eqs. (36) and (37).

#### *Case 1 - Linear Creep*

For this case $K_C \equiv 1$ in Eqs. (36) and (37) and the four describing function gains $(\delta_1, \delta_2, \lambda, a_1)$ are tabulated functions of $\sigma_{\Delta x}$, where $\Delta x$ is the relative wheelset excursion, $\Delta x = x_w - x_a$. For the numerical examples in this paper the Heumann profile on standard gage, new rail was used. Tabulated wheel/rail geometric constraints for this profile can be found in [11]. The rolling radii difference for this profile is shown in Figure 3. The computed describing functions for this profile are shown in Figures 4a and 4b.

The transfer function for the wheelset-excursion can be found from Eqs. (36) and (37), i.e.:

$$\frac{\Delta x(j\omega)}{x_a(j\omega)} \overset{\Delta}{=} H_1(j\omega) \tag{38}$$

The variance of $\Delta x$ is found from:

$$\sigma^2_{\Delta x} = \int_0^\infty |H_1(j\omega)|^2 S_{xa}(\omega)\,d\omega \tag{39}$$

For this paper the spectral density of the random alignment irregularities was assumed to have the form:

$$S_{xa}(\omega) = \frac{AV}{\omega^2 + (V\Omega_o)^2} \tag{40}$$

where A is the roughness parameter, V is the forward speed, $\Omega_o$ is the spatial break frequency, and $\omega$ is the temporal frequency.

Equation (39) represents a nonlinear integral equation since $H_1(j\omega)$ is a nonlinear function of $\sigma_{\Delta x}$ through the describing function gains. An iterative solution technique described in [7] was used to converge to a solution.

#### *Case 2 - Nonlinear Creep*

In this case the equivalent linear equations have the additional describing function gain, $K_C(\sigma_{\bar{F}'_R})$, defined in Figure 2, thus the additional parameter $\sigma_{\bar{F}'_R} = \sqrt{E[\bar{F}'_R)^2]}$ needs to be computed. From Eq. (35) it is seen that it is necessary to compute $E[(F'_x)^2]$ and $E[(F'_z)^2]$, this is done by finding the appropriate transfer functions and integrating the corresponding spectral densities.

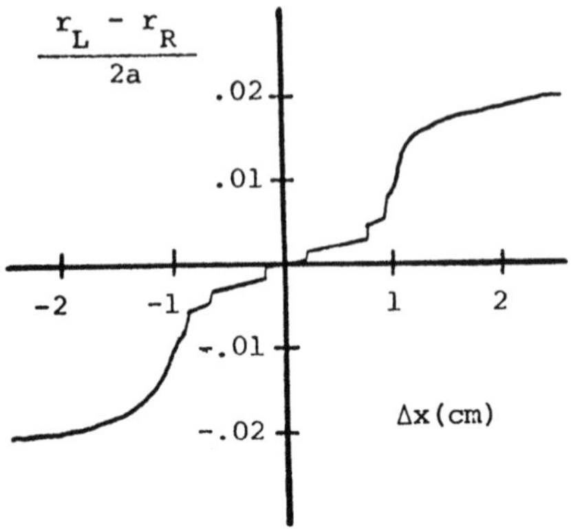


Figure 3. Rolling Radii Difference (Heumann Wheel, New Rail, Standard Gage)

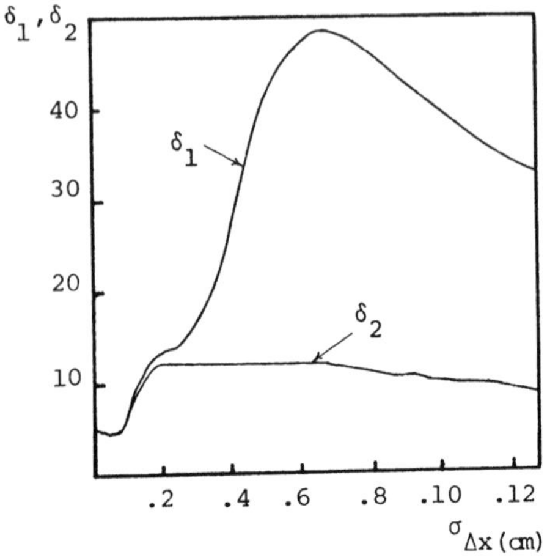


Figure 4a. $\delta_1, \delta_2$ vs. $\sigma_{\Delta x}$

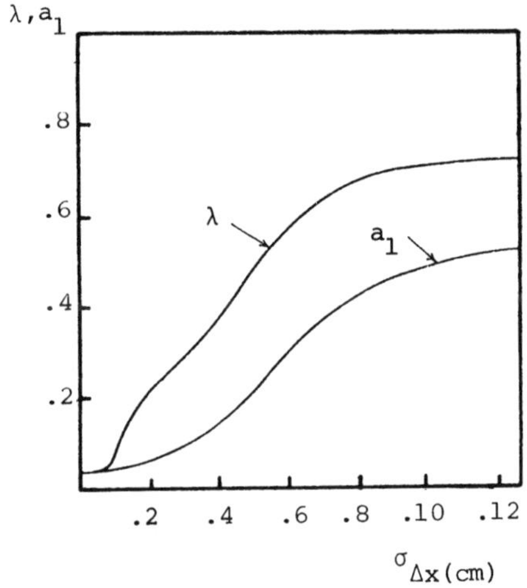


Figure 4b. $\lambda$,a, vs. $\sigma_{\Delta x}$

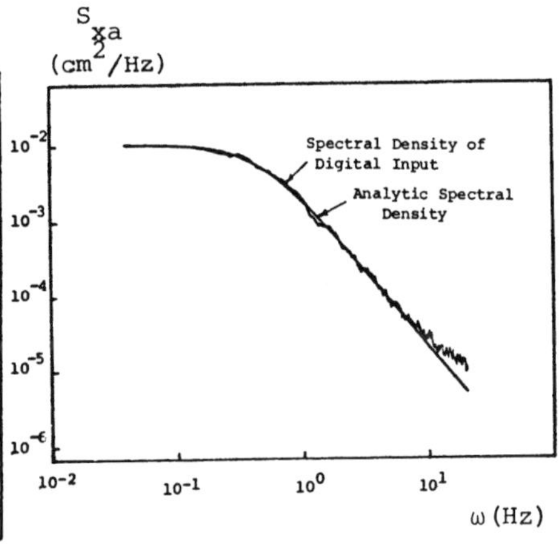


Figure 5. Alignment Spectral Density

*Digital Simulation*

In order to compare the statistical linearization results with a nonlinear digital simulation it was necessary to create a sequence of random lateral alignment inputs that had a spectral density corresponding to Eq. (40). This was done by passing a Gaussian purely random sequence through an appropriate digital filter. A comparison of the spectral density of this sequence with the analytic spectrum is shown in Figure 5.

## 5. NUMERICAL RESULTS

A representative set of wheelset parameters were chosen for the numerical simulation, several of the parameters were varied to allow reduced integration time steps, i.e., to reduce the spread between the maximum and minimum eigenvalues.

*Linear Creep*

Figure 6 shows a comparison between the statistical describing function method, nonlinear digital simulation, and the results of a linear analysis if the describing function gains at the centered position were used throughout the analysis. The rms wheelset excursion is plotted as a function of the irregularity roughness parameter (A in Eq. (40)). The first value of this parameter, $A_1$, was small enough so that the wheelset always remained in the linear region, hence all three methods agree. Figure 6 shows that as A is increased the linear analysis considerably overestimates $\sigma_{\Delta x}$, whereas the describing function underestimates but provides a more accurate prediction (10% error at 16A, as opposed to 26% for the linear analysis). This can be partially explained by the predicted effective conicity, $\lambda$, in each case. The linear analysis uses the centered value of $\lambda = .037$ while the describing function analysis converged at a value of $\lambda = .14$, the higher conicity yields lower wheelset excursions.

Figure 7 shows a comparison of the wheelset excursion spectral densities computed by the describing function method and from the digital simulation for a value of A that results in a nonlinear response. It should be pointed out that the computation time is an order of magnitude less for the descrbing function method if rms values are desired and several orders of magnitude less if spectral densities are required.

Figure 8 shows the damping ratio of the least damped mode as a function of the wheelsets forward speed. It is interesting to note that the describing function method predicted that the critical speed would be 5.75 $V_1$ while the linear analysis predicted 12$V_1$. The digital simulation yields a critical speed of 5.65 $V_1$ thus the describing function was within 2% while the linear analysis was off by 112%.

*Nonlinear Creep*

Figure 9 shows the comparison between the digital simulation results, describing function method with linear creep, and the describing function method with the creep force saturation describing function.

The digital simulation used in this case was of the complete wheelset model (Eqs. (4) and (5)) and thus the differences in Figure 9 are due to model differences as well as the equivalent linearization. It is interesting to note that the creep force saturation gain helps to improve the accuracy for larger values of A where creep force saturation becomes important.

## 6. ACKNOWLEDGEMENT

The authors would like to thank Professor N.K. Cooperrider for his communications during the course of this research.

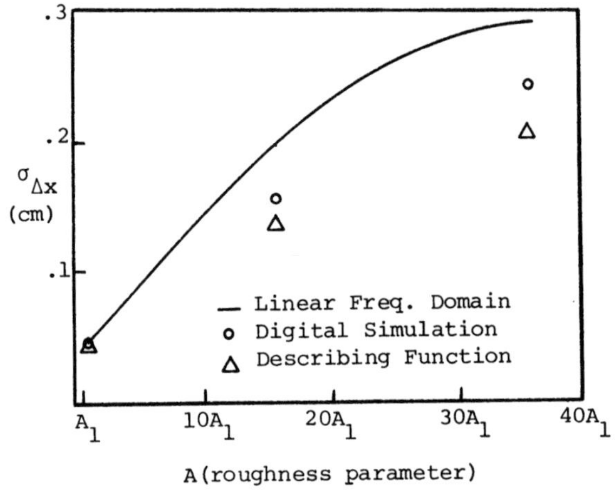


Fig. 6. RMS wheelset excursion as a function of track roughness.

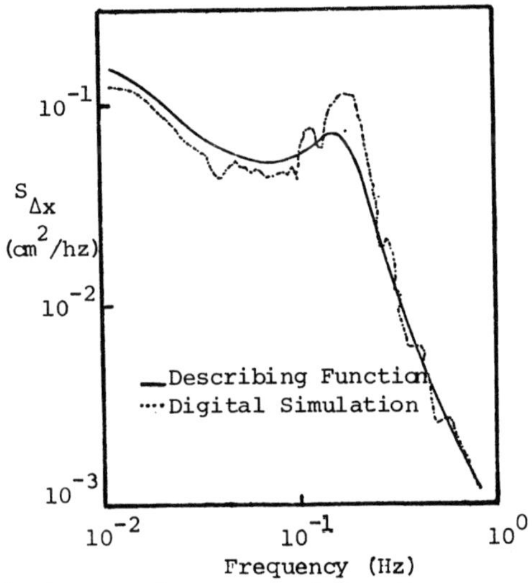


Fig. 7. Wheelset excursion spectral density.

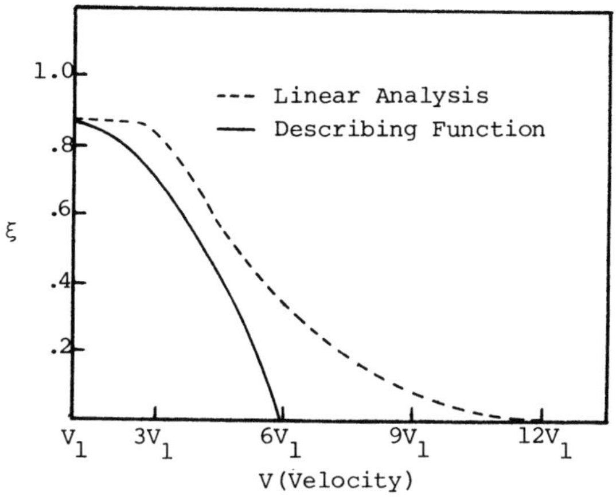


Fig. 8. Damping ratio of least damped mode as a function of forward speed.

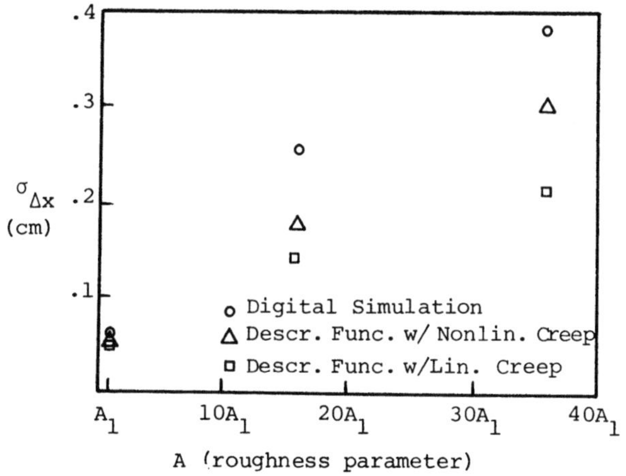


Fig. 9. RMS wheelset excursion as a function of track roughness.

## 7. REFERENCES


1. Krylov, N. and Boboliubov, N., Introduction to Nonlinear Mechanics, (Trans. by S. Lefschetz) Princeton Press, 1974.

2. Booton, R.C., "Nonlinear Control Systems with Random Inputs", IRE Trans. Circuit Theory, Vol. CT-1, No. 1, March 1954, pp. 9-17.

3. DePater, A.D., "The Approximate Determination of the Hunting Movement of a Railway Vehicle by the Aid of the Method of Krylov and Bogoliubov", Xth International Congress of Applied Mechanics, Stressa, Sept. 1960.

4. Stassen, H.G., "Random Lateral Motions of Railway Vehicles", WTHD. No. 49, Laboratory for Measurement and Control, Delft University of Technology, The Netherlands, 1973.

5. Hedrick, J.K., Cooperrider, N.K., and Law, E.H., "The Application of Quasi-Linearization to Rail Vehicle Analysis", Final Report, U.S. Department of Transportation, Report No. FRA/ORD-78-56, Nov. 1978.

6. Rus, L., "Running Stability, and Railway Vehicle Transfer Functions, Solved by the Method of Statistical Linearization", 5th VSD-2nd IUTAM Symposium, Vienna, Austria 1977.

7. Hedrick, J.K., and Arslan, A.V., "Nonlinear Analysis of Rail Vehicle Forced Lateral Response and Stability", ASME Trans., Journal of Dynamic Systems, Measurements, and Control, September 1979.

8. Gelb, A. and Vanderwelde, W.E., Multiple-Input Describing Functions and Nonlinear System Design, McGraw-Hill, New York, 1968.

9. Hedrick, J.K., and Cooperrider, N.K., "Equivalent Linearization of the Nonlinear Rail Vehicle Wheelset", 1979 IAVSD Extensive Summaries, Vehicle System Dynamics, Volume 8, Number 2-3, September 1979.

10. Vermeulen, P. and Johnson, K., "Contact of Nonspherical Elastic Bodies Transmitting Tangential Forces", Journal of Applied Mechanics, Trans. ASME, Vol. 86, No. 2, June 1964, pp. 338-340.

11. Cooperrider, N.K., et al., "Analytical and Experimental Determination of Nonlinear Wheel/Rail Geometric Constraints", Report No. FRA-OR&D-76-244 (PB 24290), December 1975.

# OPTIMIZING THE WHEEL PROFILE TO IMPROVE RAIL VEHICLE DYNAMIC PERFORMANCE

Rainer Heller and E. Harry Law

College of Engineering, Clemson University, Clemson, S.C., U.S.A.

SUMMARY

A procedure for designing wheel profiles is presented which incorporates an analysis of the stability and curving performance of an idealized nonlinear rail vehicle model and the tendency of a profile to change shape with wear. The procedure is a closed-loop process implemented under operator control on a hybrid computer system. The vehicle performance for an initial profile is analyzed and the profile is systematically adjusted until dynamic performance and wear tendency objectives are met or have reached a successful compromise.

Several studies were conducted to (1) compare the performance of existing wheel profile designs, (2) to investigate various conically tapered wheel designs, (3) to develop optimum wheel profile designs having a single circular arc tread, and (4) to compare performance of optimum single tread circle designs with existing profiles.

The objectives of the work were (1) to develop an analytically-based wheel profile design procedure, and (2) to use the procedure to demonstrate the feasibility of the process.

## 1. INTRODUCTION

The lateral profiles of wheels play a fundamental role in the interaction between rail vehicle and track. The profile shape strongly influences the lateral dynamic response of a vehicle through the conicity and gravitational stiffness effects, directly affecting the safe operating speed, curving performance, vibration response, and the progression of wheel-rail surface wear.

The effective values of conicity and gravitational stiffness strongly influence the critical speed for hunting. Hunting is a violent limitcycle oscillation in which wheelsets slam from flange to flange. For good curve negotiation, it is desirable that contact of the wheel flanges against the rails does not occur and that wheel tread creep forces guide the vehicle through the curve. During flange contact, high lateral forces cause excessive wheel and rail wear, high vibration levels and the possibility of wheel climb and track structure failures which may lead to derailment. For conventional rail vehicles, low conicity profiles facilitate high speed stable operation on tangent track while, within limits, high conicity promotes flange-free curving ability.

A major objective in the design of rail vehicles and their profiles must be a satisfactory resolution of the conflicting requirements of stability and curving performance in terms of both conicity and suspension design while avoiding flange contact.

The standard coned profile used on many existing vehicles has some well-known disadvantages. Its low conicity leads to flange guidance on all but the widest curves. A major fault of coned treads is their propensity to change shape with use. The contact regions on wheel and rail surfaces are small and the contact stresses are high leading to rapid wear and hollowing of the profile shape and, as a consequence, long-term changes in vehicle lateral dynamic behavior. To reestablish the original running properties, the traditional cure is to reprofile the wheel to the original coned shape, which is costly.

As early as 1934, Heumann [1] pointed out the irrationality of using the wear-prone standard coned design. He maintained that the initial contour design should fit closely to the curved rail head. Thus, contact stresses and the wear rate would be decreased and a more complete use of the wheel surface would result. Using this idea, there has been a reawakening of interest in the last 15 years in wheel profile design. Various hollow or worn designs have been proposed [1-6] based on approximately matching "typical" wheel and rail shapes found in service. The hollow designs do not wear as rapidly as coned profiles and their high conicity also significantly improves curving performance. The major drawback is the lowered critical velocity on straight track for existing equipment.

The design procedure for the hollow profiles reported in [1-6] did not consider the dynamics of the vehicle while designing the profile. The performance of the profiles were experimentally tested on existing equipment, but, because of the expense involved, were not immediately refined to further improve wheel profile and vehicle performance.

A design method is needed that directly evaluates the vehicle dynamic performance within the wheel profile design stage. This is the approach reported in this paper. The wheel profile is analytically studied and through a closed-loop process is optimized for a given vehicle and track geometry. The objective is to minimize the tendency of a profile to change shape by wear while providing an optimum resolution of the conflicting requirements of stability and curving.

The overall objectives of the work were (1) to develop an analytically based wheel profile design procedure and (2) to use the procedure to demonstrate the feasibility of the process.

In the following sections, our approach in characterizing an adjustable wheel profile is outlined. The nonlinear vehicle model chosen within the design procedure and methods of evaluating the performance of the model as the wheel design is changed are discussed. A closed-loop design process as implemented on a computer is then described, and the use of the procedure to design optimum profiles is presented in the final section.

## 2. THE WHEEL PROFILE AND WHEEL-RAIL GEOMETRY

### *The Wheel Profile*

To be readily adaptable to existing and future conventional rail vehicles, any new wheel profile should have the usual flange, flange throat, and tread regions. For current wheel profiles, these regions are designed with a series of circular arcs and linear segments as shown in Fig. 1. The AAR 1:20 tapered contour is recommended for use for all conventional North American freight vehicles [7]. The coned "Modified Heumann" design, proposed by Eck and Berg [8], is a modification of the profile designed by Heumann for British Railways [9] to meet "average U.S. worn rail conditions." The hollow CN 'A' profile was developed by Canadian National Railways [5] to closely approximate typical service-worn wheels.

The general wheel profile characterization in Fig. 2 was developed to

1. represent existing profile designs.
2. provide a basis for adjusting the profile during profile design.

The general profile has n connected circular arcs of radius $r_i$ extending from point G (the reference point for most North American designs) to the field side. The tread contact region is then completed by a linear segment connected to arc $r_n$. The segmented profile curve is continuous through the first derivative and, at the points of intersection between adjacent segments, the angles with respect to the baseline are $\alpha_i$.

As we are not concerned with flange design in the present work, but with tread design to avoid flanging, the flange is represented by a simple arc. The end regions, represented by dashed lines in the figure, do not come in contact with the rail during normal operation and can later be added, as required, to complete the design.

Fig. 2 identifies the profile parameters for an adjustable wheel contour. For a fixed flange, a profile with n flange throat/tread radii and a field-side linear segment is completely defined by the 2n parameters, $\alpha_i$ and $r_i$. With the appropriate algebra, the coordinates of the circular arc centers and the connecting points $(x_i, y_i)$ may be calculated. For the profile design process, the calculations are performed by a numerical algorithm.

With the proper selection of the 2n parameters, the construction in Fig. 2 can be specialized to represent most current profile designs. The parameters of the designs in Fig. 1 are given in Tab. 1. In the table, the profiles are classified (1) by the number of arcs and (2) as hollow or coned depending whether the linear segment has a zero or positive slope, respectively.

### *Rails and Track Geometry*

Rail profiles are also, in general, designed with a connected series of circular arcs. A new rail profile commonly used in the U.S. and used in this study is shown in Fig. 3 along with sample worn rail profiles [10]. Nominal values for the track gauge and rail cant (inclination angle) are given in Fig. 4.

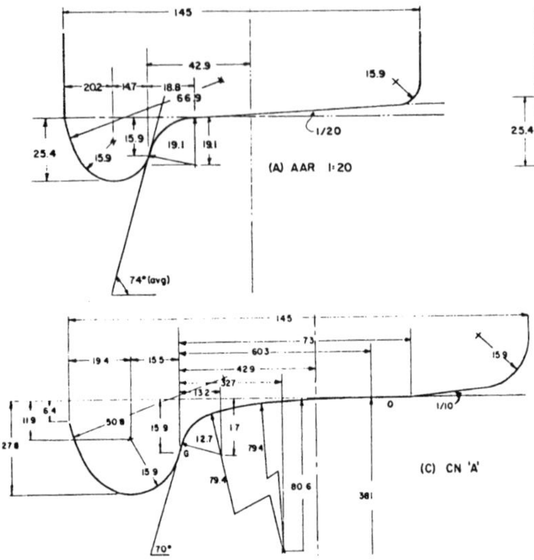


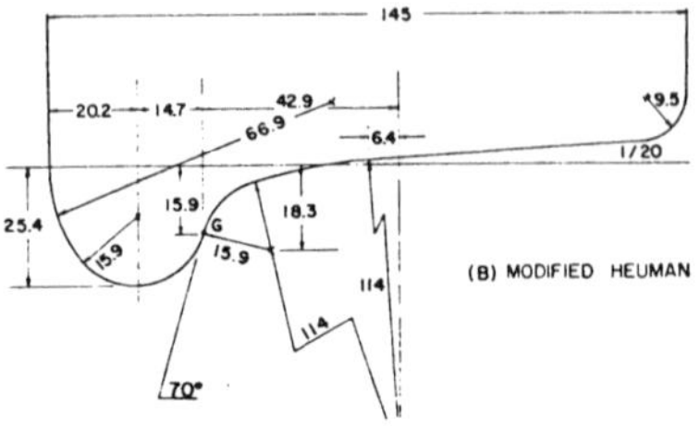


Fig. 1. Sample wheel profile designs for wide-flange steel wheels for freight service in North America.

| | AAR 1:20 Coned n=1 | Mod Heu Coned n=2 | CN 'A' Hollow n=3 | |
|---|---|---|---|---|
| $\alpha_f$ | 80.4 | 70.0 | 70.0 | deg |
| $r_f$ | 15.9 | 15.9 | 15.9 | mm |
| $r_1$ | 19.1 | 15.9 | 12.7 | mm |
| $r_2$ | — | 114 | 79.4 | mm |
| $r_3$ | — | — | 381 | mm |
| $\alpha_1$ | 2.86 | 16.0 | 18.1 | deg |
| $\alpha_2$ | — | 2.86 | 5.25 | deg |
| $\alpha_3$ | — | — | 0.00 | deg |

Tab. 1. Profile parameters

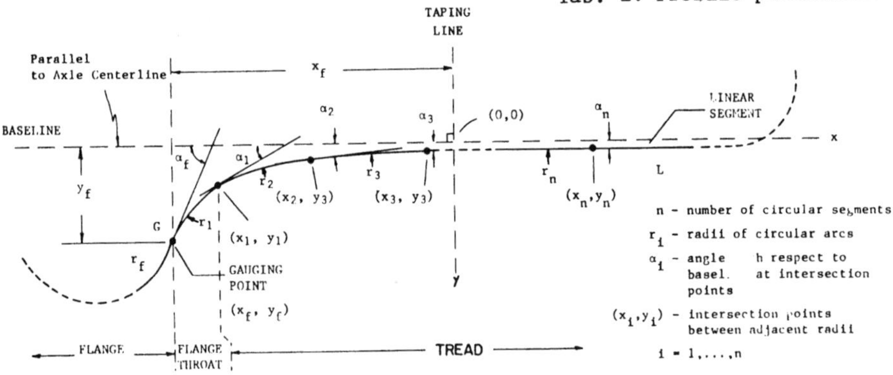


Fig. 2. General adjustable wheel profile design.

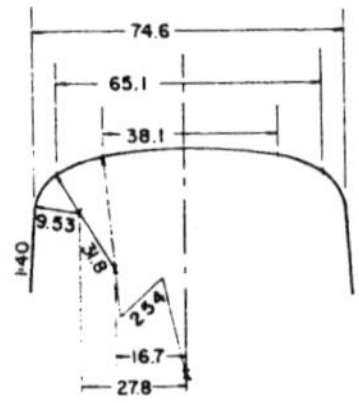

AREA 132 lb/yd design

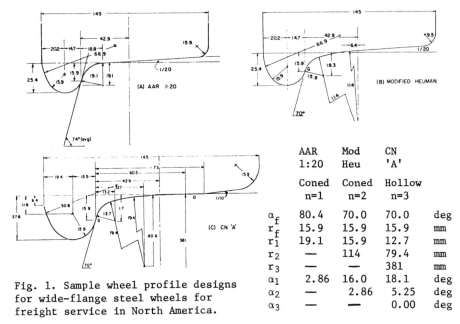

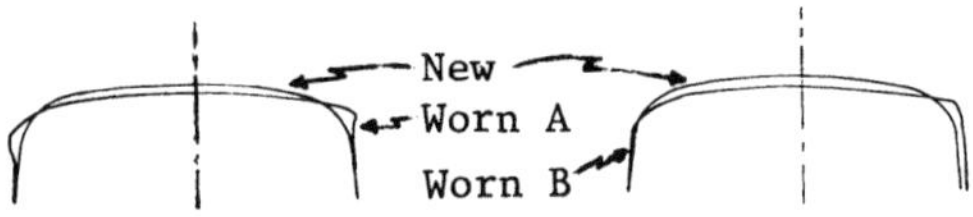


Fig. 3. New and worn rail profiles.

*Geometric Constraint Functions*

For the general case of curved profiles, the wheel rolling radii and contact angles as defined in Fig. 4 are nonlinear functions of the wheelset lateral displacement $x_w$, the dependence on wheelset yaw being negligible for practical cases. A number of numerical approaches have been developed recently to determine the locations of contact points and the kinematic parameters for arbitrarily-shaped wheel and rail profiles [10-12]. We shall use the approach of [10], which was validated by full-scale experiments.

Results of this numerical procedure are plotted in Fig. 5 for symmetric configurations of the wheel profiles of Fig. 1 and new rail profile of Fig. 3.

The slope of the $(r_L-r_R)/2$ function at a particular value of $x_w$ determines the local effective conicity at that value. The slope of the $(\delta_L-\delta_R)/2$ function affects the magnitude of the lateral resultant of the normal forces applied to the wheels, i.e. the lateral gravitational stiffness. These functions directly affect the vehicle dynamics and, in other studies, are commonly linearized: the values shown in the figure are obtained by fitting least squares straight lines to the curves for all $x_w$ within flange clearance. The $x_w$ at which the wheel contact point moves to the flange is clearly evident by the slope discontinuities in the curves.

## 3. VEHICLE MODEL

The railway vehicle model employed in the wheel profile design process was intentionally chosen to be simple to facilitate development of the process. The rail vehicle was modeled as a "pseudo" car body connected to a single wheelset through linear suspension elements. The model describes the lateral dynamics of an idealized vehicle translating uniformly along tangent or curved track. Lateral alignment irregularities can be used to force the vehicle. External forces due to aerodynamic, coupler, braking, and traction effects are neglected.

The wheelset is assumed rigid, is fitted with arbitrarily-shaped profiles, and runs on arbitrarily-shaped rail profiles. Wheel and rail profiles are assumed identical on left and right sides. Rail cant and track gauge remain fixed along the track and the track structure as well as the rails are rigid. Wheelset equations of motion were derived [13] assuming small perturbations from equilibrium or steady-state. Wheelset degrees of freedom correspond to lateral and yaw displacements as well as axle angular velocity.

The car body was represented by a symmetrical two-dimensional mass in a vertical plane with lateral and roll degrees of freedom as shown in Fig. 6. The vertical suspension consists of two fixed-end linear springs which resist relative lateral, vertical, and roll motions between car body and wheelset. Relative yaw motions are resisted by a simple yaw stiffness. In a curve, the model is assumed to be located in a position corresponding to either the front or rear of a full car body whose centerline is perpendicular to the curve radius vector at the car center of gravity. Thus, for a curve of radius R, a prescribed yaw moment $k_\theta \ell/R$ is applied to the wheelset (Fig. 6), where $k_\theta$ is the yaw stiffness and $\ell$ the semi-car length.

Nominal Track Parameters:

New AREA rail profile
Rail gauge 1435 mm
Rail cant 1:40

Nominal Wheelset Parameters:

Taping line diameter 838 mm
Wheelset gauge 1346 mm

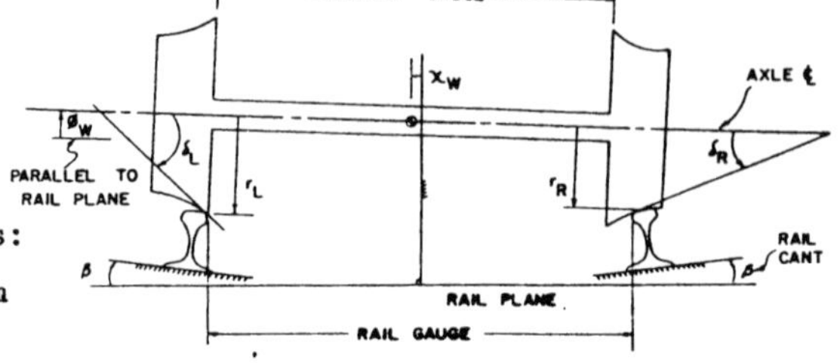


Fig. 4. Wheel/rail kinematic parameters, plan view [10].

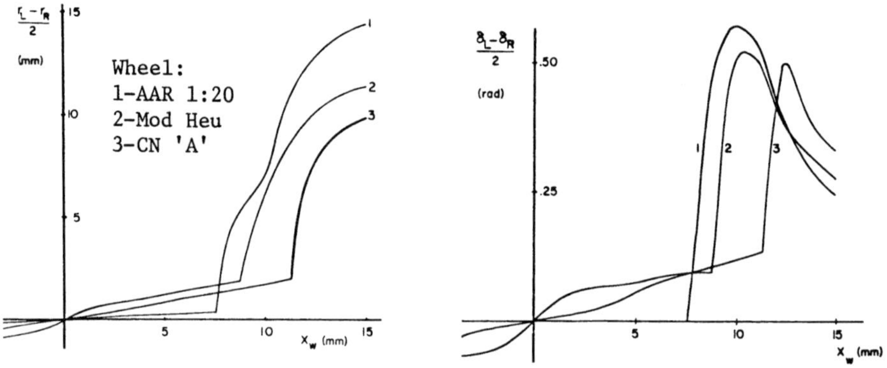


Rail Profile: New AREA
Rail Cant: 1:40
Track Gauge: 1435 mm

Linearized Parameters:

| | (1) | (2) | (3) |
|---|---|---|---|
| $\lambda = (r_L - r_R)/2x_w$ | 0.05 | 0.23 | 0.17 |
| $\Delta = (\delta_L - \delta_R)a/2x_w$ | 0.0 | 9.5 | 8.8 |

Fig. 5. Rolling radii difference and contact angle difference plots for 3 wheel profile designs, symmetric case.

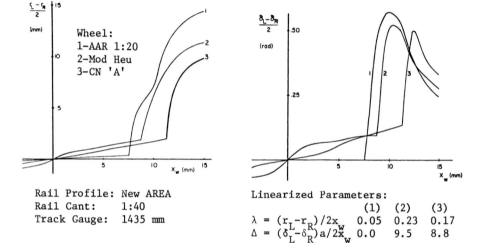

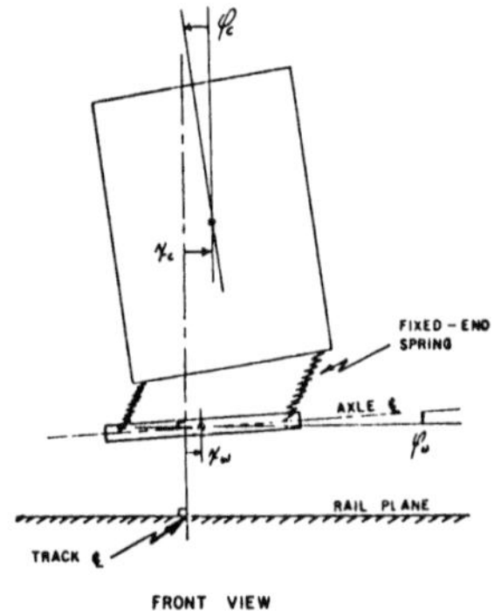


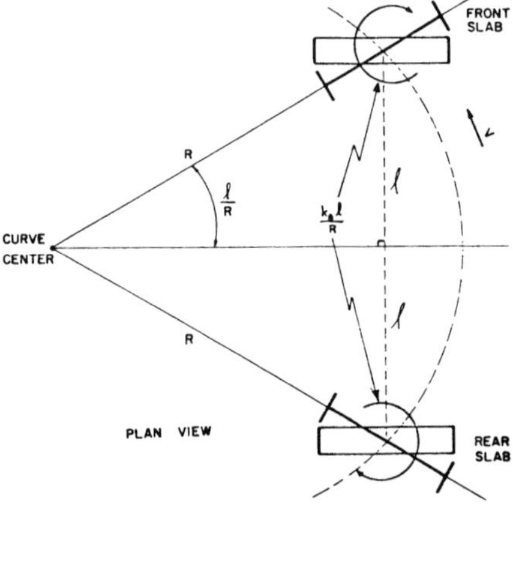


Fig. 6. Vehicle model.

In order to closely assess the effects of profile design changes on vehicle dynamics, special attention is given to an accurate modeling of contact forces. The creepages for the left wheel are [13]:

$$\gamma_{lat_L} = \frac{1}{V}\{(\dot{x}_w + r_o\dot{\phi}_w - V\theta_w)\cos\delta_L + a\dot{\phi}_w \sin\delta_L\} \quad (1)$$

$$\gamma_{long_L} = \frac{1}{V}\{-r_o\dot{\beta}_w - a(\dot{\theta}_w + \frac{V}{R}) + V(1 - \frac{r_L}{r_o})\} \quad (2)$$

$$\gamma_{spin_L} = \frac{1}{V}\{(\dot{\theta}_w + \frac{V}{R})\cos\delta_L - \frac{V}{r_o}\sin\delta_L\} \quad (3)$$

where $\delta_L$, $r_L$, and $\dot{\phi}_w$ are kinematic functions of $x_w$. The creep force/creepage relationship was formulated using Kalker's linear theory [14], which was considered valid for the small creepages found during tread contact. The creep coefficients were allowed to vary with curvatures of the wheel and rail profiles and variations in the normal forces as contact points changed with wheelset lateral displacement. Checks were made during solutions to determine when the vector creep force exceeded the friction available.

The normal forces of the rails on the wheels were obtained from simultaneous solution of the roll and vertical equations for the complete vehicle. Effects of large contact angles, cant deficiency, and car body lateral shift were included. Inertial effects were neglected as small for the conditions considered.

## 4. PERFORMANCE MEASURES, PERFORMANCE INDICATORS, AND PROFILE DESIGN GOALS

### *Performance Measures*

To permit comparisons between alternate profile designs, simple measures were developed for vehicle curving and stability performance and wheel tread wear tendency. Although possible with the developed model, performance measures for vehicle response to random rail irregularities were not formulated.

The performance measure chosen for lateral stability was the critical velocity $V_c$, determined by examining the vehicle response on smooth, tangent track to an initial condition at increasingly larger values of speed until hunting occured.

Flange-free curving ability was measured by the minimum radius curve that the vehicle can traverse in the steady-state at balance speed without flange contact or gross slip at any wheel. For profiles with relatively small tread contact angles ( $<5^\circ$), initial slip at the front outside wheel at balance speed occurred at a curve radius slightly larger than

$$R_s \cong k_\theta \ell/(\mu W a), \quad (4)$$

the approximate value of $R_s$ for a linear wheelset with zero lateral suspension stiffness. For larger tread contact angles, $R_s$ was generally significantly larger (a factor of 2 for extreme cases) due to the effects of increased spin creep. A good estimate of the curve radius below which flange contact occurred at the front outside wheel was found to be

$$R_f \cong \frac{r_o}{x_f\lambda}\left[\frac{k_\theta \ell}{2af_{33}} + a\right], \qquad (5)$$

the value for a linear wheelset with zero lateral stiffness at balance speed. In the step-by-step process of determining the actual values of $R_s$ and $R_f$, the expressions in Eq. 4 and 5 were used to choose initial estimates. Non-zero cant deficiency increased $R_s$, while $R_f$ remained essentially unaffected. For simplicity, $R_s$ and $R_f$ were evaluated only at balance speed.

Quantitatively assessing the propensity of a wheel profile to change shape by wear and/or plastic flow is very difficult, if not impossible. Wear depends on many factors such as temperature, material chemistry, load, regions of contact, presence of contaminants, braking operations, and creepage levels. Recent studies [15,16] suggest that the major wear factors in the mechanical environment (which exclude metallurgical and external factors such as humidity and contaminants) are due to the geometry of wheel-rail contact, creepage levels, creep forces, and Hertzian stress levels. Beagley [16] showed experimentally that there exists a critical combination of normal stress and creepage levels that can suddenly increase wear from a mild to a very high rate involving plastic flow at the surface of contact. From his results, an approximate boundary can be calculated for the normal stress level at low creepage that would initiate severe wear. This stress level is approximately 3.5 times the value of the yield stress in shear, $\tau_{yp}$. For railway wheels, $\tau_{yp} \cong 480$ MN/m$^2$ and thus the value for the normal stress at the "severe wear boundary" is about 1700 MN/m$^2$.

We evaluated the possibility of using for tread wear criteria the distribution of the rms levels of creepages, creep forces, and the normal forces over the contact regions while running the vehicle model over tangent track with typical random lateral alignment irregularities. It was found that the distribution of the normal tread forces stayed within a few percent of wheel load for different wheel profiles. No correlation could be established between a change in profile and a change in the distribution of tread creepages or creep forces. The level of rms creepages and creep forces depended more strongly on forward velocity, wheel load, and the rms level of the alignment irregularity than on the specific profile design. Contact forces and creepages were thus not used as wear criteria to compare the wear tendency of alternate profile designs.

For the purposes of this study, the magnitudes of maximum normal contact stresses over the wheel contact regions and the extent of these contact regions were used as measures to assess the tendency of the profile to change shape. For example, a wheel profile with high contact stresses and small contact regions would very probably change shape faster than one with lower stresses over large contact regions.

*Indicators for Performance Comparisons*

Indicators to compare the wear tendency, stability, and curving performance of alternate wheel profile designs were formulated as follows. The wear tendency indicator for wheel tread contact is

$$f_w = 100\,[.5\;C_G/\Sigma + .5\;\bar{\sigma}_m/S_G] \qquad (6)$$

where $\Sigma$ is the sum of the widths of the contact bands (excluding jumps) on both the wheel and rail profiles for all positions of the wheelset within flange clearance. $\bar{\sigma}_m$ is the mean of the maximum values of the normal contact stresses developed at all discrete locations of the wheelset within flange clearance. $C_G$ and $S_G$ are arbitrary nominal contact width and stress goals and 100 is a convenient reference number.

To assess both vehicle curving and stability performance, a combined dynamic performance indicator was formulated as

$$f_p = \max(R_f, R_s)/V_c \qquad (7)$$

where $R_f$ and $R_s$ are curving indicators discussed before and $V_c$ the critical speed on straight track. An overall performance indicator containing stability, curving, and wear measures

$$f = \frac{100}{w_1+w_2+w_3}\left[w_1 \frac{V_G}{V_c} + w_2 \frac{\max(R_f, R_s)}{R_G} + \frac{w_3}{2}\left(\frac{C_G}{\Sigma} + \frac{\bar{\sigma}_m}{S_G}\right)\right] \qquad (8)$$

was constructed that contains goals $V_G$ and $R_G$ for critical speed and curving, respectively, and weighting parameters $w_i$.

The performance of existing profile designs (as in Fig. 1) can be evaluated using these indicators, as will be discussed in Sec. 6. A small $f_w$ indicates low wear tendency, a small $f_p$ suggests good dynamic performance, and a low value of f indicates good overall performance. For profile design, the major objective is to adjust the profile to minimize one or more of these indicators.

### *Wheel Profile Design Goals*

In the profile design process, the resulting profile shape for a given vehicle depends on the goals of the design, which can be stated in terms of the performance indicators of Eq. 6, 7, and 8. The wheel profile can be designed and optimized for

- Goal 1 The best overall performance: $f \to \min$
- Goal 2 The best compromise performance between stability and curving without regard to wear tendency: $f_p \to \min$
- Goal 3 The least wear tendency without regard to vehicle dynamic performance: $f_w \to \min$
- Goal 4 A specified high speed $V_o$ while minimizing wear tendency: $f_w \to \min$, subject to $V_c \geq V_o$
- Goal 5 A specified minimum curve radius $R_o$ that can be traversed without flange contact while minimizing wear tendency: $f_w \to \min$, subject to $\max(R_f, R_s) \leq R_o$.

## 5. THE DESIGN PROCESS

The overall closed-loop wheel profile design process is illustrated in Fig. 7. It is implemented on a hybrid computer system that performs the necessary tasks under operator control via a computer terminal. Digital programs sequence the steps in the process and provide operator interaction in directing the profile design and analysis. Two modes of operation are possible: (1) An open-loop evaluation of a given profile, or (2) the closed-loop adjustment of a profile to meet a stated design goal.

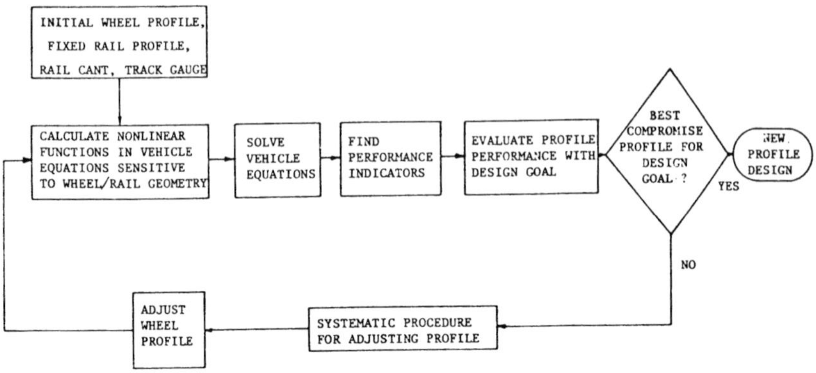


Fig. 7. Overall wheel profile design process.

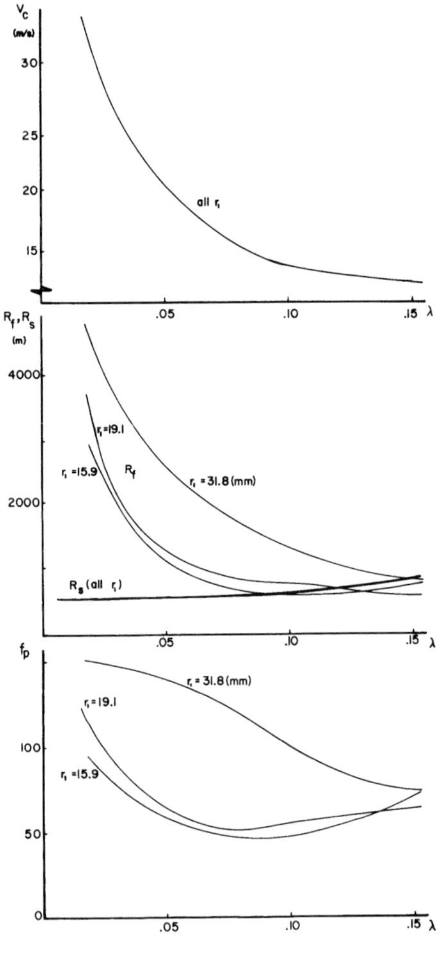


Empty vehicle
Axle load 56.5 KN
Full Kalker creep
$\mu = 0.2$

New AREA rail profile
Track gauge 1435 mm
Rail cant 1:40

Fig. 8. Coned profile design results.

The sections below discuss the details involved in each step of the process as presented in Fig. 7.

*Fix Rail Profile and Track Parameters and Select Initial Wheel Profile*

To initiate the profile design procedure, a fixed rail profile, rail cant, and track gauge are selected. Either an algorithm for a new rail design or data points for a worn rail are accepted. For a worn rail, the data is first smoothed, then fit with a cubic spline function which provides continuous curvatures for creep coefficient and contact stress calculations.

The flange parameters and the 2n profile parameters discussed in Sec. 2, are then initially specified. For an open-loop evaluation of an existing profile, the parameters are readily obtainable from a design drawing. For a closed-loop profile design, the initial parameters are also generally specified as those of an existing design.

*Calculate Nonlinear Functions and Wear Tendency Measure*

The contact points as a function of wheelset lateral position $x_w$ are then numerically calculated for the wheel-rail profile combination, from which the nonlinear wheel-rail geometric constraint functions are computed. The wheel and rail curvatures are calculated and, using these, the creep coefficients and Hertzian stresses at each wheel are numerically determined as functions of $x_w$. The levels of tread contact stresses and the extent of the wheel-rail contact regions are used to numerically formulate the wear tendency measure $f_w$. The creep coefficient and constraint functions are stored in tables for subsequent use in the solution of the equations of motion.

*Solve Vehicle Equations for Performance Measures and Determine Indicators*

For the next step, the analog part of the hybrid computer solves the equations of motion while the digital part dynamically updates the geometric constraint functions and the creep coefficients. The solutions yield time histories in real time for the vehicle response variables, wheel-rail forces, and creepages.

The stability and curving performance of the vehicle on smooth track is evaluated by integrating the equations of motion. The lateral stability performance measure, the critical velocity $V_c$, is determined by examining the response to an initial wheelset lateral displacement equal to the flange clearance at increasingly larger values of forward speed until hunting occurs. The curving performance measures, the minimum radius curves at which slip and flange contact occur at any wheel ($R_s$ and $R_f$, respectively), are evaluated during steady-state conditions at balance speed. From the values of $V_c$, $R_s$, and $R_f$, the vehicle dynamic performance indicator $f_p$ is formulated. With the wear measures obtained in the previous step, the overall performance indicator f is computed.

For an open-loop evaluation of an existing profile, the procedure stops here with the evaluation of the performance as indicated by $f_w$, $f_p$, and f (see Sec. 6 for an example). For a closed-loop profile design, the process is continued to the next step.

*Evaluate Performance with Design Goal and Systematically Adjust Wheel Profile*

The profile performance is now evaluated with respect to the previous value of the indicator for the selected design goal (assuming at least one complete cycle through the process). If the performance indicator for the goal has not reached a minimum or a constraint is violated (Goals 4 and 5 of Sec. 4.), the profile parameters are then readjusted for another pass through the process. If the indicator has converged to a minimum (determined when the current value has a small difference with the previous value), we have an optimum profile based on the design goal. Usually, but not always (see Sec. 6 for example), the final profile parameters differ for different goals.

Closed-loop profile design is conducted in two overall steps. First, no matter what the final design goal is, the initial profile parameters are systematically adjusted to minimize the wear tendency indicator $f_w$. This involves sequential iterations of the process without conducting vehicle simulation runs. Second, after $f_w$ is minimized, the resulting profile is readjusted for the final design goal.

For simplicity, sequential single-dimensional searches of the profile parameters were conducted. The number of parameters involved is reduced by selecting the end parameters $r_1$ and $\alpha_n$ (see Fig. 2). On a fixed track geometry, $r_1$ affects the flange clearance and is selected for the desired clearance. $\alpha_n$ is selected on the basis of whether a coned or hollow profile is desired.

For the remaining 2(n-1) parameters, a relationship is determined among the parameters such that only one adjustable parameter is involved. An example of this procedure is given in Sec. 6.

## 6. STUDIES USING THE DESIGN PROCESS

The studies that follow are presented to demonstrate the use of the wheel profile design process.

*Performance of Existing Designs*

The vehicle model discussed previously was used to compare the performance of existing wheel profile designs. The profiles, shown in Fig. 1, were evaluated using the design process in an open-loop mode. Results are compared in Tab. 2. The first two columns give the flange clearance $x_f$ and the least-squares trend conicity $\lambda$. The next two columns are the profile wear tendency measures for $x_w \leq x_f$. The vehicle dynamic performance measures on straight and curved track are in the next three columns. The indicators of relative performance (Eq. 6-8) are tabulated in the last three columns; the numbers for the indicators express relative merits only and have no physical meanings. The lower the number, the better the corresponding performance.

In Tab. 2, although the AAR 1:20 coned profile has the lowest mean tread contact stress level, it has the highest wear tendency indicator $f_w$ due to the very small wheel-rail regions of contact. The low conicity of the AAR profile allows a comparatively high critical

| | Wheel | $x_f$ | $\lambda$ | Wear Tendency Measure $\Sigma$ | $\overline{\sigma}_m$ | Dynamic Perf. Measure $V_c$ | $R_f$ | $R_s$ | Performance Indicators $f_w$ | $f_p$ | $f$ |
|---|---|---|---|---|---|---|---|---|---|---|---|
| 1) | AAR 1:20 | 7.6 | .05 | 16.0 | 876 | 20.9 | 1844 | 640 | 302 | 88 | 207 |
| | Mod Heu | 8.9 | .23 | 24.6 | 1300 | 13.3 | 396 | 680 | 249 | 51 | 166 |
| | CN 'A' | 11.4 | .17 | 72.6 | 993 | 14.2 | 372 | 665 | 124 | 47 | 121 |
| 1) | 483 mm | 11.4 | .11 | 73.7 | 682 | 15.1 | 542 | 588 | 101 | 36 | 107 |
| | 330 mm | 11.4 | .05 | 73.4 | 751 | 23.0 | 804 | 552 | 106 | 37 | 110 |
| 2) | AAR 1:20 | 7.6 | .05 | 16.0 | 1420 | 33.7 | 1585 | 350 | 341 | 47 | 197 |
| | Mod Heu | 8.9 | .23 | 24.6 | 2100 | 20.9 | 335 | 390 | 307 | 19 | 154 |
| | CN 'A' | 11.4 | .17 | 72.6 | 1610 | 24.2 | 305 | 427 | 169 | 18 | 104 |
| | | mm | | mm | $MN/m^2$ | m/s | m | m | | | |

1) Vehicle Empty (Axle load 56.5 KN); 2) Loaded (Axle Load 242 KN)

Track Parameters:

Cant 1:40
Rail Gauge 1435 mm
Rail Profile New AREA (Fig. 3)

Indicator Constants (Eq. 6-8):

$C_G$ = 76 mm  $V_G$ = 22.9 m/s
$S_G$ = 690 $MN/m^2$  $R_G$ = 873 m
$w_1 = w_2 = w_3 = 1$

Tab. 2. Performance comparison of profile designs.

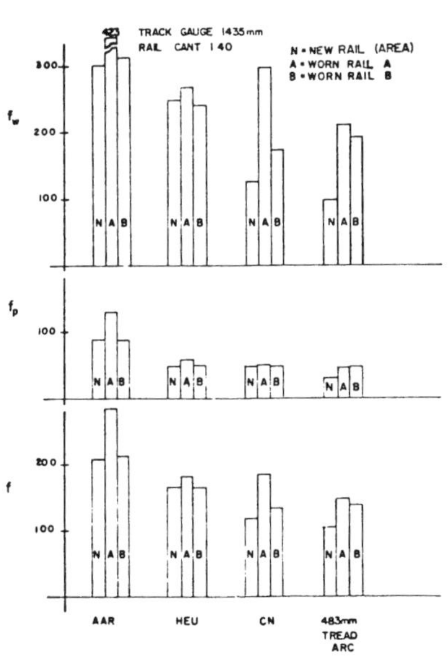


Fig. 9. Sensitivity of performance to rail profile.

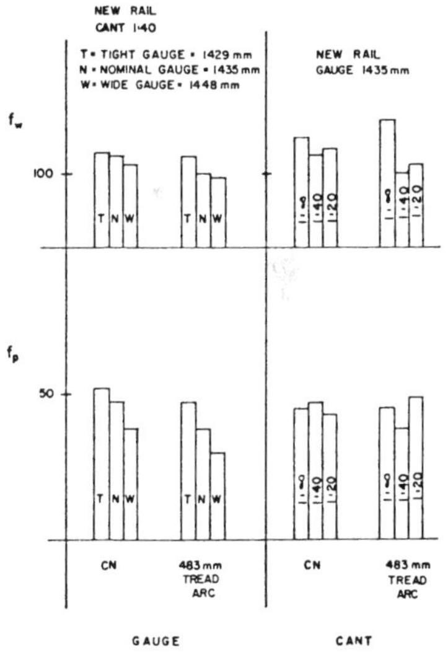


Fig. 10. Sensitivity to track gauge and rail cant.

velocity, but leads to a flange-free curving ability that is substantially inferior to the other profiles. Consequently, its dynamic performance and overall performance as measured by $f_p$ and f, respectively, ranks as poorest of the three profiles. The coned Modified Heumann design ranks as second in performance. Its high conicity leads to the lowest critical velocity. The critical velocities and curving measures of the Heumann and CN 'A' profiles are close and they thus have similar dynamic performance indicators $f_p$. However, the Heumann tread develops a higher mean stress level, has much smaller contact regions and would wear more rapidly than the CN profile. The closer matching of the hollow CN design to the new rail profile leads to an extensive use of the wheel tread and rail profiles. Judging from the performance indicators, the CN profile has by far the least wear tendency and best overall performance.

Unlike the AAR profile, both Heumann and CN profiles develop contact point jumps on their tread surfaces. This results in unequal wear across the tread and occurs when the contact point on the new rail shifts from the rail crown to the adjacent radius; for the Heumann design the jump is a relatively large 15 mm.

For a loaded vehicle, all of the vehicle dynamic characteristics, particularity $V_c$ and $R_s$, improve proportionally for each profile while, as expected, contact stresses are 62% higher, increasing the wear tendency.

*Conical Wheel Design*

The simplest case of wheel profile design is that of the standard coned contour. The adjustable profile parameters for this n=1 design (Fig. 2) are the flange throat radius $r_1$ and the angle $\alpha_1$ which defines the tread taper or the conicity $\lambda$.

Fig. 8 shows the effects of the two profile parameters on the stability and curving performance of the vehicle for a particular case on nominal track geometry. The critical velocity $V_c$ on smooth, tangent track is unaffected by changes in $r_1$ and is approximately inversely proportional to $\sqrt{\lambda}$ , a well-known result. For a given $r_1$, increasing $\lambda$ improves the curving performance by reducing the minimum radius curve $R_f$ before wheelset lateral displacement exceeds flange clearance at the front outside wheel (for steady-state conditions at balance speed on smooth track). Reducing $r_1$ opens up the flange clearance which allows more wheelset lateral movement before flange contact occurs, thus decreasing and improving $R_f$. As $\lambda$ is increased, the conflict between the requirements of stability and flange-free curving is clearly evident.

The minimum radius curve $R_s$ before gross slipping occurs at the front outside wheel tread is independent of $r_1$. For conicities below 0.10, $R_s$ is approximately constant near the value predicted by Eq. 4. As $\lambda$ increases beyond 0.10, slip occurs at increasingly larger curve radii due primarily to increases in the spin creepage contribution to the lateral creep forces with the higher contact angles at larger $\lambda$. This contribution becomes rapidly larger with increases in $\lambda$ above 0.15 and above a value of $\lambda$ of about 0.20 the treads continuously slip on the rails on tangent track.

Increases in track gauge improved curving ability by shifting the

$R_f$ curves in Fig. 8 to the left, but did not affect $V_c$ and $R_s$. Increasing creep coefficients had the same effect, also slightly increasing $V_c$ for all $\lambda$. Reductions in $\mu$ reduced curving ability by shifting the $R_s$ curve proportionally up, while $R_f$ and $V_c$ remained unaffected. Higher axle loads improved all three dynamic performance measures for all $\lambda$.

The dynamic performance indicator $f_p$ plot shows that decreasing $r_1$ improves overall dynamic performance. In terms of the best compromise between stability and curving (minimum $f_p$) for this vehicle model, the optimum $\lambda$ with $r_1$ = 15.9 mm is about 0.085 or a taper of 1:11.8.

For coned profiles, as the wheelset moves within flange clearance, the contact width on the tread is small and contact with the rail remains practically at a single point. On the tread, the maximum Hertzian stress magnitudes $\sigma_m$ remain constant. For small tread tapers, the contact point on the new rail is on the large rail crown radius. As $\lambda$ is increased, however, the rail contact point can move to the radius adjacent to the crown radius. For the track geometry case in Fig. 8, the rail contact point moves to the adjacent radius for $\lambda$ larger than 0.10, increasing $\sigma_m$ two-fold and moving $\sigma_m$ into a potentially high wear regime (Sec. 4). Thus, tread tapers above 1:10 should be avoided for the coned profile design.

*Tread Arc Wheel Profile Design*

The design process was applied to design optimum hollow wheel profiles having a single circular arc for the wheel tread. Two design examples are given, one for the goal of best overall performance, and the second for high speed operation. The profiles were designed for the new AREA rail profile and nominal track parameters. The vehicle was empty, the creep coefficients were factored down to half Kalker's values, and $\mu$ = 0.2.

The single tread arc design is an n=2 type profile having the 2n profile parameters $r_1$, $\alpha_1$, $r_2$, and $\alpha_2$. For a fixed track geometry, the flange throat radius $r_1$ affects the flange clearance and was fixed at 12.7 mm to give an approximate flange clearance of 11 mm. The angle $\alpha_2$ determines the slope of the field-side linear tread segment and was set at $\alpha_2$ = 0 to give a hollow profile with a cylindrical wheel segment.

Now, to result in a single parameter profile adjustment, $\alpha_1$ can be fixed and the tread radius $r_2$ varied. Conversely, $r_2$ can be fixed at a radius larger than the rail crown radius and $\alpha_1$ varied. However, a better procedure is to establish a relationship between $\alpha_1$ and $r_2$ based on considerations of the wheel tread profile contact conditions existing with the new rail profile. From the standpoint of distributing wear across the wheel and rail profiles, it is desirable to use as much of the tread radius and rail crown radius as possible for contact within flange clearance. However, if contact occurs between the tread radius and the radius on the rail adjacent to the crown radius, the contact stresses can increase by a factor of 2 or more over stresses occurring between the wheel tread radius and rail crown radius. To avoid this potentially high stress condition, wheel tread contact can be constrained to remain on the rail crown for displacements of the wheelset within flange clearance. This is done by requiring that the point where the contact jumps from the rail crown to the gauge corner is that point where rail crown and the adjacent radius meet. This establishes a relationship between the $\alpha_1$ and $r_2$ profile parameters and a single variable then remains for the adjustment of the profile.

For the n=2 profile design, restricting the contact to only the rail crown results in the following characteristics. Assuming contact only on $r_2$ within flange clearance, the Hertzian stress levels and creep coefficients remain constant along the profile. In addition, the conicity and gravitational stiffness remain constant. This is because for all points of tread contact, the radii of curvature of the wheel and rail are constant. In contrast, contact between the new rail and either the Heumann or CN profiles (Fig. 1) results in variable stress levels and creep coefficients as well as nonlinear functions for the conicity and gravitational stiffness.

Using the relationship between $\alpha_1$ and $r_2$, the profile is first adjusted to minimize the wear tendency indicator. A single-dimensional, half-interval search procedure involving several cycles of the design process is used to find this minimum. The optimum profile parameters found for the vehicle model used were:

$$r_2 = 483 \text{ mm}, \quad \alpha_1 = 6.53° \qquad (9)$$

Coincidently, these parameters also minimized the dynamic performance indicator $f_p$ and the overall performance indicator f. The results of this particular study are shown in Tab. 2. The new 483 mm profile resulted in a significant improvement over the performance of the existing designs. The sum of the widths of the wheel and rail contact regions of the 483 mm profile are about the same as those for the CN profile. However, since the average stresses are lower, the 483 mm profile has a lower wear tendency. The 483 mm profile resulted in a higher critical velocity a[illegible]ller radius curve before slip occurred. Thus, the dynamic performance indicator shows an improvement over that of the CN profile.

A disadvantage of the 483 mm design is its low critical velocity as compared to the AAR profile. Using the design process, a redesign of the tread circle profile was undertaken to increase the critical velocity to at least 23 m/s while minimizing the wear tendency. (Goal 4 of Sec. 4). This resulted in a low-conicity profile having a tread radius of 330 mm. The results in Tab. 2 show that compared to the AAR profile, this profile has a higher critical velocity while retaining significantly less wear tendency and superior overall performance. Compared to the 483 mm design, the wear tendency is only slightly increased and the overall performance only slightly degraded.

*Sensitivity to Variations in Track Parameters*

In practice, the rail profile shape, track gauge, and rail cant vary along the track. It is therefore of interest to analyse the sensitivity of the performance of the vehicle with a given wheel profile to changes in the track geometry. We examined the sensitivity of the developed 483 mm tread circle profile and the existing profiles of Fig. 1 to changes in rail profile, track gauge, and rail cant. The same vehicle parameters as in the tread circle design were used.

Fig. 9 shows the sensitivity of the vehicle performance with the different wheel profiles to changes in rail profile. The track gauge and rail cant are nominal. The variations in vehicle performance with the new and two worn rail heads of Fig. 3 are analysed in terms of the developed performance indicators $f_w$, $f_p$, and f. Low values for the indicators indicate good performance.

The results show that the standard AAR 1:20 coned profile exhibited the worst overall performance of the four wheel profiles. In general, the best performance is shown by the 483 mm profile, particularly for the new rail profile.

The results indicate that profile wear, vehicle dynamic, and overall performance as measured by $f_w$, $f_p$, and f, respectively, can be very sensitive to the rail profile. In particular, all wheel profiles performed better on new rail than on worn rail. The wear tendency of the AAR, CN, and 483 mm profiles depended more strongly on the rail profile than did the Heumann design. Except for the CN on the worn A profile, the wear tendency of the hollow CN and 483 mm profiles was lower than the coned profiles. The wear tendency of the 483 mm profile was lower for new and worn A rails than the CN profile, but higher for the worn B rail.

Since the 483 mm profile was designed to minimize $f_p$ with the new rail, it exhibited the best dynamic performance on new rail. However, on the worn rail profiles, it had only a slightly better dynamic performance than the CN and Heumann profiles. In terms of the overall performance indicator f, the 483 mm profile design generally ranked the best of the wheel profiles on the various rails. The only exception is a slightly worse performance on the worn B rail compared to the CN profile performance on the same rail.

Fig. 10 shows the sensitivity of the performance to changes in track gauge and rail cant for the CN and 483 mm wheel profiles. Results are shown in terms of the wear tendency and dynamic performance indicators. For both profiles, the wear tendency is lowered as the gauge is widened since larger gauge widens the tread and rail profile contact regions. For both profiles, the dynamic performance indicator is also improved with increases in gauge, since the conicity is lowered with increasing gauge. Thus, the critical velocity is significantly increased. For both wear tendency and dynamic performance, the 483 mm profile had the better performance with changes in gauge.

For both wheel profiles, variations from nominal rail cant (1:40) increased the wear tendency. Reductions in rail cant increased the wear tendency of the 483 mm profile over that of the CN profile, while the opposite occured for increases in rail cant. The dynamic performance of the CN profile improved slightly with variations in rail cant from nominal, but cant variations degraded the performance of the 483 profile.

## 7. CONCLUSIONS

Although the vehicle model in the study was very much simplified, we feel that the feasibility of the wheel profile design procedure has been demonstrated. The extension to more realistic vehicle models depends mainly on the availability of a large, state-of-the-art computing facility. This may be either hybrid, as used in this investigation, or a large digital computer.

The procedure developed may be used to address the design of more complex profiles ($n \geq 2$) at the expense of increasing the computation time for and increasing the complexity of the search procedure. Different vehicle performance indicators and/or design goals may also be formulated and used to address the needs of a particular user.

It has been demonstrated that the approach developed may be used to analyze and compare the dynamic performance of rail vehicles equipped with various wheel profiles. It has also been shown that the closed-loop design procedure may be used to develop a profile that is optimum with respect to stated design goals.

NOMENCLATURE

| | |
|---|---|
| $a$ | nominal semi-distance between contact points |
| $f_{33}$ | longitudinal creep coefficient |
| $k_\theta$ | primary yaw stiffness |
| $\ell$ | car body semi-length |
| $r_L, r_R$ | wheel rolling radii (see Fig. 4) |
| $r_o$ | wheel rolling radius at the taping line |
| $R$ | curve radius |
| $V$ | vehicle forward velocity |
| $x_w$ | wheelset lateral displacement |
| $W$ | axle load |
| $\dot{\beta}_w$ | axle rotational velocity |
| $\gamma_{lat}$ | lateral creepage |
| $\gamma_{long}$ | longitudinal creepage |
| $\gamma_{spin}$ | spin creepage |
| $\delta_L, \delta_R$ | contact angles (see Fig. 4) |
| $\theta_w$ | wheelset yaw displacement |
| $\lambda$ | effective conicity |
| $\mu$ | coefficient of friction |
| $\phi_w$ | axle roll angle (see Fig. 4) |

REFERENCES


1. Heumann, H., Zur Frage des Radreifen-Umrisses. Organ für die Fortschritte des Eisenbahnwesens, v. 89, n. 18, Sep. 1934, pp. 336-342.
2. Müller, C.H., Radreifenverschleiss und Fahrzeuglauf. Österreichische Ingenieur-Zeitschrift, v. 7, n. 7, Jul. 1964, pp. 215-224.
3. King, B.L., New Tyre Profiles for British Railways. The Railway Gazette, v. 124, n. 2, Jan. 1968, pp. 60-64.
4. Matusi, N., On the Tyre Profile of Freight Car Wheels. JNR Railway Technical Research Institute Quarterly Reports, v. 11, n. 2, Mar. 1970, pp. 61-65.
5. Marcotte, P., Test Report on the Comparative Curving Performance of Freight Car Trucks on Special Wheel Profiles. Report No. 122, Canadian National Railways, St. Laurent, Quebec, Mar. 1975.
6. Anon., Recommendations for a Uniform Worn Wheel Profile, Independent of Wheel Diameter and Vehicle Design. Report No. S1002/RP2, ORE, Utrecht, the Netherlands, Apr. 1973.
7. Anon., Wheel and Axle Manual, 11 ed. Association of American Railroads, Washington, D.C., Oct. 1975, p. 73.

8. Eck, B.J. and Berg, N.W., Looking for Tomorrow's Wheel Contour. Presented at ASME Winter Annual Meeting, Detroit, Michigan, Nov. 1973.
9. Koffman, J.L., Heumann Tyre Profile Tests on British Railways. The Railway Gazette, Vol. 121, No. 7, Apr. 1965, pp. 279-283.
10. Cooperrider, N.K., et al, Analytical and Experimental Determination of Nonlinear Wheel/Rail Geometric Constraints. Report No. FRA-OR&D 76-244, NTIS Pub. No. PB 252290, U.S. Department of Transportation, Washington, D.C., Dec. 1975.
11. Anon., Geometry of Contact Between Wheelset and Track, Part 1: Methods of Measurement and Analysis. Question C116, Interaction Between Vehicles and Track, Report No. 3, ORE, Utrecht, the Netherlands, Oct. 1973.
12. Hauschild, W., Die Kinematic des Rad-Schiene-Systems. Institute für Mechanik, Technische Universität Berlin, Mar. 1977.
13. Heller, R., Rail Vehicle Wheel Profile Optimization Using Hybrid Computation. Ph.D. Dissertation, Clemson University, to be published 1980.
14. Kalker, J.J., On the Rolling Contact of Two Elastic Bodies in the Presence of Dry Friction. Doctoral Thesis, Technical University of Delft, 1967.
15. Jamison, W.E., Mechanical Wear of Railroad Components. Report prepared for the Association of American Railroads, Chicago, Mar. 1978.
16. Beagley, T.M., Severe Wear of Rolling/Sliding Contacts. Wear, n. 36, 1976, pp. 317-335.

# GENERIC RAIL TRUCK CHARACTERISTICS


A. K. Kar, D. N. Wormley and J. K. Hedrick

Department of Mechanical Engineering, Massachusetts Institute of Technology, Cambridge, MA, U.S.A.



SUMMARY

A six degree of freedom, linear rail truck model is analyzed to identify six independent stiffness and one geometric parameter which can be selected in the wheelset-truck interconnection stiffness matrix. This generic truck model is shown to reduce to a two stiffness parameter conventional truck configuration and a four stiffness parameter radial truck configuration. Improvement in the stability and curving performance of a selected conventional and radial truck configuration augmented by an additional independent stiffness parameter is illustrated.


## 1.0 INTRODUCTION

In conventional rail vehicles lateral guidance is achieved with pairs of tapered wheels mounted on a common axle to form a wheelset. The design of the suspension elements coupling the wheelsets to a rail truck represents a compromise in performance between stability and curve negotiation. A free wheelset is theoretically capable of performing pure rolling motion to negotiate a constant radius curve, however, an uncontrained wheelset is dynamically unstable and elastic and damping elements are required to couple wheelsets to the truck to provide dynamic stability. These constraints tend in general to limit the ability of wheelsets to sustain pure rolling motion on curves with the result that curve negotiation is accomplished with flange contact and/or slip resulting in wheel and rail wear.

A number of studies [1-7] have been conducted of wheelset-truck suspension design to achieve a good compromise between curving and stability. Parametric studies of conventional truck suspensions which provide direct coupling of each wheelset to a truck frame [1,6,7] as well as studies of radial truck configurations [2,3] which allow direct connections between the two wheelsets in a truck have been conducted. In addition studies of the fundamental characteristics of interconnected wheelsets [4,5] have identified a number of the fundamental limits to curving and stability.

In this paper the coupling of two wheelsets to a truck is studied using a six degree of freedom truck model incorporating lateral and yaw motion of the truck and lateral and yaw motion of each wheelset. This planar model is used to identify using symmetry and rigid body motion constraints the maximum number of independent linear elastic elements which can be used to couple the wheelsets and the truck. Conventional and radial truck configurations are special cases of the generic configuration which contains six independent stiffness parameters and a wheelset spacing parameter. The stability and curving performance of the generic truck is explored to determine if the additional stiffness parameters allow an improvement in curving/stability performance compared with conventional and radial configurations.

## 2.0 TRUCK REPRESENTATION

A planar model of a rail truck is formulated to identify the number of independent wheelset-truck stiffness elements required to characterize the truck completely. The model is based upon a linear representation of interconnection stiffness and damping elements and wheel-rail interactions. The representation consists of two degrees of freedom for each wheelset and the truck frame as shown in Fig. 1 with:

Truck frame - lateral displacement ($y_t$), yaw ($\psi_t$)

Leading wheelset - lateral displacement ($y_1$), yaw ($\psi_1$)

Trailing wheelset-lateral displacement ($y_2$), yaw ($\psi_2$)

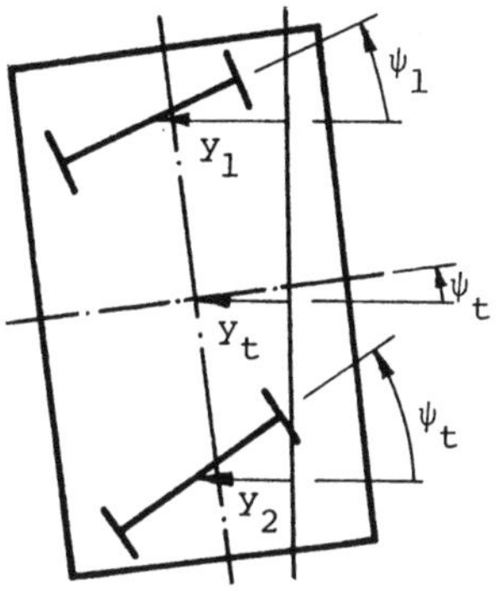


Figure 1. Truck degrees of freedom.

The general linear equation of motion for the six D-O-F isolated truck traveling with constant velocity V on rigid, smooth track may be written for dynamic operation on tangent track or steady-state operation in a constant radius curve as:

$$\underline{M}\,\ddot{\underline{X}} + (\underline{B}_s + \underline{B}_c)\dot{\underline{X}} + (\underline{K}_s + \underline{C})\underline{X} = \underline{Q} \qquad (1)$$

where: $\underline{X}$ = coordinate vector $[y_t, \psi_t, y_1, \psi_1, y_2, \psi_2]^T$

$\underline{M}$ = mass matrix

$\underline{B}_s$ = interconnection damping matrix

$\underline{B}_c$ = wheel-rail interaction damping matrix

$\underline{K}_s$ = interconnection stiffness matrix

$\underline{C}$ = wheel-rail interaction stiffness matrix

$\underline{Q}$ = external and body force vector

For stability studies on tangent track $\underline{Q}$ is set to zero while in steady-state negotiation of a constant radius curve $\underline{Q}$ is finite.

*Wheel-Rail Interactions*

The wheel-rail interaction model is based upon linear creep theory. The forces for each wheelset due to lateral, longitudinal and spin creep as well as gravitational stiffness and gyroscopic effects are summarized in [7]. This wheel-rail interaction model results in terms which are velocity and displacement dependent ($\underline{B}_c$ and $\underline{C}$ matrices) and in effective forces generated in steady-state curving ($\underline{Q}$ matrix). In this investigation only the primary terms related to lateral and longitudinal creep are retained for the exploratory studies described below. The spin creep, gyroscopic and gravitational effects are in general small compared with these primary creepages. When only lateral and longitudinal creep are considered the wheel-rail interaction forces may be summarized as:

Matrix $\underline{B}_c$: all $b_{ij} = 0$ except for

$$b_{33} = \frac{2f_{11}}{V} = b_{55} \qquad b_{44} = \frac{2a^2 f_{33}}{V} = b_{66}$$

Matrix $\underline{C}$: all $c_{ij} = 0$ except for

$$c_{34} = -2f_{11} = c_{56} \qquad c_{43} = 2\,af_{33}\frac{\lambda}{r_o} = c_{65}$$

Vector $\underline{Q}$:

$$Q_1 = M_T g\phi_D \qquad Q_2 = 0$$

$$Q_3 = \frac{2f_{11}b}{R} + M_w g\phi_D \qquad Q_4 = \frac{2f_{33}a^2}{R}$$

$$Q_5 = \frac{2f_{11}b}{R} + M_w g\phi_D \qquad Q_6 = Q_4$$

where:

$f_{11}(f_{33})$ = lateral (longitudinal) creep coefficient

$a$ = wheel spacing

$b$ = leading to trailing wheelset half spacing

$\lambda$ = wheel conicity

$r_o$ = wheel radius

$R$ = curve radius

$\phi_D$ = cant deficiency

$g$ = acceleration due to gravity

$M_T(M_w)$ = truck (wheelset) mass

*Inertial Matrix*

For the two wheelsets and truck, a translateral mass is associated with each lateral D-O-F and an inertia with each angular D-O-F. The truck mass matrix is diagonal with:

$$\text{Diagonal } [\underline{M}] = [M_T,\ I_T,\ M_w,\ I_w,\ M_w,\ I_w]$$

where: $I_T(I_w)$ = truck (wheelset) inertia

*Truck Interconnection Stiffness Matrix*

The truck frame-wheelset interconnection stiffness matrix is derived in a general manner to identify the number of independent stiffness parameters which exist for the planar truck model. The interconnection elements are assumed linear and passive. In addition, the constraints of elastic symmetry and rigid body motion are also satisfied.

For a set of linear, passive, elastic interconnections between a set of N bodies with each body possessing a single translational and rotational degree of freedom, a general interconnection stiffness matrix may be defined which is 2Nx2N. The matrix is symmetric since the elements are derived from a general potential energy function.

When 2 bodies (N=2), i.e. two wheelsets spaced 2b apart, are considered, the general 4x4 stiffness matrix may be shown [4] with the additional assumptions of lateral plane symmetry and rigid body motion constraints to contain two independent stiffnesses-lateral shear stiffness ($k_s$) and bending stiffness ($k_b$) and one geometric parameter-the wheelset spacing (2b). Thus for two isolated wheelsets spaced 2b apart only two stiffness parameters may be set independently.

When two wheelsets and a truck frame are considered (N=3) symmetry requires 21 distinct terms in the 2Nx2N interconnection stiffness matrix. Simplification of this matrix using the conditions of elastic plane symmetry and rigid body motion may be conveniently accomplished by considering the motions in terms of the generalized coordinates $\underline{\Phi}$ shown in Fig. 2.

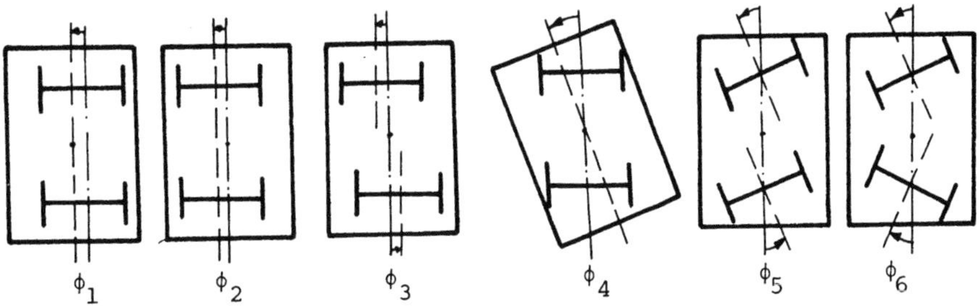


Figure 2. Truck generalized coordinates.

The generalized coordinates $\underline{\Phi}$ are related to the original coordinates $\underline{X}$ through a similarity transformation $\underline{T}$:

$$\underline{\Phi} = \underline{T}\ \underline{X} \tag{2}$$

$$\underline{T} = \frac{1}{\sqrt{2}} \begin{bmatrix} \sqrt{2} & 0 & 0 & 0 & 0 & 0 \\ 0 & 0 & 1 & 0 & 1 & 0 \\ 0 & 0 & 1 & 0 & -1 & 0 \\ 0 & \sqrt{2} & 0 & 0 & 0 & 0 \\ 0 & 0 & 0 & 1 & 0 & 1 \\ 0 & 0 & 0 & 1 & 0 & -1 \end{bmatrix} \tag{3}$$

The stiffness matrix expressed in generalized coordinates is symmetric and represented as $\underline{K}$ with each element represented as $k_{ij}$. The relationship among the stiffness parameters in $\underline{K}$ may be identified by considering elastic plane symmetry and rigid body motion constraints.

Lateral plane elastic symmetry requires that the elements of the interconnection matrix $\underline{K}$ be invariant for a symmetric displacement in the coordinates $\underline{\Phi}$ with respect to a lateral plane of symmetry passing through the truck center and requires:

$$k_{13} = k_{14} = k_{15} = 0 \tag{4}$$

$$k_{23} = k_{24} = k_{25} = 0 \tag{5}$$

$$k_{36} = k_{46} = k_{56} = 0 \tag{6}$$

The truck assembly is capable of two rigid body motions - a pure lateral displacement of all bodies, a combination of $\phi_1$ and $\phi_2$ and a pure rotation, a combination of $\phi_3, \phi_4$ and $\phi_5$.

Under rigid body motion no net forces are generated in the interconnection elements. Rigid body translation requires:

$$k_{12} = -\frac{1}{\sqrt{2}} k_{11} \tag{7}$$

$$k_{22} = \frac{1}{2} k_{11} \tag{8}$$

$$k_{26} = -\frac{1}{\sqrt{2}} k_{16} \tag{9}$$

while rigid body rotation requires:

$$k_{34} = -\sqrt{2}\ k_{35} - \sqrt{2}\ b\ k_{33} \tag{10}$$

$$k_{44} = -\sqrt{2}\; k_{45} + 2bk_{35} + 2b^2 k_{33} \qquad (11)$$

$$k_{55} = -\frac{1}{\sqrt{2}} k_{45} - b\, k_{35} \qquad (12)$$

When eqs. (4)-(12) are used to reduce the 21 distinct terms in $\underline{K}$ to the minimum number of independent terms, the following form for $\underline{K}$ may be derived:

$$\underline{K} = \begin{bmatrix} k_{11} & -\frac{k_{11}}{\sqrt{2}} & 0 & 0 & 0 & k_{16} \\ & \frac{k_{11}}{2} & 0 & 0 & 0 & -\frac{k_{16}}{\sqrt{2}} \\ & & k_{33} & -\sqrt{2}(k_{35}+bk_{33}) & k_{35} & 0 \\ & & & -\sqrt{2}\, k_{45}+2bk_{35}+2b^2k_{33} & k_{45} & 0 \\ & \text{SYMMETRIC} & & & -\frac{k_{45}}{\sqrt{2}} - bk_{35} & \\ & & & & & k_{66} \end{bmatrix} \qquad (13)$$

the $\underline{K}$ matrix contains six independent stiffness elements - $k_{11}, k_{33}, k_{66}, k_{16}, k_{35}$ and $k_{45}$ - and a geometric parameter - b, the wheelset half spacing. The interconnection stiffness matrix $\underline{K}_s$ in physical coordinates may be derived in terms of these six stiffness and wheelset spacing as:

$$\underline{K}_s = \underline{T}^T \underline{K}\; \underline{T} \qquad (14)$$

Thus the wheelset-truck interconnection stiffness matrix $\underline{K}_s$ has a maximum of six independent stiffness elements and one geometry parameter which may be used to specify the possible interconnections.

*Interconnection Damping Matrix*

In this study the damping due to wheelset-truck coupling is assumed to be related directly to the stiffness elements with:

$$\underline{B}_s = \underline{T}^T (\gamma\; \underline{K}) \underline{T} \qquad (15)$$

where $\gamma$ is a constant.

## 3.0 TRUCK CONFIGURATION

A number of truck configurations may be developed in which the six stiffness elements of the interconnection matrix may be independently specified. One configuration is illustrated in Fig. 3 in which the springs $k_x$ and $k_y$ represent the lateral and longitudinal stiffness elements associated with a conventional truck. With all other stiffnesses set to zero a conventional truck is represented and only two terms in the stiffness matrix may be set for fixed wheelset spacing. The springs $k_s$ and $k_b$ represent direct shear and bending stiffnesses between the wheelsets such as employed in radial trucks with elastic interwheelset links. With only $k_x, k_y, k_s$ and $k_b$ four terms in the stiffness matrix may be set independently - $k_{11}$, $k_{35}$, $k_{45}$ and $k_{66}$. With the addition of the spring $k_o$ between the wheelset and truck, the stiffness $k_{16}$ may be independently specified and with the spring $k_\psi$, the stiffness $k_{33}$ may be independently specified and all six elements of the stiffness matrix may be specified. The springs $k_o$ and $k_\psi$ relate yaw to lateral force and lateral displacement to torque and thus represent lateral-rotational coupling.

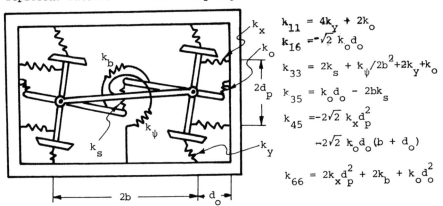

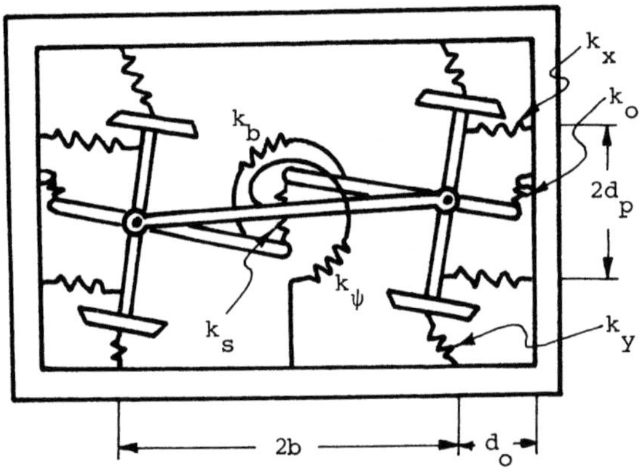


$$k_{11} = 4k_y + 2k_o$$

$$k_{16} = -\sqrt{2}\, k_o d_o$$

$$k_{33} = 2k_s + k_\psi/2b^2 + 2k_y + k_o$$

$$k_{35} = k_o d_o - 2bk_s$$

$$k_{45} = -2\sqrt{2}\, k_x d_p^2 - 2\sqrt{2}\, k_o d_o (b + d_o)$$

$$k_{66} = 2k_x d_p^2 + 2k_b + k_o d_o^2$$

Figure 3. A generic truck configuration.

## 4.0 CHARACTERISTICS OF A SINGLE TRUCK

A number of the characteristics of the generic truck may be established in a general form. The stability of the truck on tangent track is determined from solution of the equation:

$$\underline{M}\,\ddot{\underline{X}} + (\underline{B}_s + \underline{B}_c)\dot{\underline{X}} + (\underline{K}_s + \underline{C})\underline{X} = 0 \qquad (16)$$

In Eq. (16) the creep terms in $\underline{B}_c$ and $\underline{C}$ are functions of forward velocity. The critical speed is determined by calculating the eigenvalues of the equation for increasing values of speed until an unstable eigenvalue is found and the corresponding speed is the truck critical speed.

The steady-state curving performance of the truck is determined by solution of the equation:

$$(\underline{K}_s + \underline{C})\underline{X} = \underline{Q} \tag{17}$$

where $\underline{Q}$ is a function of the creep coefficients, the curve radius, and the cant deficiency $\phi_D$. For the balanced running condition ($\phi_D = 0$), eqn. (17) may be solved analytically with results that are readily interpreted:

$$y_t = \frac{ar_o}{\lambda R} + \frac{2}{\sqrt{2}}\,\frac{k_{16}}{k_{11}}\,\frac{b}{R} \cdot \frac{1}{1 + \dfrac{k_{ps}k_{pb}r_o}{af_{11}f_{33}\lambda}} + \frac{b^2 k_{ps}k_{pb}r_o}{af_{33}\cdot\lambda R} \cdot \frac{f_{11}}{k_{ps}k_{pb} + af_{11}f_{33}\dfrac{\lambda}{r_o}} \tag{18}$$

$$\psi_t = \frac{2(k_{35} + bk_{33})}{-\sqrt{2}\,k_{45} + 2bk_{35} + 2b^2k_{33}} \qquad \frac{k_{pb}f_{11}\dfrac{b}{R}}{k_{ps}k_{pb} + af_{11}f_{33}\dfrac{\lambda}{r_o}} \tag{19}$$

$$y_1 = \frac{ar_o}{\lambda R} + \frac{k_{pb}f_{11}\dfrac{b}{R}}{k_{ps}k_{pb} + af_{11}f_{33}\dfrac{\lambda}{r_o}} + \frac{b^2k_{ps}k_{pb}r_o}{af_{33}\,\lambda\,R} \cdot \frac{f_{11}}{k_{ps}k_{pb} + af_{11}f_{33}\dfrac{\lambda}{r_o}} \tag{20}$$

$$\psi_1 = -\frac{b}{R} \cdot \frac{1}{1 + \dfrac{k_{ps}k_{pb}r_o}{af_{11}f_{33}\lambda}} \tag{21}$$

$$y_2 = \frac{ar_o}{\lambda R} - \frac{k_{pb}f_{11}\dfrac{b}{R}}{k_{ps}k_{pb} + af_{11}f_{33}\dfrac{\lambda}{r_o}} + \frac{b^2k_{ps}k_{pb}r_o}{af_{33}\,\lambda\,R} \cdot \frac{f_{11}}{k_{ps}k_{pb} + af_{11}f_{33}\dfrac{\lambda}{r_o}} \tag{22}$$

$$\psi_2 = \frac{b}{R} \cdot \frac{1}{1 + \frac{k_{ps} k_{pb} r_o}{a f_{11} f_{33} \lambda}} \qquad (23)$$

These equations are written in terms of the total interconnection bending $k_{pb}$ and shear $k_{ps}$ stiffnesses between wheelsets which are defined as:

$$k_{pb} = \frac{1}{2} (k_{66} - \frac{k_{16}^2}{k_{11}}) \qquad (24)$$

$$k_{ps} = \frac{1}{2} \frac{(\sqrt{2}\, k_{33} k_{45} + 2k_{35}^2 + 2b\, k_{35} k_{35})}{\sqrt{2}\, k_{45} - 2bk_{35} - 2b^2\, k_{33}} \qquad (25)$$

Examination of eqs. (18)-(23) shows that the wheelset displacement and radial alignment are controlled by $k_{ps}$ and $k_{pb}$ while the truck displacement is influenced by additional parameters. The wheelsets achieve a pure radial alignment as $k_{pb}$ approaches zero and $k_{pb} = 0$ yields the perfect steering condition similar to that discussed in [5]. The interwheelset bending stiffness may be reduced to zero by setting:

$$k_{16} = -\beta\, k_{11} \qquad (26)$$

$$k_{66} = \beta^2 k_{11} \qquad (27)$$

where $\beta$ is a constant.

If $\beta = 0$, two of the six stiffness parameters are required to be zero. If $\beta \neq 0$ two of the six stiffness parameters are established by the constant $\beta$ and $k_{11}$.

## 5. TRUCK PERFORMANCE

The stability and curving performance of conventional and radial truck configurations augmented with an additional stiffness parameter have been computed to determine if improvements in performance are achieved with the use of the additional stiffness elements. In the study a set of baseline parameters summarized in Table 1 corresponding to a conventional truck have been adopted. The critical speed of a configuration is determined by solving eqn. (16). Since the steady-state curving performance of the wheelsets is determined to first order by the interwheelset total bending stiffness $k_{pb}$ and shear stiffness $k_{ps}$ as shown by eqns. (20)-(23), the curving performance of the truck is represented by the parameter $\psi R/b$ which approaches 1.0 as perfect steering is approached. This study is limited to consideration of the effects of $k_x, k_y, k_s, k_b$ and $k_o$ on performance and has not evaluated the influence of $k_\psi$.

TABLE 1. BASELINE PARAMETER LIST

| TRUCK PARAMETERS | | | BASELINE CONVENTIONAL TRUCK | | |
|---|---|---|---|---|---|
| $M_w$ | = | 120 slugs (1750 kg) | $k_x^*$ | = | $9.6 \times 10^5$ lb/ft ($1.401 \times 10^7$ N/m) |
| $I_w$ | = | 550 slug-ft$^2$ (745.9 kg-m$^2$) | $k_y^*$ | = | $4.8 \times 10^5$ lb/ft ($7.0 \times 10^6$ N/m) |
| $M_T$ | = | 250 slugs (3647.5 kg) | $\gamma$ | = | 0.004 S |
| $I_T$ | = | 2500 slug-ft$^2$ (3647.5 kg-m$^2$) | CREEP AND WHEEL PARAMETERS: | | |
| $d_o$ | = | 1.25 ft (0.38 m) | $f_{11}$ | = | $2.12 \times 10^6$ lb ($9.43 \times 10^6$ N) |
| $a$ | = | 2.35 ft (0.7163 m) | $f_{33}$ | = | $2.3 \times 10^6$ lb ($10.23 \times 10^6$ N) |
| $b$ | = | 4.25 ft (1.295 m) | $r_o$ | = | 1.5 ft (0.457 m) |
| $d_p$ | = | 2.00 ft (0.6096 m) | $\lambda$ | = | 0.1 |

A conventional truck is considered with initial baseline values of $k_x$ and $k_y$ and $k_b = k_s = k_\psi = 0$. The stiffness $k_o$ has been added to the truck suspension and is varied to determine its influence on critical speed and the curving parameter $\psi R/b$ calculated with eqn. (23). The performance of the conventional truck augmented with $k_o$ is summarized in Fig. 4 for two sets of data. In one set of data as $k_o/k_y^*$ is varied, $k_x$ and $k_y$ are adjusted to maintain $k_{pb}/k_y^* = 8.0$ and $k_{ps}/k_y^* = 0.31$, the baseline values, so that curving performance is held constant with $\psi R/b = 0.575$. ($k_y^*$ is the baseline value of $k_y$). In these data as $k_o/k_y^*$ is increased from 0.0 to 2.0, the critical speed $V_{cr}$ increases to 1.2 times its baseline value of $V_{cr}^* = 76$ m/s. In the second set of data in Fig. 4 as $k_o$ is increased, $k_x$ is altered to maintain the critical speed constant and the curving performance is computed in terms of $\psi R/b$. As $k_o/k_y^*$ is increased from 0.0 to 2.0 the curving parameter $\psi R/b$ increases from 0.575 to 0.759 and $k_{pb}$ is reduced to 67% of its initial value and $k_{ps}$ is reduced to 68% of its initial value. In an additional study using two trucks and the full carbody steady-state curving model described in [6], the degree curve which is negotiated without meeting either slip or flange limits as $k_o/k_y^*$ is increased from 0.0 to 2.0 and critical speed held constant increases from 1.1° to 1.6°, reflecting the same trend as the isolated truck.

In the second case a radial truck has been considered with values of $k_x$ and $k_y$ one fifth the baseline values of the conventional truck and with $k_s$ and $k_b$ selected to yield values of $k_{ps}$ and $k_{pb}$ identical to the baseline conventional truck. With these selections of $k_s/k_y^* = 0.25$ and $k_b/k_y^* = 6.4$ the radial truck curving parameter $\psi R/b$ and critical speed $V_{cr}$ are equivalent to those of the conventional truck. For the radial truck two sets of data illustraing the influence of $k_o/k_y^*$ on performance are summarized in Fig. 5. In one set of data as $k_o/k_y^*$ is varied for fixed $k_x$ and $k_y$ the values of $k_s$ and $k_b$ are adjusted to maintain $k_{ps}$ and $k_{pb}$ constant yielding constant curving performance and the critical speed is calculated. These data show that

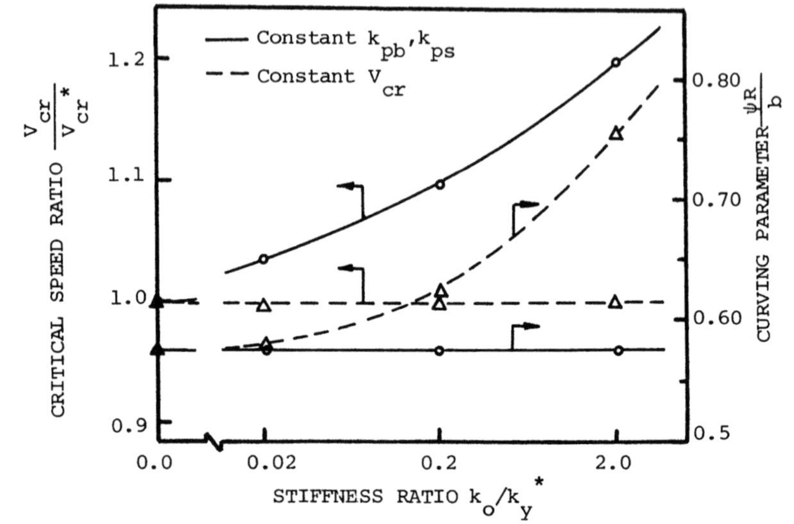


Figure 4. Influence of $k_o$ on the curving and stability of conventional truck.

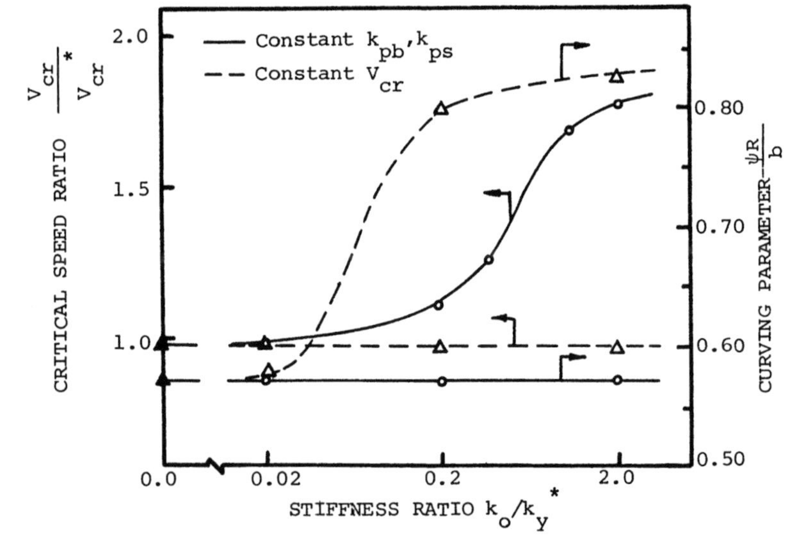


Figure 5. Effect of $k_o$ on the curving and stability of radial truck.

as $k_o/k_y^*$ increases from 0.0 to 2.0 the critical speed increases to 1.8 times the baseline value.

In the second set of data in Fig. 5 as $k_o/k_y^*$ is varied, $k_b$ is altered holding $k_x$, $k_y$ and $k_s$ constant to keep the critical speed constant and the curving performance is calculated. These data show that as $k_o/k_y^*$ increases from 0.0 to 2.0, the curving parameter $\psi R/b$ increases from 0.575 to 0.815 and $k_{pb}$ is reduced to 0.3 times its baseline value indicating an improvement in curving performance.

## 6.0 CONCLUSIONS

The analysis of a six D-O-F planar rail truck has shown that the constraints of linearity, passivity, elastic symmetry and rigid body motions lead to identification of a truck interconnection stiffness matrix with six independent stiffness parameters and one geometric parameter-the wheelset spacing. This interconnection stiffness matrix may be reduced to represent two stiffness parameter conventional and four stiffness parameter radial truck configurations.

The general conditions derived to represent steady-state negotiation of a constant radius curve have shown that wheelset lateral displacements and radial angular alignment are established by total interwheelset shear $k_{ps}$ and bending $k_{pb}$ stiffnesses which are functions of the general stiffness parameters. To achieve perfect steering two of the six stiffness parameters are required to be related by a constant $\beta$ to a third stiffness parameter.

An exploratory parametric study of a two parameter conventional truck augmented with the stiffness $k_o$ and a four parameter radial truck augmented with the stiffness $k_o$ has shown that as $k_o$ is increased either the critical speed may be increased while maintaining constant curving performance or curving performance may be improved while maintaining constant critical speed. These studies indicate that investigation using a full carbody model with two trucks is warranted to determine in detail the potential for rail vehicle performance improvement with the full six stiffness parameter truck.

## 7.0 REFERENCES


1. Law, E.H., and Cooperrider, N.K., "Literature Survey of Railway Vehicle Dynamics Research", Journal of Dynamic Systems, Measurements, and Control, June 1974.

2. Scheffel, H., "The Hunting Stability and Curving Stability of Railway Vehicles", Rail International, February 1974.

3. List, H.A., "Design System Approach to Problem Solving", 12th Annual Railroad Engineering Conference, 1975.

4. Wickens, A.H., "Steering and Dynamic Stability of Railway Vehicles", Vehicle System Dynamics, Volume 5, Number 1-2, 1975/76, pp.15-46.

5. Wickens, A.H., "Static and Dynamic Stability of a Class of Three-Axle Vehicles Possessing Perfect Steering", Vehicle System Dynamics, Volume 6, Number 1, 1977, pp. 1-20.

6. Hedrick, J.K., Wormley, D.N., Kar, A.K., Murray, W., Baum, W., "Performance Limits of Rail Passenger Vehicles: Evaluation and Optimization", Final Report, U.S. Department of Transportation, Report No. DOT-OS-70042, August 19, 1978.

7. Cox, J.J., Hedrick, J.K., Cooperrider, N.K., "Optimization of Rail Vehicle Operating Speed with Practical Constraints", Journal of Dynamic Systems, Measurements and Control, Vol. 100, No. 4, December 1978.

# MEASUREMENTS OF INTERIOR NOISE IN VEHICLES IN THE INFRASOUND AND HEARING RANGE

M. Lemke and H.-P. Willumeit

Institute of Automotive Engineering, Technische Universität Berlin, F.R.G.

SUMMARY

Beyond the hearing range from 16 Hz to 16 kHz influences on the human organism can occur, e.g. in the range below 16 Hz which is the infra-sound range. Interior noise in driving vehicles has been measured with the goal to find the spectra of the noise load on the passengers in the range from 0.4 Hz to the hearing range. Parameters were driving speed and the opening of the windows and the ventilation system. It turned out that 90 to 98% of the sound intensity was in the infra-sound range. A "resonant portion" in the frequency domain has been defined. Resonant portions higher than 10 dB have been found mostly at frequencies below 100 Hz.

The noise load on car passengers is generally measured by its RMS-value. It is restricted to the hearing range from 16 Hz to 16 kHz. For subjective rating and comparison of noises the "A-weighting" is added, a bandpass filter which is to simulate the empirical, frequency-dependent sensitivity of the human ear. However, beyond this range there can be some influence of sound on the human organism, e.g. below 20 Hz in the infrasound range. The results can be physiological reactions, and even illness. This has been established by investigations in skyscrapers with infrasound emitting air-condition. A review of the appearance of infrasound, the measurement techniques, and the influence on the human organism is found in GONO [1]. LEVENTHALL [3] established the performance loss by infrasound using a tracking task.

The laboratory simulation of the noise environment of car passengers could help to find physiological reactions due to infrasound. But it requires the analysis of the noise load in real driving situations as a first step. Investigations on this aspect are reported in this paper.

Two vehicles have been equipped for the investigations:
- Car I, 4 cylinder, 1.6 liter engine
- Car II, 6 cylinder, 3.0 liter engine

The equipment comprised a microphone (AM type, lower cut-off frequency 0,1 Hz), positioned in the head area of the front passenger, and a real-time frequency analyser. Due to the real-time analysis the results could be estimated and data reduction could be performed immediately during the tests. The frequency resolutions were 0.4 Hz for the infrasound range and 20 Hz for the hearing range, the frequency ranges were overlapping. The sound pressure levels in the diagrams were normalized to 1 Hz bandwidth so that the measurements from both frequency ranges could be compared directly. Test parameters were driving speed and different openings of windows and ventilation system; the tests were performed on the West-Berlin Autobahn which has no speed limit.

The multitude of the parameter combinations required a total of approx. 500 test runs. Therefore, only a few frequency analyses of the interior noise are shown in Fig. 1 ... 3 (the dark areas in the schematic vehicles show the opening of windows and sun roof).

The presentation of a larger number of results required further data reduction by calculation of RMS-values in certain frequency ranges from the frequency analyses. This was achieved by integrating the power spectral densities in the infrasound range from 0.4 Hz ... 20 Hz (yielding the sound pressure level $L_{PI}$). Integration over the whole range from 0.4 Hz ... 5000 Hz yielded the overall sound pressure level $L_{Pges}$ (the measurements were restricted to a frequency of 5000 Hz, since there was no significant sound pressure level above 5000 Hz compared with the range below that frequency). A schematic graph of this data reduction is shown in Fig. 4. To explain the portion of the infrasound, compared with the overall sound, an "intensity ratio" IV was defined. For reasons of clearness this energy ratio has a linear scale and not a dB-scale in the diagrams. Some of the figures show additional A-weighted sound pressure levels.

The influence of varied driving speed with closed windows and ventilation system is shown in Fig. 5, in Fig. 6 the front door window has different openings. The smaller car I and the bigger car II can be compared from Fig. 7. The open sunroof yields a steep increase of the sound pressure level from 50 to 100 km/h compared with the open front door window but no further significant increase from 100 to 150 km/h. Generally, the sound pressure

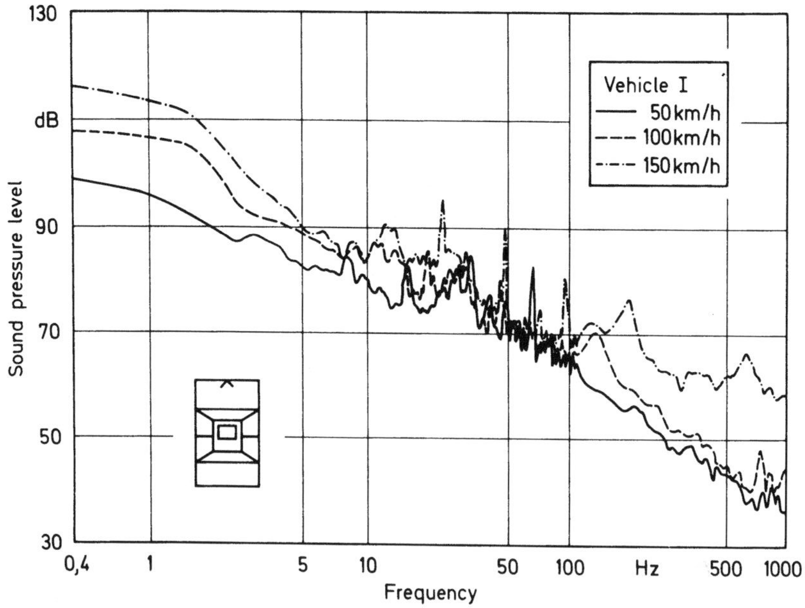


Fig. 1: Noise spectrum. Car I, ventilation closed (LO), window and sun roof closed.

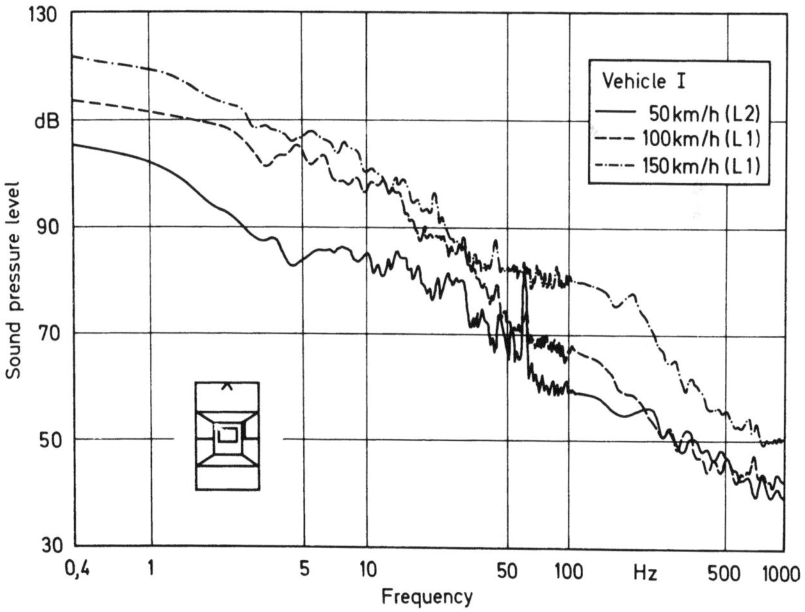


Fig. 2: Noise spectrum. Car I, ventilation (L1), fan on (L2), small opening of right window.

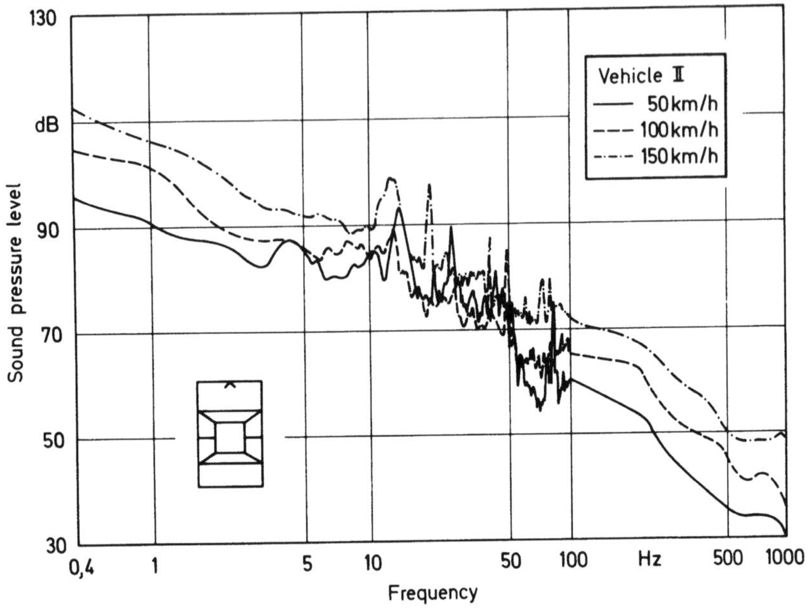


Fig. 3: Noise spectrum. Car II, ventilation and windows closed.

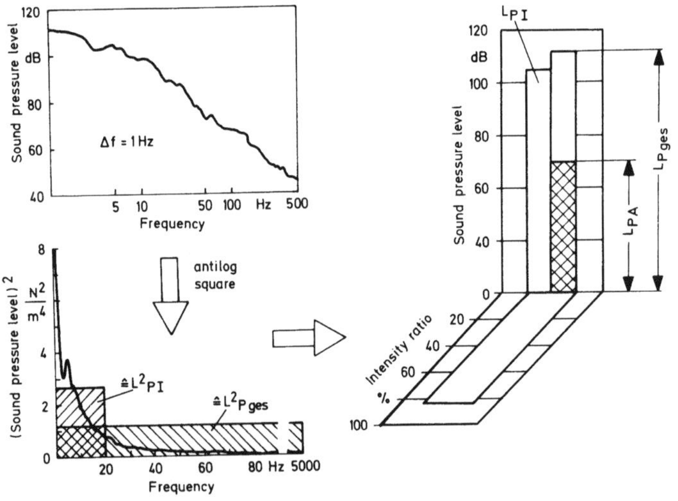


Fig. 4: Evaluation of infra-sound and over-all sound pressure level from the noise spectrum

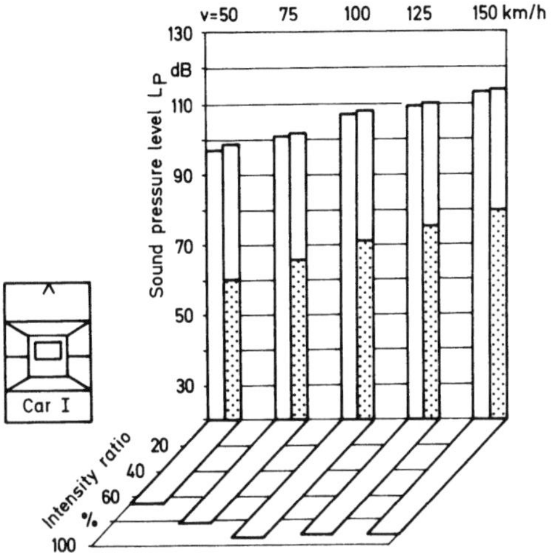


Fig. 5: Sound pressure level and intensity rate at variable speed (ventilation closed).

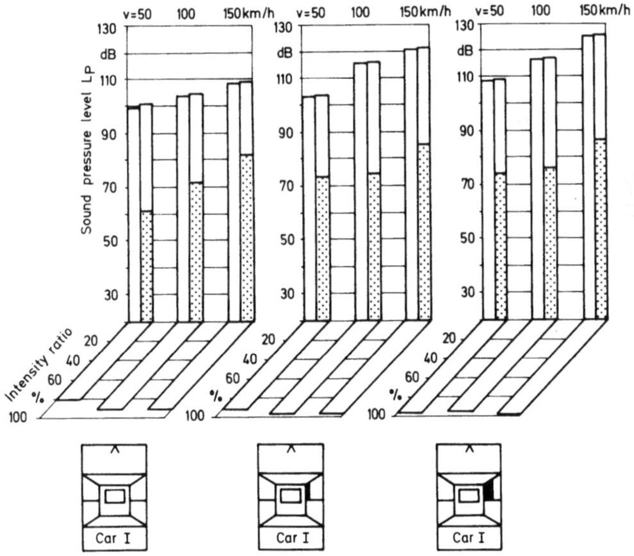


Fig. 6: Sound pressure level and intensity rate at gradually opened right window.

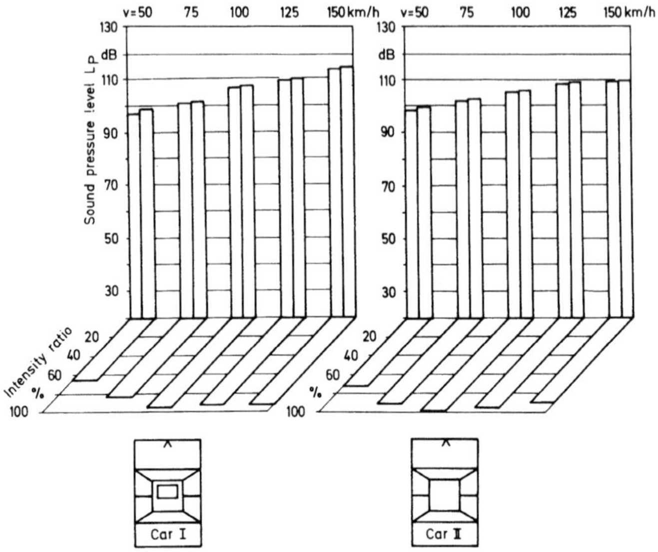


Fig. 7: Sound pressure level and intensity rate.
Comparison of both cars. Ventilation closed.

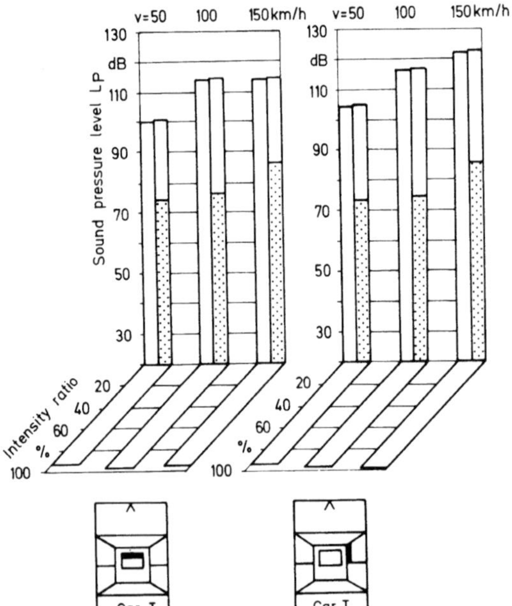


Fig. 8: Sound pressure level and intensity rate.
Comparison of open window and open sun roof.

levels of the open sun roof are below those of the open front door window (Fig. 8).

The comparison of the situations "closed ventilation" and "open ventilation" for two selected velocities of car I can be seen in Table 1: A higher infrasound portion is the result of the closed ventilation and even a higher absolute infrasound pressure level.

| Velocity / km/h | Ventilation | $L_{PI}$ / dB | IV / % |
|---|---|---|---|
| 100 | closed | 107,1 | 94,4 |
| | open | 103,6 | 90,4 |
| 150 | closed | 113,5 | 92,3 |
| | open | 108,5 | 89,7 |

Table 1: Infrasound pressure level ($L_{PI}$) and intensity ratio (IV) for closed and open ventilation (car I)

A statistical analysis of the data was performed to get an overall view of the results. It must be regarded, however, that the combination of the test parameters was not a random sample from a population with normal distribution. Therefore, the two vehicles can hardly be compared in this analysis. On the other hand, the results at various velocities for each vehicle can be examined, since the sets of test parameters are identical. Figures 9 and 10 show the sound pressure levels for the velocities v = 50, 100, 150 km/h and their absolute class frequencies in the infrasound and the overall range. The general impression is an increasing sound pressure level as a function of increasing velocity. At a velocity of 100 km/h the sound pressure levels form a narrow cluster, obviously independent of the opening area. This could be the indication of a resonance situation which is not or only little influenced by the opening of windows, sun roof, and ventilation. The histograms show outliers at low sound pressure levels which were measured with closed windows or only open sun roof at 100 and 150 km/h. The mathematical representation of sound pressure level versus velocity results in a potency regression (Fig. 11) having correlation coefficients from r = 0.86 to r = 0.92, the graphs of both vehicles being remarkably similar.

In another potency regression the total open area of windows and sun roof was the variable and the velocity the

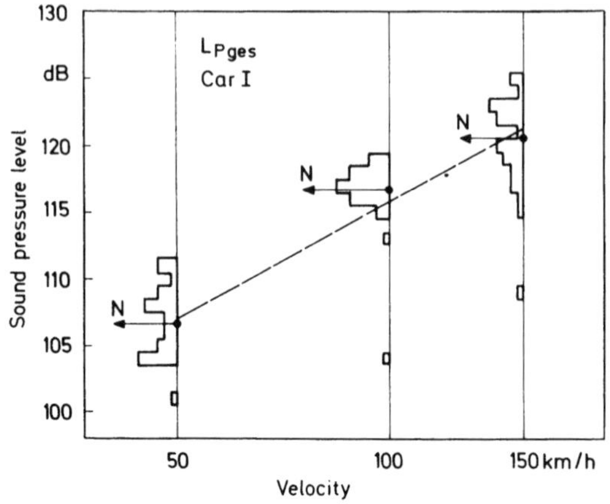


Fig. 9: Histogramm of over-all sound pressure level at variable speed. N is counting parameter for sound pressure level.

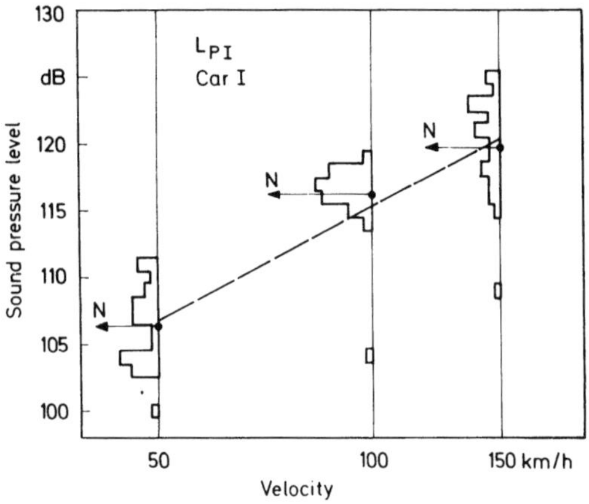


Fig. 10: Histogram of infra-sound pressure level at variable speed. N is counting parameter for sound pressure level.

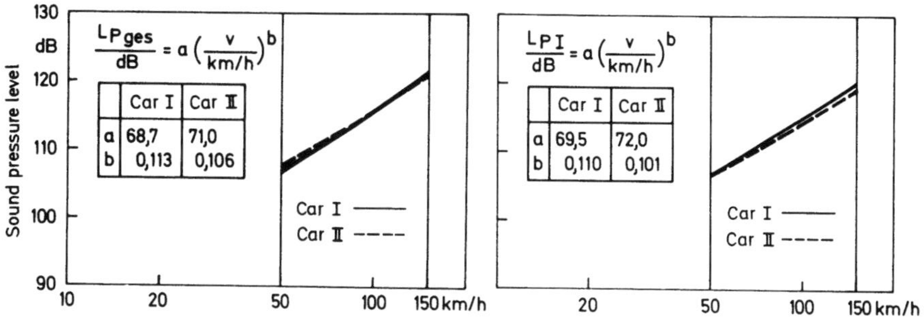


Fig. 11: Potency regression of the sound pressure levels for variable speed, evaluated from all measurements.

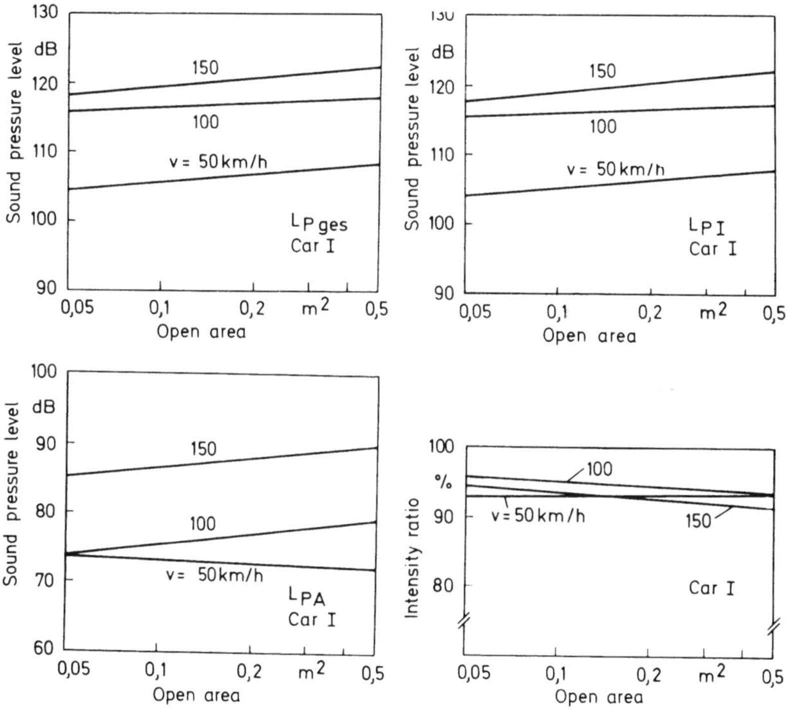


Fig. 12: Potency regression of the sound pressure levels $L_{Pges}$, $L_{PI}$, $L_{PA}$, and intensity rates for variable opening areas, evaluated from all measurements with car I.

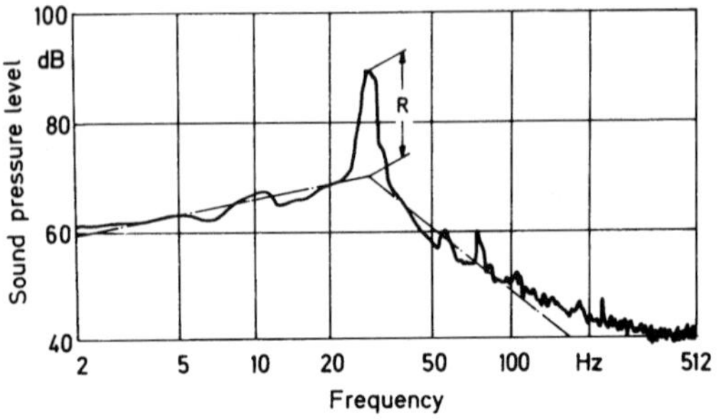


Fig. 13: Definition of the resonant portion R in the frequency domain.

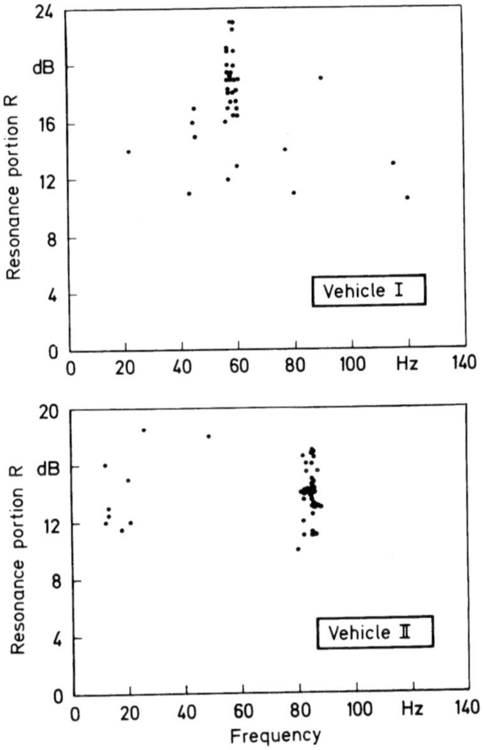


Fig. 14: Measured resonant portions in the frequency domain and resonant frequencies. Only resonant portions higher than 10 dB have been considered.

parameter. The results for car I are shown in Fig. 12. The sound pressure levels in the infrasound range and in the overall range are increasing with larger openings, the A-weighted sound pressure level, however, is decreasing at a velocity of v = 50 km/h. The graphs of the intensity ratio show a relative decrease of the infrasound energy for velocities v = 100 km/h and v = 150 km/h and a slight increase for v = 50 km/h as a function of increasing open window areas.

For subjective rating of broadband noise it is most important to consider resonances (discrete frequency peaks respectively). The VDI-draft 2574 (rating of interior noise of vehicles) comprises such a resonance portion. It does not, however, define discrete frequency peaks but the variation of the A-weighted sound pressure levels at various velocities and can therefore measure e.g. a general increase of the noise spectrum for all frequencies at certain velocities. In contrast to this draft a "resonant portion R" was defined in the frequency domain: It measures discrete (or narrowband) frequency components exceeding the sound pressure level spectrum which dominate the subjective impression of the noise as pure tones (Fig. 13). Only resonant portions exceeding 10 dB from all measured noise spectra were taken into consideration and only the highest resonant portion from each spectrum. Together with their respective frequencies they are shown in Fig. 14. As it turns out there is a concentration of resonances at vehicle specific frequencies (car I: 60 Hz, car II: 80 Hz) and the resonance portions of car I are higher than those of car II. In the virtual infrasound range only a few resonance peaks can be found; some resonance peaks in the 20 Hz range occur at the parameter combinations "all windows closed", "rear windows open", and some diagonal combinations.

The measurement of infrasound pressure levels in vehicles can, of course, only be the first step in the investigation of the infrasound problem. There is only little knowledge till now if there are any relevant physiological reactions due to these rather moderate sound pressure levels. If they do occur, the sources of infrasound in vehicles must be investigated to achieve improvement of the noise load on the passengers.

The interior noise is only one aspect of the problem, the other is the noise load on the vehicle's environment: windows in buildings seem to have natural frequencies not in the infrasound range but in the lower hearing range about 30 Hz. The noise spectra of standing vehicles with idling engines show high energy peaks in that range representing the second order vibrations of the engines. Thus, high sound pressure levels in the living rooms are the result which could possibly cause disturbances of the inhabitants.

The analysis of the interior noise of two passenger cars with open windows, sun roof, and ventilation yielded the following results:

- Despite the large number of experimental parameters and the considerable variability of the measured values a "smooth" relation between velocity and interior noise was found.
- There was no significant difference between the sound pressure level of both vehicles.
- Both vehicles had characteristic "resonance peaks" (frequency and level). The peaks were in the lower hearing range below 100 Hz.
- The energy ratio between infrasound and overall range comprised values from 90 ... 98%. The absolute level of the sound pressure in the infrasound range has average values of 105, 115, 120 dB for velocities of 50, 100, 150 km/h (normalized to a bandwidth of 1 Hz).
- Future investigations of physiological reactions due to infrasound could be limited in the number of experimental parameters. Thus, a larger number of vehicles could be examined with less effort, since some selected test parameters yield the whole range of results.

REFERENCES


1. GONO, F., Infraschall und seine Wirkung auf Menschen. Arbeitsmedizin, Sozialmedizin, Präventivmedizin 13 (1978) 7, 137 - 142.
2. HEISS, P., Schallfrequenzanalyse im fahrenden Personenkraftwagen bei geöffneten Fenstern und Lüftungsklappen. Diplomarbeit, TU Berlin, 1975.
3. LEVENTHALL, H.G., Man-made Infrasound Occurance and Some Subjective Effects. Colloques Internationaux du Centre National de la Recherche Scientifique, Paris 24 - 27 Sep. 1973, No. 232, 1974, 129 - 152.
4. TEMPEST, W., Loudness and Annoyance Due to Low Frequency Sound. Acustica 29, 1973, 205 - 209.

# PRELIMINARY INVESTIGATION OF A VARIABLE BRAKING VEHICLE

Stan Lukowski

Technical University, Wroclaw, Poland

SUMMARY

A computer study of variable braking vehicle was made in which the time required to apply and release the brake torque were considered. The equations of motion were derived in general terms for a simplified four-wheel vehicle and then solved on the computer for particular vehicle parameter values. The solution indicated that the application of variable braking torque which operated satisfactorily in the presence of relatively high or medium coefficient of friction could not operate satisfactorily at low coefficient of friction between the tire and the road. It was also found that stopping distances are not significantly dependent on the tire longitudinal stiffness.

## 1. INTRODUCTION

There is a need to investigate the problem of vehicle control during emergency conditions. A good starting point in attacking this problem is the commercially available braking control system. Such systems called anti-lock braking control systems are used in many recently designed vehicles. In general, operation of these systems is based on the fact that both the maximum lateral retaining force and the maximum frictional braking force between the tire and the road usually occur when the angular velocity of the wheel is greater than zero. The performance of the systems results in improved vehicle yaw stability and in stopping distances not greater than those obtained with the four wheels locked.

A particular vehicle braking control sytem which is sensitive to the angular velocity of wheels is considered. The basic principle of the system is as follows: The application of the brakes causes the angular velocity of the wheel to decrease. When imminent wheel lock is sensed, a self-adaptive control module signals to a vacuum powered brake actuator. The actuator causes the brake pressure to decrease which increases the angular velocity of the wheel, and thus the wheel lock is avoided. When the angular velocity has increased enough, the control module again signals the actuator and the brake pressure is then increased. This, in turn, causes the angular velocity to decrease, thus completing the control cycle. The cycle is repeated until the vehicle reaches a low speed. At a specific value of speed the system cuts off and the normal braking system takes over.

In this study dynamic modelling and simulation were employed to investigate the effectiveness of wheel angular velocity-sensitive braking torque application. In addition a different longitudinal tire stiffness was investigated.

## 2. THE VEHICLE MODEL

The simplified model used in this study is a two-axle vehicle considered in the longitudinal plane only /Fig. 1/.

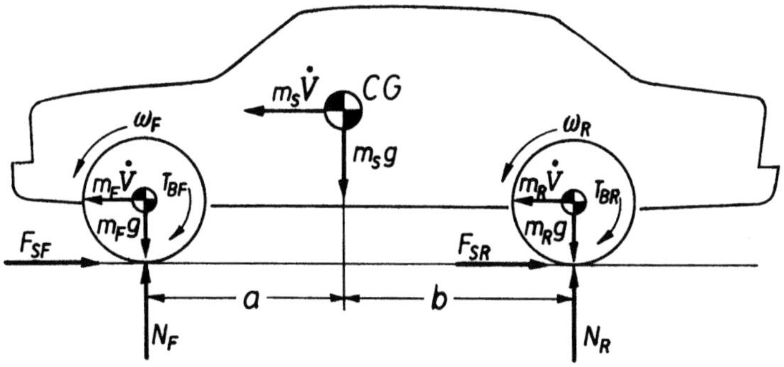


Fig. 1. Vehicle model diagram

The vehicle model results in three coupled first order differential equations; one equation for each of the state variables which are - the vehicle forward velocity, the front wheel rotational velocity, and the rear wheel rotational velocity. There are three control variables - the front wheel braking torque, the rear wheel braking torque, and the longitudinal tire stiffness. The model of the tire longitudinal friction force is taken from Reference 2. The values of all vehicle parameters involved in the simulation were selected to be representative of a medium-size passenger car.

## 3. VEHICLE MOTION EQUATIONS

The mathematical equations describing the behaviour of the vehicle under braking are as follows:

Vehicle forward motion

$$\frac{dV}{dt} = \frac{\sum F_{si}}{m} \tag{1}$$

Front wheel rotation

$$\frac{d\omega_{Fi}}{dt} = \frac{F_{SFi}\, R_{Fi}}{J_{Fi}} - \frac{T_{BFi}}{J_{Fi}} \tag{2}$$

Rear wheel rotation

$$\frac{d\omega_{Ri}}{dt} = \frac{F_{SRi}\ R_{Ri}}{J_{Ri}} - \frac{T_{BRi}}{J_{Ri}} \tag{3}$$

Road/tire friction force

$$F_{Si} = \frac{C_{Si}\ S_i}{1 - S_i}\ f\ (\lambda_i) \tag{4}$$

where:

$$f(\lambda_i) = 1 \qquad , \text{ when } \lambda_i \geqslant 1 \text{ (low slip)}$$

$$f(\lambda_i) = \lambda_i(2 - \lambda_i) \qquad , \text{ when } \lambda_i < 1 \text{ (high slip)}$$

$$\lambda_i = \frac{\mu_o\ N_i}{2C_{Si}\ S_i}\ (1 - A_S\ V_{Wi}\ S_i)(1 - S_i)$$

Tire normal load

$$N_i = K_{oi}(R_{ui} - R_i) \tag{5}$$

Rolling radius of front wheel

$$R_{Fi} = R_{ui} - \frac{g}{2K_{oi}} \left[ m_F + \frac{b}{a + b}\ m_S \right] \tag{6}$$

Rolling radius of rear wheels

$$R_{Ri} = R_{ui} - \frac{g}{2K_{oi}} \left[ m_R + \frac{a}{a + b}\ m_S \right] \tag{7}$$

Fron wheel % slip

$$S_{Fi} = 1 - \frac{\omega_{Fi}\ R_{Fi}}{V_{WFi}} \tag{8}$$

Rear wheel % slip

$$S_{Ri} = 1 - \frac{\omega_{Ri}\ R_{Ri}}{V_{WRi}} \tag{9}$$

With brakes and load adjusted symmetrically left and right, one obtains

$$F_{S1} = F_{S2} = F_{SF} \ , \quad F_{S3} = F_{S4} = F_{SR} \ , \quad N_1 = N_2 = N_F$$

$$N_3 = N_4 = N_R \ , \quad R_1 = R_2 = R_F \ , \quad R_3 = R_4 = R_R$$

$$S_1 = S_2 = S_F \ , \quad S_3 = S_4 = S_R \ , \text{ and } V_{W,F,R} = V$$

Vehicle stopping distance

$$V(t) = \int_{t_o}^{t} \dot{V}\, dt + V(t_o) \qquad (10)$$

$$X(t) = \int_{t_o}^{t_o} V\, dt + X(t_o) \qquad (11)$$

State variables: $V$, $\omega_F$, $\omega_R$. Control variables: $T_{BF}$, $T_{BR}$, $C_S$.

## 4. GENERATION OF BRAKE TORQUE

In this study the brake was considered to be torque limited and the limiting value is denoted as $T_{B,max}$. It was further assumed that, when brake torque was applied or released, it varied linearly with time. The brake torque-time variation used is illustrated schematically in Fig. 2. The time for torque to increase from zero to its limited value $T_{B,max}$ is called $t_1$ and the time to decrease from $T_{B,max}$ to zero called $t_2$. The values of $t_1$ and $t_2$ may be selected in order to simulate brakes having various response times.

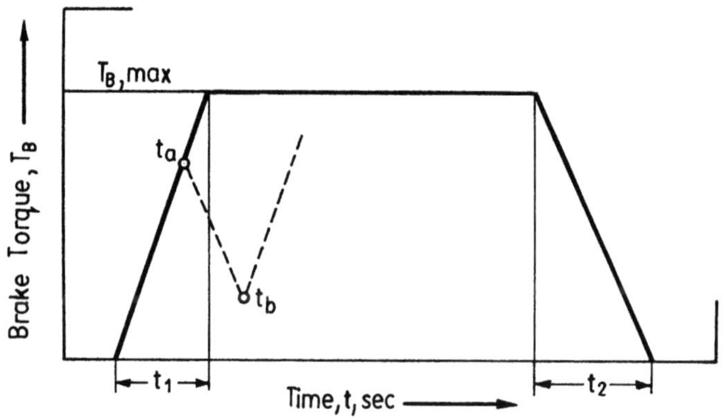


Fig. 2. Variation of brake torque with time used in computer braking study.

The computer was programmed either to permit the brakes to be released prior to reaching $T_{B,max}$ or to allow the brakes to be reapplied during the release interval before reaching zero torque. The input to the program was arranged to be inter-active so as to allow the wheel to be locked under certain operating conditions, i.e. if the response of the brakes and/or the control system was not fast enough to reduce the brake torque in order to prevent the wheel from being locked.

In order to eliminate prolonged periods of wheel locking, the computer was programmed to generate a brake release signal when the wheel angular velocity was less than 10 radians per second. Even so, some wheel locking did occur, and during that time the brake torque applied to the wheel was a function of the applied ground drag torque and the tire inertia. The locked-wheel brake torque $T_{B,L}$ is then

$$T_{B,Li} = F_{Si}R_i - \dot{\omega}_i J_i \tag{12}$$

When wheel locking occured, the computer was programmed to compute both $T_B$ and $T_{B,L}$ and to compare the two values. As long as $T_B$ was greater then $T_{B,L}$, the wheel remained locked and the brake torque used in the equations was $T_{B,L}$. When, however, $T_B$ became less than $T_{B,L}$, the value of $T_B$ was used in the equations and the wheel began turning again.

Studies were also made to compare stopping distances obtained with use of both variable and fixed braking torques.

## 5. SIMULATED RESULTS

The results obtained from the computer study of the variable braking vehicle are presented in Tables 1 to 4.

As the presented results have indicated, the stopping distances with the application of variable braking torque are considerably shorter as compared with stopping distances obtained by using fixed braking torque. The differences between these stopping distances can particularly be observed when the friction coefficient between the tire and the road is relatively high.

In the case of a low friction coefficient the stopping distances with the variable braking torque application are not shorter, but on the contrary, they are in some cases longer than those obtained with the use of fixed braking torque.

The results presented in table 3 have shown that brakes having different torque decay rates have no significant effect on stopping distances.

In addition, as can be seen from the results presented in Table 4, the tire longitudinal stiffness does not considerably affect stopping distances. Under some circumstances, however, the differences occuring in stopping distances with different tire longitudinal stiffness, may be of great importance.

TABLE 1

STRAIGHT LINE BRAKING SIMULATION

| $C_S$ = 90,000 N/unit slip | | | | |
|---|---|---|---|---|
| Initial Velocity km/h | Stopping Distance m | | | Remarks |
| | $\mu_o$ = 0.75 | $\mu_o$ = 0.5 | $\mu_o$ = 0.25 | |
| 16 | 1.38 | 2.15 | 4.75 | Variable braking torque |
| 32 | 5.65 | 7.65 | 17.76 | |
| 64 | 24.55 | 32.25 | 73.75 | |
| 96 | 55.45 | 88.75 | 178.35 | |

TABLE 2

STRAIGHT LINE BRAKING SIMULATION

| $C_S$ = 90,000 N/unit slip | | | | |
|---|---|---|---|---|
| Initial Velocity km/h | Stopping Distance m | | | Remarks |
| | $\mu_o$ = 0.75 | $\mu_o$ = 0.5 | $\mu_o$ = 0.25 | |
| 16 | 1.57 | 2.20 | 4.20 | Fixed braking torque |
| 32 | 6.70 | 8.69 | 17.28 | |
| 64 | 26.57 | 34.45 | 73.55 | |
| 96 | 59.16 | 92.35 | 179.70 | |

TABLE 3

STRAIGHT LINE BRAKING SIMULATION
INITIAL VELOCITY, ALL RUNS - 64 km/h

| $C_S$ = 90,000 N/unit slip, $\mu_o$ = 0.5 | | |
|---|---|---|
| Brake Torque Decay Rates sec | | Stopping Distance m |
| $t_1$ = 0.1 | $t_2$ = 0.1 | 32.25 |
| $t_1$ = 0.1 | $t_2$ = 0.5 | 31.64 |
| $t_1$ = 0.1 | $t_2$ = 0.75 | 32.75 |

TABLE 4

STRAIGHT LINE BRAKING SIMULATION

INITIAL VELOCITY, ALL RUNS - 64 km/h, $\mu_o = 0.5$

| Longitudinal Tire Stiffness $C_S$ N/unit slip | Stopping Distance m | Remarks |
|---|---|---|
| 23,000 | 38,15 | Fixed |
| 90,000 | 34,45 | braking torque |
| 184,000 | 36.54 | |

## 6. CONCLUDING REMARKS

A preliminary computer study has been made of the vehicle braking process in which the variable braking torque application was taken into account in developing the system equations of motion. These equations were derived in general terms for a simplified four-wheel vehicle and then solved on the computer for a particular case of vehicle parameters values. The brake simulation used was the torque limited, and the variation of brake torque with time during its application and release was assumed to be linear.

Studies were made of several brakes having a different variation of torque decay with time. The results have not shown a significant influence of a different variation of torque decay with time on stopping distances.

The results have also shown that the application of variable braking torque considerably improves vehicle performance in stopping distances under the conditions of a relatively high or medium coefficient of friction between the tire and the road. In the presence of a low coefficient of friction the use of variable braking torque gives negative results. It seems that in order to obtain the required effect of minimum stopping distance by using variable braking torque under the conditions of a low friction coefficient between the tire and the road it would be desirable to choose the appropriate value of wheels deceleration to generate a brake release signal.

Studies were also made of the effect of tire longitudinal stiffness on stopping distances. The results obtained indicate that stopping distances are only slightly affected by the tire longitudinal stiffness.

Generally speaking, the results of this preliminary study cannot, however, be considered as a precise description of the control system operation braking process since many factors affecting the operation of the braking system are not taken into consideration. In order to verify the results obtained further research is needed which would approach the factors affecting the vehicle braking process in a wider range.

NOTATIONS

| | |
|---|---|
| $m$ | total vehicle mass (kg) |
| $m_S$ | spring mass (kg) |
| $m_F$ | mass of front wheels (kg) |
| $m_R$ | mass of rear wheels (kg) |
| $a$ | distance of front axle from body CG (m) |
| $b$ | distance of rear axle from body CG (m) |
| $A_S$ | friction reduction factor |
| $C_S$ | longitudinal tire stiffness (N/unit slip) |
| $F_S$ | longitudinal tire friction force (N) |
| $g$ | gravitational acceleration ($m/s^2$) |
| $J_F$ | moment of inertia of front wheels ($kg.m^2$) |
| $J_R$ | moment of inertia of rear wheels ($kg.m^2$) |
| $K_o$ | tire spring rate (N/m) |
| $N$ | tire normal load (N) |
| $R_u$ | undeflected radius of wheels (m) |
| $R_F$ | rolling radius of front wheels (m) |
| $R_R$ | rolling radius of rear wheels (m) |
| $S$ | slip ratio of tire (%) |
| $T_{BF}$ | braking torque of front wheels (Nm) |
| $T_{BR}$ | braking torque of rear wheels (Nm) |
| $V$ | vehicle forward velocity (m/s) |
| $\omega_F$ | angular velocity of front wheels (rad/s) |
| $\omega_R$ | angular velocity of rear wheels (rad/s) |
| $\mu_o$ | nominal friction coefficient |
| $i$ | wheel number, 1 - left front, 2 - right front, 3 - left rear, 4 - right rear |

REFERENCES


1. Aurora B.R., Simulation of Vehicle Dynamic Braking Characteristics. The Journal of Automotive Engineering, September 1972.
2. Dugoff, H., Fancher P.S., and Segel, L., An Analysis of Tire Traction Properties and Their Influence on Vehicle Dynamic Performance Paper 700377 presented at SAE Mid-Year Meeting, Detroit, May 1970.
3. Lukowski, S., Simulation of the Dynamics of Automobile During Braking KONMOT'78 Conference, Krakow, 1978.

4. Segel, L., Murphy R.W., The Variable Braking Vehicle Concept and Design. Proceedings of the First International Conference on Vehicle Mechanics. Detroit, 1968.
5. Yim, B., Olsson, G.R., and Fielding, P.G., Highway Vehicle Stability in Braking Maneuvers. SAE Paper 700515.

# RESONANCE TESTING OF RAILWAY VEHICLES

Derek Lyon and Norman C. Remedios

British Rail Research and Development Division, Derby, United Kingdom

SUMMARY

Most papers dealing with resonance testing have concentrated on lightly damped modes of vibration. Resonance tests for rail vehicles must also measure the damped modes if the results are to be relevant to vehicle performance on the track. Recent developments in testing and analysis have made such tests possible but by no means simple. This paper presents a background to the techniques, briefly discusses the practical problems and presents some results of the tests carried out on the Power Car of British Rail's Advanced Passenger Train (APT).

## 1. INTRODUCTION

To achieve the objectives of low capital costs and low maintenance costs, the railway suspension designer must have a thorough understanding of vehicle dynamics combined with an ability to measure the performance of suspensions both on the track and under controlled laboratory conditions. In this paper it is intended to deal with those aspects of the latter termed resonance testing; that is, the excitation of the vehicle or vehicle structure with external forces so that its responses can be measured and analysed to yield the vehicle parameters.

The practice and development of resonance testing spans the past four decades. Although originally used in the aircraft industry it now has wide application in the transportation and power generation industries and in other fields where significant dynamic effects occur. Its development has been governed by advances in many associated technologies and reflects the general, rapid advance in the application of science. The earliest work had to employ rotating out-of-balance force generators and it was developments in electronics and servo-mechanisms leading to the widespread use of servo-hydraulic and electro-magnetic force generators which made possible the variety of testing techniques now available. These techniques also depend on accurate measurements of force, acceleration and displacement over a wide frequency range which in turn require precision measuring transducers and high performance electronic amplifiers. Finally, the introduction of digital computers has enabled large quantities of data to be collected and analysed in much greater depth with comparative ease.

All these factors have influenced the way tests are conducted. Because instrumentation was rudimentary and data handling non-existent, many of the early techniques concentrated, by the use of several force generators, on producing system responses which were easy to interpret and required little analysis. In some instances these responses were the normal modes of the system,which lead to a direct measure of natural frequency and damping. In other techniques completely artificial responses were introduced - such as the "characteristic modes" of Veubeke [1] - from which the system parameters could be determined. Considerable effort was involved at the experimental stage in positioning the force generators and adjusting their relative phases so as to induce the prescribed motions. It is important to grasp that this effort was necessitated by the limitations and deficiencies in the available measurement and analysis techniques.

With the development of instrumentation, it was possible to improve analysis techniques. The main advance came with the application of the method developed by Kennedy and Pancu [2] which recognises that the responses of lightly damped modes can be approximated by circular arcs in the complex plane. With strongly coupled modes, however, the inter-action can so distort the data as to make analysis extremely difficult and consequently multi-shaker techniques still offered many advantages particularly when combined with vector analysis methods.

Many papers have been published dealing with the Kennedy-Pancu technique [3, 4] and exploring its applications and limitations. Much of this detail has now been superseded by the introduction of digital analysis methods which enable curves of the expected theoretical shape to be fitted to the experimental data using non-linear regression techniques. The relevance of Kennedy-Pancu is now not so much the techniques themselves but rather the insight they provide which enable the computer methods to be used in a sensible way.

There already exists a considerable body of literature dealing with the testing and analysis of fairly lightly damped structural systems. Consequently, although tests are conducted at British Rail to measure the structural modes of vehicle bodies, this paper will concentrate exclusively on those aspects which are peculiar to rail vehicle testing; namely low frequency rigid body modes having fairly heavy damping and with significant damper series stiffness.

## 2. A BROAD OUTLINE OF THE SUBJECT

The mathematical model used to study the behaviour of railway vehicles will take the form of the matrix equation

$$[M]\{\ddot{q}\} + [C]\{\dot{q}\} + [K]\{q\} = \{P(t)\} \qquad (1)$$

where $\{q\}$ is a column vector of n generalised co-ordinates and $[M]$, $[C]$, $[K]$ are square matrices of order n which represent the "inertia, damping and stiffness" of the vehicle, respectively.

The only restrictions placed on the system matrices are that they are real with constant coefficients and that the roots of the characteristic equation have negative real parts. The former requirement ensures that we are dealing with linear systems while the latter merely guarantees that the responses are bounded - a prudent requirement for any laboratory test. There is no requirement for the matrices to be symmetric so the methods would be applicable to the resonance testing of a vehicle running on a roller rig where the wheelset contact forces give rise to asymmetric terms in the stiffness matrix. A departure from similar systems of equations usually presented for structural analysis, is that the mass matrix will frequently be either singular or ill-conditioned because some suspension elements take the form of a series-spring and dash-pot with a **zero or small** mass at the junction.

Equation 1. can always be expressed as an equivalent set of first order equations, even when $[M]$ is singular.

$$[I_m]\,\{\dot{z}\} + [U]\,\{z\} = \{\Omega(t)\} \tag{2}$$

The relationships between $[U]$, $\{\Omega\}$ and $[M]$, $[C]$, $[K]$, $\{P\}$ and also between $\{q\}$ and $\{z\}$ can be established easily and will be omitted; $[I_m]$ is an mxm unit matrix.

A receptance $\alpha_{rs}(i\omega)$ is defined in the usual way, as the steady state response of the r'th generalised co-ordinate $q_r$ to a harmonic force applied at the s'th co-ordinate.

It is shown in the Appendix that

$$\alpha_{rs}(i\omega) = \sum_{j=1}^{m} \frac{A_j}{i\omega - \lambda_j}$$

where $A_j$ are constants determined from the eigenvectors and $\lambda_j$ are the roots (eigenvalues) of the system.

The roots will often occur in complex conjugate pairs which can be combined to give

$$\alpha_{rs}(i\omega) = \sum_{j=1}^{m-2N} \frac{A_j}{i\omega - \lambda_j} + \sum_{r=1}^{N} \frac{B_r.iw + G_r}{(-\omega^2 + 2\zeta_r.\omega_r.i\omega + \omega_r^2)}$$

N is the number of complex conjugate roots $\lambda_r$ ; $\lambda^*_r$

$\omega_r^2 = \lambda_r . \lambda^*_r$ ; $-(\lambda_r + \lambda^*_r) = 2\zeta_r \omega_r$

and $\omega_r$ ; $\zeta_r$ are the undamped natural frequency and damping ratio of the r'th resonant mode.

Each term of this series closely resembles the receptance of either a first or second order single-degree-of-freedom system. All the analysis techniques make use of this property but before dealing with these it is appropriate to discuss the actual test methods used to acquire the receptance data.

A "direct" laboratory resonance test consists of applying at each frequency in turn, sinusoidal forces which induce a steady state harmonic response. System responses are measured at various locations and the receptances are calculated directly from the amplitude and phase relationships between responses and forces. Testing must be carried out over the frequency range of interest and the spacing of the test frequencies must be sufficiently close to define the receptances adequately, particularly in those regions around the natural frequencies.

With the advent of digital techniques in general and the development of the Fast Fourier Transform algorithm in particular, it has become possible to determine system receptances by what can be termed "indirect" methods. Arbitrary force histories such as impulses, swept sine-waves, random and pseudo-random processes can be employed and the receptances determined by manipulation of the Fourier Transforms of the applied force histories and induced responses [5].

In practice the choice of test technique will depend upon many factors including the type of force generation equipment available, the time available for the test, the instrumentation and analysis facilities and also the type of vehicle to be tested. The latter is often a major consideration because although the theories are derived for linear systems, all vehicles exhibit some degree of non-linearity. This may take the form of non-linear viscous damping characteristics, clearances, non-linear springs or it may be due to presence of dry friction.

Having determined experimentally the complex receptances by the most appropriate method, the final stage is the analysis of this data to yield the system parameters. Since the coefficients in the series form of the receptance are related to the eigenvalues and eigenvectors of the system, an accurate determination of these coefficients will - in theory at least - provide a complete description of the system.

Experimental receptances can be classified on the basis of their constituent parts.

i. All terms second order with real numerators.

ii. Some terms first order, second order terms with real numerators.

iii. All terms second order but with complex numerators.

iv. Some first order terms, second order terms with complex numerators.

When all the roots are complex conjugates then all the modes will have less than critical damping, that is they will be resonant to some degree and all the terms of the receptance series will be of second order. The significance of real numerators is that the modal vectors must be real and identical with the modes of the undamped system. This occurs when the damping matrix is formed from linear combinations of the mass and stiffness matrices. In practice this will rarely be exactly so but it may be a close approximation in which case the number of coefficients is halved and it follows that the receptance vector will always be parallel to the imaginary axis at the undamped natural frequency of the mode in question.

For each of the above classes of receptance one can define three sub-categories:-

A. Each term of the series is distinct. This requires the frequency separation of the modes to be large compared with the band width of the modes. Additionally, the resonant response of each mode must be large when compared with the sum of all the other off-resonant responses. Under these conditions analysis reduces to the determination of the characteristics of a number of virtually independent single-degree-of-freedom systems (that is providing the data is of class i or iii).

B. The modes are well separated in frequency but for some modes the off-resonant response is of the same order or larger than the resonant response. Although the terms must be added vectorially, the separation of the modes ensures that the off-resonant contribution is essentially constant over the bandwidth of the resonant mode. Under these conditions the vector plot resembles that of a single-degree-of-freedom system with a displaced origin.

C. The modal separation is less than the modal band width and two or more modes interact strongly to make analysis extremely difficult.

## 3. ANALYSIS PROGRAMMES

The non-linear regression analysis to extract modal information from complex receptance plots uses programs developed at RAE Farnborough [6]. The latest version is self-starting, can be used when combinations of first and second order terms are present and will successfully analyse closely interacting modes even with quite large amounts of damping. For each resonant mode analysed, the programme calculates the mean values of the four coefficients of a second order term, and the 95% confidence limits. A response synthesised from the calculated data is also plotted to enable visual comparisons to be made with the original experimental data. A number of similar analysis programs are commercially available and without programs of this type none of the analysis described in this paper would have been possible.

## 4. PRACTICAL CONSIDERATIONS

Several factors such as damper series stiffness, friction and number of actuators have been discussed during the presentation of experimental results (Sec. 5 & 6).

### *Limitations of servo-hydraulic equipment*

Two types of control are available on proprietary equipment, one uses actuator displacement, the other actuator forces. Because a servo-valve generates a flow proportional to the applied current, the characteristics of the vehicle being tested enter into the control equations. Also, because the vehicle body masses are large and the actuators have fairly long strokes to enable a variety of tests to be performed, oil compressibility and its associated resonance cannot be neglected.

Summarized this means that with displacement feedback for body resonance tests, low gain must be used to ensure stability of the oil column resonance. True displacement control is only possible at low frequencies and the system operates as an open loop velocity system over much of the frequency range. With force feedback, effective control is possible at frequencies away from the resonant frequencies of the vehicle. Around resonance however, the closed loop behaviour is very dependant on the vehicle damping and for light damping the control system becomes virtually an open loop system.

### *Choice of Test Technique*

Providing care is taken it should be possible to measure the parameters of a vehicle with any of the available techniques. The choice depends mainly on practical considerations which to some extent depend upon the equipment available.

### *Sinusoidal Testing*

This provides a simple and effective means of measuring receptances with the minimum of post-test processing of results. One disadvantage is that with the equipment available, testing is slow - half a day may be required for a single test over the frequency range 0.1 Hz to 3 Hz. However, it is possible with this type of testing to overcome the deficiencies of the servo-control system by using external iterative control at each test frequency.

### *Random Testing*

Testing times are very much shorter because response channels can be recorded in parallel. Extensive manipulation of the data is required to derive the receptances. The main advantage is that the vehicle can be subjected to amplitudes and frequencies which resemble those experienced on the track and in the case of non-linearities the best linear transfer function - in the least squares sense - can be derived.

## 5. TESTS ON A PANTOGRAPH ANTI-TILT MECHANISM

Electric trains which tilt, such as the Advanced Passenger Train (APT) require a mechanism to maintain the pantograph in the correct relationship to the overhead wire. On APT this is a passive mechanism which is heavily damped and provides an ideal illustration of the effect of damper series stiffness on a mode of vibration.

The mechanism is illustrated in (Fig 1) where it is shown mounted on the Power Car body. The pantograph assembly (A) is supported on two frames which form an inverted "pendulum" (B). The pantograph is maintained parallel to the track by the torsion bar (C) and drive arms (D). Damping is provided by a pair of viscous dampers (E). For resonance test purposes the mechanism was attached to a concrete plinth and excited using a servo-hydraulic actuator. Sinusoidal excitation was used and tests were carried out both with and without damping.

*Undamped Behaviour*

With dampers removed the mechanism behaves as a lightly damped single-degree-of-freedom system with the pantograph assembly oscillating on the torsion bar. Analysis of the measured receptances gave values of 1.9 Hz for the undamped natural frequency and 6% damping. The latter arises mainly from the losses in the spherical bearings of the supporting frames and the drive arms. The modal mass and stiffness were calculated to be 240 kg $m^2$ and 35 KN.m/rad.

*Damped Behaviour*

The dampers were connected and the system was re-tested. The measured receptance is shown in (Fig.2) and is clearly not that of a simple heavily damped system. The data was analysed in three separate ways:-

(a) A second order transfer function was fitted to the data in the resonant region.

(b) The data was transformed to velocity/force data as shown in (Fig.3) and the analysis repeated.

(c) The data was analysed as a combination of first and second order systems.

The derived values of natural frequency and damping are listed below for the various methods of analysis and compared with the values to be expected if an infinite series stiffness and the design rate of the damper were to be assumed.

FIG. 1

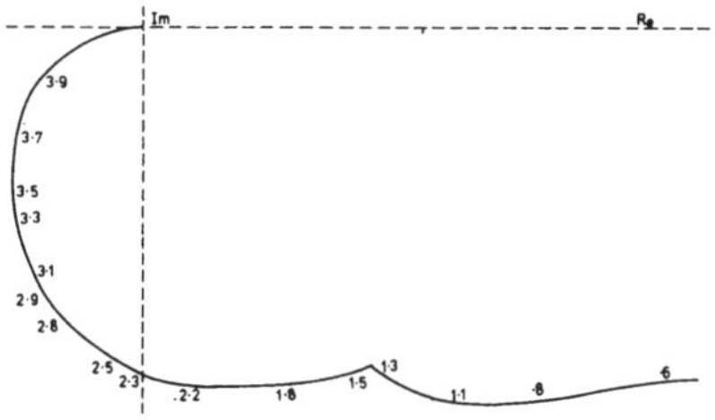

FIG. 2. DISPLACEMENT RECEPTANCE. DAMPERS FITTED.

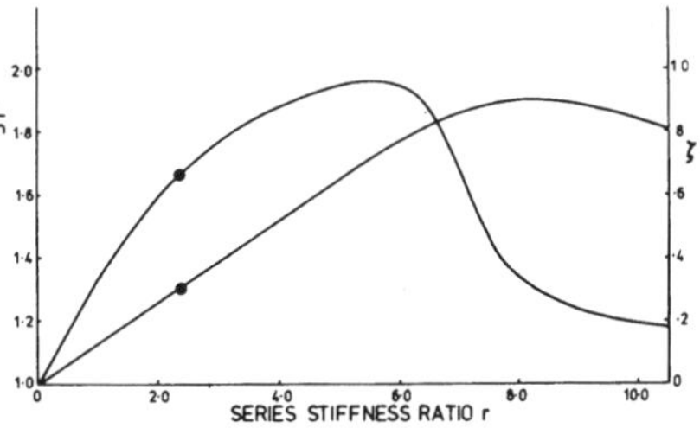


FIG. 4.

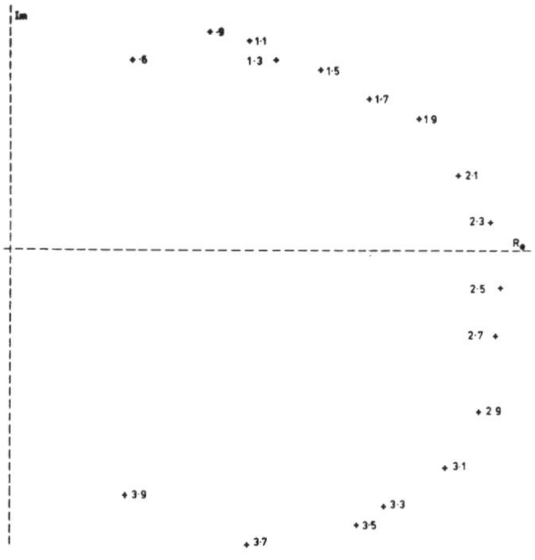


FIG. 3. VELOCITY RECEPTANCE. DAMPERS FITTED.

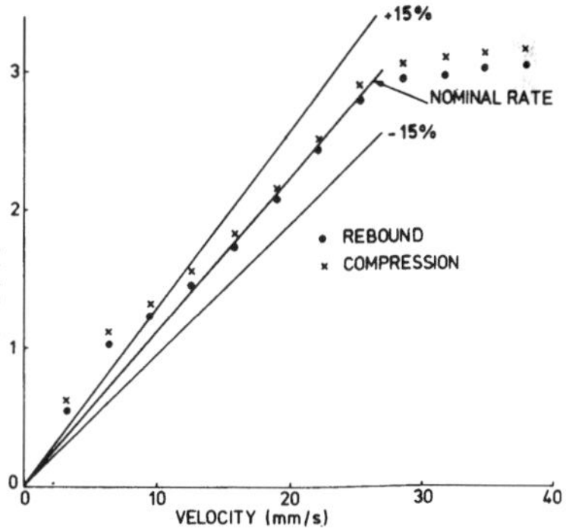


FIG. 5. DAMPER CHARACTERISTICS.

| | Frequency | Damping |
|---|---|---|
| Method (a) | 3.2 Hz | 48% |
| Method (b) & (c) | 3.15 Hz | 30% |
| Calculated, infinite series stiffness | 1.9 Hz | 75% |

The system is clearly not accurately represented by the simple model but if a damper series spring is included it can be shown that the transfer function takes the form

$$\frac{x(\bar{S})}{x_{St}(\bar{S})} = \frac{A}{\bar{S} + \frac{1}{\bar{\omega}_n^{\,2}} \cdot \frac{r}{2a}} + \frac{B.\bar{S} + C}{\bar{S}^2 + 2\zeta\bar{\omega}_n\bar{S} + \bar{\omega}_n^{\,2}}$$

$$\text{with } \omega_o^{\,2} = \frac{k}{m}\ ;\ 2a.\omega_o = \frac{c}{m}\ ;\ \bar{S} = \frac{i\omega}{\omega_o}\ ;\ \bar{\omega}_n = \frac{\omega_n}{\omega_o}\ ;\ x_{St} = \frac{P}{k}$$

m,c,k are the mass, damping and stiffness of the system, rk is the damper series spring stiffness and P the applied force.

For a particular value of a (the ratio of critical damping for $r = \infty$ ) it is possible to plot the variation of $\bar{\omega}_n$ and $\zeta$ of the second order term as a function the series stiffness ratio r. An example, approximately that of the anti-tilt mechanism, is shown in (Fig.4). At extreme values of series stiffness, the undamped natural frequency of the second order term tends to the undamped natural frequency of the simple spring-mass system while for intermediate values the natural frequency of the second order term can be as much as twice this value. The values of $\bar{\omega}_n$ and $\zeta$ from the table of results above yield a stiffness ratio r= 2.4 and a damper rate of 134 KN.s/m.

Shown in (Fig.5), are the nominal design and measured rates of the damper in question. It can be seen that the measured rate at low velocities (typical of those used for the test) is approximately 130 KN.s/m compared with the nominal design rate of 112 KN.s/m.

Because the system is very simple it has been possible to explain the experimental results fully. In many cases with complex systems such agreement is not always possible. The results show that modal data can be influenced significantly by both the series stiffness of the damper and the non-linear characteristics of a damper at low velocities.

## 6. TESTS ON THE APT POWER CAR

The bogies of the APT Power Cars are representative of modern designs for high performance vehicles. Both the stiffness and damping elements of the suspension are designed, as far as possible, to have linear values within the normal working range. The lateral rigid body tests serve to illustrate the problems encountered and the methods which can be employed. A three dimensional arrangement of the vehicle suspension is shown in (Fig.6).

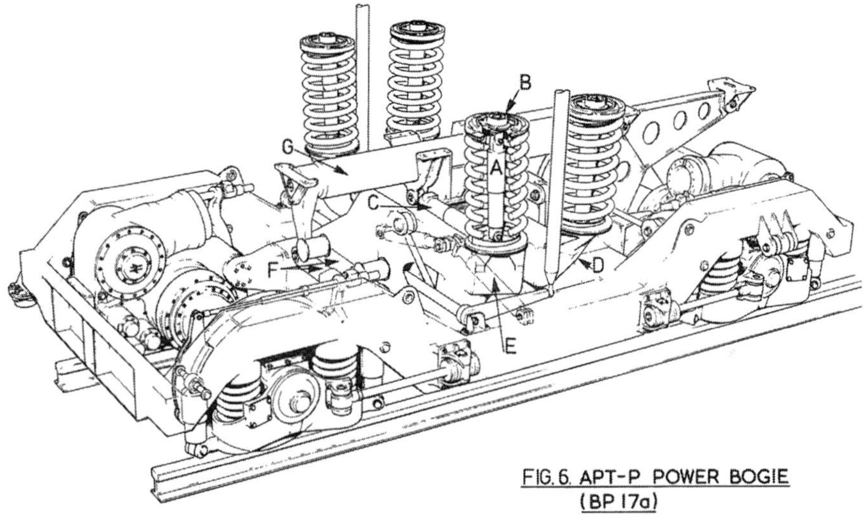


FIG.6. APT-P POWER BOGIE
(BP 17a)

The principal difficulties experienced during testing are due to :-

(a) The limited number of positions on the vehicle body at which hydraulic actuators can be attached.

(b) The need to measure the damped modes of the vehicle, and the coupling which exists between modes.

(c) The non-linearities, in this case principally friction when the dampers are disconnected. For other vehicle types, non-linearity might also be present in the form of clearances and non-linear stiffnesses.

(d) The effects of damper series stiffness which for the vehicle in question arise in a number of distinct ways and are worth elaborating.

(i) The vertical dampers marked (A) (Fig.6) are designed to have a soft series rubber spring (B).

(ii) The lateral dampers (C) act between the tilting bolster (D) and the body. In this case the series stiffness is the oil compressibility of the tilt jack (E) and associated pipework.

TABLE I

| | | ONE ACTUATOR | | TWO ACTUATORS IN-PHASE | | TWO ACTUATORS ANTI-PHASE | |
|---|---|---|---|---|---|---|---|
| | | fo (Hz) | % Damping | fo (Hz) | % Damping | fo (Hz) | % Damping |
| ALL DAMPERS DISCONNECTED | Lower Sway | 0.79 ±0.02 | 4.5 ±0.4 | 0.80 ±0.02 | 5.0 ±0.4 | — | — |
| | Upper Sway | 1.47 ±0.06 | 5.7 ±0.3 | 1.49 ±0.05 | 6.0 ±0.2 | — | — |
| | Yaw | 1.13 ±0.02 | 5.0 ±0.4 | — | — | 1.13 ±0.01 | 5.0 ±0.3 |
| YAW DAMPERS CONNECTED | Lower Sway | 0.80 ±0.03 | 8.1 ±0.4 | 0.79 ±0.02 | 8.7 ± .6 | — | — |
| | Upper Sway | 1.31 ±0.05 | 4.0 ±1 | 1.48 ±0.03 | 13.0 ±2 | — | — |
| | Yaw | 1.27 ±0.02 | 23.0 ±5 | — | — | 1.25 ±.03 | 39.0 ±2 |
| YAW + VERTICAL DAMPERS CONNECTED | Lower Sway | 0.84 ±.02 | 20.0 ±2 | 0.81 ±0.03 | 17.0 ±1 | — | — |
| | Upper Sway | 1.32 ±.03 | 9.7 ±2 | 1.61 ±0.04 | 25.0 ±1.7 | — | — |
| | Yaw | 1.42 ±.07 | 32.0 ±5 | — | — | 1.4 ±.08 | 47.0 ±3 |
| ALL DAMPERS CONNECTED | Lower Sway | 0.90 ±.10 | 26.0 ±7 | 0.82 ±0.04 | 23.0 ±1 | — | — |
| | Upper Sway | — | — | 1.66 ±0.03 | 43.0 ±4 | — | — |
| | Yaw | 1.51 ±.4 | 47.7 ±12 | — | — | 1.39 ±.04 | 82.0 ±6 |

N.B. The lower sway mode consists of in-phase body lateral displacement and roll.

The upper sway mode consists of out-of-phase body lateral displacement and roll.

(iii) The high rate yaw dampers (F) act in series with the torsion bar stiffness (G) to facilitate low frequency rotation of the bogie in curves but stiff restraint at bogie kinematic frequencies.

There is also the general problem of devising a test which will yield results which are amenable to analysis, provide data which is relevant to the performance of the vehicle on track, enables comparison to be made with design calculations and illuminates suspension performance and possible deficiencies.

*Tests using Random Excitation*

For the results to be presented random testing was employed using two actuators mounted one at each end of the vehicle; the tests were also repeated with one actuator disconnected. The vehicle was tested with various configurations of suspension.

(i) All dampers removed.

(ii) Secondary yaw damping only.

(iii) Secondary yaw and secondary vertical damping.

(iv) All dampers connected.

The results for the four configurations have been summarised in Table I for tests with both a single actuator and two actuators. Mean values of the undamped natural frequencies and damping ratios are presented and the 95% confidence limits are shown.

*Discussion of results : undamped vehicle*

The modal behaviour was distinct and similar results were obtained from all the tests. A receptance plot from a single actuator shows all the modes (Fig. 7) whereas with two actuators in-phase, only the sway modes are present (Fig. 8). The separation of modes using two actuators obviously makes analysis easier but in this case the analysis of single actuator results was well within the capabilities of the programs. Even with a single actuator, adding or subtracting response channels can be employed to suppress some modes and enhance other modes as an aid to analysis.

*Secondary yaw dampers connected.*

For two actuator anti-phase excitation, the yaw frequency is increased to 1.25 Hz and the damping to 39% when the secondary yaw damping is added. The damper acts in series with a torsion bar and the results show the expected increase in undamped frequency for a system of this type.

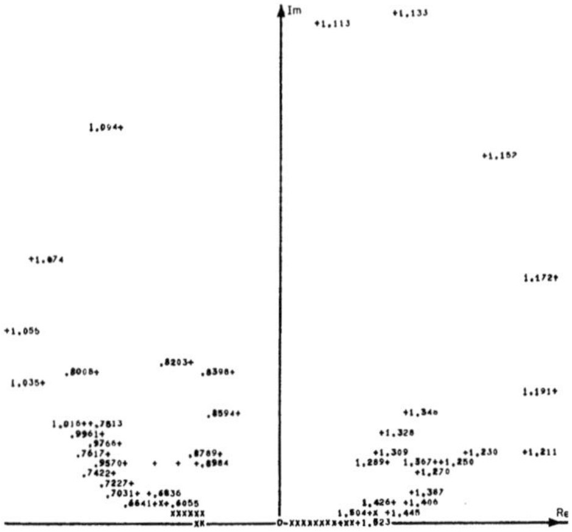


FIG. 7 ONE ACTUATOR. NO DAMPERS.

FIG. 8. TWO ACTUATORS. NO DAMPERS.

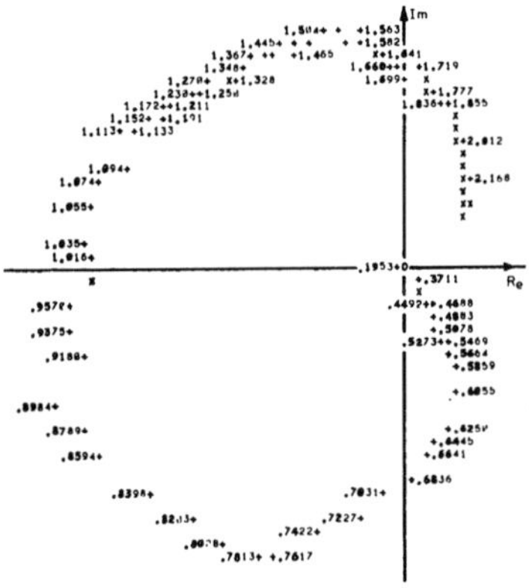


FIG. 9. TWO ACTUATORS IN-PHASE
ALL DAMPERS.

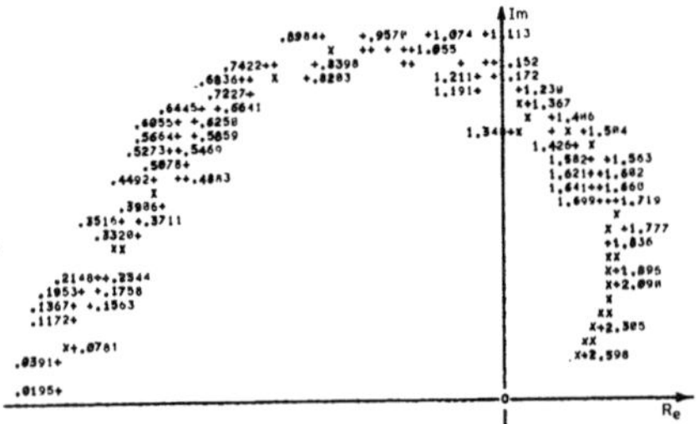


FIG. 10. TWO ACTUATORS OUT-OF-PHASE
ALL DAMPERS.

The addition of these dampers should not affect the sway frequencies, however from the two actuator in-phase tests it is apparent that although the undamped frequency is unchanged the damping has increased significantly. This can be explained by the presence of some 16 spherical bearings in the secondary yaw restraints of both bogies. These bearings are PTFE - glass fibre which when interference fitted have high friction break-out torques during their initial lives.

The effect of this friction is to make testing and analysis with a single actuator extremely difficult. At high levels of excitation, the results tend to those for two actuators but as the excitation level is reduced the modes become strongly coupled and the frequencies change accordingly. Friction is often linearised as an amplitude dependent viscous damper. Tests with a single actuator produce dissimilar amplitudes at the two bogies and consequently dissimilar damping rates. This asymmetry in the suspension, couples the yaw and sway modes and in the extreme, at low excitation levels, the body tends almost to pivot about the bogie at the end remote from the actuator.

*Secondary yaw and vertical dampers connected, and all dampers fitted.*

Only the results from the two actuator tests yield realistic results with the expected variations in frequencies and dampings. The same applies to the results for the case with all the dampers connected. Receptances for the fully damped condition, two actuator excitation with in-phase and out-of-phase excitation are shown in (Fig 9,10).

APPENDIX

Consider a set of m first order equiations

$$[I_m]\{\dot{Z}\} + [U]\{Z\} = \{\Omega(t)\}$$

$[I_m]$ is an mxm unit matrix

By application of Laplace transform theory, the solution in terms of transformed variables can be written

$$\{Z(s)\} = ([I_m]\, s + [U])^{-1}.\ (\{Zo\} + \{\Omega(s)\})$$

where $\{Zo\}$ is a vector of initial conditions and $\{\Omega(s)\}$ a vector of transformed forces.

The matrix $([I_m]\, s + [U])^{-1}$ defines the properties of the system and for steady state harmonic excitation can be considered as the system receptance matrix $\{Z(i\omega)\} = ([I_m]\, i\omega + [U])^{-1}\{\Omega(i\omega)\}$

It will now be shown that the receptance matrix can be expressed in series form and that the coefficients of the series are determined by the eigenvalues and eigenvectors of the system.

Let $[f(s)] = ([I_m]s + [U])$ then $[f(s)]^{-1} = [F(s)] / \Delta(s)$
$[F(s)]$ is the adjoint of $[f(s)]$ and $\Delta(s)$ is the determinant.

If the roots of $[f(s)]$ are distinct then $\Delta(s) = (s - \lambda_1)\ (s - \lambda_2) \ldots \ldots (s - \lambda_m)$ and $[f(s)]^{-1}$ can be expanded as
$[F(s)] / \Delta(s) = [A_1] / (s - \lambda_1) + [A_2]/(s - \lambda_2) + \ldots [A_m] (s - \lambda_m)$

A typical term can be evaluated as

$$[A_r] = \left( \frac{[F(s)]\ (s - \lambda_r)}{\Delta(s)} \right)_{s = \lambda_r} = F[(\lambda_r)] / \Delta^{(1)}(\lambda_r).$$

Giving $$[f(s)]^{-1} = \sum_{r=1}^{m} \frac{[F(\lambda_r)]}{\Delta^{(1)}(\lambda_r).(s - \lambda_r)}$$

It can be shown, see chapter III [7], that when the roots are distinct $[f(\lambda_r)]$ is simply degenerate. The adjoint $[F(\lambda_r)]$ therefore has unit rank and can be expressed as the product

$$[F(\lambda_r)] = \{\Phi_r\} \lfloor L_r \rfloor$$

Also from the matrix properties

$[f(s)]\ [F(s)] = [F(s)]\,[f(s)] = [I]\,\Delta(s)$ it can be shown that the

scalar quantity $\Delta^{(1)}(\lambda_r) = \lfloor L_r \rfloor \{\Phi_r\}$

Finally, it can be shown that $\{\Phi_r\}$ ; $\lfloor L_r \rfloor$ are the eigenvectors associated with the root $\lambda_r$.

Other derivations can be found which start by assuming solutions. Thus $\Delta^{(1)}(\lambda_r)$ corresponds to the "Generalised Coefficient" of Halfman [8] and since the scaling is arbitrary, it is used by Wahed & Bishop [9] to "normalise" the modes by choosing $\lfloor L_r \rfloor \{\Phi_r\} = 1$

For steady state response, the receptance matrix can be expressed in series form

$$([I_m]\, i\omega + [U])^{-1} = [\alpha(i\omega)] = \sum_{r=1}^{m} \frac{\{\Phi_r\} \lfloor L_r \rfloor}{\lfloor L_r \rfloor \{\Phi_r\}\ (i\omega - \lambda_r)}$$

When the system matrices are symmetric $\{\Phi_r\} = \lfloor L_r \rfloor'$

When the system is undamped or the damping matrix is formed from linear combinations of symmetric mass and stiffness matrices then $\{\Phi_r\}$ are real.

When complex conjugate roots occur, pairs of terms can be combined to form second order terms.

Any measured receptance can therefore be expressed in the form

$$\alpha_{rs}(i\omega) = \sum_{j=1}^{m-2N} \frac{A_j}{i\omega - \lambda_j} + \sum_{r=1}^{N} \frac{B_r.i\omega + G_r}{(-\omega^2 + 2\zeta_r.\omega_r.i\omega + \omega_r^2)}$$

REFERENCES


1. Fraeijs de Veubeke, B.M. Déphasages Charactéristiques et Vibrations Forcées d'un Systeme Amorti. Bull. Acad. Belg. Sci. 1948
2. Kennedy, C.C., Pancu, C.D.P., Use of Vectors in Vibration Measurement and Analysis, Jnl. Aeronautical Sciences. Vol.14 No.11 Nov 1946.
3. Bishop, R.E.D., Pendered, J.W., A Critical Introduction to some Industrial Resonance Testing Techniques. Jnl. Mech. Eng. Science. Vol.5 No.4. 1963.
4. Bishop, R.E.D., Fawzy, I., On the nature of resonance in non-conservative systems. Jnl. Sound and Vibration. Vol.55 No.4. Dec 1977.
5. Ramsay, A.K., Effective Measurements for Structural Dynamics Testing. Sound & Vibration Nov. 1975
6. Gaukroger, D.R., Skingle, C.W., Heron, K.H., Numerical Analysis of Vector Response Loci. Jnl. Sound and Vibration. Vol. 29. No.3 August 1973.
7. Frazer, R.A., Duncan, W.J., Collar, A.R. Elementary Matrices. Cambridge Univ. Press. 1960.
8. Halfman, R.L. Dynamics. Vol II. Addison-Wesley Co. Inc. 1962.
9. Bishop, R.E.D., Wahed, I.F.A., On the equations governing the free and forced vibrations of a general non-conservative system. Jnl. Mech. Eng. Science. Vol.18 No.1 Feb 1976.


ACKNOWLEDGEMENTS


The authors thank the British Railways Board for permission to publish this paper, and would also like to acknowledge the assistance of colleagues in carrying out the laboratory tests.

# DETERMINATION OF VEHICLE PARAMETERS FROM TRACK TESTING DATA

Derek Lyon and Norman C. Remedios

British Rail Research and Development Division, Derby, United Kingdom

SUMMARY

Using standard instrumentation to measure accelerations and relative displacements during track testing, it has been possible to determine accurate vehicle transfer functions. Manipulation of these transfer functions and the application of modal analysis techniques have enabled the modal and suspension parameters of the vehicles to be determined. The methods have been applied to the vertical response of 2 axle and 4 axle passenger vehicles and results are presented. The general methods have been discussed and the problems - which arise because the input excitation is of arbitrary form and consists of multiple correlated inputs - have been outlined.

## 1. INTRODUCTION

The vertical ride of a railway vehicle is determined by the roughness of the track and the effectiveness of the vehicle suspension. Although the contribution of each is of no concern to the passenger it is of paramount importance to the vehicle designer. By separating the suspension system from the inputs it becomes possible to ascertain whether design values have been realised in terms of suspension components and in cases of disappointing performance, design changes can be pursued in a rational manner.

In the past such questions have been answered by a combination of laboratory and track testing with the latter being used to assess the ride quality due to the combined effects of track roughness and suspension transmisibility. Laboratory testing has generally been employed to measure suspension parameters under controlled conditions and to study the effects of design changes. Although laboratory testing used to be conducted - of necessity - by sinusoidal excitation, the development of digital techniques over the past decade has provided a variety of test methods using random or transient excitations. It is therefore a logical step to let the track provide the excitation and to work directly with the measured responses.

Work is currently being undertaken to study the problems and to develop suitable practical techniques. Initially, only the vertical ride has been considered in detail although the technique can be applied directly to some lateral problems. If the method is to have wide application in development testing then it must be capable of using standard instrumentation which would normally be fitted to monitor ride and suspension performance. Although still at an early stage in the development, the techniques are already being applied to suspension development problems where the direct evaluation of suspension parameters with known confidence offers considerable advantages over existing techniques.

## 2. TRANSFER FUNCTION MEASUREMENT

*Determination of the transfer function of a simple system.*

The problem of determining the transfer function of a linear system having a single input and a single output but in the presence of an extraneous incoherent noise source has been dealt with at length. [1, 2]. As this forms the basis of the subsequent techniques, it will be briefly summarised.

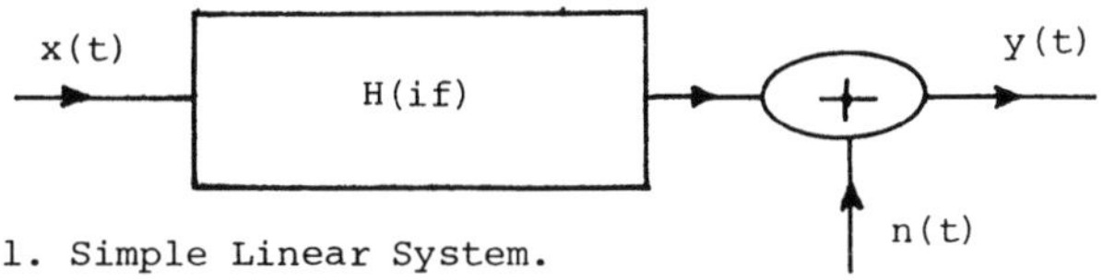


Fig 1. Simple Linear System.

x(t) is the time history of the system input
n(t) is the time history of the incoherent noise source
y(t) is the time history of the combined output
H represents the linear system and
H(if) is the system transfer function which is to be evaluated.

The linear Fourier Spectrum of a time history, say x(t) is defined as Sx(if).

The auto Power Spectrum of a time history, say x(t) is defined as $Gxx = Sx(if)S^{*}x(if)$ where * denotes the complex conjugate.

The cross Power Spectrum between two time histories, say x(t) and y(t) is defined as $Gyx = Sy(if)S^{*}x(if)$

The average of a number of estimates of a Power Spectrum is denoted by $\overline{Gyx}$

The transfer function to be determined can in fact be defined in the frequency domain as H(if) = Sy(if)/Sx(if).

For practical reasons, transfer functions are in fact determined from the relationship

$$H(if) = \overline{Gyx}/\overline{Gxx}$$

since the use of the Power Spectra permits averaging a number of estimates of H(if) to obtain more accurate results. Averaging also reduces the effects of noise since in this case an erroneous transfer function will be calculated which is given by

$$H(if) = \overline{Gyx}/\overline{Gxx} - \overline{Gnx}/\overline{Gxx}$$

but the term $\overline{Gnx}$ which is the cross Power Spectrum between two uncorrelated sources will tend to zero when averaged over a large number of estimates.

Finally the calculation of the coherence function, which can be used to estimate how much of the output is due to the input, enables the quality of the transfer function measurement to be assessed. Coherence is defined as

$$\gamma^2 = |\overline{Gxy}|^2/\overline{Gxx}.\overline{Gyy}$$

and must lie in the range $0 \leq \gamma^2 \leq 1$.

The use of these techniques in laboratory testing with random or transient inputs is now well established to the extent that it is possible to buy hard-wired digital analysers which perform the tasks at the touch of a button. In order to apply the techniques to the analysis of track data it is necessary to deal with inputs which are neither simple random nor pseudo random and to deal with multiple correlated inputs to a fairly complex system.

*Application to Railway Vehicle Analysis.*

The vertical input at the wheelsets of a rail vehicle contains components which resemble a Gaussian random process. In addition other components of the input arise from periodic effects due either to the joints or to the welds which on BR are spaced at just over 18 m intervals, even on continuous welded rail. Finally, at infrequent intervals, there occur large excursion transient inputs due to discrete track features such as points, crossings, catch points and insulated joints.

In general, the methods used in laboratory testing can be applied directly to the arbitrary mixture of track inputs provided that the Fourier Transforms can be calculated, which will always be the case. The following observations are worth noting.

a) Since one is no longer working with samples from a Gaussian random process, the methods for estimating variance and transfer function accuracy which are based on stationary random process theory can no longer be used.
b) The calculations will invariably be based on digital Fourier Transforms and the length of each transform must be such that errors due to truncation or end effects are acceptably small.
c) The coherence function can be used to assess the results and therefore averaging can be used to improve signal to noise ratios. However, averaging cannot be used to overcome errors which may be introduced by using too short a Fourier Transform.

The work is still at a relatively early stage in its development and it is not possible to give hard rules for the choice of transform and record length. The amount of averaging required can be estimated from the coherence function while choice of record length is clearly determined by the lowest frequency and damping present in the system impulse response function.

*Multiple correlated inputs.*

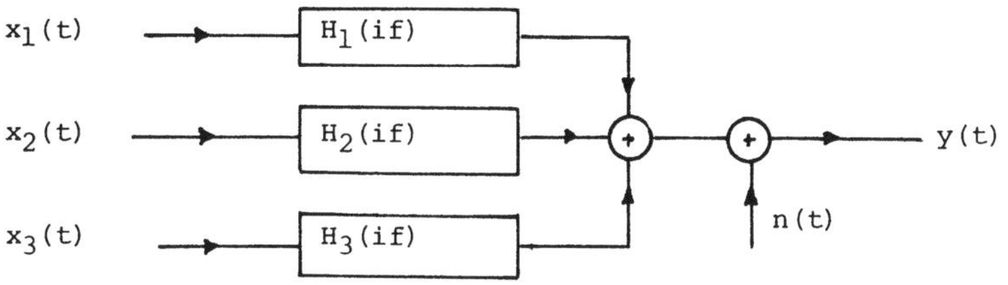


Fig 2. Linear System with multiple inputs.

Multiple inputs are represented diagrammatically in the above figure, and can be expressed using matrix notation as

$$Sy = \lfloor H(if) \rfloor \{Sx\}$$

Sy is the linear Fourier Spectrum of the input
$\lfloor H(if) \rfloor$ is a row vector of the transfer functions $H_1$ (if), $H_2$(if)... etc.
$\{Sx\}$ is a column vector of the linear Fourier Spectra of the inputs $x_1(t)$, $x_2(t)$... etc.

Post-multiplying by $\lfloor S^*x \rfloor$ gives $Sy\lfloor S^*x \rfloor = \lfloor H(if) \rfloor [Gxx]$

where

$$[Gxx] = \begin{bmatrix} Sx_1.S^*x_1 & Sx_1.S^*x_2 & Sx_1.S^*x_3\ldots \\ Sx_2.S^*x_1 & Sx_2.S^*x_2 & \ldots\ldots\ldots\ldots \\ \ldots\ldots\ldots\ldots & \ldots\ldots\ldots\ldots & \ldots\ldots\ldots\ldots \end{bmatrix}$$

and is a matrix formed from the input auto and cross Power Spectra.

For the case of vertical excitation, all the inputs $x_1(t)$, $x_2(t)$ are due to the same track irregularity but occur with time delays $T_1$,$T_2$, etc, depending on the vehicle speed and the distances between the various wheelsets of the vehicle. Consequently, the matrix [Gxx] will be singular and the transfer functions $\lfloor H(if) \rfloor$ cannot be evaluated by extension of the simple methods.

By manipulation of the inputs and exploiting the symmetry of the vehicle it is possible to reduce the problem to what is in effect a number of single input systems which can then be used to determine practical transfer functions. This will be dealt with in the next section.

## 3. VEHICLE MODELS AND TRACK INPUTS

The equations governing the vertical and longitudinal motions of a rail vehicle can be expressed as

$$[M]\{\ddot{q}\} + [C]\{\dot{q}\} + [K]\{q\} = [Cf]\{\dot{z}\} + [Kf]\{z\}$$

where $[M]$ $[C]$ $[K]$ are nxn matrices which define the mass, damping and stiffness of the vehicle.

$\{q\}$ is a column vector of generalised co-ordinates.
$\{z\}$ is a column vector of the vertical displacement inputs at each wheelset.

$[Cf]$, $[Kf]$ are the primary suspension elements which act at the inputs i.e. at each wheelset.

For vertical suspension modelling in the frequency range 0 - 15 Hz the wheelset masses and track stiffness and damping will normally be neglected; for passenger vehicles the unsprung mass - track resonance occurs at 60 - 70 Hz. In general all suspension dampers will be modelled by a series dashpot-spring arrangement. This is particularly important for the secondary suspension where the choice of a suitable series stiffness enables a compromise between resonant damping and high frequency transmission to be made. Even when dampers are designed with stiff end mountings, the effects of oil compressibility and expansion of the pressure tube are significant and laboratory tests suggest that stiffnesses of greater than 5 - 10 MN/m are rarely achieved.

For the 4 axle vehicle there will be an input at each axle due to the vertical track irregularity profile.

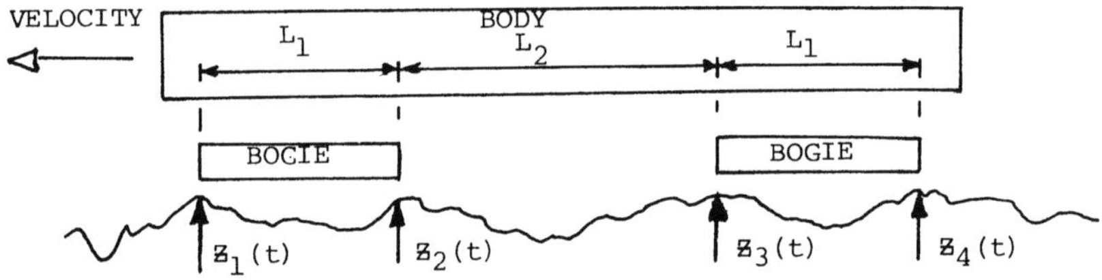


Fig 3. Vertical Inputs for 4 axle bogie vehicle.

If the time delays are given by $T_1 = L_1/V$; $T_2 = L_2/V$ then the four inputs become

$$z_1(t) = z(t) \qquad z_2(t) = z(t+T_1)$$
$$z_3(t) = z(t+T_1+T_2) \qquad z_4(t) = z(t+2T_1+T_2)$$

By making use of the symmetry of the vehicle it is possible to define four inputs which simplify the determination of transfer functions and give physical insight into vehicle behaviour.

(a) Symmetric Body Modes. $W_1 = 1/4\ (z_1+z_2+z_3+z_4)$

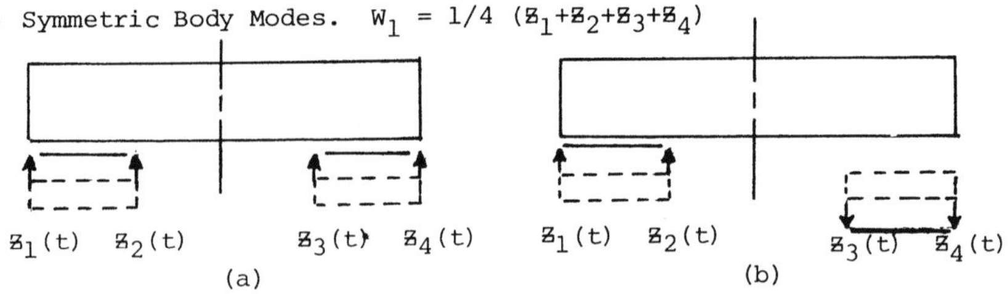

(b) Asymmetric Body Modes. $W_2 = 1/4\ (z_1+z_2-z_3-z_4)$

(c) Longitudinal Body Modes. $W_3 = 1/4\ (z_1-z_2+z_3-z_4)$

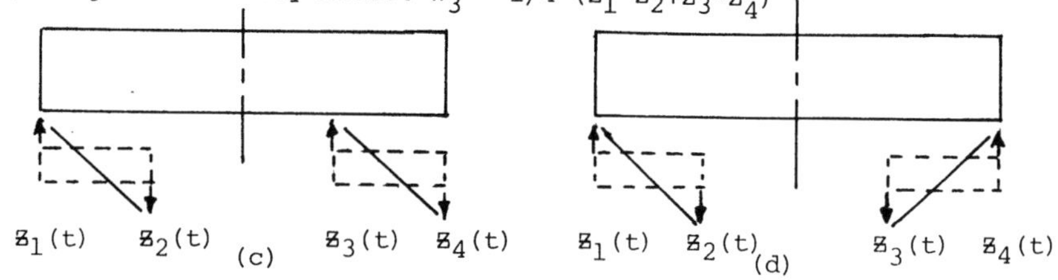


(d) Body Bending Modes. $W_4 = 1/4\ (z_1-z_2-z_3+z_4)$

The spectral content of the four transformed inputs has been shown in Fig. 4 for typical values of $L_1$ = 3m and $L_2$ = 12 m with a vehicle speed of 50 m/s. The frequency (about 1½ Hz) at which the bounce input $W_1$ falls to zero and the pitch input $W_2$ reaches a first peak is clearly shown. Peaks in the other inputs which tend to produce longitudinal body vibration and body bending are also evident.

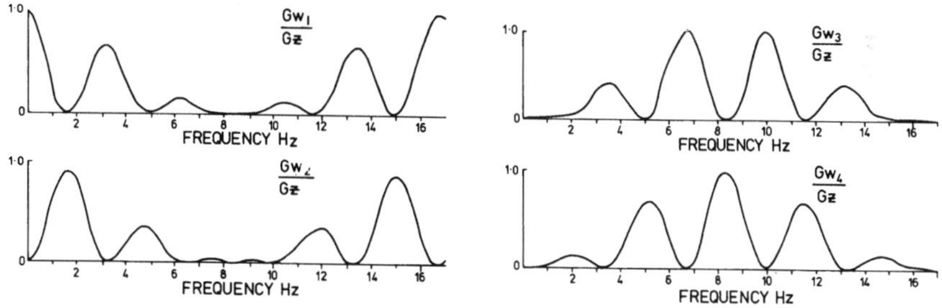


FIG. 4. SPECTRAL CONTENT OF TRANSFORMED INPUTS.

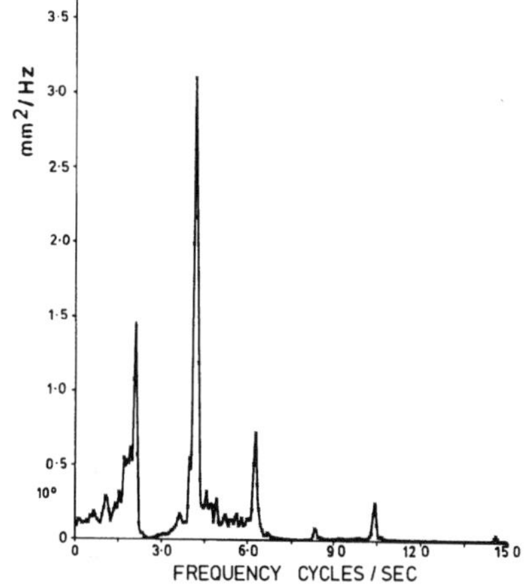


FIG. 5. P.S.D. OF INPUT. T = 100 SECS.

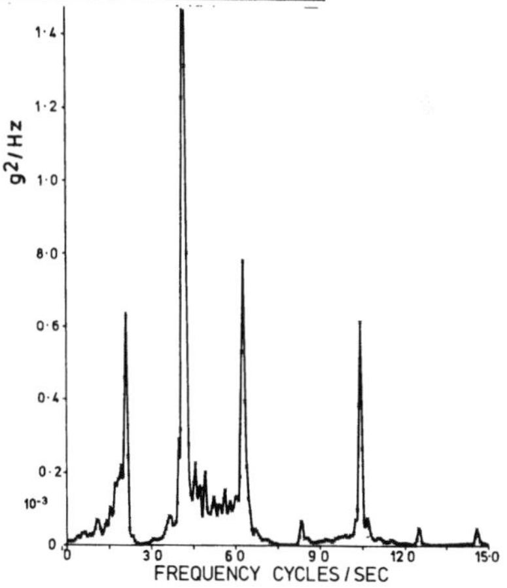


FIG. 6. P.S.D. OF OUTPUT. T = 100 SECS.

## 4. TESTS WITH 2 AXLE RAIL CAR

The testing of the 2 axle rail car was not planned as an exercise to determine transfer functions, in fact the experimental data were accumulated as part of the normal development testing programme and were only subsequently used for the transfer function determination.

It therefore demonstrates the important requirement that the method should be applicable to standard instrumentation.

Experimental data were available in the form of five accelerations

$\ddot{x}_1;\ \ddot{x}_2;\ \ddot{x}_3;\ \ddot{x}_4;\ \ddot{x}_5$

mounted in the vehicle body and four displacement potentiometers

$\delta_1;\ \delta_2;\ \delta_3;\ \delta_4$

one mounted at each axlebox. One of the accelerometers was positioned at the body centre and the others were symmetrically disposed over each wheelset and at the vehicle ends. The measured responses were added and subtracted to yield bounce and pitch mode inputs and responses.

| | |
|---|---|
| Bounce Input | $\delta b = 1/4\ (\delta_1+\delta_2+\delta_3+\delta_4)$ |
| Bounce Outputs | $\ddot{x}b_1 = \frac{1}{2}(\ddot{x}_1+\ddot{x}_5)$; $\ddot{x}b_2 = \ddot{x}_3$; $\ddot{x}b_3 = \frac{1}{2}\ (\ddot{x}_2+\ddot{x}_4)$ |
| Pitch Input | $\delta p = 1/4\ (\delta_1+\delta_2-\delta_3-\delta_4)$ |
| Pitch Outputs | $\ddot{x}p_1 = \frac{1}{2}\ (\ddot{x}_1-\ddot{x}_5)$; $\ddot{x}p_2 = \frac{1}{2}\ (\ddot{x}_2-\ddot{x}_4)$ |

The power spectral densities of the inputs and outputs for the bounce mode are shown in Fig 5-8 for record lengths of 100 seconds and 3500 seconds. Although giving the distribution of power in the input and output signals these are of little direct use in assessing vehicle performance and both input and output is dominated by periodic effects.

Vehicle transfer functions were determined using the normal techniques of single input - single output systems as outlined in Section 2. Space limitations permit only a small selection of results to be presented and a typical acceleration-displacement transfer function is shown in Fig. 9. Because of the form of this transfer function the resolution of the rigid body mode is not very good but the first symmetric vertical bending mode is clearly defined.

At low frequencies the three bounce transfer functions are identical since at rigid body frequencies all the accelerations will be the same. At high frequencies the transfer functions diverge as each accelerometer responds to the vertical bending mode in accordance with its position along the body. The idealised rigid body model for the bounce mode is shown in Fig. 10 as a spring (k), mass (m), damper (c) and damper series spring (Nk). The acceleration/relative displacement transfer function for this system will be

$$g(s) = \frac{\ddot{x}(s)}{\underset{.}{S}(s)} = \frac{\ddot{x}(s)}{x(s)\ \delta(s)} = \frac{-(1+N)k}{m} + \frac{N^2k^2}{m(cs+Nk)}$$

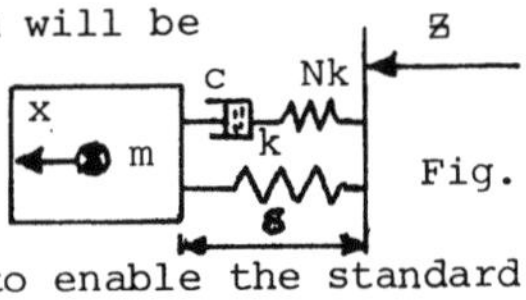

Fig. 10

To assist in low frequency analysis and to enable the standard techniques of modal analysis to be applied, the acceleration transfer function was manipulated to yield the displacement transfer function.

$$\frac{x(s)}{\delta(s)} = \frac{g(s)}{g(s)-s^2}$$

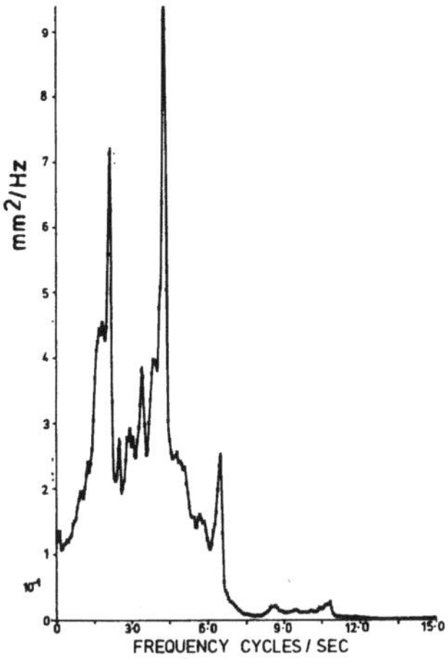


FIG. 7. P.S.D. OF INPUT. T=3500 SECS.

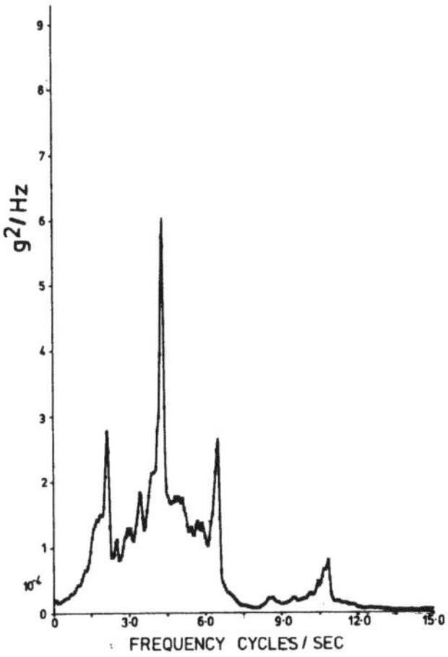


FIG. 8. P.S.D. OF OUTPUT. T=3500 SECS.

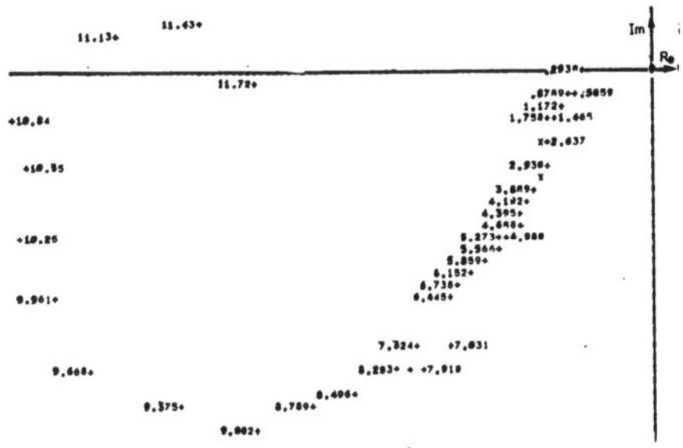


FIG. 9. TRANSFER FUNCTION OF BODY ACCELERATIONS / SUSPENSION DISPLACEMENTS.

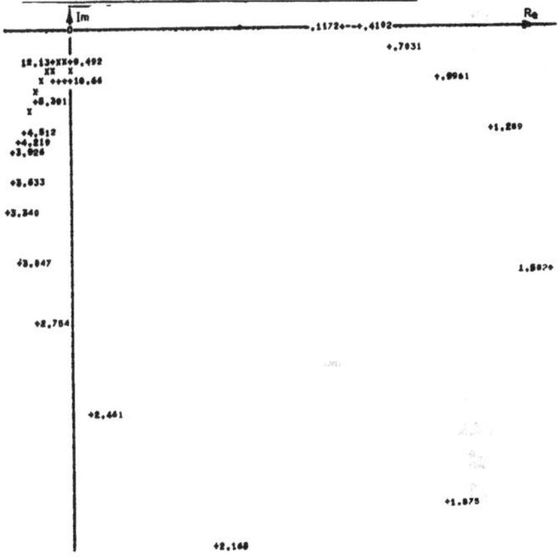


FIG. 11. TRANSFER FUNCTION OF BODY DISPLACEMENTS / SUSPENSION DISPLACEMENTS.

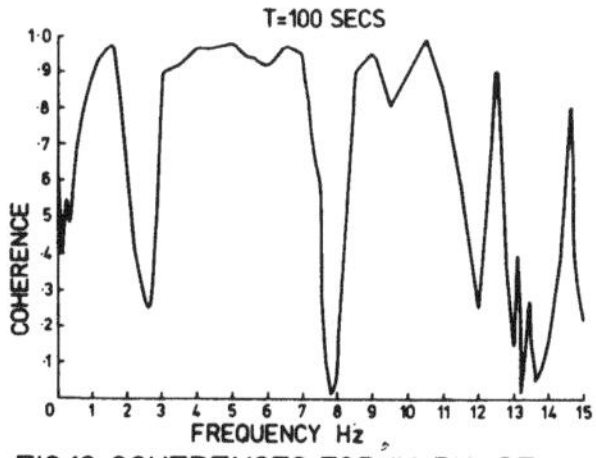


FIG. 12. COHERENCES FOR IN PHASE TRANSFER FUNCTIONS.

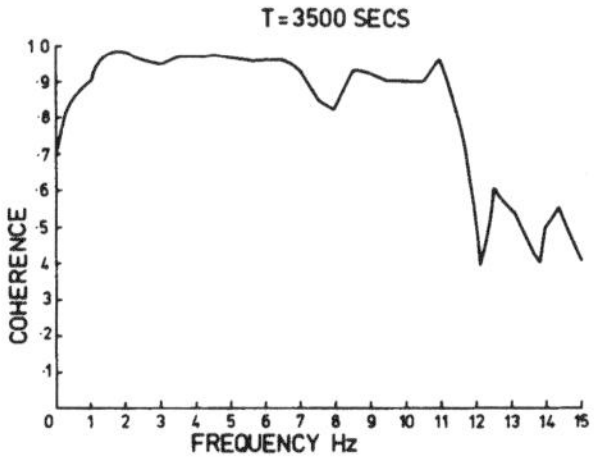


FIG. 13. COHERENCES FOR IN PHASE TRANSFER FUNCTIONS.

A plot of this transfer function is shown in Fig. 11 and the application of modal analysis techniques gave the natural frequency and damping of the bounce mode. The analysis of the pitch mode was carried out in a similar manner and the complete suspension parameters of the vehicle were determined as listed below.

| | Natural Frequency | Damping |
|---|---|---|
| Bounce Mode | 2.0 Hz ± 0.6 Hz | 29.7% ± 2% |
| Pitch Mode | 2.1 Hz ± 0.04 Hz | 21.8% ± 2.0% |
| Body Bending | 10.52 Hz ± 0.15 Hz | 9.0% ± 2.0% |

Suspension Stiffness (k) 2.18 MN/m.
Damper Coefficient (c) 128 KN.s/m
Damper Series Stiffness (Nk) 12.8 MN/m.

As a measure of the accuracy of the results coherences are shown in Fig. 12, 13 for record lengths of 100 seconds and 3500 seconds. The coherence is generally high for the longer record while the troughs in the coherence of the shorter record coincide with the frequency at which the bounce input (for the particular vehicle speed) is virtually zero i.e. the input at this frequency is wholly out of phase at the two wheelsets. It is this low level of input which leads to poor signal to noise ratios at these frequencies and low values of coherence when short records are analysed.

## 5. TESTS WITH 4 AXLE BOGIE VEHICLE

The testing of this vehicle was carried out specifically to study the application of transfer function techniques. Acceleration measurements were taken on both the body and the bogie and relative displacements were measured between axle-boxes and bogie frame and between bogie frames and body. All the instrumentation was standard and testing was conducted during normal running on main line. Several tests were conducted and changes to the suspension were introduced between each test run. Thus it was possible to study :

(i) The variation of secondary suspension damping which for the vehicle in question is provided by orifices between the air springs and the auxiliary reservoirs.

(ii) The variation of primary suspension damping by removing one damper of each pair of viscous dampers fitted at every axle-box.

(iii) The variation of secondary traction rod stiffness which for many vehicles is the principal coupling between bogie pitch and both body longitudinal and body bending modes.

Secondary suspension parameters were determined by computing transfer functions from the body accelerations and the relative bogie to body displacements. This was necessary because at the rigid body frequencies the primary displacements are too small to use effectively. Therefore for the secondary suspension measurements, the bogie vehicle is treated in an exactly similar manner to the two axle vehicle.

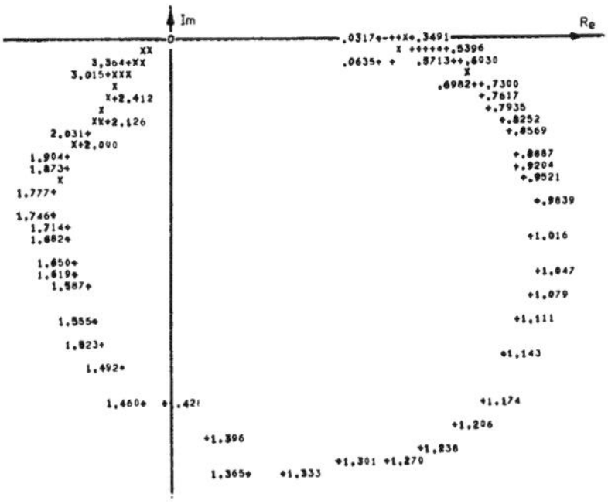


FIG. 14. BODY PITCH RESPONSE - BOGIE VEHICLE FULL SECONDARY DAMPING.

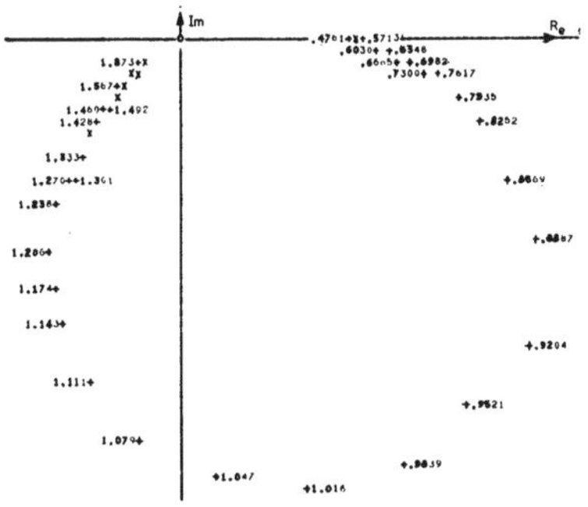


FIG. 15. BODY PITCH RESPONSE - BOGIE VEHICLE REDUCED SECONDARY DAMPING.

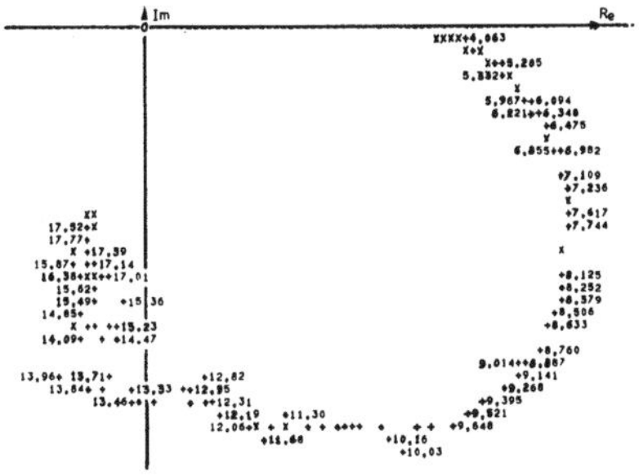


FIG. 16. BOGIE PITCH MODE - FULL PRIMARY DAMPING.

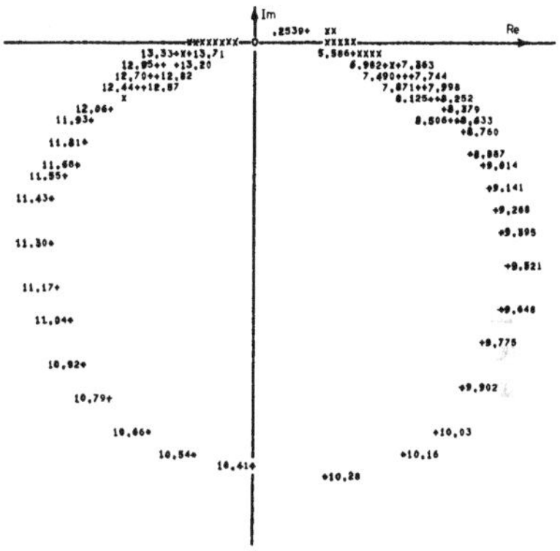


FIG. 17. BOGIE PITCH MODE - REDUCED PRIMARY DAMPING.

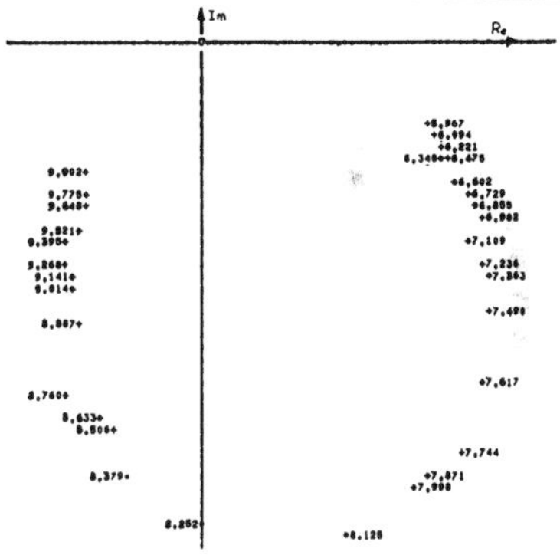


FIG. 18. BOGIE LATERAL MODE.

Transfer functions for the body pitch mode are shown in Figs. 14 & 15 for two levels of secondary suspension damping. The bogie pitch transfer functions for two levels of primary suspension damping are shown in Figs. 16 & 17. Although the work reported here has dealt exclusively with vertical inputs it was mentioned that the techniques could be applied to some lateral problems. As an illustration Fig.18 shows the transfer function for the lateral mode of a bogie.

The modal parameters determined from the transfer functions shown in Figs 14-17 are listed below.

*Body pitch.*

Soft secondary traction rod, 10 mm orifice in air spring pipe.
Frequency 1.3 Hz ± 0.1 Hz Damping 25% ± 1%.

Stiff secondary traction rod, no orifice in air spring pipe.
Frequency 1.0 Hz ±0.05 Hz Damping 14% ± 0.5%

*Bogie pitch*

All primary dampers fitted.
Frequency 10.4 Hz ± 0.4 Hz Damping 45% ± 5%
Half primary dampers fitted
Frequency 10.4 Hz ± 0.3 Hz Damping 8.5%± 0.3%

It is clear from these results that the bogie pitch damping has reduced by much more than the anticipated 50%. Analysis of the results is still being made but the large reduction is thought to be influenced by :

A reduction in damper series stiffness. With 2 dampers per axle-box the series stiffness is due to the oil flexibility and the end mountings of the damper and is estimated to be about 10 MN/m. But when one damper is removed, the single damper is able to rotate the axle-box against the stiffness of the primary springs and therefore has a much lower effective series stiffness.

Non-lineararities in the damper characteristics. Dampers have non linear characteristics and the damping rates for small velocities can be much larger than the nominal damping rate. Removing one damper will obviously increase the velocity of the remaining damper and this may well produce a lower effective rate.

REFERENCES.

1. Bendat, J.S., Piersol, A.G., Random Data : Analysis and Measurement Procedures, Wiley-Interscience, 1971.
2. Newland, D.E., An Introduction to Random Vibration and Spectral Analysis, Longman, 1975.

ACKNOWLEDGEMENTS

The authors thank the British Railways Board for permission to publish this paper. In full scale track testing many people participate and the authors would like to thank all who contributed to the collection of the data presented.

# COMPUTER SIMULATION AND PARAMETER SENSITIVITY STUDY OF A COMMERCIAL VEHICLE DURING ANTISKID BRAKING

Charles C. MacAdam

Highway Safety Research Institute, The University of Michigan, Ann Arbor, MI, U.S.A.

SUMMARY

A large-scale digital computer model, used for simulating the dynamical response of commercial vehicles, was employed in a parameter sensitivity study which focused on the antiskid braking performance of a heavy truck. Full-scale vehicle test results and laboratory measurements were used in developing a baseline computer representation of the test vehicle. Subsequent to the development of the baseline computer representation, the computer simulation was exercised to study the influence of various vehicle and antiskid system parameters. Basic categories examined in the parameter sensitivity study were: brake system properties, vehicle mass and inertia properties, tire traction characteristics, suspension properties, wheel dynamics, and antiskid system characteristics.

The antiskid braking performance of the examined baseline vehicle was primarily dominated by the adverse effects deriving from rear brake torque imbalances and hysteresis precipitated during wheel cycling. The most noteworthy findings of the parameter sensitivity study for improving the baseline stopping performance, outside of reducing the side-to-side imbalances and rear brake hysteresis, were found to be: (a) reduction in rear brake effectiveness or use of load proportioning valves, (b) modification of antiskid system operation, and (c) modification of rear suspension properties.

## 1. INTRODUCTION

Promulgation of Federal Motor Vehicle Safety Standard FMVSS 121 in the U.S. has spawned considerable discussion and controversy over the past decade regarding antiskid braking performance of commercial vehicles. Largely out of need to objectively address its own concerns over the impending government regulation, the truck industry members of the Motor Vehicle Manufacturers Association (MVMA) initiated a long-term motor truck braking and handling performance study at the Highway Safety Research Institute (HSRI) of The University of Michigan in 1971. The purpose of this study was to develop computer-based methods, principally time-domain dynamical simulations, for representing and predicting the braking and handling performance of heavy trucks and tractor-trailers. Since that time, a comprehensive set of computer models has evolved under MVMA sponsorship [1-3]. Examples of previous studies which have employed these models are given in References [4 and 5]. The material presented in this paper represents a recent study performed at HSRI to (1) simulate the baseline dynamical response of a specific problem vehicle during straight-line, antiskid braking and (2) perform a subsequent parameter sensitivity study using the baseline vehicle computer representation.

The antiskid braking study discussed here focused on the acquisition and examination of test data obtained from both full-scale vehicle tests and laboratory measurements. The intent of this effort was to gather parameter information for constructing a representative digital simulation of the test vehicle. Upon development of the representative baseline vehicle performance, the digital simulation would then be exercised to study the influence and importance of various vehicle and antiskid system parameters.

## 2. COMPUTER MODEL

The straight truck vehicle model employed in this study is known as the Phase III MVMA/HSRI braking model [3]. It can be characterized as essentially a pitch-plane "bicycle" model providing pitch and bounce degrees of freedom for the sprung mass, and vertical degrees of freedom for each of the front and rear unsprung masses. Each wheel is allowed its own rotational degree of freedom. Separate side-to-side wheel rotation degrees of freedom are provided for representing the effects due to brake torque imbalances on each axle.

The basic dynamical model is augmented by a set of specialized subprogram models used to represent single- and tandem-axle suspensions [1, 3, 6], antiskid braking systems [3, 7, 8], mechanical friction brake characteristics [1, 3], and longitudinal tire force properties [1, 3]. These subprogram models provide a convenient and flexible means for simulating either simple or more complex operating characteristics exhibited by these individual components.

The block diagram of Figure 1 shows the basic inter-relationships between the major components of the vehicle simulation.

## 3. TEST VEHICLE/MEASUREMENTS

The test vehicle used in this study was a loaded, 51,000-lb straight truck equipped with a beam axle, single-leaf front suspension, rear "walking-beam" tandem suspension, and an axle-by-axle antiskid brake control system. The front axle was equipped with dual-wedge brakes, each rear axle with S-cam brakes. The center-of-gravity height above ground was 64 in. for the loaded vehicle configuration. The payload weight of 26,000 lbs was positioned 23 in. in front of the center line of the rear suspension.

Full-scale vehicle tests were performed by White Motors, Inc., at the Bendix Automotive Proving Ground. The tests were straight-line braking stops from 60 mph using a full treadle valve application (100 psi). The primary data variables measured in each test are listed in Table 1. "Torque wheels" were used on the right front wheel and right rear-tandem wheel to measure the instantaneous wheel torques.

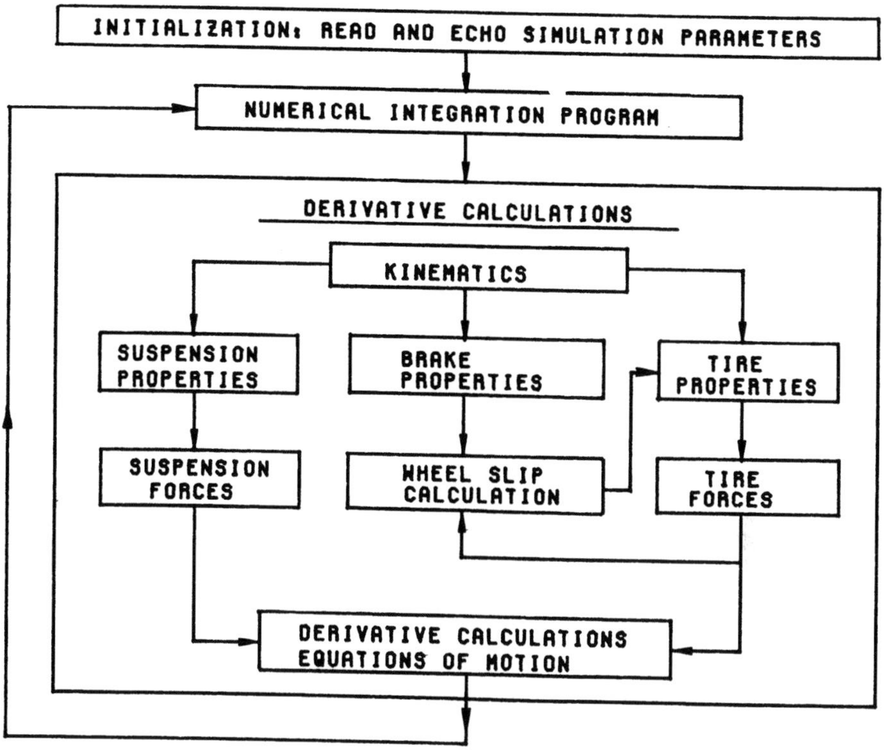


Fig. 1. Vehicle simulation block diagram.

Following the series of full-scale braking tests, the vehicle inertial properties and component system characteristics were measured by HSRI. Table 2 lists the principal vehicle components and their sources of measurement.

The tire force measurements were performed on the same surface used for the full-scale vehicle tests. Data was collected at three speeds (20, 40, and 60 mph) and three vertical loads (rated load ± 50%). Brake dynamometer data provided initial estimates for the torque effectiveness of each brake. Subsequent examination of the full-scale test records for the front and rear torque wheels revealed significant side-to-side brake imbalances and considerable levels of hysteresis in the rear S-cam actuated brakes. Suspension measurements showed a particularly complex force-deflection relationship for the rear walking-beam suspension. The laboratory analog computer tests of the anti-skid system and the full-scale vehicle test records both indicated significant differences in the cycling operation of the front and rear antiskid systems. The antiskid system used on the vehicle employed "worst wheel" (slower running) side-to-side control for each axle and a "pneumatic logic" mechanism within the modulator valve to control the rate of pressure build-up during each cycle.

Table 1. Listing of Data Variables Measured During Full-Scale Vehicle Tests.

Left Front Wheel Speed
Right Front Wheel Speed
Left Middle Wheel Speed
Right Middle Wheel Speed
Left Rear Wheel Speed
Right Rear Wheel Speed
Right Front Wheel Torque
Right Rear Wheel Torque
Front Axle Vertical Position
Rear Axle Vertical Position
Front Axle Brake Pressure
Middle Axle Brake Pressure
Rear Axle Brake Pressure
Driver Treadle Pressure
Vehicle Deceleration
Fifth Wheel Vehicle Velocity
Stopping Distance

Table 2. Sources of Measurement for the Vehicle Parameters and Components.

| | |
|---|---|
| Tire Force Properties: | Mobile Truck Tire Dynamometer (HSRI) |
| Brake System: | Brake Dynamometer Data and Full-Scale Vehicle Test Records |
| Vehicle Mass and Inertia Properties: | Inertial Swing Facility (HSRI) |
| Suspension Characteristics: | Suspension Measurement Facility (HSRI) |
| Antiskid System: | Laboratory Analog Computer Tests (HSRI) and Full-Scale Vehicle Test Records |

## 4. COMPUTER MODEL REFINEMENTS

Review of the full-scale vehicle and component tests suggested the need for certain refinements in the brake and suspension subprogram models. Several preliminary computer runs were performed to substantiate and highlight major differences between vehicle test data and the initial computer representation. Rear brake release times which occurred during vehicle tests were much greater than those predicted by the initial simulation runs. Vehicle tests showed average brake release times of approximately 200 milliseconds; simulation release times were less than 100 milliseconds. These differences were reflected in stopping distances—250 feet for the initial computer representation, 300 feet (average) for the vehicle tests. The principal items requiring improved representation within the computer model were found to be:

•Brake pressure-torque hysteresis of the rear S-cam brakes

•Air system transport time lag effects at low pressure levels during antiskid cycling

and

•Improved representation of the rear tandem suspension force-deflection characteristics.

Study of the brake pressure, torque wheel, and wheel speed traces from the vehicle tests indicated significant levels of hysteresis (27,000 in-lb) in the rear S-cam brakes during antiskid cycling. A typical hysteresis loop is shown in Figure 2. (No hysteresis was

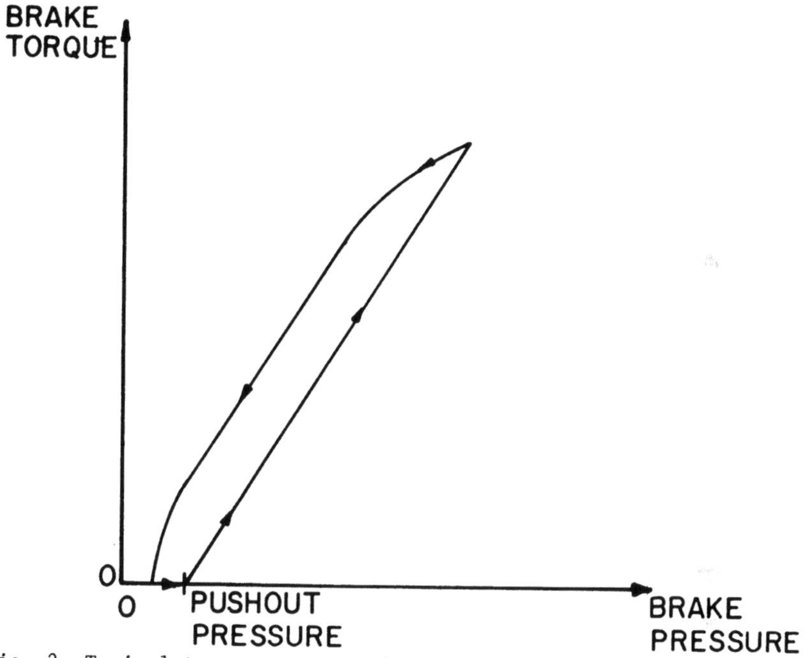


Fig. 2. Typical torque-pressure hysteresis loop; rear S-cam brake.

observed in the front wedge brakes.) The amount of hysteresis present in the rear brakes acted in combination with side-to-side wheel imbalances (caused by asymmetric vehicle loading and/or side-to-side differences in brake effectiveness) to produce significantly greater brake release times than would occur otherwise. Furthermore, the lengthening of brake release times by the hysteresis and imbalance allowed rear brake pressures to decrease to relatively low levels, often below the brake pushout pressure. Hence, additional lags were incurred from transport time delays due to volumetric changes in the air supply system for line pressures less than brake pushout. Such transport time lags in the air supply system were approximately proportional to how far the line pressure dropped during each cycle below the pushout pressure of the brake.

The manner in which the rear brake hysteresis, air supply delays, and side-to-side imbalance all combined to produce a significant stopping performance sensitivity is detailed by the following discussion and reference to Figure 3. During an antiskid pressure

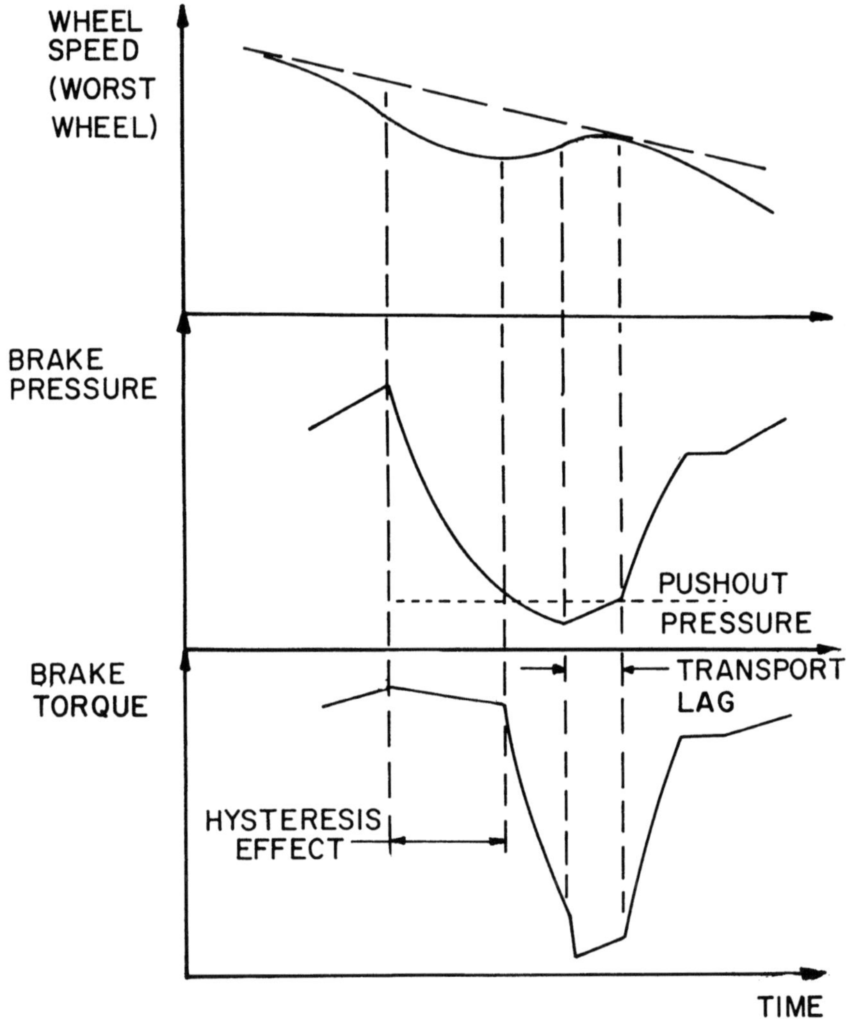


Fig. 3. Representative cycle showing effects of brake hysteresis and air supply transport lag.

application, the "worst" wheel (more effective brake and/or smaller side-to-side vertical load) on an axle is driven toward the peak of the tire-road μ-slip curve with the "best" wheel lagging behind as a result of the side-to-side imbalance. When the antiskid system interrupts the pressure application, the "worst" wheel is generally in the vicinity of the tire force μ-slip peak (15-25% slip). As the brake pressure drops, the brake torque tends to remain high, or lags, due to the hysteresis in the brake. This residual or lagging torque causes the "worst" wheel to remain in a moderate slip regime (25-50% slip) for an additional period of time, forcing the antiskid system to lower the pressure to a level below the pushout pressure of the brake. Meanwhile, the "best" wheel is returning to a relatively low slip condition, providing little braking. With the pressure now decreased to a very low level, the "worst" wheel accelerates back to a low slip condition causing the antiskid system to generate an "ON" command, applying air to the brake again. Depending on how far below pushout the line pressure has fallen in the last release, a transport lag occurs due to the air volume change required in pushing the brake shoes back out to the drum. (The data for this truck indicated a maximum time lag of 40 milliseconds to go from 0 psi to the pushout pressure of 7 psi.) The net effect of the brake hysteresis and air supply delay is to permit free-rolling wheel conditions that otherwise would not occur.

In order to include the hysteresis and air supply delay effects exhibited by this vehicle within the computer model, (1) a small but significant modification was made to the hysteresis computer algorithm in order to represent more accurately the hysteresis-torque relationship at low pressure, and (2) the air supply transport delay effect, as presently represented, was modified and permitted to occur for any line pressure falling below pushout during an antiskid discharge cycle.

The remaining item requiring computer model refinements was the representation of force-deflection characteristics of the tandem suspension. The tandem suspension measurements made at HSRI (see Fig. 4) indicated a somewhat more complex relationship between vertical load and spring deflection than was presently assumed within the computer model. To better represent the suspension characteristics shown in Figure 4, the following equation was developed to approximate the force-deflection characteristics.

$$F_i = F_{ENV_i} + \left(F_{i-1} - F_{ENV_i}\right) e^{-\beta\left|\delta_i - \delta_{i-1}\right|} \qquad (1)$$

where

$F_i$ is the suspension force at the current simulation time step

$F_{i-1}$ is the suspension force at the last simulation time step

$\delta_i$ is the suspension deflection at the current simulation time step

$\delta_{i-1}$ is the suspension deflection at the last simulation time step

$F_{ENV_i}$ is the force corresponding to the deflection, $\delta_i$, of the outer envelopes of the measured suspension characteristic. $F_{ENV}$ is represented in the simulation by two force vs. deflection tabular functions (upper and lower) input by the program user.

and $\beta$ is an input parameter used for describing the rate at which the suspension force within an envelope loop approaches the outer envelope, $F_{ENV}$. Different values of $\beta$ for increasing vs. decreasing load may be specified.

The representation of suspension force-deflection characteristics by the above method provided a convenient means within the subsequent

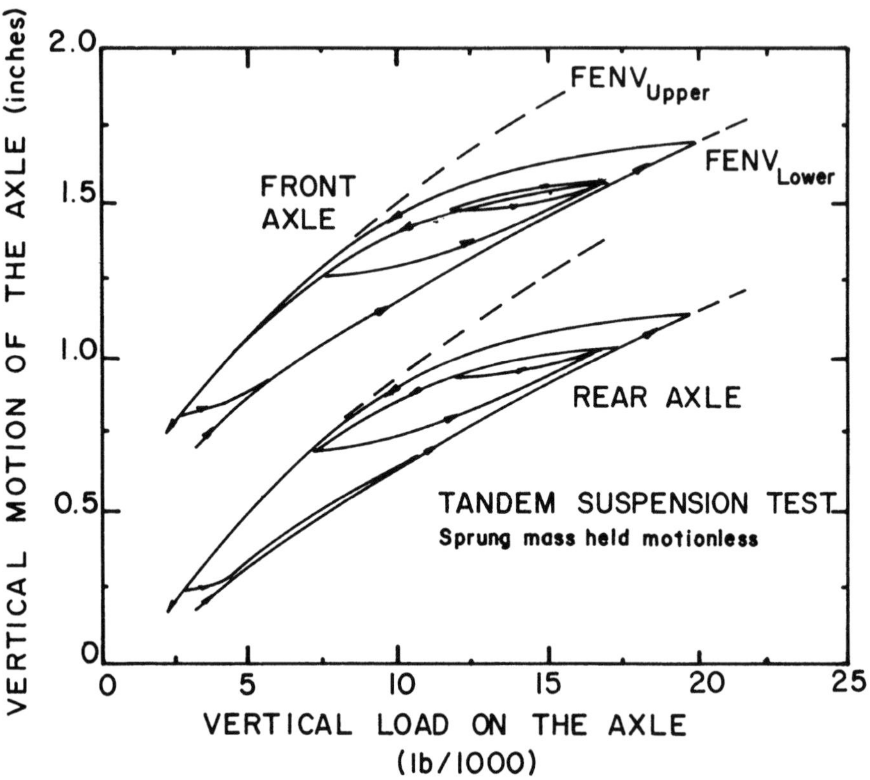


Fig. 4. Rear tandem suspension force-deflection measurement.

parameter sensitivity study for examining realistic variations in the rear suspension properties. For example, reduction in size of the outer envelope loops, $F_{ENV}$ (closer together), corresponds physically to a reduction in the suspension coulomb friction, yet remains consistent with the overall qualitative nature of the measured data.

## 5. VALIDATION OF THE BASELINE COMPUTER MODEL

Following the program modifications discussed above, several computer runs were performed to observe the influence of these modifications and to determine a final baseline set of parameters representative of the average 60-mph vehicle stop. Vehicle test data showed considerable variation in stopping distances for repeated 60-mph stops (280 ft.-340 ft.), thereby confusing any definition of a baseline performance. However, since the majority of these test repeats did produce stopping distances between 285 ft-310 ft. and displayed similar cycling and dynamic behavior, a representative test result from this majority was selected as the baseline measured response. The final baseline computer representation of the average vehicle response included side-to-side brake torque imbalances, hysteresis in the rear brakes, air system transport lags at low line pressures, and the above-described rear suspension force-deflection characteristics—all selected to closely reflect the measured component characteristics discussed in the previous sections, and to further the agreement between initial simulation attempts and full-scale vehicle test results.

Figure 5 shows a comparison of the baseline measured vehicle response and the final baseline computer model representation for front and rear (leading tandem) axle brake pressures, front and rear axle "worst" wheel speeds, and vehicle velocity and acceleration. As can be seen, the level of agreement between the measured and simulated vehicle response is quite good, particularly in light of the variability displayed by the measured data for repeated tests. The level and frequency of measured brake pressures and wheel speeds during cycling are closely approximated by the simulation results. Furthermore, the baseline computer representation now predicted a stopping distance of 295 feet as compared to 296 feet for the baseline vehicle test.

Figure 6 illustrates the intensity of severe pitching and bouncing that an antiskid braking maneuver can evoke in a commercial vehicle of this class. The variables shown here are from the same baseline computer simulation run as Figure 5. The predicted front axle load on this vehicle routinely exceeds 30,000 lbs. during antiskid cycling with several cycles producing maximum loadings of nearly 40,000 lbs. Likewise, the predicted sprung mass pitch and bounce excursions during cycling have peak-peak amplitudes of approximately 1.5 deg. and 1 in., respectively.

The severe dynamic response which is normally associated with most commercial vehicle antiskid braking is significantly amplified when side-to-side brake torque imbalances and hysteresis effects are also present. Such additional factors, as discussed in the previous section, promote greater opportunities for wheels to free-roll during cycling, thereby magnifying the variations in tire braking forces from maximum levels to near zero. The pulse-like character of the measured and simulated vehicle acceleration traces directly reflects these extreme variations in rear tire braking forces. Figure 7 demonstrates the reduction in pitch and bounce motions predicted by the computer model when hysteresis is removed from the rear brakes. In addition, the predicted stopping distance is significantly decreased from 295 feet to 253 feet.

As will be indicated in the next section, the presence of hysteresis alone, without side-to-side brake torque imbalances and air system transport lags, is not, in itself, sufficient to cause a significant degradation in braking performance. Rather, it is the dynamical interaction amongst these less desirable brake properties during antiskid cycling that precipitates the markedly diminished braking performance in this vehicle.

## 6. PARAMETER SENSITIVITY STUDY USING THE BASELINE COMPUTER MODEL

Findings of the parameter sensitivity study performed on the baseline computer model are presented and discussed in this section. A summary of selected braking performance and descriptive numerics is provided in tabular form for each parameter variation examined.

The following categories were examined in the sensitivity analysis:

- •Brake System Properties •Tire Traction Characteristics
- •Mass and Inertia Properties •Suspension Properties

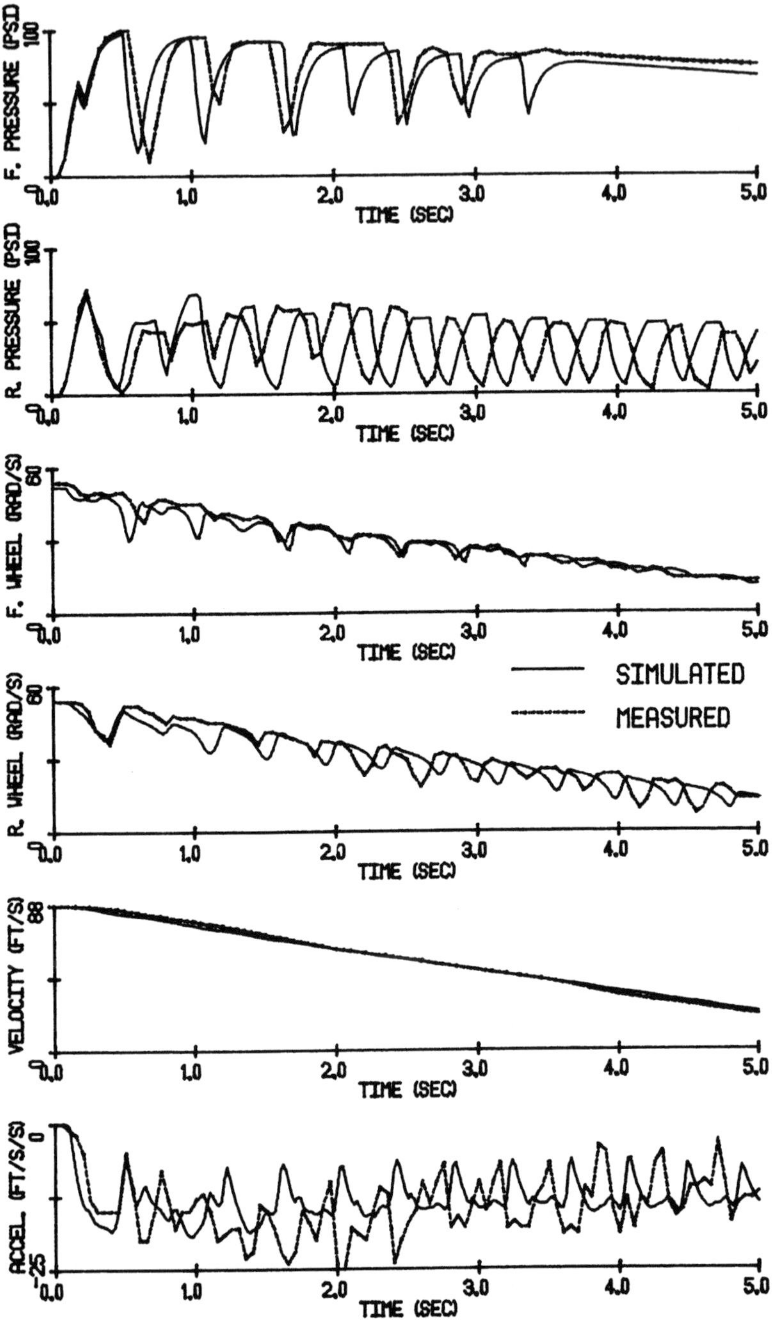


Fig. 5. Comparison of simulated and measured baseline vehicle response.

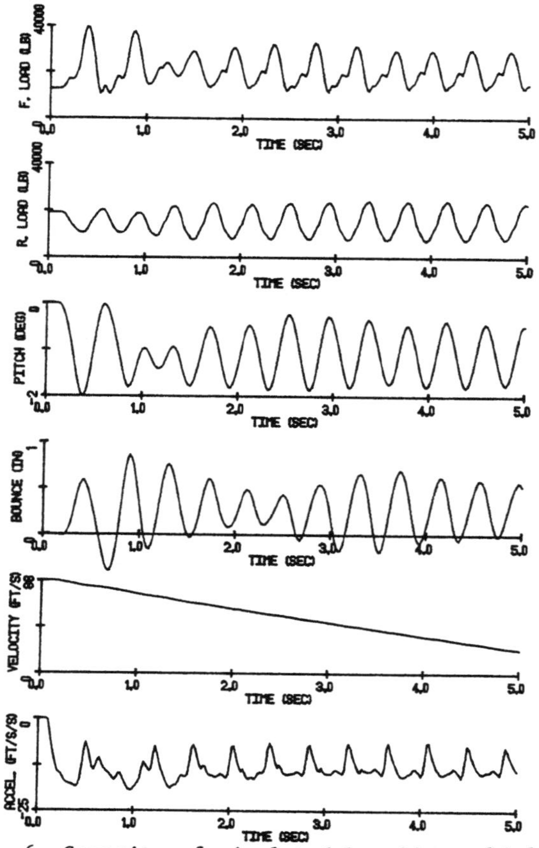


Fig. 6. Severity of simulated baseline vehicle response.

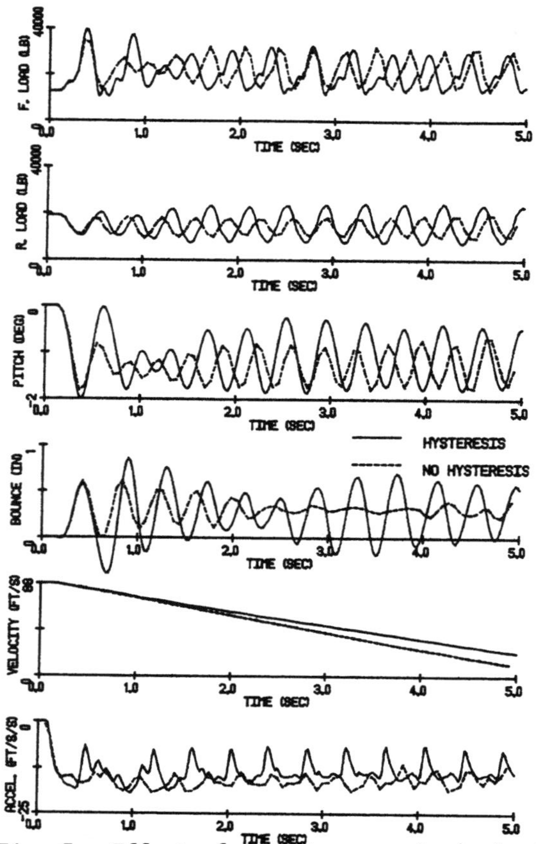


Fig. 7. Effect of removing rear brake hysteresis

•Wheel Dynamics •Antiskid System Characteristics

Table 3 presents the specific parameter variations studied and a summary of results showing predicted stopping distance and time, number of antiskid pressure cycles occurring for each axle, and the maximum vertical load experienced by the front axle. The baseline computer representation results are listed first, followed by the results obtained for each of the described parameter variations from the baseline case.

*Brake System Variations* - The brake system results shown in Table 3 clearly indicate a strong sensitivity of braking performance to the combined effects of rear brake hysteresis and side-to-side brake torque imbalances. The presence of both these factors were necessary for accurate representation of the dynamic response of the baseline vehicle. As discussed in the previous section, the synergistic mechanism involving rear brake hysteresis, side-to-side imbalances, and air system transport lags, is the principal brake-related reason for the reduced level of braking performance displayed by the baseline vehicle. The parameter sensitivity results further suggest that interruption of this synergism by removal or reduction of either brake hysteresis (Items 1.3, 1.4) or side-to-side brake torque imbalance (Item 1.5) can lead to significantly improved braking performance.

The other brake-related items of interest in Table 3 are the effects of reduced rear brake effectiveness (Item 1.2) and load proportioning (Items 1.11, 1.12). Each of these influences are similar in effect and are advantageous for two reasons: (1) the reduced torque output of the rear brake, whether deriving from a reduction in brake effectiveness or the indicated proportioning valve mechanisms, results in improved brake torque distribution on the vehicle, thereby lessening the redistribution requirements performed by the antiskid system, and (2) the reduced rear torque raises the minimum brake pressures experienced during cycling, thereby reducing the influences of hysteresis and air supply time lags, as discussed above.

The proportioning Items 1.11 and 1.12 demonstrate the effects of including load-sensitive brake proportioning valves, in addition to antiskid, on each axle. Item 1.11 represents an ideal valve which modulated treadle pressure near optimally at each axle in proportion to the prevailing vertical load, with an upper pressure limit determined by the demanded treadle pressure. Item 1.12 was an attempt to emulate the ideal proportioning scheme, 1.11, by use of a physically realistic method which employed suspension deflection as the mechanical means for estimating vertical axle loads.

Finally, Item 1.10 shows the predicted antiskid braking results for this vehicle with an ideal brake system having no hysteresis, imbalance, or air system transport lag effects.

*Mass and Inertia Variations* - The effects of payload location are demonstrated by Items 2.1 and 2.2 in Table 3. Lowering of the payload produces less pitching and bouncing motion with less fore-aft load transfer and correspondingly fewer antiskid cycles. Forward movement of the payload worsens the vehicle brake torque distribution.

Items 2.3 and 2.4 indicate an improvement in braking performance with vehicle pitch inertia modified from its baseline value. Presumably, the pitch and bounce dynamics associated with the pitch inertia of the baseline vehicle were more susceptible to antiskid cycling frequencies than the pitch and bounce dynamics resulting from increased or decreased values of inertia.

*Tire Property Variations* - Items 3.1 and 3.2 represent decreases in peak tire-road friction and the total tire-road friction characteristic, respectively. These results are interesting in that they do not demonstrate a proportional increase in stopping distance accompanying the lowered friction. Due to faster brake pressure releases during each antiskid cycle caused by the reduced tire traction, improvement in antiskid cycling efficiency offset much of the loss in braking performance expected from the reduced friction condition.

Variations in the tire vertical stiffness (Item 3.3) and tire traction sensitivity to large vertical loads (Item 3.4) demonstrate little or no effect upon braking performance. The negligible influence played by tire traction sensitivity for large vertical loads occurs because heavy tire loading during cycling corresponds to a low wheel slip condition, thereby minimizing any traction losses/gains obtained from the load sensitive property of the tire.

*Suspension Property Variations* - Variation of front suspension characteristics demonstrated minimal influence upon stopping performance. However, variation of rear suspension characteristics did have significant impact, particularly Items 4.4 and 4.5, rear Coulomb friction and spring rate changes. Lowering of the rear suspension Coulomb friction level promoted increased pitching and bouncing motion of the vehicle and correspondingly poorer braking performance. The number of front-axle antiskid cycles more than doubled for this variation. Reduction in the stiffness of the rear spring rate altered the pitch and bounce motions in the opposite manner, producing a more controlled stop with fewer antiskid cycles on all axles.

Item 4.6 indicates minimal effects on stopping performance for this vehicle from increasing rear tandem suspension interaxle load transfer from 5% (baseline) to 20%.

*Wheel Dynamics* - Item 5.1, wheel inertia, was the only wheel variation examined in this category and demonstrated negligible influence upon stopping performance.

*Antiskid System Variations* - Item 6.1 shows the effects on stopping performance for a faster prediction (brake release) by the antiskid logic module, and Item 6.2 for a faster re-selection (brake reapplication) by the logic module.

Items 6.3-6.5 are related to the operating characteristics of the antiskid pressure modulation valve. Item 6.3 shows improvement in stopping performance from increased solenoid valve actuation delays for both release and re-application during each cycle. The small increase in braking performance indicated here is principally a result of deeper wheel slip excursions for the "best" wheel on each axle promoted by the increased valve lags.

Item 6.4 demonstrates the effect of altering the "pneumatic logic" of the modulator valve. During each pressure increase portion of a cycle, the "pneumatic logic" mechanism of the valve causes the rate of pressure increase to abruptly switch to a slow, linear rise rate. This mechanism is active during the modulation of rear axle pressure shown in Figure 5. By lowering the switching point 20% from the baseline operation, a significant improvement in braking performance is

Table 3. Parameter Sensitivity Results

| Parameter Variation | Description | Stopping Distance (Ft.) | Stopping Time (Sec.) | Number of Cycles Front | Number of Cycles Middle | Number of Cycles Rear | Maximum Front Load (lb) |
|---|---|---|---|---|---|---|---|
| | Baseline | 295 | 6.55 | 8 | 14 | 14 | 39,600 |
| 1.1 | Front Brake Effectiveness Increased 25% | 279 | 6.04 | 13 | 12 | 12 | 40,400 |
| 1.2 | Rear Brake Effectiveness Decreased 25% | 259 | 5.68 | 3 | 7 | 7 | 37,600 |
| 1.3 | Rear Brake Hysteresis Decreased 50% | 272 | 6.08 | 8 | 14 | 14 | 39,000 |
| 1.4 | Rear Brake Hysteresis Removed | 253 | 5.72 | 8 | 14 | 14 | 38,000 |
| 1.5 | Rear Side-to-Side Brake Imbalances Removed | 252 | 5.49 | 6 | 11 | 11 | 40,400 |
| 1.6 | Air System Transport Lag Removed | 287 | 6.34 | 8 | 14 | 14 | 40,000 |
| 1.7 | Air System Losses During Cycling Decreased 50% | 292 | 6.54 | 15 | 15 | 15 | 39,600 |
| 1.8 | Front Brake Fade Sensitivity Increased 50% | 306 | 6.82 | 6 | 14 | 15 | 39,600 |
| 1.9 | Rear Brake Hysteresis Increased 25% | 296 | 6.59 | 8 | 14 | 14 | 39,600 |
| 1.10 | Ideal Brakes (No Hysteresis, Imbalances, or Air System Transport Lag) | 230 | 5.12 | 10 | 14 | 14 | 39,200 |
| 1.11 | Ideal Vertical Load Proportioning | 259 | 5.68 | 3 | 9 | 9 | 38,300 |
| 1.12 | Suspension Deflection Load Proportioning | 260 | 5.64 | 4 | 6 | 6 | 39,000 |
| 2.1 | Payload Lowered 2.0 ft. | 259 | 5.72 | 6 | 13 | 13 | 34,600 |
| 2.2 | Payload Moved Forward 1.5 ft. | 312 | 6.90 | 5 | 17 | 17 | 39,000 |
| 2.3 | Vehicle Pitch Inertia Decreased 25% | 269 | 5.90 | 8 | 14 | 14 | 38,200 |
| 2.4 | Vehicle Pitch Inertia Increased 25% | 282 | 6.23 | 8 | 13 | 13 | 40,600 |
| 3.1 | Peak Tire/Road Friction Decreased 10% | 303 | 6.62 | 11 | 15 | 15 | 39,000 |
| 3.2 | Total Tire/Road Friction Decreased 10% | 304 | 6.65 | 11 | 15 | 15 | 38,600 |
| 3.3 | Tire Vertical Stiffness Decreased 20% | 300 | 6.81 | 9 | 14 | 14 | 39,200 |
| 3.4 | Load Sensitivity of Tires Removed for Tire Loads Exceeding 10,000 lb | 296 | 6.59 | 8 | 14 | 14 | 39,600 |

Table 3 (Cont.)

| Parameter Variation | Description | Stopping Distance (Ft.) | Stopping Time (Sec.) | Number of Cycles | | | Maximum Front Load (lb) |
|---|---|---|---|---|---|---|---|
| | | | | Front | Middle | Rear | |
| 4.1 | Front Suspension Coulomb Friction Decreased 50% | 297 | 6.77 | 8 | 15 | 15 | 42,200 |
| 4.2 | Front Suspension Coulomb Friction Increased 50% | 273 | 5.99 | 8 | 13 | 13 | 38,400 |
| 4.3 | Front Suspension Spring Rate Doubled | 288 | 6.48 | 3 | 14 | 14 | 31,200 |
| 4.4 | Rear Suspension Coulomb Friction Decreased 50% | 325 | 6.90 | 19 | 14 | 14 | 39,400 |
| 4.5 | Rear Suspension Spring Rate Decreased 50% | 253 | 5.51 | 6 | 11 | 11 | 44,000 |
| 4.6 | Rear Suspension Interaxle Load Transfer Increased from 5% (Baseline) to 20% | 287 | 6.34 | 8 | 13 | 13 | 39,000 |
| 5.1 | All Wheel Inertias Decreased 25% | 296 | 6.60 | 9 | 14 | 14 | 39,800 |
| 6.1 | Antiskid Logic Module Prediction Level Quickened by ~5% of Wheel Slip | 304 | 6.78 | 9 | 15 | 15 | 38,000 |
| 6.2 | Antiskid Logic Module Re-Selection Level Quickened by ~5% of Wheel Slip | 285 | 6.33 | 8 | 14 | 14 | 39,600 |
| 6.3 | Antiskid Solenoid Valve Delays Increased 50% | 287 | 6.28 | 9 | 13 | 13 | 39,000 |
| 6.4 | Antiskid "Pneumatic Logic" Level Decreased by 20% | 263 | 5.78 | 4 | 11 | 11 | 39,600 |
| 6.5 | Antiskid Pressure Rise and Exhaust Rates Reduced by 20% | 278 | 6.28 | 7 | 13 | 13 | 39,400 |
| 6.6 | Antiskid "Average Wheel" Option | 265 | 5.80 | 6 | 12 | 12 | 40,000 |

predicted. The principal effect of this valve modification was to cause a faster brake release during each cycle, resulting in improved wheel cycling with increased minimum wheel slip levels.

Item 6.5 shows the influence of reducing the pressure discharge and application rates of the modulator valve. The resulting slow-down in pressure release and build-up produced improvement in wheel cycling, attained primarily from higher levels of minimum wheel slip.

Lastly, Item 6.6 demonstrates the improvement in stopping performance which can be expected by employement of "average wheel" axle control on this vehicle. "Average wheel" antiskid systems simply utilize the average of left and right wheel speeds for each axle within the logic module in place of the axle's "worst" wheel speed. While such systems usually provide improved braking from higher average wheel slip conditions, the likelihood of diminished vehicle directional stability is generally enhanced.

Finally, it should be reiterated that the mechanism having primary influence over the stopping performance of this vehicle was the aforementioned interactive effects of side-to-side rear brake imbalance and hysteresis. Many of the other parameter variations which displayed significant influence upon stopping performance, did so indirectly, by modifying or lessening the adverse effects of the baseline brake imbalance-hysteresis mechanism. Primarily for this reason, the results of the parameter sensitivity study presented here may not be applicable to other vehicles, particularly ones not having similar brake system properties.

## 7. CONCLUSIONS

The antiskid braking performance of the examined baseline vehicle was primarily dominated by the adverse effects deriving from rear brake torque imbalances and hysteresis precipitated during wheel cycling. The most noteworthy findings of the parameter sensitivity study for improving the baseline stopping performance, outside of reducing the side-to-side imbalances and rear brake hysteresis, were found to be: (a) reduction in rear brake effectiveness or use of load proportioning valves, (b) modification of antiskid system operation, and (c) modification of rear suspension properties.

This study has also presented the opportunity to demonstrate the utility of detailed computer models in representing and studying the frequently complicated and inter-active properties of complex physical systems. While it is usually desirable to employ as simple a mathematical model as possible to represent a dynamical system under study, in certain cases the complexity of the dynamical system precludes use of simple mathematical approximations which would fail in representing important detailed mechanisms. Such was the case in this study where relatively obscure brake system properties combined to strongly influence the predicted braking performance during antiskid cycling.

## ACKNOWLEDGEMENT

The support of this work by the Motor Vehicle Manufacturers Association of America is greatly appreciated.

REFERENCES


1. Murphy, R.W., et al. "A Computer-Based Mathematical Method for Predicting the Braking Performance of Trucks and Tractor-Trailers." Phase I Report, Motor Truck Braking and Handling Performance Study, Highway Safety Res. Inst., Univ. of Michigan, Sept. 15, 1972.
2. Bernard, J.E., et al. "A Computer-Based Mathematical Method for Predicting the Directional Response of Trucks and Tractor-Trailers." Phase II Report, Motor Truck Braking and Handling Performance Study, Highway Safety Res. Inst., Univ. of Michigan, Rept. No. UM-HSRI-PF-73, June 1, 1973.
3. Winkler, C.B., et al. "Predicting the Braking Performance of Trucks and Tractor-Trailers." Phase III Report, Truck and Tractor-Trailer Braking and Handling Project, Highway Safety Res. Inst., Univ. of Michigan, Rept. No. UM-HSRI-76-26, June 1976.
4. Fancher, P.S. and MacAdam, C.C. "Computer Analysis of Antilock System Performance in the Braking of Commercial Vehicles." Proceedings of Inst. of Mech. Engrs. Conference on Braking of Road Vehicles, March 1976.
5. Fancher, P.S. "Pitching and Bouncing Dynamics Excited During Antilock Braking of a Heavy Truck." 5th VSD-2nd IUTAM Symposium on Dynamics of Vehicles on Roads and Tracks, Vienna, Sept. 19-23, 1977.
6. Winkler, C.B. "Analysis and Computer Simulation of the Four Elliptical Leaf Spring Tandem Suspension." SAE Paper No. 740136, Feb. 25-Mar. 1, 1974.
7. MacAdam, C.C. "A General-Purpose Mathematical Model for Simulating Antilock Systems." Report to Motor Vehicle Manufacturers Association, by Highway Safety Res. Inst., Univ. of Michigan, Feb. 15, 1973.
8. MacAdam, C.C. "A General-Purpose Simulation for Antiskid Braking Systems." Proceedings of a Symposium on Commercial Vehicle Braking and Handling, Highway Safety Res. Inst., Univ. of Michigan, Rept. No. UM-HSRI-PF-75-6, May 5-7, 1975.

# A PRACTICAL CALCULATION METHOD OF QUASI-STATIC CURVING PERFORMANCE OF RAILWAY BOGIE VEHICLES

Nobuo Matsui

Technical Research Center, Tokyu Car Corporation,
Yokohama, Japan

SUMMARY

This study is intended to obtain a calculation method of quasi-static curving performance of railway bogie vehicles consisting of both 6 and 14 degrees of freedom systems. In this analysis, the non-linear characteristics of contact forces between wheel and rail are taken into consideration, and the successive approximation method is applied to carry out the numerical calculations.

Although there are still areas which are in need of more detailed study, an experimental evidence suggests that this method is adequate for use in the predictions on curving performance of railway bogie vehicles with linearly tapered wheel tread.

The analytical solutions of the fundamental equations are derived to enable us the numerical calculations of curving performance even with a portable calculator.

The optimum values of the bogie design parameters such as the stiffnesses of primary and secondary suspension systems, the frictional or linear resisting moment to vehicle body/bogie yaw should be determined from the point of view of both the running safety on curved tracks with various radii of curvature and the hunting stability on straight track, and this author considers that this newly developed method is readily applicable to the former study.

## 1. INTRODUCTION

The quasi-static lateral forces such as the side thrust and the flange force on curved track are the most important physical quantities in connection with the vehicle derailment, the track destruction and the wheel flange/rail wear. Therefore, various studies mainly based on the friction theory and the linear creep theory have been carried out by a number of researchers, but these results are not necessarily satisfactory from the point of view of the practical application because of the lack of generality and accuracy caused by the inadequate assumptions with regard to the bogie construction and the contact phenomena between wheel and rail.

Stimulated by the fact mentioned above, this author has tried to analyse the entitled subject by applying the non-linear creep theory to the following three cases:

category 1. no resisting moment is applied (6 degrees of freedom),
category 2. dry frictional moment is applied (6 degrees of freedom),
category 3. linear resisting moment is applied (14 degrees of freedom) to the vehicle body/bogie yaw.

## 2. TRADITIONAL THEORIES OF VEHICLE CURVING

Heumann's friction center theory [1] is the most famous method to analyse the curving performance, but this theory involves the following group of assumptions:

1. that the wheel tread is cylindrical,
2. that the wheelsets are rigidly connected with a given frame,
3. that the tangential force at each wheel-tread obeys Coulomb's law irrespective of the magnitude of wheel/rail slip.

Despite of these impractical assumptions, this theory has been widely used, especially in Europe, for a long period due to the lack of the alternatives.

A friction theory for the case of coned wheel-tread and of elastically connected wheelsets with a bogie frame was developed by Kunieda [2], but, even in this case, an over-estimation of the tangential force at each wheel-tread with small slip against the rail (i.e., for the trailing axle wheels) is inevitable.

Recently, the analysis applying creep theory to the tangential force is being remarkable with keen interest, but the traditional theories [3, 4], assumed the linear creep force, have an impractical limitation that they are available only when the creep force is fairly smaller than the friction force or, in other words, when the radius of track curvature is at least larger than a range of 1,000-1,200 meters.

From the reasons mentioned, this author considers that the appropriate estimation of vehicle curving performance, especially on sharp curves, is difficult by applying the traditional theories, and that an establishment of the new calculation method is being required.

## 3. ASSUMPTIONS AND ANALYTICAL MODEL

To analyse the quasi-static curving performance of a bogie vehicle, the following group of assumptions is involved:

1. each wheel has a linearly tapered tread,
2. in the case of wheel/rail two points contact (on wheel tread and on wheel flange), the wheel load and the creep forces in longitudinal and lateral directions act at the first (wheel tread) contact point and only a lateral flange force acts at the second (wheel flange) contact point.

The vehicle model on curved track is shown in Fig.1. In two cases of categories 1 and 2, it is sufficient to consider a half of vehicle (6 degrees of freedom), but, in a case of category 3, a complete vehicle should be considered (14 degrees of freedom) because of the interaction between leading and trailing bogies.

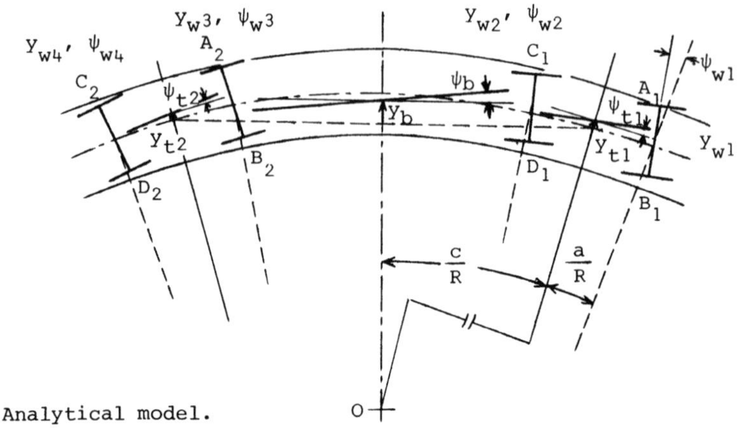


Fig. 1. Analytical model.

## 4. NON-LINEARITY OF CREEP FORCES AND EQUIVALENT CREEP COEFFICIENTS

The characteristics of creep forces acting between wheel and rail have been made fairly clear by the theoretical and the experimental studies [5, 6]. The relation between tangential force and creepage is schematically shown in Fig.2, and the tangential force is approximately given by the following equation:

$$\frac{1}{(T/\mu N)^{\beta}} = 1 + \frac{1}{(f_k \nu/\mu N)^{\beta}} \tag{1}$$

The value of β can be given by 1 to 2 and β=1.5 is considered appropriate on the basis of the experimental results [6].

It is clear from Fig.2 that the equivalent creep coefficient can be obtained as a function of the creepage as follows:

$$\frac{f}{f_k} = \frac{T}{f_k \nu} = 1/\{1+(f_k \nu/\mu N)^{\beta}\}^{1/\beta} \tag{2}$$

and this relation is shown in Fig.3.

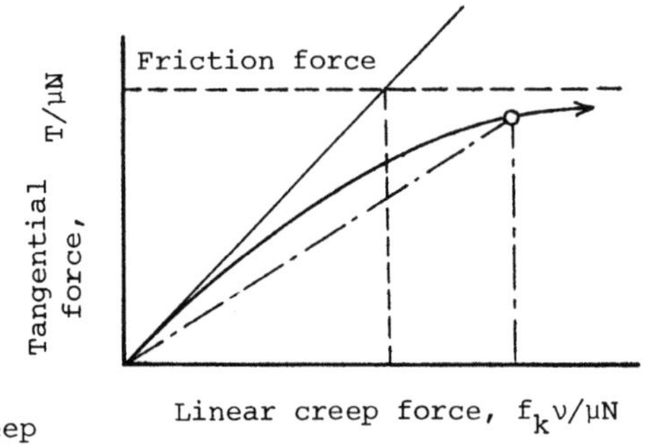


Fig.2. Non-linearity of creep force.

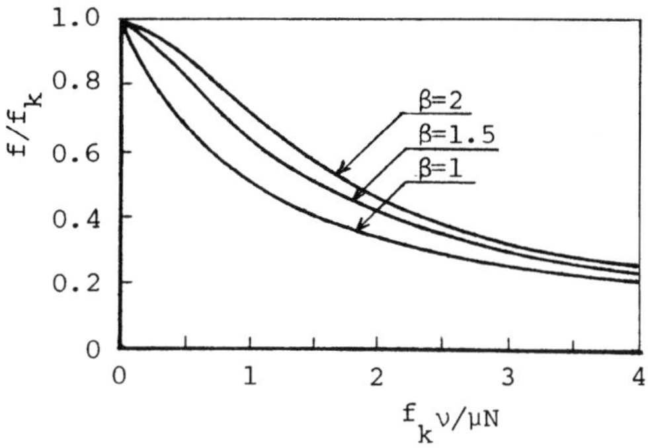


Fig.3. Equivalent creep coefficient.

The creepages between wheel and rail, on which the equivalent creep coefficients depend, are approximately given as follows:

| | longitudinal direction (positive forwards) | lateral direction (positive outwards) |
|---|---|---|
| A, B wheels | $\pm(\frac{b}{R} - \gamma\frac{y_{w1}}{r})$, | $\psi_{w1}$ |
| C, D wheels | $\pm(\frac{b}{R} - \gamma\frac{y_{w2}}{r})$, | $\psi_{w2}$ |

Therefore, the equivalent creep coefficients are unable to be determined unconditionally.

The longitudinal and lateral creep forces acting on each wheel and on each wheelset and the moment around the vertical axis on each wheelset are given by the following expressions:

| | longitudinal creep force | lateral creep force | moment ar. vert. axis |
|---|---|---|---|
| leading outside wheel; | $-f_{11A}(\frac{b}{R} - \gamma\frac{y_{w1}}{r})$, | $f_{22A}\psi_{w1}$, | - |
| leading inside wheel; | $f_{11B}(\frac{b}{R} - \gamma\frac{y_{w1}}{r})$, | $f_{22B}\psi_{w1}$, | - |
| trailing outside wheel; | $-f_{11C}(\frac{b}{R} - \gamma\frac{y_{w2}}{r})$, | $f_{22C}\psi_{w2}$, | - |
| trailing inside wheel; | $f_{11D}(\frac{b}{R} - \gamma\frac{y_{w2}}{r})$, | $f_{22D}\psi_{w2}$, | - |
| leading wheelset; | $(f_{11B}-f_{11A})(\frac{b}{R} - \gamma\frac{y_{w1}}{r})$, | $2f_{22}\psi_{w1}$, | $2f_{11}b(\frac{b}{R} - \gamma\frac{y_{w1}}{r})$ |
| trailing wheelset; | $(f_{11D}-f_{11C})(\frac{b}{R} - \gamma\frac{y_{w2}}{r})$, | $2f'_{22}\psi_{w2}$, | $2f'_{11}b(\frac{b}{R} - \gamma\frac{y_{w2}}{r})$ |

with

$f_{11A} + f_{11B} = 2f_{11}$, $\quad f_{11C} + f_{11D} = 2f'_{11}$

$f_{22A} + f_{22B} = 2f_{22}$, $\quad f_{22C} + f_{22D} = 2f'_{22}$

## 5. FUNDAMENTAL EQUATIONS OF STEADY STATE CURVING AND THEIR SOLUTIONS

The external forces acting on the wheelsets, on the bogie frames and on the vehicle body consist of the above mentioned creep forces, the spring forces, the uncompensated centrifugal forces and the flange forces (only when the flanges contact the rail). The fundamental equations of steady state curving can be obtained from the equilibrium conditions of the external forces in lateral direction and their moments around the vertical axis.

*Cases of Categories 1 and 2*

The fundamental equations in each case of category 1 (M=0) and category 2 (M≠0) are given as follows, for example, when the leading outside wheel flange contacts the rail:

$$\begin{pmatrix} -2f_{22} & -k_y & -ak_y & 0 & 0 & 1 \\ k_\psi & 0 & -k_\psi & 0 & 0 & 0 \\ 0 & 2k_y & 0 & -k_y & 0 & 0 \\ -k_\psi & 0 & 2(k_\psi+a^2k_y) & ak_y & -k_\psi & 0 \\ 0 & -k_y & ak_y & k_y & -2f_{22}' & 0 \\ 0 & 0 & -k_\psi & 2f_{11}'\gamma\frac{b}{r} & k_\psi & 0 \end{pmatrix} \begin{pmatrix} \psi_{w1} \\ y_t \\ \psi_t \\ y_{w2} \\ \psi_{w2} \\ F_A \end{pmatrix}$$

$$= \begin{pmatrix} 0 & m_w g & -k_y & 0 \\ 2f_{11}b^2+ak_\psi & 0 & -2f_{11}\gamma\frac{b}{r} & 0 \\ 0 & (\frac{1}{2}m_b+m_t)g & k_y & \frac{1}{2s} \\ 0 & 0 & ak_y & 1 \\ 0 & m_w g & 0 & 0 \\ 2f_{11}'b^2-ak_\psi & 0 & 0 & 0 \end{pmatrix} \begin{pmatrix} \frac{1}{R} \\ \alpha \\ \delta \\ M \end{pmatrix} \quad (3)$$

Therefore, the solutions of Eq.(3) are obtained as follows:

$$y_{w1}=\delta, \quad y_{w2}=a\Delta_{yw2}/\Delta, \quad y_t=a\Delta_{yt}/\Delta$$
$$\psi_{w1}=\Delta_{\psi w1}/\Delta, \quad \psi_{w2}=\Delta_{\psi w2}/\Delta, \quad \psi_t=\Delta_{\psi t}/\Delta, \quad F_A=\Delta_{FA}/\Delta \quad (4)$$

where,

$$\Delta = f_{22}'+f_{11}'\{2(\frac{1}{ak_y}+\frac{2a}{k_\psi})f_{22}'-1\}\gamma\frac{b}{r} \quad (5.1)$$

$$\Delta_{yw2} = \left(\frac{2}{ak_y}f_{22}'-1\right)\{f_{11}\left(\frac{b}{R}-\gamma\frac{\delta}{r}\right)+f_{11}'\frac{b}{R}\}\frac{b}{a}+f_{22}'\left(\frac{4b}{k_\psi}f_{11}'\frac{b}{R}-\frac{2a}{R}+\frac{\delta}{a}\right)+\frac{1}{4}mg\alpha$$

$$+\frac{1}{2}\{\left(\frac{2}{ak_y}f_{22}'-1\right)+\frac{a}{2c}\}\frac{M}{a} \qquad (5.2)$$

$$\Delta_{yt} = \frac{1}{2}[f_{11}\left(\frac{2}{ak_y}f_{22}'-1\right)\frac{b}{a}\left(\frac{b}{R}-\gamma\frac{\delta}{r}\right)+f_{11}'\{\left(\frac{1}{ak_y}+\frac{2a}{k_\psi}\right)f_{22}'-1\}\frac{b}{a}\left(\frac{b}{R}+\gamma\frac{\delta}{r}\right)]$$

$$-f_{22}'\left(\frac{a}{R}-\frac{\delta}{a}\right)+\frac{1}{8}\{m+\frac{2\Delta}{ak_y}(m_b+2m_t)\}g\alpha$$

$$+\frac{1}{4}\{\left(\frac{2}{ak_y}f_{22}'-1\right)+\frac{a}{2c}\left(1+\frac{2\Delta}{ak_y}\right)\}\frac{M}{a} \qquad (5.3)$$

$$\Delta_{\psi w1} = \frac{1}{2}\{(f_{11}+f_{11}')+\frac{4a}{k_\psi}f_{11}\Delta-\frac{4a}{k_\psi}f_{22}'f_{11}'\left(1-\frac{2}{ak_y}f_{11}\gamma\frac{b}{r}\right)\}\frac{b}{a}\left(\frac{b}{R}-\gamma\frac{\delta}{r}\right)$$

$$+\{\Delta+f_{22}'\left(1+\frac{2}{ak_y}f_{11}'\gamma\frac{b}{r}\right)\}\frac{a}{R}-\frac{1}{8}\left(1+\frac{2}{ak_y}f_{11}'\gamma\frac{b}{r}\right)mg\alpha$$

$$+\frac{1}{4}\{\left(1+\frac{8}{k_yk_\psi}f_{22}'f_{11}'\gamma\frac{b}{r}\right)-\frac{a}{2c}\left(1+\frac{2}{ak_y}f_{11}'\gamma\frac{b}{r}\right)\}\frac{M}{a} \qquad (5.4)$$

$$\Delta_{\psi w2} = \frac{1}{2}\{(f_{11}+f_{11}')+\frac{4a}{k_\psi}f_{11}f_{11}'\gamma\frac{b}{r}\}\frac{b}{a}\left(\frac{b}{R}-\gamma\frac{\delta}{r}\right)+f_{11}'\gamma\frac{b}{r}\cdot\frac{a}{R}$$

$$-\frac{1}{8}\{1+2\left(\frac{1}{ak_y}+\frac{2a}{k_\psi}\right)f_{11}'\gamma\frac{b}{r}\}mg\alpha$$

$$+\frac{1}{4}[\left(1+\frac{4a}{k_\psi}f_{11}'\gamma\frac{b}{r}\right)-\frac{a}{2c}\{1+2\left(\frac{1}{ak_y}+\frac{2a}{k_\psi}\right)f_{11}'\gamma\frac{b}{r}\}]\frac{M}{a} \qquad (5.5)$$

$$\Delta_{\psi t} = \frac{1}{2}\{(f_{11}+f_{11}')-\frac{4a}{k_\psi}f_{22}'f_{11}'\left(1-\frac{2}{ak_y}f_{11}\gamma\frac{b}{r}\right)\}\frac{b}{a}\left(\frac{b}{R}-\gamma\frac{\delta}{r}\right)$$

$$+f_{22}'\left(1+\frac{2}{ak_y}f_{11}'\gamma\frac{b}{r}\right)\frac{a}{R}-\frac{1}{8}\left(1+\frac{2}{ak_y}f_{11}'\gamma\frac{b}{r}\right)mg\alpha$$

$$+\frac{1}{4}\{\left(1+\frac{8}{k_yk_\psi}f_{22}'f_{11}'\gamma\frac{b}{r}\right)-\frac{a}{2c}\left(1+\frac{2}{ak_y}f_{11}'\gamma\frac{b}{r}\right)\}\frac{M}{a} \qquad (5.6)$$

$$\Delta_{FA} = [(f_{22}+f_{22}')(f_{11}+f_{11}')+\frac{4a}{k_\psi}f_{22}f_{11}\Delta$$

$$-\frac{4a}{k_\psi}f_{22}'f_{11}'\{f_{22}-f_{11}\left(\frac{2}{ak_y}f_{22}+1\right)\gamma\frac{b}{r}\}]\frac{b}{a}\left(\frac{b}{R}-\gamma\frac{\delta}{r}\right)$$

$$+2\{f_{22}(\Delta+f_{22}')+f_{22}'f_{11}'\left(\frac{2}{ak_y}f_{22}+1\right)\gamma\frac{b}{r}\}\frac{a}{R}$$

$$+\frac{1}{4}\{\Delta-f_{22}-f_{11}'\left(\frac{2}{ak_y}f_{22}+1\right)\gamma\frac{b}{r}\}mg\alpha$$

$$+\frac{1}{2}[(f_{22}+f_{22}')+\frac{4a}{k_\psi}f_{22}'f_{11}'\left(\frac{2}{ak_y}f_{22}+1\right)\gamma\frac{b}{r}$$

$$+\frac{a}{2c}\{\Delta-f_{22}-f_{11}'\left(\frac{2}{ak_y}f_{22}+1\right)\gamma\frac{b}{r}\}]\frac{M}{a} \qquad (5.7)$$

The side thrust of leading outside wheel with flange contact can be obtained as follows:

$$Q_A = F_A - f_{22A}\psi_{w1} \tag{6}$$

*Case of Category 3*

In this case, the fundamental equations consist of 14 simultaneous linear equations as shown by Eq.(7), for example, when the leading outside wheel of each bogie contacts the rail on its tread and flange:

$$\begin{pmatrix}
1 & -2(f_{22})_1 & 0 & 0 & -k_y & -ak_y & 0 & 0 & 0 & 0 & 0 & 0 & 0 & 0 \\
0 & k_\psi & 0 & 0 & 0 & -k_\psi & 0 & 0 & 0 & 0 & 0 & 0 & 0 & 0 \\
0 & 0 & k_y & -2(f_{22})'_1 & -k_y & ak_y & 0 & 0 & 0 & 0 & 0 & 0 & 0 & 0 \\
0 & 0 & 2(f_{11})'_1\gamma\frac{b}{r} & k_\psi & 0 & -k_\psi & 0 & 0 & 0 & 0 & 0 & 0 & 0 & 0 \\
0 & 0 & -k_y & 0 & k_{y0}+2k_y & 0 & -k_{y0} & -ck_{y0} & 0 & 0 & 0 & 0 & 0 & 0 \\
0 & -k_\psi & ak_y & -k_\psi & 0 & k_{\psi0}+2(k_\psi+a^2k_y) & 0 & -k_{\psi0} & 0 & 0 & 0 & 0 & 0 & 0 \\
0 & 0 & 0 & 0 & -k_{y0} & 0 & 2k_{y0} & 0 & 0 & -k_{y0} & 0 & 0 & 0 & 0 \\
0 & 0 & 0 & 0 & -ck_{y0} & -k_{\psi0} & 0 & 2(k_{\psi0}+c^2k_{y0}) & -k_{\psi0} & ck_{y0} & 0 & 0 & 0 & 0 \\
0 & 0 & 0 & 0 & 0 & 0 & 0 & -k_{\psi0} & k_{\psi0}+2(k_\psi+a^2k_y) & 0 & -k_\psi & ak_y & -k_\psi & 0 \\
0 & 0 & 0 & 0 & 0 & 0 & -k_{y0} & ck_{y0} & 0 & k_{y0}+2k_y & 0 & -k_y & 0 & 0 \\
0 & 0 & 0 & 0 & 0 & 0 & 0 & 0 & -k_\psi & 0 & k_\psi & 2(f_{11})'_2\gamma\frac{b}{r} & 0 & 0 \\
0 & 0 & 0 & 0 & 0 & 0 & 0 & 0 & ak_y & -k_y & -2(f_{22})'_2 & k_y & 0 & 0 \\
0 & 0 & 0 & 0 & 0 & 0 & 0 & 0 & -k_\psi & 0 & 0 & 0 & k_\psi & 0 \\
0 & 0 & 0 & 0 & 0 & 0 & 0 & 0 & -ak_y & -k_y & 0 & 0 & -2(f_{22})_2 & 1
\end{pmatrix}
\begin{pmatrix}
F_{A1} \\ \psi_{w1} \\ y_{w2} \\ \psi_{w2} \\ y_{t1} \\ \psi_{t1} \\ y_b \\ \psi_b \\ \psi_{t2} \\ y_{t2} \\ \psi_{w4} \\ y_{w4} \\ \psi_{w3} \\ F_{A2}
\end{pmatrix}$$

$$= \begin{pmatrix}
0 & -k_y & m_w g \\
2(f_{11})_1 b^2+ak_\psi & -2(f_{11})_1\gamma\frac{b}{r} & 0 \\
0 & 0 & m_w g \\
2(f_{11})'_1 b^2-ak_\psi & 0 & 0 \\
0 & k_y & m_t g \\
ck_{\psi0} & ak_y & 0 \\
0 & 0 & m_b g \\
0 & 0 & 0 \\
-ck_{\psi0} & ak_y & 0 \\
0 & k_y & m_t g \\
2(f_{11})'_2 b^2-ak_\psi & 0 & 0 \\
0 & 0 & m_w g \\
2(f_{11})_2 b^2+ak_\psi & -2(f_{11})_2\gamma\frac{b}{r} & 0 \\
0 & -k_y & m_w g
\end{pmatrix}
\begin{pmatrix}
\frac{1}{R} \\ \delta \\ \alpha
\end{pmatrix} \tag{7}$$

Eq.(7) can be solved numerically by the use of any electronic computer with moderate capacity.

It is apparently troublesome to derive the analytical solutions of Eq.(7), but this author dared to do it by applying Shur-Frobenius' law and obtained the following approximate results which enable us to carry out the numerical calculations even with a portable calculator:

$$
\begin{aligned}
&y_{w1}=\delta, \quad y_{w2}=a\Delta_{yw2}/\Delta, \quad y_{w3}=\delta, \quad y_{w4}=a\Delta_{yw4}/\Delta,\\
&\psi_{w1}=\Delta_{\psi w1}/\Delta, \quad \psi_{w2}=\Delta_{\psi w2}/\Delta, \quad \psi_{w3}=\Delta_{\psi w3}/\Delta, \quad \psi_{w4}=\Delta_{\psi w4}/\Delta,\\
&F_{A1}=\Delta_{FA1}/\Delta, \quad F_{A2}=\Delta_{FA2}/\Delta
\end{aligned}
\tag{8}
$$

where,

$$
\begin{aligned}
\Delta = \Delta_1\Delta_2\Big(&\{1+(\tfrac{1}{c})^2 k_{\psi 0}(\tfrac{1}{k_{y0}}+\tfrac{1}{2k_y})\}+\tfrac{1}{4a}k_{\psi 0}[\{1+\tfrac{1}{2}(\tfrac{a}{c})^2\}(\tfrac{1}{\Delta_1}+\tfrac{1}{\Delta_2})\\
&+\tfrac{8}{k_y k_\psi}\{(f_{22})'_1(f_{11})'_1\tfrac{1}{\Delta_1}+(f_{22})'_2(f_{11})'_2\tfrac{1}{\Delta_2}\}\gamma\tfrac{b}{r}]\Big)
\end{aligned}
\tag{9.1}
$$

$$
\Delta_i=(f_{22})'_i+(f_{11})'_i\{2(\tfrac{1}{ak_y}+\tfrac{2a}{k_\psi})(f_{22})'_i-1\}\gamma\tfrac{b}{r}
\tag{9.2}
$$

$$
\begin{aligned}
\left.\begin{matrix}A_1\\A_2\end{matrix}\right\}=\ &\{1+(\tfrac{1}{c})^2 k_{\psi 0}(\tfrac{1}{k_{y0}}+\tfrac{1}{2k_y})\}\Delta_i+\tfrac{1}{4a}k_{\psi 0}\{1+(\tfrac{a}{c})^2+\tfrac{8}{k_y k_\psi}(f_{22})'_i(f_{11})'_i\gamma\tfrac{b}{r}\}\\
&\pm\tfrac{1}{4c}k_{\psi 0}\{1+\tfrac{2}{ak_y}(f_{11})'_i\gamma\tfrac{b}{r}\}
\end{aligned}
\tag{9.2}
$$

$$
\begin{aligned}
\left.\begin{matrix}\Delta_{yw2}\\\Delta_{yw4}\end{matrix}\right\}=\ A_j\Big(&\{\tfrac{2}{ak_y}(f_{22})'_i-1\}\{(f_{11})_i(\tfrac{b}{R}-\gamma\tfrac{\delta}{r})+(f_{11})'_i\tfrac{b}{R}\}\tfrac{b}{a}\\
&+(f_{22})'_i\{\tfrac{4b}{k_\psi}(f_{11})'_i\tfrac{b}{R}-\tfrac{2a}{R}+\tfrac{\delta}{a}\}+\tfrac{1}{4}mg\alpha\Big)\\
&+\tfrac{1}{2}(\tfrac{1}{a})^2 k_{\psi 0}\tfrac{1}{k_y}\Delta_j\Big(2(f_{22})'_i\{\tfrac{2b}{k_\psi}(f_{11})'_i\tfrac{b}{a}-1\}\tfrac{a}{R}+\tfrac{1}{2}k_y\delta+\tfrac{1}{4}mg\alpha\\
&\pm\{(f_{11})_i(\tfrac{b}{R}-\gamma\tfrac{\delta}{r})+(f_{11})'_i\tfrac{b}{R}+\tfrac{1}{2}k_y(\tfrac{a}{b})\delta\}\tfrac{b}{c}\pm\{2(f_{22})'_i-ak_y\}\tfrac{c}{R}\Big)
\end{aligned}
\tag{9.4}
$$

$$
\begin{aligned}
\left.\begin{matrix}\Delta_{\psi w1}\\\Delta_{\psi w3}\end{matrix}\right\}=\ A_j\Big(&\tfrac{1}{2}[(f_{11})_i+(f_{11})'_i+\tfrac{4a}{k_\psi}(f_{11})_i\Delta_i\\
&-\tfrac{4a}{k_\psi}(f_{22})'_i(f_{11})'_i\{1-\tfrac{2}{ak_y}(f_{11})_i\gamma\tfrac{b}{r}\}]\tfrac{b}{a}(\tfrac{b}{R}-\gamma\tfrac{\delta}{r})\\
&+[\Delta_i+(f_{22})'_i\{1+\tfrac{2}{ak_y}(f_{11})'_i\gamma\tfrac{b}{r}\}]\tfrac{a}{R}-\tfrac{1}{8}\{1+\tfrac{2}{ak_y}(f_{11})'_i\gamma\tfrac{b}{r}\}mg\alpha\Big)\\
&+\tfrac{1}{4a}k_{\psi 0}\Delta_j\Big(\{1+\tfrac{8}{k_y k_\psi}(f_{22})'_i(f_{11})'_i\gamma\tfrac{b}{r}\}\{\tfrac{2b}{k_\psi}(f_{11})_i(\tfrac{b}{R}-\gamma\tfrac{\delta}{r})+\tfrac{a}{R}\pm\tfrac{c}{R}\}\\
&\pm\{1+\tfrac{2}{ak_y}(f_{11})'_i\gamma\tfrac{b}{r}\}\{\tfrac{2b}{k_\psi}(f_{11})_i(\tfrac{b}{R}-\gamma\tfrac{\delta}{r})+\tfrac{a}{R}\}\tfrac{a}{c}\Big)
\end{aligned}
\tag{9.5}
$$

$$\left.\begin{matrix}\Delta_{\psi w2}\\ \Delta_{\psi w4}\end{matrix}\right\}= A_j\Big(\frac{1}{2}\{(f_{11})_i+(f_{11})'_i+\frac{4a}{k_\psi}(f_{11})_i(f_{11})'_i\gamma\frac{b}{r}\}\frac{b}{a}(\frac{b}{R}-\gamma\frac{\delta}{r})+(f_{11})'_i\gamma\frac{b}{r}\cdot\frac{a}{R}$$

$$-\frac{1}{8}\{1+2(\frac{1}{ak_y}+\frac{2a}{k_\psi})(f_{11})'_i\gamma\frac{b}{r}\}mg\alpha\Big)$$

$$+\frac{1}{2a}k_{\psi 0}\Delta_j\Big(\frac{b}{k_\psi}(f_{11})'_i(\frac{b}{R}-\gamma\frac{\delta}{r})-\frac{a}{2R}-\frac{1}{2k_yk_\psi}(f_{11})'_i\gamma\frac{b}{r}\,mg\alpha$$

$$\pm[\frac{b}{k_\psi}(f_{11})'_i\{1-\frac{2}{ak_y}(f_{11})_i\gamma\frac{b}{r}\}(\frac{b}{R}-\gamma\frac{\delta}{r})-\frac{1}{2}\{1+\frac{2}{ak_y}(f_{11})'_i\gamma\frac{b}{r}\}\frac{a}{R}]\frac{a}{c}$$

$$\pm\frac{1}{2}\{1+\frac{4a}{k_\psi}(f_{11})'_i\gamma\frac{b}{r}\}\frac{c}{R}\Big) \tag{9.6}$$

$$\left.\begin{matrix}\Delta_{FA1}\\ \Delta_{FA2}\end{matrix}\right\}= A_j\Big(\Big\{\{(f_{22})_i+(f_{22})'_i\}\{(f_{11})_i+(f_{11})'_i\}+\frac{4a}{k_\psi}(f_{22})_i(f_{11})_i\Delta_i$$

$$-\frac{4a}{k_\psi}(f_{22})'_i(f_{11})'_i[(f_{22})_i-(f_{11})_i\{\frac{2}{ak_y}(f_{22})_i+1\}\gamma\frac{b}{r}]\Big\}\frac{b}{a}(\frac{b}{R}-\gamma\frac{\delta}{r})$$

$$+2[(f_{22})_i\{\Delta_i+(f_{22})'_i\}+(f_{22})'_i(f_{11})'_i\{\frac{2}{ak_y}(f_{22})_i+1\}\gamma\frac{b}{r}]\frac{a}{R}$$

$$+\frac{1}{4}[\Delta_i-(f_{22})_i-(f_{11})'_i\{\frac{2}{ak_y}(f_{22})_i+1\}\gamma\frac{b}{r}]mg\alpha\Big)$$

$$+\frac{k_{\psi 0}}{2a}\Delta_j\Big(\frac{2b}{k_\psi}[(f_{22})'_i(f_{11})'_i+(f_{22})_i(f_{11})_i\{1+\frac{8}{k_yk_\psi}(f_{22})'_i(f_{11})'_i\gamma\frac{b}{r}\}](\frac{b}{R}-\gamma\frac{\delta}{r})$$

$$-[(f_{22})'_i-(f_{22})_i\{1+\frac{8}{k_yk_\psi}(f_{22})'_i(f_{11})'_i\gamma\frac{b}{r}\}]\frac{a}{R}$$

$$+\frac{1}{4}\{1+\frac{4}{k_yk_\psi}(f_{22})'_i(f_{11})'_i\gamma\frac{b}{r}\}mg\alpha$$

$$\pm[(f_{11})_i+(f_{11})'_i-\frac{2a}{k_\psi}\{(f_{22})'_i(f_{11})'_i-(f_{22})_i(f_{11})_i\}](\frac{b}{R}-\gamma\frac{\delta}{r})\frac{b}{c}$$

$$\pm\{(f_{22})_i+(f_{22})'_i\}[\frac{4}{k_yk_\psi}(f_{11})_i(f_{11})'_i\gamma\frac{b}{r}\cdot\frac{b}{a}(\frac{b}{R}-\gamma\frac{\delta}{r})+\{1+\frac{2}{ak_y}(f_{11})'_i\gamma\frac{b}{r}\}\frac{a}{R}]\frac{a}{c}$$

$$\pm[(f_{22})_i+(f_{22})'_i+\frac{4a}{k_\psi}(f_{22})'_i(f_{11})'_i\{\frac{2}{ak_y}(f_{22})_i+1\}\gamma\frac{b}{r}]\frac{c}{R}\Big) \tag{9.7}$$

In each of Eqs.(9), i=1, j=2 and upper signs of the right side correspond to the upper term of the left side, and i=2, j=1 and lower signs of the right side to the lower term of the left side. Therefore, the upper term of the left side can be transformed into the lower term by changing i from 1 to 2, j from 2 to 1 and the sign of c from (+) to (-).

## 6. PRACTICAL CALCULATION METHOD

Six unknowns in Eq.(3) and fourteen unknowns in Eq.(7) can formally be obtained by solving these equations, but they contain the equivalent creep coefficients which depend on themselves and, therefore, it is difficult to obtain these unknowns analytically. Then, this author tried to apply the following successive approximation method:

1. assume first the adequate creep coefficients (for example, Kalker's) as the equivalent ones,
2. calculate the longitudinal and lateral creepages on each wheel by Eq.(3) (or Eqs.(4), (5)) in the cases of categories 1 and 2, and by Eq.(7) (or Eqs.(8), (9)) in the case of category 3 with the assumed creep coefficients,
3. calculate the equivalent creep coefficients of second approximation by Eq.(2) with above obtained creepages,
4. repeat the above procedure until the creepages on each wheel converge to some definite values,
5. calculate finally the necessary unknowns such as the flange forces by Eq.(3) (or Eqs.(4), (5)) in the cases of categories 1 and 2 and by Eq.(7) (or Eqs.(8), (9)) in the case of category 3 with the definitive creepages on each wheel.

This method seems to be apparently complicated, but, in almost all cases, one time repeat gives the satisfactory results with practical accuracy.

## 7. NUMERICAL CALCULATION RESULTS

Some examples of numerically calculated results for the cases of categories 1 and 2 with the following fundamental data are shown in Figs.4 and 5.

$a = 1.1$ m, $b = 0.56$ m, $c = 6.9$ m, $r = 0.43$ m, $R = 600$ m,
$\delta = 0.006$ m, 0.008 m, 0.014 m, $\gamma = 1/20$, $\alpha = 0$, $\mu = 0.25$,
$N = 34.6$ kN, $k_y = 30{,}820$ kN/m, $k_\psi$, $M$ = variables.

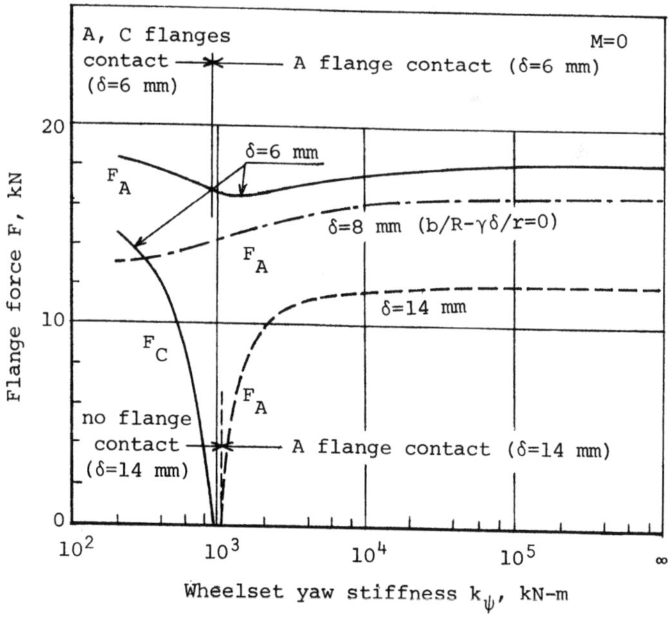


Fig.4.Relation between flange forces and wheelset yaw stiffness.

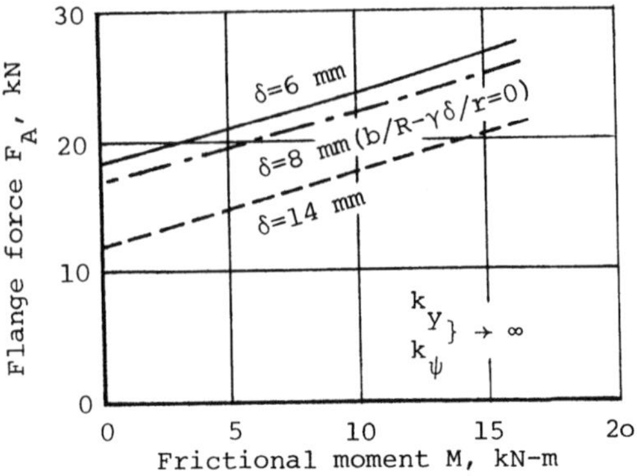


Fig.5. Effect of frictional moment on flange force.

## 8. COMPARISON BETWEEN THEORETICAL AND EXPERIMENTAL RESULTS

In order to validate convincingly the theoretical predictions, the theoretically calculated results are compared with the experimental data which were obtained from the running test of an electric car with bolsterless bogies (14 degrees of freedom system). In this test, the side thrust and the vertical load on individual wheel were measured by fitting load measuring wheels to the leading wheelset of each bogie. Furthermore, the lateral acceleration of vehicle body was measured on the floor just above each bogie center. The forces and the lateral accelerations were averaged to extract each quasi-static component.

The fundamental data of vehicle and track are:

$c = 5.5$ m, $\delta = 0.0105$ m, $k_y = 4{,}400$ kN/m, $k_\psi = \infty$

$k_{y0} = 583$ kN/m, $k_{\psi 0} = 510$ kN-m, $m_w g = 13.00$ kN, $m_t g = 23.00$ kN,

$m_b g = 241.16$ kN (114.72 kN on leading bogie, 126.44 kN on trailing),

$\mu$ (experimentally suggested) $= 0.4(1-0.00225v)$; v in m/s.

The remainders (a, b, r, R, γ) are same as those in §7.

The theoretical and the experimental results are compared in Figs.6-8 which show mutual good agreement.

## 9. CONCLUDING REMARKS

Although there are still areas which are in need of more detailed study, an experimental evidence suggests that this calculation method is adequate for use in the predictions on curving performance of railway bogie vehicles with linearly tapered wheel tread.

The optimum values of the bogie design parameters such as the stiffnesses of primary and secondary suspension systems, the frictional or linear resisting moment to vehicle body/bogie yaw should be determined from the point of view of both the running safety on curved tracks and the hunting stability on straight track, and this author is intending to study the related matters in more detail to contribute towards the development of new railway bogies in Japan.

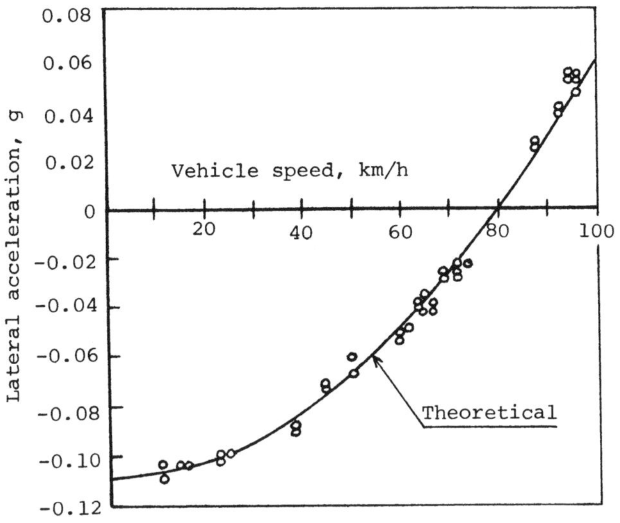


Fig.6. Comparison between theoretical and experimental results: vehicle body lateral acceleration.

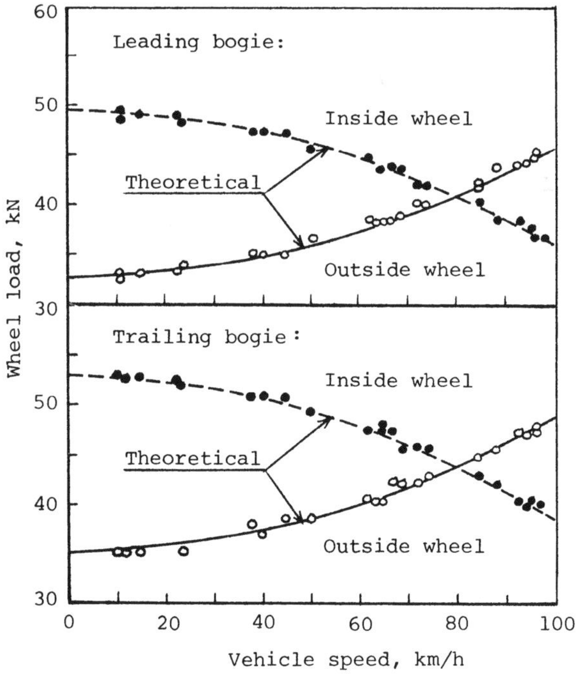


Fig.7. Comparison between theoretical and experimental results: wheel load.

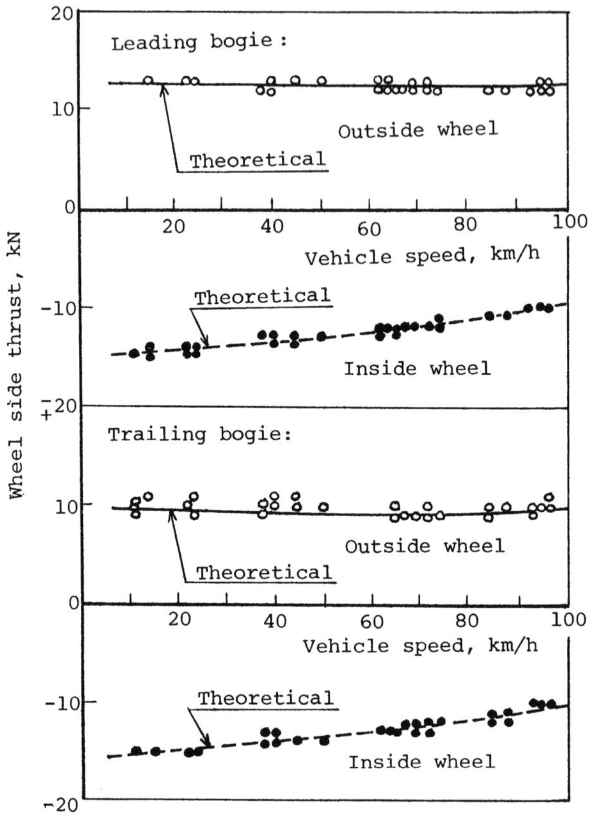


Fig.8. Comparison between theoretical and experimental results: wheel side thrust.

NOTATIONS

| | |
|---|---|
| $a$ | bogie semi-wheelbase |
| $b$ | semi-track |
| $c$ | half distance between bogie centers |
| $2f_{11}$, $2f_{22}$ | longitudinal, lateral equivalent creep coefficients per axle (suffix 1; leading bogie, suffix 2; trailing bogie, with a prime; trailing wheelset) |
| $f_k$ | Kalker's linear creep coefficient |
| $F_A$ | flange force of leading outside wheel (suffix 1; leading bogie, suffix 2; trailing bogie) |
| $g$ | gravitational acceleration |
| $k_y$, $k_\psi$ | primary lateral, yaw suspension stiffnesses per axle |
| $k_{y0}$, $k_{\psi 0}$ | secondary lateral, yaw suspension stiffnesses per bogie |
| $m_b$, $m_t$, $m_w$ | vehicle body, bogie frame, wheelset masses |

| | |
|---|---|
| M | frictional moment, vehicle body to bogie frame |
| N | wheel load |
| $Q_A$ | side thrust of leading outside wheel |
| r | mean wheel radius |
| R | mean track radius of curvature |
| T | wheel/rail tangential force |
| v | vehicle speed |
| $y_b$, $y_{ti}$, $y_{wj}$ | vehicle body, bogie frame, wheelset lateral displacements (i; bogie no., j; wheelset no.) |
| $\alpha$ | uncompensated lateral acceleration |
| $\gamma$ | wheel tread conicity |
| $\delta$ | semi-flange clearance |
| $\mu$ | coefficient of friction, wheel to rail |
| $\nu$ | creepage |
| $\psi_b$, $\psi_{ti}$, $\psi_{wj}$ | vehicle body, bogie frame, wheelset yaw (i; bogie no., j; wheelset no.) |

REFERENCES


1. Heumann, H. In Meineke, Kurzes Lehrbuch des Dampflokomotivbaues. Springer, Berlin, 1931.
2. Kunieda, M. "Analytical Study on the Side Thrust of Truck Wheels Running on Curves upon the Friction Theory." Railway Technical Research Report, No.773, 1971 (in Japanese).
3. Kunieda, M. "Theoretical Study on the Side Thrust of Truck Wheels Running on Curves." Railway Technical Research Report, No.693, 1969 (in Japanese).
4. Gilchrist, A. O. and Hobbs, A. E. W. "The Guidance of Railway Vehicles." The Dynamics of Vehicles on Roads and on Railway Tracks, 1976 pp.302-313.
5. Kalker, J. J. "On the rolling contact of two elastic bodies in the presence of dry friction." Doctoral Thesis, Delft University of Technology, 1967.
6. Arai, S. and Yokose, K. "Simulation of Lateral Motion of 2-Axle Railway Vehicle in Running." The Dynamics of Vehicles on Roads and on Railway Tracks, 1976, pp.345-368.

# TRACK DYNAMICS AT HIGH VEHICLE SPEEDS

P. Meinke[1], W. Hehenberger[2], K. Perger[1] and J. Gruber[2]

[1] M.A.N. München
[2] Bundesbahn-Zentralamt München, F.R.G.

SUMMARY

Of technical importance for trackguided traffic is the dynamic interaction vehicle and track. This fact is especially true for the wheel/rail system. This paper supplies in addition to the known comparatively extensive models of vehicles, which are in use in related theoretical investigations, appropriate models of the track. Only by this means is an entire optimization in the design of the interacting system-parts, vehicle and track, possible. Keeping in mind that a major financial investment depends upon the construction and maintenance of the track, the particular importance of the considered questions is apparent.

A dynamic model of the track is established including nonlinear influence. In an experimental/theoretical system identification on real tracks of the Deutsche Bundesbahn the dynamic parameters are determined. With the aid of hybrid simulation, technical relevant questions are considered and an "evaluation-system" of the track is presented. This offers a means for optimization and redesign of the track as well as a mobile checking device for the performance of existing tracks. Even the optimal design of rail vehicles is impossible without the knowledge of the full requirements of the track.

## 1. INTRODUCTION

The dynamic interaction between vehicle and track is of technical importance for track guided traffic. This applies especially to the wheel/rail system. Comparatively extensive models of vehicles are available while few dynamic models of tracks have been developed to date. However, when taking into account that construction and maintenance of the track require a major share of financial means for wheel/rail systems, the special importance of the questions raised, becomes obvious. Without detailed knowledge of the track even a satisfactory design of vehicle and bogies cannot be realized. Therefore the final objective of such efforts must be, within the framework of comprehensive optimization, the development of specifications for "standard vehicles" and "standard tracks".

## 2. EVALUATION SYSTEM

The above mentioned optimization of vehicle/track interaction appears possible only with the aid of an evaluation system in accordance with fig.1. This system takes into account as relevant problems the models and the evaluation criteria to be applied, to attain a prediction vector for the evaluation of the wheel/rail system.

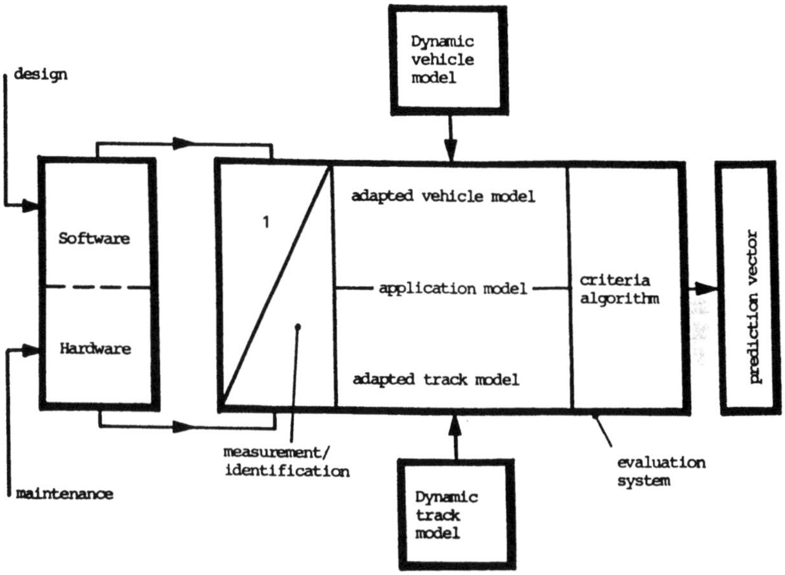


Fig.1. Evaluation System

The extensive results of theoretical and experimental investigations of vehicles available to date are suitable to supply the required inputs for an application model of the evaluation system; however, they must be supplemented by appropriate track model concepts which will be the focal point of this lecture.

Such an evaluation system may serve on the one hand, as a calculation model to contribute to the design and development of new track concepts for given operating conditions, and on the other hand as an experimental calculational model in a test device (vehicle) supplying date, in an inspection process, on the maintenance state of the track. Further objective is to realize the design and evaluation of new and conventional track concepts for speeds and vehicle configurations outside the scope of our present state of experience. Based on the knowledge of operating wheel/rail systems as well as on experiments in the laboratory and on existing high speed tracks, these calculation models will serve to assess and evaluate the influence of the parameters of track and vehicle on relevant reactions to assumed operating conditions. Theory will be addapted to practical application, in the following steps:

a) Preparation of calculation models to determine the relevant reactions of the track under the rolling vehicle regarding short term dynamics (load, safety, ride quality) and long term behavior (setting, wear).
b) Compilation of model parameters and disturbance characteristics of vehicle and track require for design,
c) Compilation of criteria algorithm and evaluation scales for the calculated reactions as a basis for the optimization for the wheel/rail system for given operating conditions.

## 3. DYNAMIC TRACK MODELS

In the following, the problems concerning track modelling are shown by the example of vertical dynamics in the ballast road bed. Discrete models treating cross-ties as rigid bodies and using "lumped parameter" description for the soil seem to be the most feasible (fig.2). Possible methods to describe a discrete system for the deflection of rails are for instants:

a) Discrete partial differential equation by means of the method of Ritz
b) Beam equation as an integral equation with consecutive numerical integration
c) Approximation of beam equation based on the difference method.

Nonlinear influences expressed as zero setting of cross-ties, nonlinear pressure-spring property of soil, and Coulomb friction of ballast, are of special importance. Irregularities in the ballast bed of the track must be introduced as well.

In accordance with fig.1 a vehicle model adapted to the evaluation purpose, should be added to complement the descripted track model. For the determination of the dynamic interaction between wheel and rail, between rail and cross-tie, as well as between cross-tie and ballast, the track model, in connection with the comparatively simple vehicle model of fig.2, can already deliver valuable results.

As degrees of freedom of the track those of the rails as elastic beams have to be added to the displacement of the cross-tie, as well as to the degrees of freedom of the lumped-parameter-modelling of the underground as far as and as long as it is decoupled by the gap. The elastic rail delivers in x- direction, according to finite element methods on degree of freedom of displacement and one of rotation per cross-tie section. For technical relevant speed range the number of degrees of freedom of the elastic rail can be reduced to half by means of condensation which, unfortunately leads to a high number of elements in the matrices. As however the condensed stiffness and mass matrices of the rail are diagonally dominant, it seems acceptable to take into account with the main diagonals only the four upper and lower side diagonals. Special attention should be paid to the fact that this model is suitable to handle track line irregularities as well as varying underground parameters, which both can be distributed statistically or deterministically. On requirement, the introduction of additional track surface irregularity in the rails is possible.

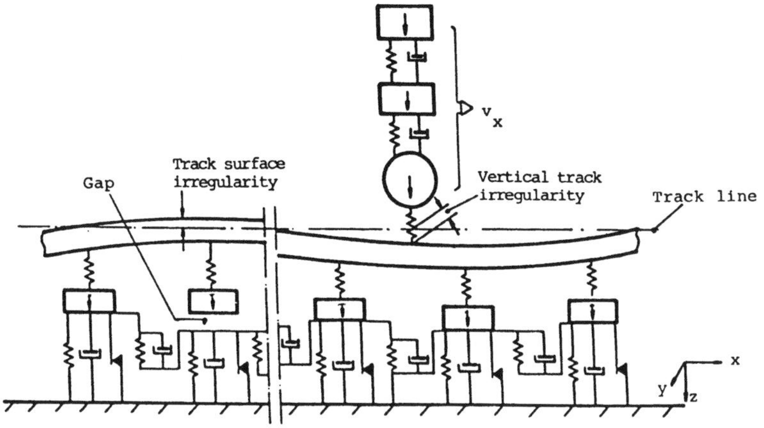


Fig.2. Vertical track model

## 4. IDENTIFICATION OF THE TRACK PARAMETERS

The identification of track model parameters is carried out of the basis of measurements on an impact excited high-speed track of the Deutsche Bundesbahn. Fig.3 gives an overview of the test rig. In fig.4 we see the impact weight and in the background the recording coach.

Fig.3. Test rig for impact exciting on a high speed track

Fig.4. Test rig and recording coach of Deutsche Bundesbahn

Fig.5. Accelerometers and force sensors on rail and cross-ties

The forces in vertical direction are measured on both rails, as accelerations on the rail and cross-ties, see fig.5.

The lay-out of sensor devices, fig.6, shows the position of all sensors on the track during the impact experiments. The variation in time of a typical exciting force and of acceleration are shown in fig.7, (the repeated striking of the impact weight is clearly discernible), while we see in fig.8 an amplitudespectrum from these experiments. We conclude that in the "upper" frequency range strong system responses are to be expected.

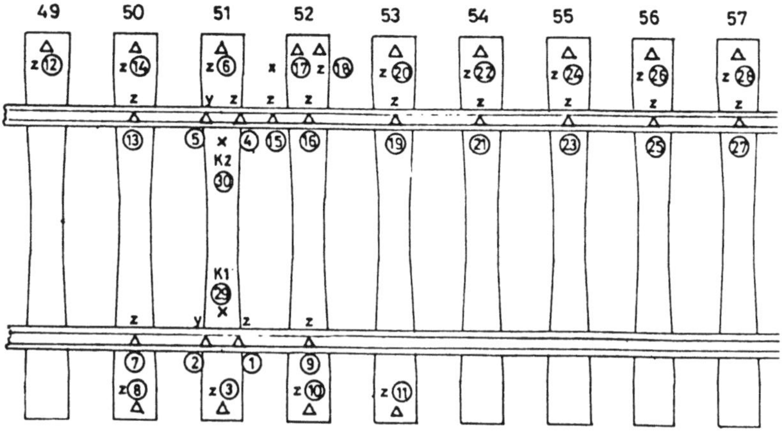


Fig.6. Lay-out of sensor devices on rail and cross-ties

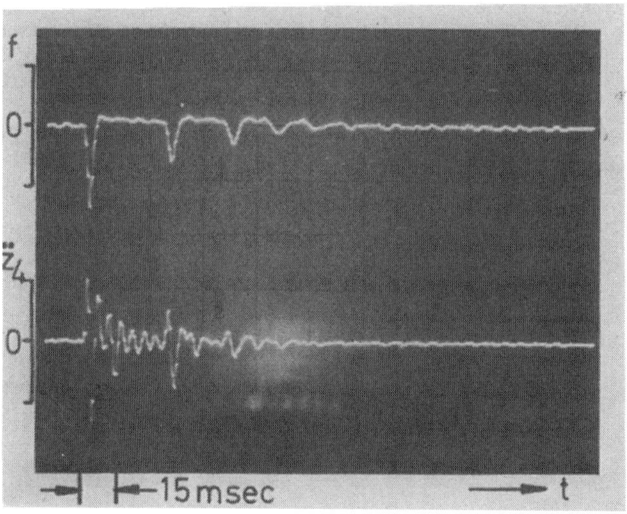


Fig.7. Typical response to impact exciting on high speed ballast track

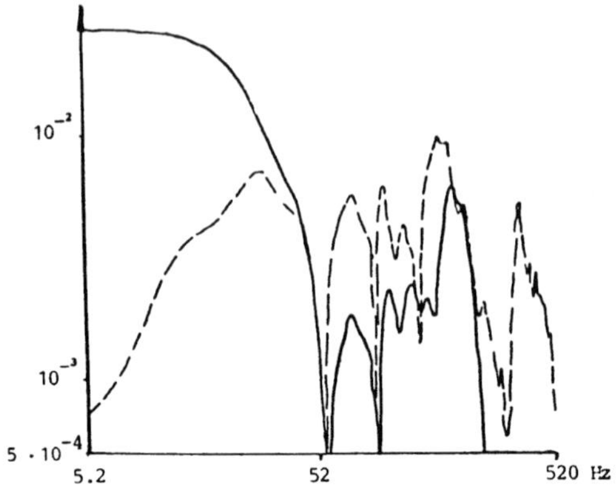


Fig.8. Amplitude spectra of exciting force (——) and rail acceleration (----) at impact position.

In the first simplified approximation step a linear track model was adapted to the measurements. A comparison of calculated and measured results for a high-speed track is shown in the frequency responses, fig.9, which indicates that the influence of the measurement and recording noises are particulary present in the high frequency range.

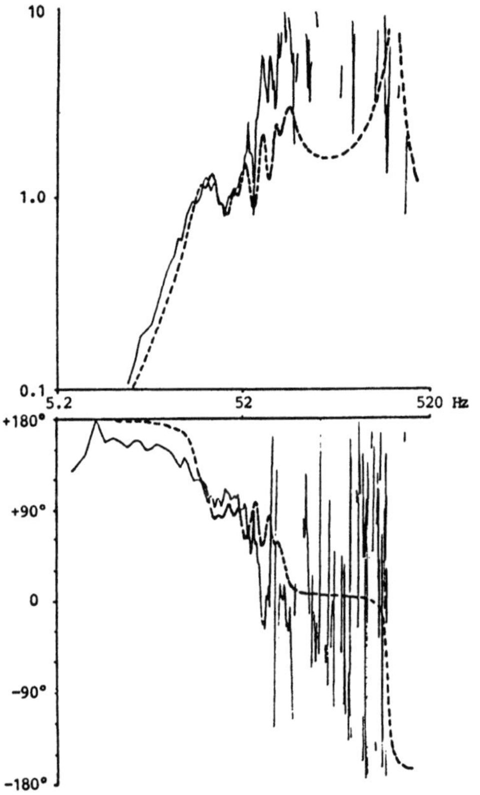


Fig.9. Frequency response of track (——) and adapted linear model (----)

The values of the parameters which were found by this method, especially the ballast-spring and -damper coefficients are used as starting values for the approximation of the nonlinear track model, described by the following equation:

$$M\,\ddot{z}(t) + D\,\dot{z}(t) + K\,z(t) + g(z,\dot{z},z_G,z_0) = f(t) \qquad (1)$$

$$z(0) = z_0,\ \dot{z}(0) = 0,$$

$g(z,\dot{z},z_G,z_0)$ : Vector function which describes nonlinear ballast-springs, nonlinear dampers, Coulomb friction, and gaps (deadbands).

The time discrete state space equation, with $x = [z\ \dot{z}]^T$ is given by

$$x(k+1) = \varphi(x(k),\ f,\ \theta),\ x(k_0) = x_0 \qquad (2)$$

$$y(k) = Cx(k) + v(k)$$

and serves as a basis for the Maximum Likelihood parameter estimation problem:

$$p\left[x(k_N)\ \middle|\ \theta\right] = \prod_{k=k_1}^{k_N} p\left[y(k)\ \middle|\ y(k-1),\ \theta\right] \rightarrow \text{MAX} \qquad (3)$$

N : Number of samples of the output vector y
$\theta$ : Unknown parameters
v(k): Discrete unbaised independent Gauß-Markov process.

These investigations will be accompanied by nonlinear approximation in the frequency domain, which uses the method of harmonic balance.

## 5. DRIVING TESTS ON HIGH SPEED TRACK

While the previous sections were directed toward the question of a suitable model, and the response of the track of the test signals (i.e. impulses), now interest is concentrated on the track under real load due to a passing train. Additional sensors were prepared according to the plan, fig.10 in the immediate neighborhood of the identified area of the high speed track, which include rail and cross-tie accelerations as well as soil pressure in the road bed. The position of the overrolling axis results out of the signal of the rail bending reaction strain gauge. Typical results of these measurement series are shown in fig.11 for four wheel-sets, which belong to the neighbored bogies of two railroad cars.

It can be noticed, that the frequencies of the oscillation fall from the rail over the cross-tie to the road bed. Further, it can be found that the cross-ties come to vibration before the wheel-sets enter the regarded area, which is explicable as a lift-off-wave. While the vertical movement of the cross-tie shows defined maxima, the excited oscillation in x- direction is less sensitive to the position of the wheel-set.

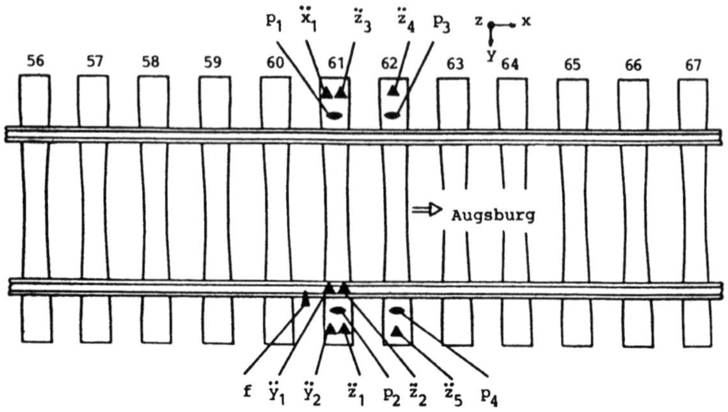


Fig.10. Lay-out of sensor devices for measurement under train with $v_x$ = 200 KM/h

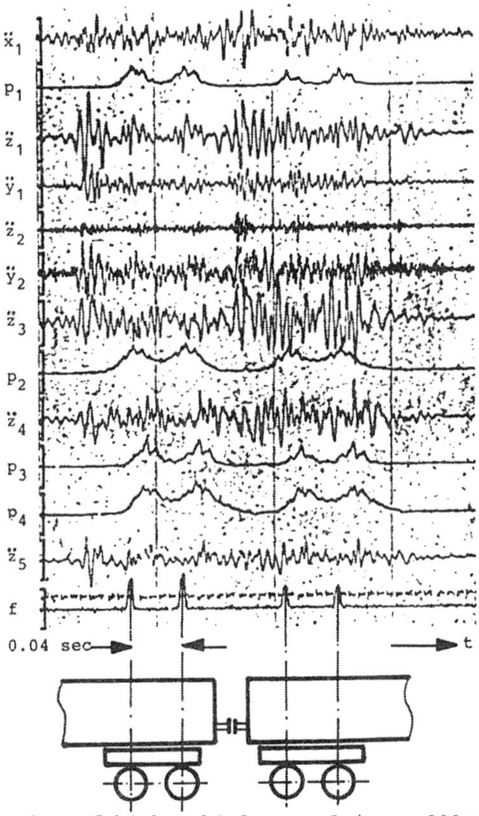


Fig.11. Track dynamics of high vehicle speed ($v_x$ = 200 KM/h)

The measurement give an interesting view on the number of cross-ties effected by the oscillation, from which important facts to the necessary size of a simulation model can be derived.

## 6. SIMULATION OF THE WHEEL-SET-MOVEMENT ON THE TRACK

To examine the influence of track- and vehicle parameters to different kinds of operating conditions computation of the relevant reaction of the system with aid of the application model is necessary. The high system order as well as the nonlinear character of that part of the model which describes the track, gives need for a simulation on the hybrid computer. Such a concept has the advantage, that a sufficient number of cross-ties can practically be closed "to a circle", so that the endless long track can be copied. By reading-in deterministically or statistically distributed track-line-irregularities or gaps, it is possible to study the movement of wheel-set and vehicle over long track distances in the simulation.

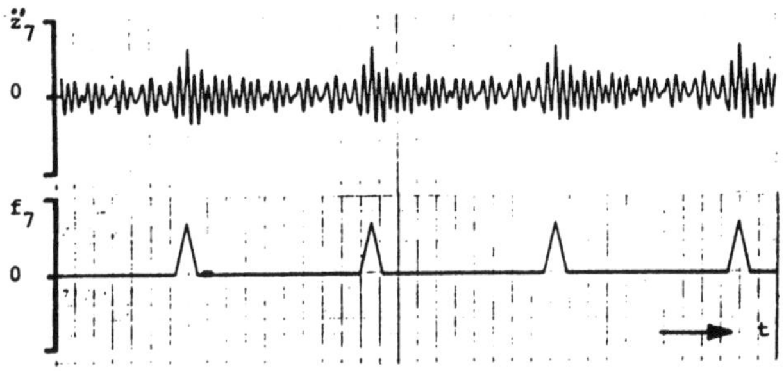


Fig.12. Simulation of forces moving along the track ($v_x$ = 100 KM/h)

Fig.12 shows an example of four successive loads moving along the track model. The represented results were obtained in real-time using parallel computation techniques. Therefore, it appears possible to practice such simulation also in a moving measurement device in order to inspect on-line the track and to check the need for maintenance.


Acknowledgement

This paper is part of a work supplied by the German Ministry of Research and Technology (FRG).

REFERENCES


1. Cottin, N., Dellinger, E., Bestimmung der dynamischen Kenngrößen linearer elastomechanischer Systeme. Zeitschrift für Flugwissenschaften 22 (Heft 8, 1974).
2. Fryba, L., Vibration of solids and structures under moving loads. Noordhoff International Publishing.
3. Koloušek, V., Dynamics in Engineering Structures, Butterworth, London 1973.
4. Moyar, G.J., Pilkey, W.D., Pilkey, B.F., Track/Train Dynamics & Design, Pergamon Press, New York, Toronto, Oxford, Sydney, Frankfurt, Paris, 1978.
5. Müller, P.C., Schiehlen, W.O., Lineare Schwingungen, Akademische Verlagsgesellschaft, Wiesbaden, 1976.
6. Sotiropoulus, G.H., Approximate Determination of the Dynamic Stiffness Coefficients of Beams, Ing. Arch. 47 (1978).
7. Richart, F.E., Woods, R.D., Hall, J.R., Vibrations of soils and Foundations, Prentice-Hall, Inc., Englewood Cliffs, N.Y. 1970
8. Sage, A.P., Melsa, J.L., System Identification, Academic Press, New York, San Francisco, London, 1971.
9. Sage, A.P., Melsa, J.L., Estimation Theory with Applications to Communications and Control, Mc Graw-Hill Book Company, New York, St. Louis, San Francisco, Düsseldorf, London, Mexico, Panama, Sydney, Toronto, 1971.

# CONTRIBUTION TO THE INVESTIGATIONS INTO THE DYNAMICS OF RAILWAY VEHICLES[1])

Wolfgang Michels and Helmut Schutzbach

Fried. Krupp GmbH, Krupp Industrie- und Stahlbau, Essen, F.R.G.

SUMMARY

The Wheel/Rail Research Program presented in this paper deals with the model-setup, the computation of linear as well as non-linear systems and the comparison of measured results with computed findings of the riding behaviour of a railway vehicle.

The fundamental investigations of the dynamics of the wheel/rail system are carried out with the basic unit of the system, i.e. with the wheelset moving on the guiding track. Thereby great importance is attached to considerations of non-linear relationships in the contact area between wheel and rail.

The wheelset and the track are described by inertia-masses coupled and interacting partly by elements following linear laws, as the axle with its torsional and bending elasticity, and partly by laws of non-linearities, as the contact geometry and the tangential forces in the wheel/rail contact area. This model is used for the investigation of low-frequency motions of the wheelset, for instance of the hunting motion, and also for the oscillations of higher frequencies such as the torsional vibrations of the axle.

The system of differential equations expressing the mathematical model is solved in their non-linear structure by numerical integration and subjected to an analysis of eigenvalues after linearizing it. The results on frequencies and modes of oscillation arrived at theoretically will be compared with the findings of measurements.

In addition to the above investigations a description of two methods of finding track data, the static disturbance variables and the dynamic track parameters, is given.

The verification of the theoretical models is attained by roller rig tests in Munich-Freimann and by field tests on the Rheine-Freren test track, which is now under construction.

1) The research work presented here is sponsored by the Federal Ministry for Research and Technology

## 1. OBJECTIVE

The major features of the wheel/rail system as used in railway technology have remained almost unchanged for more than 150 years. The basis of this system, the double cone, which is self-centring during travel, is as ingenious as it is simple.

All theoretical considerations on the running behaviour of railway vehicles must focus on the problems deriving from the dynamics of this double cone.

Theoretical considerations on the purely kinematic relationships between conicity and frequency behaviour of the wheetset have already been made by Klingel [1].

A series of papers were recently published which deal with theoretical investigations involving both kinematics and kinetics [2], [3], [4].

The aim of the investigation to be carried out must be to work out a theory which permits optimal harmony of vehicle and track parameters and the calculation of the running properties of complicated wheel/rail vehicles.

For this purpose, it must first be analysed how sensitive the stability of a wheelset reacts to the variation in the parameters which influence the motion behaviour.

## 2. SOLUTIONS / PROCEDURE

The procedure for establishing reliable data for the theory can be divided into the following steps:

- Problem definition
- Model setup
- Solutions
- Interpretation and comparison of the results.

These steps are carried out parallel to the actual theoretical work as well as to the backup tests.

The three parallel approaches are as follows:

- Theory
  As the theoretical penetration of the problem is absolutely imperative for the physical interpretation of certain processes, comprehensive theoretical fundamentals are developed for the vehicle/track dynamics with which the effects relevant to the running behaviour can be described. With this theory and the replacement models and program systems based on it, design data are to be derived for the track and high-speed railway vehicles. It is, however, imperative to verify the results of these investigations in laboratory and track running tests.
- Roller rig tests
  With the tests carried out on a roller rig under laboratory conditions, individual parameters can be varied and the other parameters kept constant in contrast to the conditions in track running tests. The roller rig guarantees good reproducibility of the test results. The tests should confirm the theory or permit a correction

of the concept underlying the model.
Track running tests
There are considerable differences between roller rig and actual track with the result that certain problems can only be investigated in track running tests.
As the theory is intended for the design of track and vehicles, it is necessary to check the theory by track running tests. The test programmme is based on the knowledge gained from the theoretical investigations and set up in accordance with the necessities of the theory.

It is well known what fundamental differences have to be taken into consideration when comparing the results from the roller rig tests with those determined in the track tests. However, here again, dependencies which can only be quantified by theoretical investigations can be specified.

Relatively sophisticated models with sometimes quite complicated coupling are studied in the theoretical work. In order that these models do not become too complicated, it is advisable to limit them initially to the basic elements wheelset and rail.

Based on the information gathered from the wheelset model, a bogie model is investigated in the second stage and a vehicle model in the third stage. This three-stage procedure not only applies to the theoretical work but, in principle, to accompanying tests on the roller rig and track.

## 3. MODELS / TEST SETUP

The model to be investigated in the first stage consists of a wheelset and a track consisting of the two rails (Fig. 1). This model is designed so that a series of conceivable and actual wheelset arrangements can be reproduced.

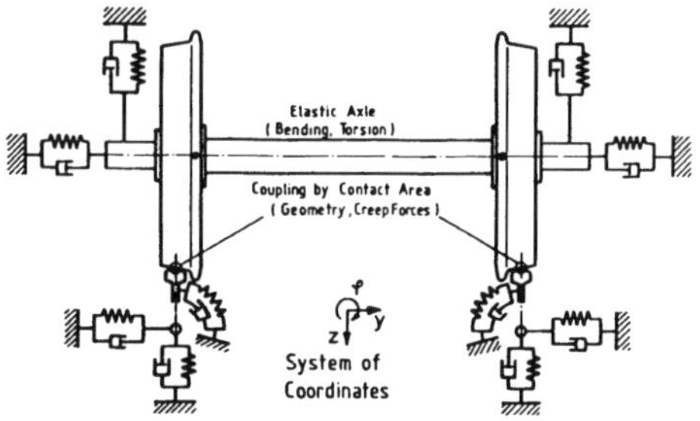


Fig. 1:
Wheelset on rails

To set up the differential equations, sections are put into the contact areas, providing three sub-models, the wheelset sub-model and two rail sub-models.

The inertias of the wheelset are reduced to 2 points which are coupled with one another via a transversely and torsionally flexible shaft. A damping coupling for the bending and torsion of the shaft can also be introduced.

The wheelset is linked to an infinite mass in longitudinal

direction in such a way that the longitudinal and torsional stiffnesses and damping can be varied independent of one another.

Spring-damper systems connected in series represent another possible arrangement in longitudinal direction. In transverse direction, the wheelset is backed against an infinitely large mass through spring/damper systems operating in parallel.

The sub-models of the track, the two rails, are situated below the wheel contact areas. Here, too, the inertias are reduced to two points. In longitudinal direction, the track can be regarded as rigid. In the transverse direction the rails are backed by parallel spring/damper systems against an infinite mass. Parallel spring/damper systems also act in the vertical direction. The centres of gravity rotate against parallel springs and dampers.

The wheelset/track model described above has 17 degrees of freedom. The wheelset sub-model has 11 degrees of freedom and the two rail sub-models 3 degrees of freedom each.

It should be noted that oscillations occur on this model whose frequencies are considerably higher than the hunting frequency. In the discussion of the results it will be shown that these frequencies occur up to 200 Hz and thus lie in the audibility range. It has been shown that in order to investigate the behaviour of the wheelset/track system such a detailed model is required so as to permit the phenomena obtaining between wheel and rail to be interpreted and quantitative information to be derived from computations.

In the bogie model two wheelset/track models are connected by a frame which in turn is coupled to an infinite mass moving with it. This frame consists of two longitudinal girders which are connected by a cross member providing torsional elasticity but transverse and compressive rigidity. This bogie frame has 7 degrees of freedom.

The vehicle model connects two of the above described bogies via a body which has five degrees of freedom of movement of a rigid body - longitudinal motion excluded. An infinite upper mass no longer exists.

In addition to the above described composition of the subsystems to form an overall system, it is possible to take into consideration modes previously calculated, for example, using the finite element method.

Moreover, a model reduction can be carried out in this way. This will be briefly explained using a single wheelset model as an example. The modes of the individual wheelset described by the eigenvectors represent the degrees of freedom of the wheelset model in the coordinate system of the modal modes. In this system the oscillation differential equations are decoupled, the eigenvector matrix is the pertinent transformation matrix. The differential equations of the bogie model can be set up by coupling the wheelset differential equations in the modal system. In this way, the number of degrees of freedom of the entire model can be limited to the essential oscillation modes without thereby neglecting those portions forming the slip vector which derive from the centre of gravity motions of the partial bodies belonging to an eigenmode. A detailed description of this procedure is contained in the following paper [5].

## 4. SOLUTIONS OF THE THEORETICAL INVESTIGATIONS

For many reasons only linear systems can be considered in parameter studies for designing vehicles. However, since the reactions between wheel and rail - geometry and adhesion - proceed non-linearly, it is not sufficient to undertake linear investigations alone. For this reason, the differential equation system representing the model is solved by

- numerical integration in non-linear form
- closed integration after linearization

### 4.1 Solutions of the non-linear equation system

The system of differential equations was built up in two steps. Firstly, the inertia, damper and spring forces were determined by the Lagrange method. In the force and moment equilibrium the wheel contact forces transmitted at the contact point also occur in addition to the forces determined by the Lagrange method. These forces, their points of application and their direction are functions of the instantaneous oscillation state of the entire system. The variables of these functions are the position and speeds of the partial masses. The functions are non-linear and together with the Lagrange terms result in a system of differential equations as shown in Fig. 2.

$$\underline{U}_1 (x, \dot{x}) \cdot \underline{\ddot{x}} + U_2 (x, \dot{x}) \cdot \underline{\dot{x}} + \underline{U}_3 (x, \dot{x}) \cdot \underline{x} = \underline{C} (x, \dot{x})$$

with

$$\underline{U_i} (x, \dot{x}) = \begin{bmatrix} u_{11}^{(i)} (x_1, \dots, x_n, \dot{x}_1, \dots, \dot{x}_n) & \dots & u_{1n}^{(i)} (x_1, \dots, x_n, \dot{x}_1, \dots \dot{x}_n) \\ \cdot & & \\ \cdot & & \\ u_{n1}^{(i)} (x_1, \dots, x_n, \dot{x}_1 \dots, \dot{x}_n) & \dots & u_{nn}^{(i)} (x_1, \dots, x_n, \dot{x}_1, \dots, \dot{x}_n) \end{bmatrix}$$

$$\underline{C} (x, \dot{x}) = \begin{bmatrix} c_1 (x_1, \dots, x_n, \dot{x}_1, \dots, \dot{x}_n) \\ \cdot \\ \cdot \\ \cdot \\ c_n (x_1, \dots, x_n, \dot{x}_1, \dots, \dot{x}_n) \end{bmatrix}$$

Fig. 2: Differential Equation System

The coefficients $u_{jk}$ (Fig. 2) cannot all be represented in the form of algebraic terms; for calculating them, however, computation rules were derived and programmed.

When solving the system of differential equations of the second order, it is transformed into a system of the first order with 2 n equations.

The variable coefficients of the differential equation system are computed with our wheel/rail contact geometry program [6] and the program for the adhesion/slip relationship according to Kalker [7].

The result of the initial value problem are functions of time which describe the state of the oscillation system in the integration interval.

The above-described model contains non-linearities which result from the coupling of wheel and rail. The condition that wheel and rail are in contact results in a geometrical relationship for the spatial position of the partial bodies of the model to one another. This condition is represented mathematically by a function which connects the centre of gravity coordinates with one another in the form of

$$f_i\ (y,\ z_i, \varphi_i, \psi_i,\ y_{oi},\ z_{oi},\ \varphi_{oi}) = 0$$

or $$z_i = z_i\ (y,\ \varphi_i, \psi_i,\ y_{oi},\ z_{oi},\ \varphi_{oi})$$

The functions are determined by the shape of the profiles which are in contact with one another.

The tangential forces which are transmitted have a functional relationship with the slip vector in the wheel contact area. They can be shown in the following form:

longitudinal force $T_{xi} = T_{xi}\ (\sigma_{xi},\ \sigma_{yi},\ \emptyset_i,\ \mu,\ E,\ N_i)$

lateral force $T_{yi} = T_{yi}\ (\sigma_{xi},\ \sigma_{yi},\ \emptyset_i,\ \mu,\ E,\ N_i)$

pivoting moment $M_i = M_i\ (\sigma_{xi},\ \sigma_{yi},\ \emptyset_i,\ \mu.\ E,\ N_i)$

where $\sigma_x$ is the longitudinal slip, $\sigma_y$ the lateral slip and $\emptyset$ the spin. The major parameters are the coefficient of friction μ, the modulus of elasticity E and the normal force N. These forces and the contact geometry functions which connect wheel and rail with one another are included in the coefficients of the equation system in Fig. 2.

A prerequisite for a closed solution of the vibration problem is a linear differential equation system. Such a system can be attained by replacing the non-linear functions with linear approximations. For example, if the $T_x$ force is plotted over the longitudinal slip $\sigma_x$ and the lateral slip $\sigma_y$, it can be shown as a curved surface.

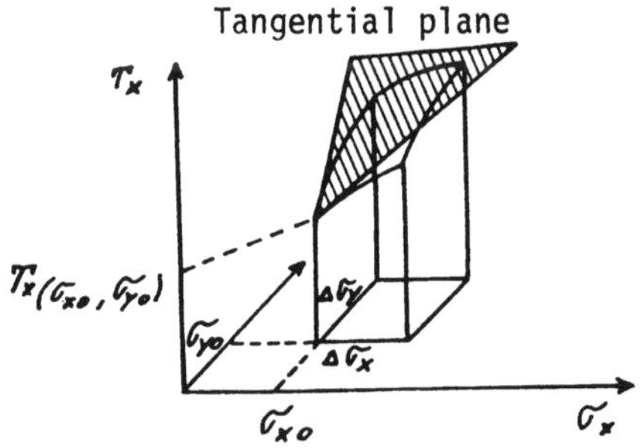


Fig. 3:
Linearization of the $T_x$-force

The tangential plane can be used as an approximation function for $T_x$:

$$T_x = T_x\ (\sigma_{xo},\ \sigma_{yo}) + \frac{\partial T_x}{\partial \sigma_x} \cdot \Delta \sigma_x + \frac{\partial T_x}{\partial \sigma_x} \cdot \Delta \sigma_y$$

The other non-linear relations are approximated in the same way as the function of the longitudinal force. The equation system (Fig. 2) is thus changed into a system with constant coefficients in the form of

$$\underline{M}\,\ddot{\underline{x}} + \underline{D}\,\dot{\underline{x}} + \underline{C}\,\underline{x} = 0$$

where $\underline{M}$ is the mass matrix
$\underline{D}$ the damping matrix
$\underline{C}$ the stiffness matrix.

The damping matrix is made up of three parts, two of which are dependent on the travelling speed V:

$$\underline{D} = \underline{D}_1 + \frac{1}{V} \cdot \underline{D}_2 + V \cdot \underline{D}_3$$

The solution of the linear equation system leads to the eigenvalue problem. The eigenfunctions dependent on the travelling speed are represented by their eigenvalues and eigenvectors.

## 5. RESULTS

The solutions to the non-linear equation system without external excitations show the transient behaviour of the system in relation to the given initial values. In order to investigate the stationary oscillation behaviour, which provides information on stability, adapted initial values are selected to avoid long computation times. The wheelset performance at irregularities in track position is examined using disturbance functions which are introduced in the form of force excitations. The solution functions contain numerous types of motions reproduced by the model. The investigation into the eigenbehaviour shows which oscillation modes are involved [8].

Hunting is the oscillation mode of the wheelset which is decisive for the running performance of vehicles. With respect to the type of motion the simulation calculation gives the amplitudes in the individual degrees of freedom. The value range which the non-linear functions assume is used as a basis for the selection of the working point for linearization.

Fig. 4 shows the oscillation mode "hunting" of the wheelset. The degree of freedom y, i.e. the lateral motion of the wheelset, has the largest amplitude. The yawing motion of the wheel disc centres of gravity $\psi_1$ and $\psi_2$ are phase-shifted by 109° relative thereto. The translatory motions, $x_1$ and $x_2$, are displaced relative by 180° to one another, $x_2$ is in phase with $\psi_1$ and $\psi_2$. The only recognizable deformation of the wheelset is torsion of the shaft - $\chi_1$, $\chi_2$ - which is approximately in phase with the yawing motion. The rails contribute to this type of motion with their degrees of freedom $y_{01}$ and $y_{02}$. The amplitudes of the other degrees of freedom are insignificant and therefore only their phase position relative to y is given.

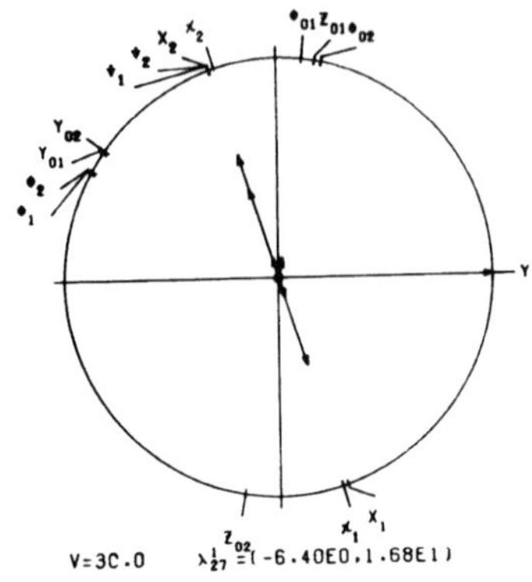

Fig. 4: Eigenmode hunting

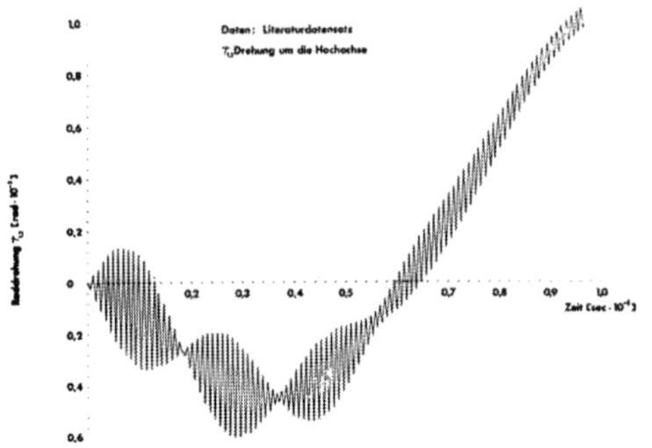

Fig. 5: Non-linear Simulation
Yaw angles $\psi_1, \psi_2$

Owing to the non-linearities in the wheel contact area, all higher-frequency modes of oscillation contained in the model are excited as the wheel rolls along the rail. Figure 5 shows a non-linear simulation computation which clearly shows how the fundamental mode "hunting" is superimposed by higher-frequency oscillation.

These oscillations can be analysed by linking the results of the simulation computation with the eigenvalue computation. In this case, flexural vibrations are involved. The eigenvalue analysis provides various bending modes. As an example, Fig. 6 shows the second bending mode of the shaft in the Y-Z plane - $\varphi_1, \varphi_2$. The rail rotations $-\varphi_{01}, \varphi_{02}$ - make a considerable contribution to this eigenmode.

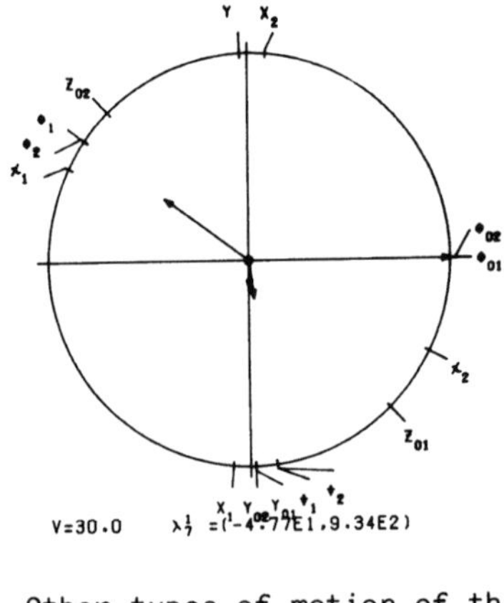

Fig. 6:
Second bending of the axle

Other types of motion of this model are, for example, the longitudinal, lateral, bouncing and rolling of the wheelset, the torsion of the shaft and oscillation modes to which the degrees of freedom of the rails are mainly contributing.

Further investigations were carried out on the bogie model in line with the proposed procedure. The modal degrees of freedom of the wheelsets remain almost unchanged in the eigenvalue analysis in this model. Of the additional modes of oscillation two are of particular interest, in-phase and anti-phase hunting. In both cases, the computation shows in the lower range an increase in the frequency proportional to the travelling speed and both can assume eigenvalues with a positive real part as the travelling speed increases. With anti-phase hunting the two wheelsets perform hunting motions that are phase-shifted by 180° relative to one another like a single wheelset coupled to a rigid mass. The bogie frame remains almost at rest. In-phase hunting is the more important of the two eigenmodes as with conventional coupling arrangements it tends towards instability.

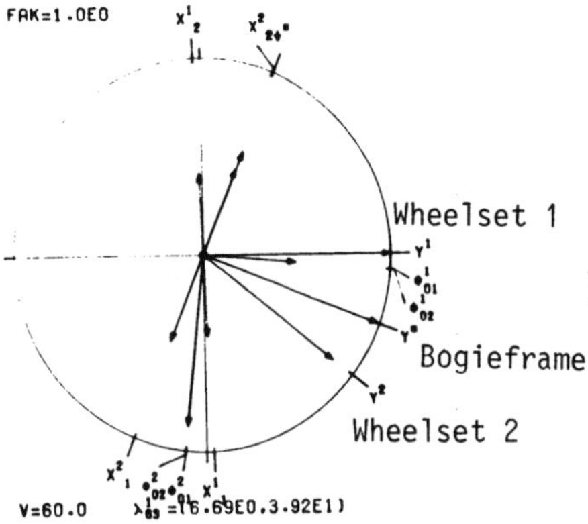


Fig. 7:
Eigenmode bogie hunting (Amplitudes)

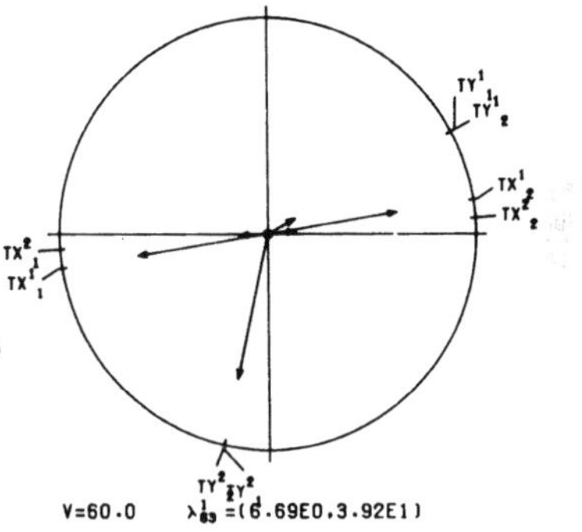


Fig. 8:
Eigenmode bogie hunting(forces)

In this case, the wheelsets together with the bogie frame describe a lateral and yawing motion of similar amplitudes and with a slight phase displacement (Fig. 7). The motion of the frame lies between the leading front wheelset and the lagging rear wheelset. The results of an investigation into the forces acting between the wheels and the rail with this mode can be compared with measurements. If the eigenvector of the motion is extended by the tangential forces acting in the wheel contact area, these can be represented as linear combinations of the shifts and the speeds. Fig. 8 illustrates the theory that on the front wheelset the $T_x$-forces predominate and on the rear wheelset the $T_y$-forces, i.e. that the front wheelset is subjected to a yawing moment from the anti-phase longitudinal forces whereas the rear wheelset produces the forces in lateral direction. A measuring trip, during which the accumulated $\Sigma$ Y-forces were measured, is shown in Fig. 9.

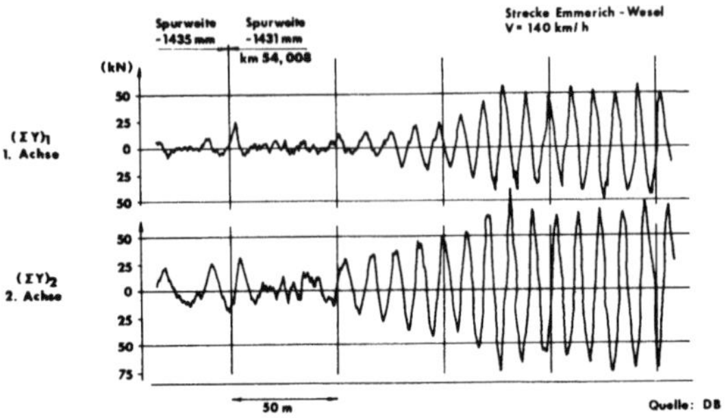


Fig. 9: $\Sigma$Y-Forces

In the following example the influence of the rail elasticities is discussed. The effect of vertical stiffness on the critical speed shall by no means be neglected. The critical travelling speed increases as the vertical stiffness of the rail decreases (Fig. 10). This is due to the fact that the rail elasticities are connected with the types of motion of the wheelset via the coupling in the wheel contact areas. The effect of lateral rail stiffness on the critical speed is not great in those areas which are representative for existing rails. Torsional stiffness has a considerably greater influence (Fig. 11). In this case, the critical speed also increases as torsional elasticity increases. If, as in the case of the rigid track, infinite torsional stiffness is assumed in the computation, this assumption involves a stability risk which cannot be expressed numerically.

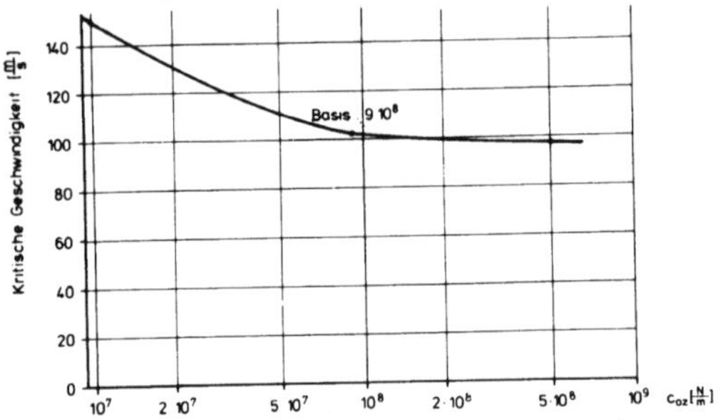


Fig. 10:
Critical speed versus
vertical rail elasticity

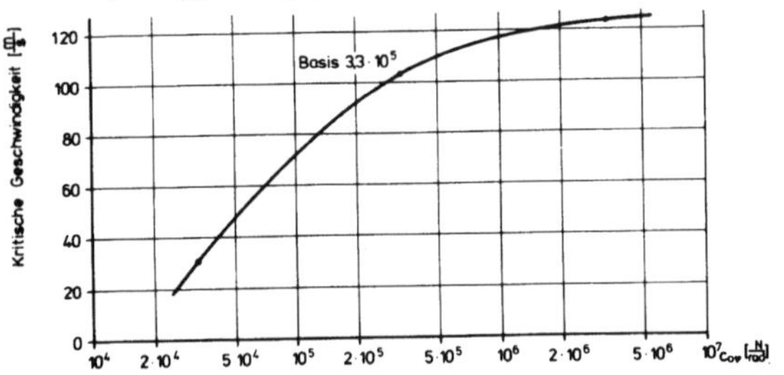


Fig. 11:
Critical speed versus
torsional rail elasticity

It is known from the operation of railway vehicles that the vehicles run more quietly with a low coefficient of friction than with a high one. The sensitivity analyses which were carried out permit the following interpretation (Fig. 12). The critical speed is increased by changing the coefficient of friction from μ = 0.35 to μ = 0.2. The actual difference, however, is that with the lower coefficient of friction the real parts are considerably greater below the critical speed and thus the system is substantially damped. Every disturbance introduced into the system from the track decreases more rapidly if the coefficient of friction is low than when it is high. This means that, in the interesting speed range, the wheelset runs more quietly with a lower coefficient of friction. Moreover, since vehicle instabilities usually occur below the critical speed of the wheelset an even greater effect on the critical speed than is shown here can be expected owing to the increased damping effect existing at a low coefficient of friction.

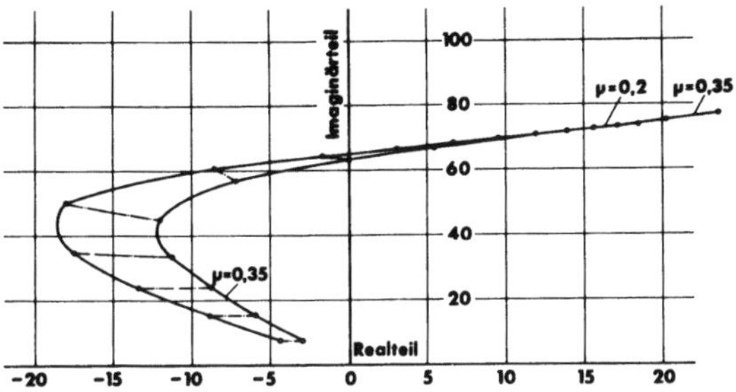


Fig. 12:
Eigenvalues, Variation of Friction

The effect of track gauge clearance is demonstrated using the example of the running performance of a DB - class 181.2 locomotive. This vehicle was the first standard locomotive to be fitted with a "worn profile" - DB II. This profile, especially when new, reacts very sensitively to a narrowing of the gauge as you can see in this picture. Fig. 9 shows the effect of the transition from one gauge to another using as an example the lateral forces ($\Sigma Y$). It can be seen that the forces increase very quickly to undesirable levels when the locomotive runs into the track section with a gauge of 1431 mm. As the gauge becomes narrower, the effective conicity $\lambda_{eff}$ increases.

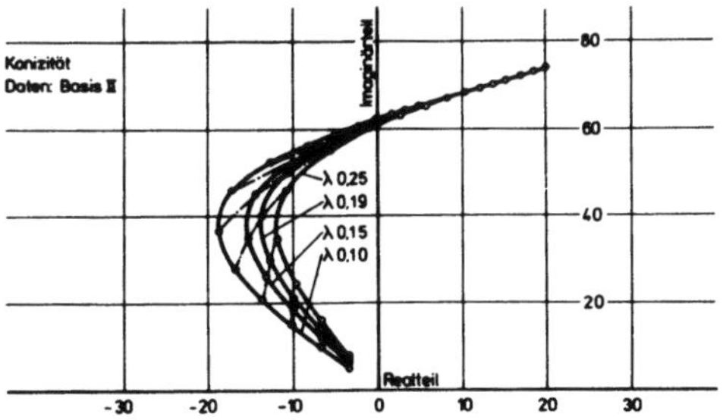


Fig. 13:
Eigenvalues, Variation of effective conicity

If the parameter $\lambda_{eff}$ is varied, which with this profile corresponds to a gauge variation, the calculation shows that damping of the hunting motion decreases sharply as $\lambda_{eff}$ increases, particularly in the higher speed range (Fig. 13). The hunting frequency at which the critical speed for a given conicity is reached remains, however, almost unchanged.

## 6. FURTHER PROBLEMS

In addition to data of the vehicle, the actual characteristic values of the track are also required for the above-described approach. Whilst the parameters of the vehicle are relatively easy to obtain, measurement of the track data causes considerable difficulties, even for the simple track model described above [9].

The excitation of oscillations in the vehicle/track system is possible by changes over distance

- of the rail head profile,
- of the position of the rails
- of the dynamic track parameters.

If the track characteristics change on the basis of the first two possibilities, we speak of "static disturbance variables". Variations in the dynamic track parameters are called "dynamic disturbance variables".

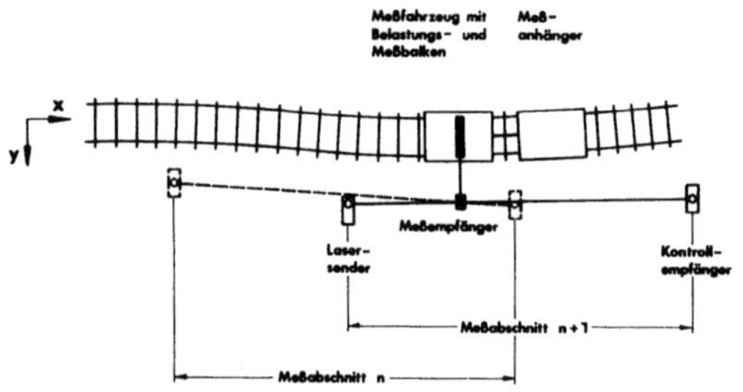


Fig. 14:
Rail disturbance variables, Test arrangement

For determining the static disturbance variables, a procedure and the attendant software and hardware (Fig. 14) were developed, with the aid of which it is possible,

- to record rail profiles, particularly in the wheelbase area, with an accuracy of 0.1 mm;
- to measure the track position with a "neighbourhood accuracy" also of 0.1 mm, where the neighbourhood is defined as an area of ± 5 m;
- to determine deviations in track position in an area of 500 m and more (Fig. 15);
- to determine the track positions under a vertical, adjustable preload, where the displacement and twisting of the rails under load is only slightly hindered.

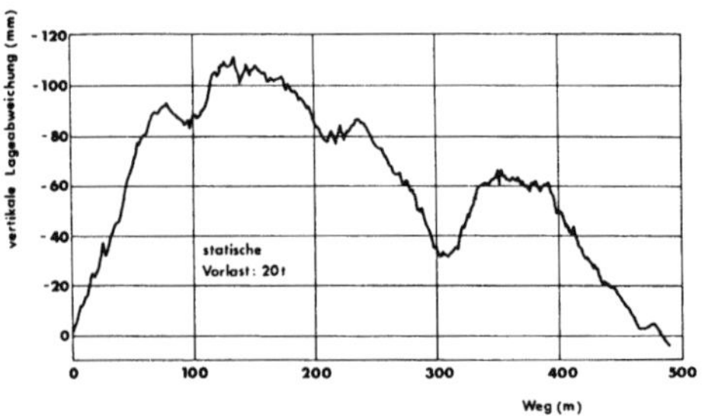


Fig. 15:
Vertical rail disturbances

The main components of the procedure are a laser beam system, with which a coordinate reference system is produced, and a measuring vehicle, whose hydraulically loaded measuring beam is linked to this coordinate system.

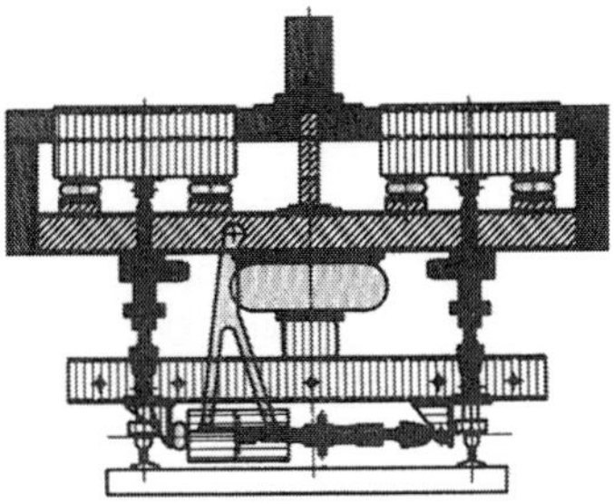

Fig. 16:
Dynamic track parameters,
Basis setup

- Frame of Car Body
- Cross Member
- Vertical Excitation
- Horizontal Excitation
- Components for Dyn. Decoupling

To determine the dynamic track parameters by experiment, a procedure was also developed and the appropriate hardware produced. In this case, excitation is produced either by an electro-dynamic shaker or a hydraulic servo-cylinder for the vertical and lateral direction.

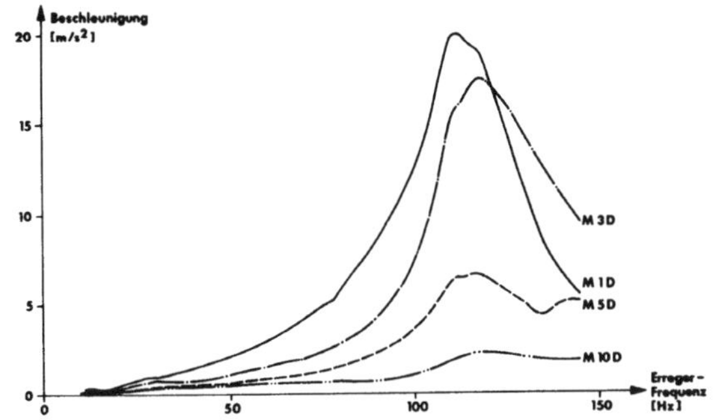


Fig. 17:
Dyn. track parameters,
Frequency function

Fig. 16 shows the basic setup of the excitation system, Fig. 17 shows a typical frequency function measurement for several measuring points on a DB main line on which speeds of 200 km/h are currently attained. Fig. 18 shows a mode of oscillation recorded at about 110 Hz which clearly shows the only slight extension in the longitudinal direction.

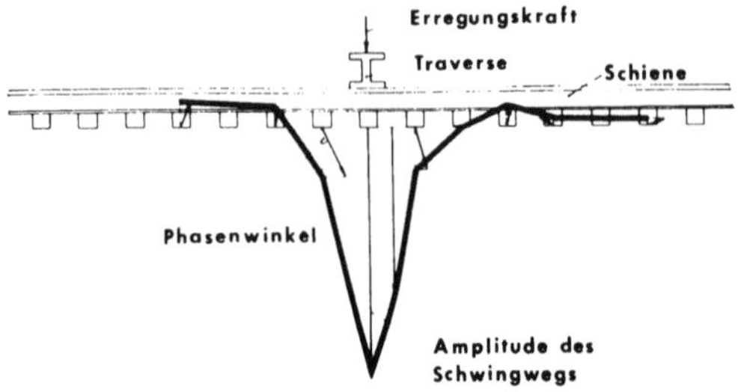


Fig. 18:
Dyn. track parameters,
Eigenmode

The roller rig tests in Munich-Freimann, intended to verify the theoretical work presented here, have just commenced.

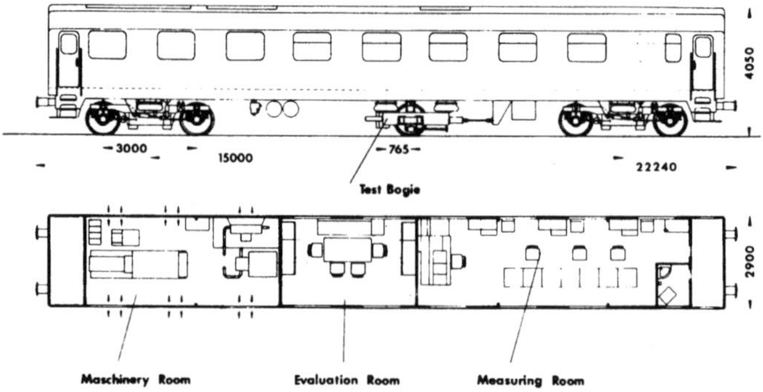


Fig. 19: Rheine-Freren - Experimental vehicle

The experimental test track Rheine-Freren is currently being built for the track running tests [10]. In the first phase single and double-axle test running gears under a special-purpose vehicle (Fig. 19) are to travel at speeds of up to 250 km/h. The test running gears, whose arrangement parameters can also be variably adjusted, can be excited by hydraulic cylinders so that certain modes, which do not normally occur owing to high damping, can be excited as desired.

From the frequency responses of the test running gears measured on the roller rig and during the track running tests the quality of the models of certain elements is checked in the mass, stiffness and damping matrix, using special evaluation and transformation programs.

## 7. REFERENCES

[1] Klingel
"Über den Lauf der Eisenbahnwagen auf gerader Bahn"
Organ für die Fortschritte des Eisenbahnwesens 38 (1883)

[2] Gerdsmeier, H.
"Status of Research and Development of High-Speed Trains within the Program of the Federal Ministry of Research and Technology"
Proceedings, Session B, International Symposium on Traffic and Transportation Technologies, Hamburg, 1979

[3] A. H. Wickens
"The Dynamics of Railway Vehicles on Straight Track: Fundamental Considerations of Lateral Stability"
Proc. Inst. Mech. Eng. London, Vol. 180 (1965)

[4] R. Joly:
"Untersuchung der Querstabilität eines Eisenbahnfahrzeuges bei hohen Geschwindigkeiten"
Schienen der Welt 3 (1972) Nr. 3

[5] Richter, R., Jaschinski A.
"Coupling and Order Reduction of Combined Linear State-Space and Multibody Systems with Application to the Dynamics of the Wheel-Rail Vehicle".
6. IAVSD-IUTAM-Symposium. Berlin, 1979

[6] Schutzbach H.
"Berührungsverhältnisse bei elastischem Radsatz und bei elastischem Gleis"
Forschungsbericht T 74-42 des Bundesministeriums für Forschung und Technologie

[7] Kalker J.J.
"On the Rolling Contact of Two Elastic Bodies in the Presence of Dry Friction"
Dissertation T. H. Delft, 1967

[8] Michels W.
"Problemkreis Zusammenwirken von Fahrzeug und Fahrweg, Übersicht über den Problemkreis"
Statusseminar V, Spurgeführter Fernverkehr, Rad-Schiene-Technik Willingen, 1978

[9] Budde U., Michels W.
"Neue Gleismeßverfahren"
ZEV - Glasers Annalen 102 (1978) No. 11

[10] Waldeck H.
"Rheine-Freren-Versuchsfahrzeug I"
Eisenbahntechnische Rundschau (28) 1/2 - 1979

# COMPARISON OF THE STABILITY OF FRONT WHEEL AND REAR WHEEL STEERED MOTOR VEHICLES

M. Mitschke

Institut für Fahrzeugtechnik, Technische Universität
Braunschweig, F.R.G.

SUMMARY

For the stability of front steered motor vehicles a caster, for the rear steered a negative caster is necessary. Beside this a vehicle with rear wheel steering is to be driven more carefully.

The usual view in vehicle technology to front wheel steered road vehicles shall be applied to rear wheel steered motor vehicles, which are often used as slow running working vehicles.

[1] describes the handling of a motor vehicle as follows (see fig.1): To drive on a road with the assigned bend $\kappa_{Soll}$, the driver will first adjust the steering wheel angle $\beta_{L,S}$ according to his "anticipating steering". By that the vehicle drives on the nominal bend $\kappa_{Ist}$. Because the nominal bend is in general different from the assigned bend $\kappa_{Soll}$, the driver uses the bend difference

$$\Delta\kappa = \kappa_{Soll} - \kappa_{Ist} \tag{1}$$

by his "compensating control" as an information for an additional steering angle $\beta_{L,R}$, in order to keep the motor vehicle on the road.

The input of the motor vehicle is the steering wheel angle $\beta_L$, which consists therefore of a steering part and a control part according to

$$\beta_L = \beta_{L,S} + \beta_{L,R} \quad . \tag{2}$$

The relation between the nominal bend $\kappa_{Ist}$ and the assigned bend $\kappa_{Soll}$ using the transfer functions of fig.1 is

$$\kappa_{Ist} = F\ \beta_L = F(\beta_{L,S}+\beta_{L,R})$$
$$= F[M_S\ \kappa_{Soll} + M_R\ (\kappa_{Soll}-\kappa_{Ist})] \tag{3}$$

That yields

$$\frac{\kappa_{Ist}}{\kappa_{Soll}} = \frac{F(M_S + M_R)}{1 + F\ M_R} \quad . \tag{4}$$

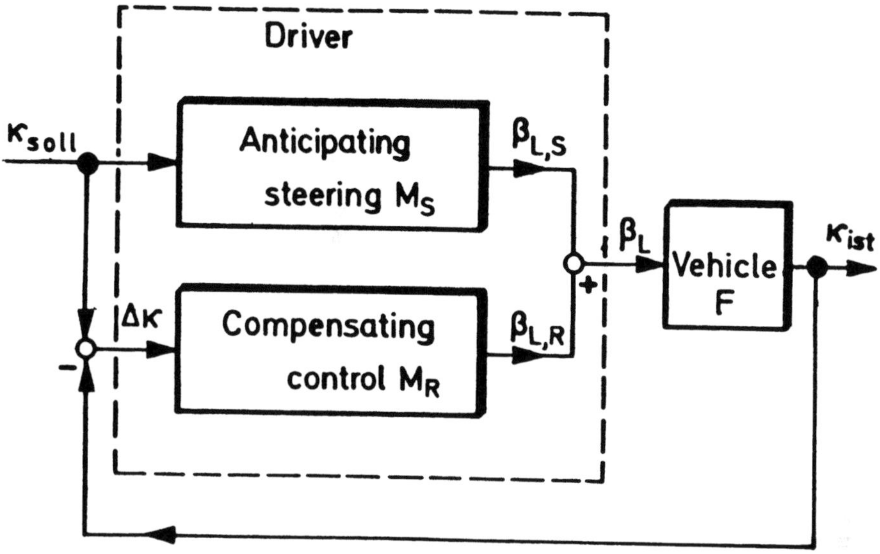


Fig. 1 Loop system driver-vehicle

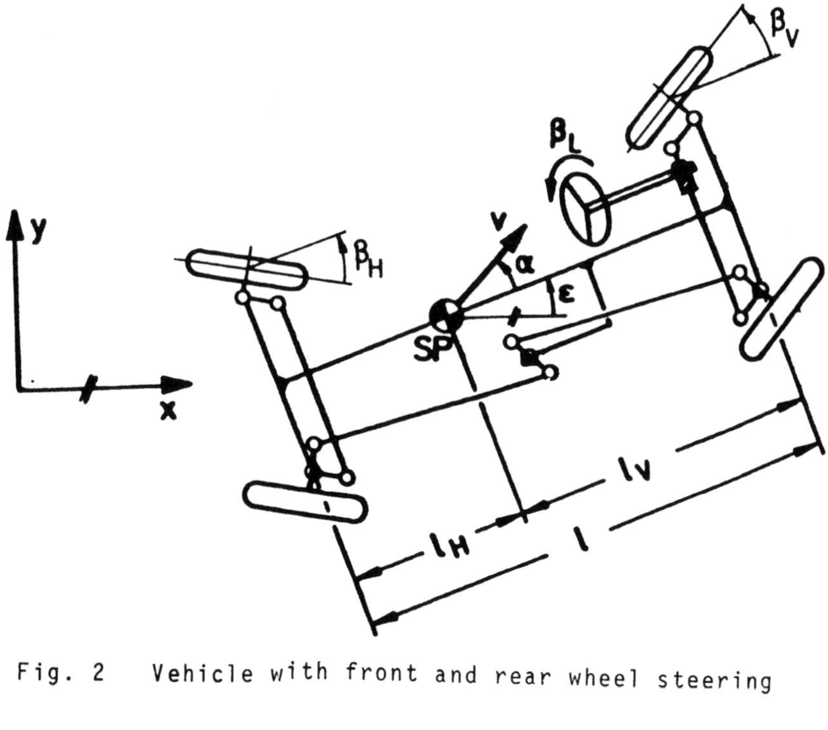


Fig. 2 Vehicle with front and rear wheel steering

For the discussion of equation (4) the human frequency functions $M_S$ and $M_R$ as well as the frequency function F of the vehicle must be known. Because there are only few results of $M_S$ and $M_R$ in consideration, the view will be limited on F.

EQUATIONS OF MOTION

To get the "response" $\kappa_{Ist}$ of the vehicle to the "input" $\beta_L$, the equations of motion have to be arranged.

$$F = \frac{\kappa_{Ist}}{\beta_L} \tag{5}$$

NOMENCLATURE

| | |
|---|---|
| $\varepsilon$ | yaw angle |
| $\alpha$ | slip angle, sideslip angle |
| $\beta_L$ | steering wheel angle |
| m | mass of vehicle |
| $\theta$ | yawing moment of inertia |
| $l_V$, $l_H$ | distance from center of gravity to front and rear axle |
| l | wheelbase, $l=l_V+l_H$ |
| $\delta_V$, $\delta_H$ | cornering stiffness of front and rear tires, including roll steer and elasticity of suspension (two wheels) |
| $i_L$ | overall steering ratio |
| n | caster (sum of constructional caster and tire caster) |
| $C_L$ | stiffness of steering system |
| v | forward speed |
| S | tire side force (two wheels) |

indices: V = front
H = rear

Assuming a two-dimensional model for the motor vehicle (see fig.2) the following equations of motion are valid [2]:

$$mv^2\kappa = S_V + S_H \tag{6}$$

$$\theta\ddot{\varepsilon} = S_V l_V - S_H l_H \ . \tag{7}$$

In a linear case the tire side forces are proportional to the slip angles $\alpha_i$

$$S_V = \delta_V\ \alpha_V \tag{8}$$

$$S_H = \delta_H\ \alpha_H \ . \tag{9}$$

The slip angles $\alpha_i$ result from the sideslip angle $\alpha$, the yaw velocity $\dot{\varepsilon}$, the front wheel angle $\beta_V$ and the rear wheel angle $\beta_H$

$$\alpha_V = -\alpha + \beta_V - l_V \frac{\dot{\varepsilon}}{v} \tag{10}$$

$$\alpha_H = -\alpha - \beta_H + l_H \frac{\dot{\varepsilon}}{v} \; . \tag{11}$$

Considering the steering ratios $i_{L,V}$ and $i_{L,H}$, the casters $n_V$ and $n_H$ and the stiffnesses of the steering systems $C_{L,V}$ and $C_{L,H}$, there are the following connections between the steering wheel angle $\beta_L$ and the wheel angle $\beta_V$ and $\beta_H$

$$\beta_L = i_{L,V} \left(\beta_V + \frac{S_V n_V}{C_{L,V}}\right) \tag{12}$$

$$\beta_L = i_{L,H} \left(-\beta_H + \frac{S_H n_H}{C_{L,H}}\right) \; . \tag{13}$$

Using the equations (8) to (13) and regarding

$$v\kappa = \dot{\alpha} + \dot{\varepsilon} \; , \tag{14}$$

the equations of motions result from the equations (6) and (7)

$$(\delta'_H l_H - \delta'_V l_V)\frac{\dot{\alpha}}{v} + (\delta'_V + \delta'_H)\alpha + \left[mv^2 - (\delta'_H l_H - \delta'_V l_V)\right]\kappa = \left(\frac{\delta'_V}{i_{L,V}} + \frac{\delta'_H}{i_{L,H}}\right)\beta_L \tag{15}$$

$$-\Theta\ddot{\alpha} - (\delta'_V l_V^2 + \delta'_H l_H^2)\frac{\dot{\alpha}}{v} - (\delta'_H l_H - \delta'_V l_V)\alpha + \Theta v\dot{\kappa} + (\delta'_V l_V^2 + \delta'_H l_H^2)\kappa = \left(\frac{\delta'_V l_V}{i_{L,V}} - \frac{\delta'_H l_H}{i_{L,H}}\right)\beta_L \; . \tag{16}$$

The equations (15) and (16) include

$$\delta'_V = \frac{\delta_V}{1 + \frac{\delta_V n_V}{C_{L,V}}} \tag{17}$$

$$\delta'_H = \frac{\delta_H}{1 + \frac{\delta_H n_H}{C_{L,H}}} \tag{18}$$

meaning, that the cornering stiffness and the stiffness of the steering system can be taken as a connexion in series.

## STABILITY OF THE MOTOR VEHICLE

The homogenous equations (15) and (16) yield the characteristic equation

$$\Theta m v^2 \lambda^2 + v[\Theta(\delta_V' + \delta_H') + m(\delta_V' l_V^2 + \delta_H' l_H^2)]\lambda + [\delta_V' \delta_H' l^2 + m v^2 (\delta_H' l_H - \delta_V' l_V)] = 0. \qquad (19)$$

This equation does not contain the steering ratios. Therefore there are no differences between front wheel and rear wheel steered vehicles. This changes by looking at the term of $\lambda^0$. Motor vehicles get unstable above a certain speed - in the literature depending on vehicles called critical speed - if

$$\delta_H' l_H - \delta_V' l_V < 0 \quad . \qquad (20)$$

This will now being explained by the example of a car having the center of gravity in the middle of the wheelbase ($l_V = l_H$) and tires with equal cornering stiffness at the front and the rear wheels ($\delta_V = \delta_H$). The equation (20) is simplified by the equations (17) and (18)

$$\frac{n_V}{C_{L,V}} > \frac{n_H}{C_{L,H}} \quad . \qquad (21)$$

For a front wheel steered car therefore having no rear wheel steering and the stiffness of the rear wheel steering being infinite, $C_{L,H} \to \infty$, the following equation is valid

$$\frac{n_V}{C_{L,V}} > 0 \quad . \qquad (22)$$

This condition is satisfied if $n_V > 0$, that means, if the vehicle has a caster at the front wheels. As this is always the case, a front wheel steered vehicle is stable [2, 3].

For the rear wheel steered vehicle $C_{L,V}$ is infinite, $C_{L,V} \to \infty$, and the condition for stability is

$$0 > \frac{n_H}{C_{L,H}} \quad . \qquad (23)$$

These vehicles are only stable for all speed ranges, if a negative caster exists at the rear wheels [4]! As the caster consists of the constructional caster and the generally positive tire caster, for rear wheel steered motor vehicles a large negative constructional caster has to be realized. If that does not succeed, a greater watchfulness of the driver is nessesary.

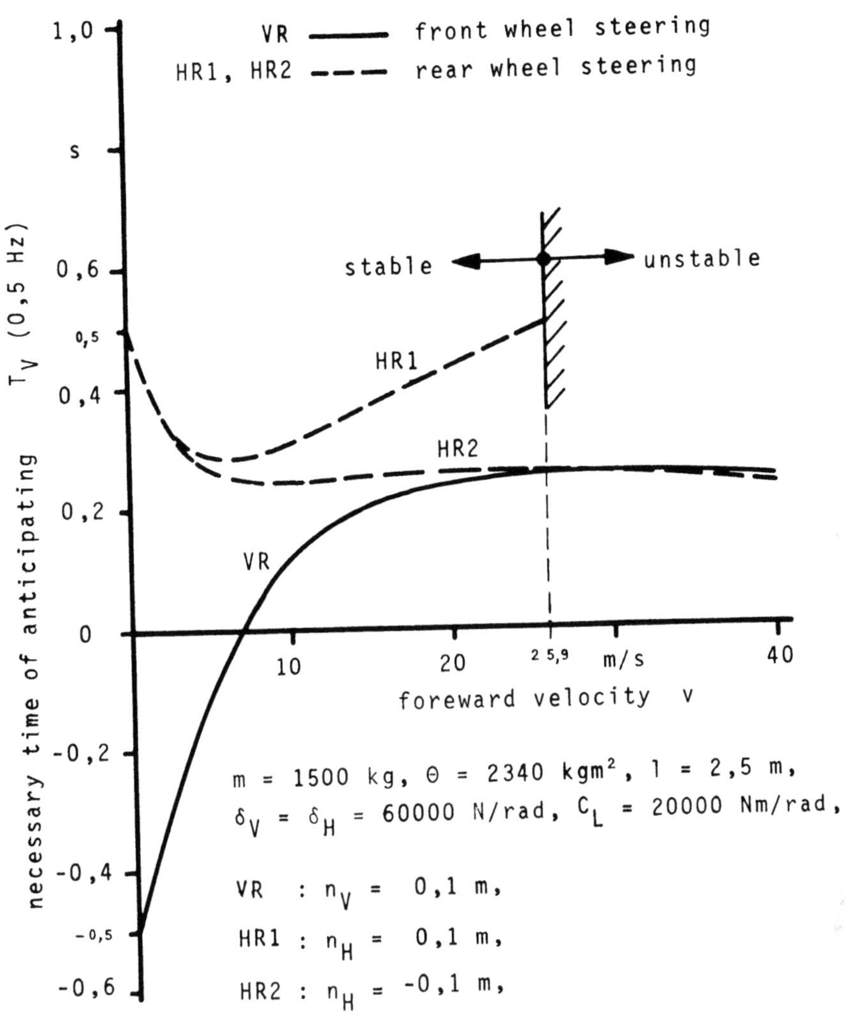


Fig. 3 Necessary time of anticipating for an ideal driver

ANTICIPATING TIME FOR THE IDEAL DRIVER

To be not forced to deal with the human transfer function, it is assumed, that the driver is able to hold the car exactly on the nominal course $\kappa_{Soll}$, that means, following equation (4),

$$\frac{\kappa_{Ist}}{\kappa_{Soll}} = 1 \quad , \tag{24}$$

and with equation (24)

$$F\,M_S = 1 \quad \rightarrow \quad M_S = \frac{1}{F} \; . \tag{25}$$

Splitting the complex frequency function $M_S$ in amplitude $|M_S|$ and phase angle $\varphi$ yields

$$M_S = |M_S| \; e^{i\varphi} \tag{26}$$

respectively

$$M_S = |M_S| \; e^{i \frac{\varphi}{\omega} \omega} = M_S \; e^{i T_V \omega} \tag{27}$$

with the human anticipating time $T_V$. It is determined from the equations (15) and (16), that means from the vehicle data.

In figure 3 the functions of $T_V$ versus the forward speed v are compared for a front wheel steered vehicle VR and two with rear wheel steering HR1 and HR2. The vehicles do not differ in mass, yawing moment of inertia, position of the c.o.g., wheelbase, stiffness of steering system. VR is stable, HR1 is stable until v = 25,9 m/s and HR2 at any speed. The comparision shows exactly, that the anticipating time for the rear wheel steered motor vehicles and especially for that ones, that are only stable for certain speed ranges, is greater than for the front wheel steered vehicles.

Hence the result follows:

1. Driving a rear wheel steered car the driver has to concentrate himself more, he has to think and to handle in advance.
2. A rear wheel steered vehicle is to be driven slowly, the same is prescribed by some legislators.

REFERENCES


1. Donges, E. Experimentelle Untersuchung und regelungstechnische Modellierung des Lenkverhaltens von Kraftfahrern bei simulierter Straßenfahrt
Dissertation Darmstadt (1977)

2. Mitschke, M. Dynamik der Kraftfahrzeuge
Springer-Verlag Berlin, Heidelberg, New York (1972)

3. Pacejka, H.B. Simplifide Analysis of Steady-State Turning of Motor Vehicles
Vehicle System Dynamics (1973), S.161-204

4. Perret, W. Lenkkräfte und Lenkwege am Lenkrad von Kraftfahrzeugen und ihr Einfluß auf die Lenksicherheit
Dissertation Stuttgart (1964)

# COMPUTATION OF LIMIT CYCLES OF A WHEELSET USING A GALERKIN METHOD

Dirk Moelle, Harald Steinborn and Robert Gasch

Institut für Luft- und Raumfahrt, Technical University of Berlin, F.R.G.

SUMMARY

The limit cycles for the nonlinear wheelset equations are determined by Galerkin's method. This method allows to consider not only the first but also the higher order Fourier terms for approximating the periodic hunting motion. Therefore, the solution can be calculated with any degree of accuracy.

As numerical results, the relations between amplitude and speed of the hunting wheelset will be presented. The time history of the state variables and of the contact forces will be demonstrated as well.

## 1. INTRODUCTION

In the lateral dynamics of railway vehicles moving on a straight track, almost all physical and kinematic connections are of a nonlinear nature.

That raises the following question: what errors result when these relations are regarded as purely linear, as has been done in almost all investigations to date. The following nonlinear analysis of the movement of a wheelset offers an answer to this question, based on a variational method of Galerkin [1,2].

By applying higher Fourier-Terms in the periodic approximation used in this calculation, the degree of accuracy may be increased at will.

The following nonlinearities are considered to be the main influences on the lateral dynamics of a wheelset on straight track.
- Coordinates of the wheel-rail-contact point,
- Geometric data of the contact area and creep,
- Creep forces.

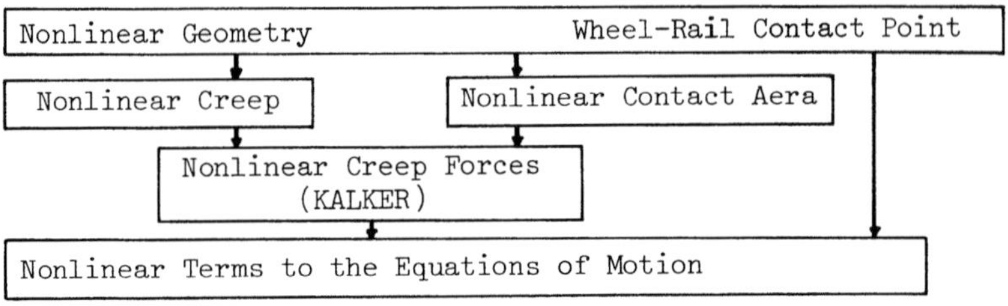


Fig. 1: Nonlinear effects

Apart from the described nonlinear effects the equations of motion are nonlinear and include second order terms [3]. The equations of motion are set up in moduls. Therefore there are no restrictions in the degrees of freedom. We have, however, until now analysed a three-degrees-of-freedom (3-DOF) wheelset.

## 2. THE PRINCIPLE OF THE GALERKIN METHOD

The equation of motion is of the following type:

$$\omega_s^2 \underline{M} (\underline{x},t)\ddot{\underline{x}} + \omega_s \underline{D} (\underline{x},t)\dot{\underline{x}} + \underline{S} (\underline{x},t)\underline{x} + \underline{r} (\underline{x},t) = 0 \tag{1}$$

The factor $\omega_s$ is used to make the independent variable t dimensionless.

Spliting up the equation (1) into a linear and a nonlinear part results in

$$\underline{g}^{LIN}(\ddot{\underline{x}},\dot{\underline{x}},\underline{x},\omega_s) + \underline{g}^{NON}(\ddot{\underline{x}},\dot{\underline{x}},\underline{x},t,\omega_s) = 0 \tag{2}$$

In order to find periodic solutions (limit cycles), a higher order Fourier-approximation is introduced for the nonlinear part of the differential equation as well as for the unknown sta. variables $\underline{x}$.

The unknown $\underline{x}$ are described by a Fourier approach

$$\underline{x} \simeq \underline{x}_m = \underline{a}_o + \sum_{k=1}^{m} (\underline{a}_{2k-1} \text{ sinkt} + \underline{a}_{2k} \text{ coskt}). \tag{3}$$

The unknown amplitudes $\underline{a}$ are also used in order to approach the nonlinearities $\underline{g}^{NON}$ of the differential equation (2).

$$\underline{g}^{NON} \simeq \underline{g}_m^{NON} = \underline{w}_o (\underline{a},\omega_s) + \sum_{k=1}^{m} (\underline{w}_{2k-1}(\underline{a},\omega_s) \text{ sinkt} + \underline{w}_{2k}(\underline{a},\omega_s) \text{ coskt}). \tag{4}$$

with

$$\underline{w}_0(\underline{a},\omega_s) = \frac{1}{2\pi}\int_0^{2\pi} \underline{g}^{NON}\left[\underline{x}_m(s),\ x_m(s),\ \underline{x}_m(s),\ \omega_s,\ s\right] ds$$

$$\underline{w}_{2k-1}(a,\omega_s) = \frac{1}{\pi}\int_0^{2\pi} \underline{g}^{NON}\left[\underline{x}_m(s),\ \underline{x}_m(s),\ \underline{x}_m(s),\ \omega_s,\ s\right] \sin ks\, ds$$

$$\underline{w}_{2k}(\underline{a},\omega_s) = \frac{1}{\pi}\int_0^{2\pi} \underline{g}^{NON}\left[\underline{x}_m(s),\ \underline{x}_m(s),\ \underline{x}_m(s),\ \omega_s,\ s\right] \cos ks\, ds$$

$$k = 1,2,\ldots,m$$

Substitution of (3) and (4) into (2) yields a system of nonlinear algebraic equations in terms of $\underline{a}$ and $\omega_s$:

$$\underline{u}(\underline{a},\omega_s) = 0 \tag{5}$$

Equation (5) is solved iteratively by a Newton-procedure

$$\underline{a}_{NEW} = \underline{a}_{OLD} - \left(\frac{\partial \underline{u}}{\partial \underline{a}}\right)^{-1}_{OLD} \underline{u}_{OLD} \ . \tag{6}$$

If there is any solution at all, then Urabe [1] has proved that the exact periodic solution of the differential equation can be calculated with an arbitrary degree of accuracy by the approach described before.

In general it is difficult to find criteria of stability for the given type of nonlinear differential equation of the second order. In our case the chosen method by Galerkin has proved to be helpful, because it is possible to replace the solution for a limit cycle in the differential equation (1). By this we obtain a linear homogeneous differential equation of first order with periodic coefficients. The stability of the limit cycle can thus be determined via the eigenvalues of the matrizant.

## 3. MECHANICAL MODEL OF A THREE-DEGREES-OF-FREEDOM WHEELSET

The motion of the wheelset is described in an inertial system that moves with the undisturbed forward speed $v_o$ (fig. 2). The degrees of freedom to be considered are

- the lateral displacement $u_y$,
- the angular displacement about the axis of revolution $\phi_y$,
- the angular yaw displacement $\phi_z$.

The angular displacement $\phi_y$ is refered to the undisturbed rotation $\omega_o$ of the wheelset. By this it is guaranteed that all state variables are small in relation to the main motion $(v_o,\omega_o)$. In addition to the state variables the dependent displacements $u_z$ (vertical) and $\phi_x$ (roll-angle) are to be considered. The kinematic constraints between

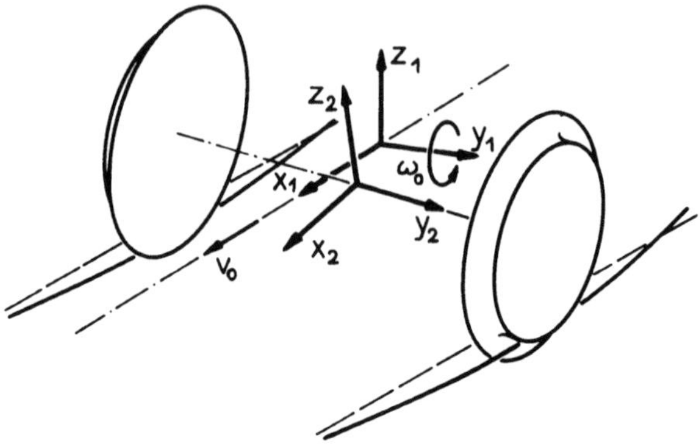


Fig. 2: Coordinate systems

$u_z$ and $\phi_x$ and the independent displacements $u_y$ and $\phi_z$ are nonlinear of high order. They are evaluated numerically by a computer-program [5].

The normal forces $N_L$ and $N_R$ (fig. 3) in the contact patches between wheel and rail have an influence on the nonlinear force/creepage relationship. They affect the slip forces $T_\xi$ and $T_\eta$ as well as the spin moment $M_\zeta$

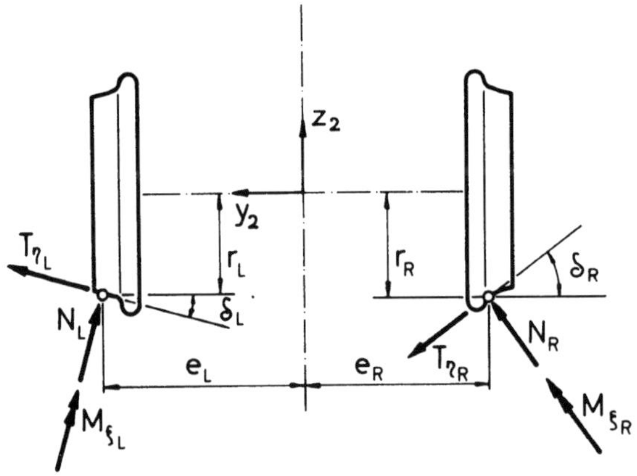


Fig. 3: Wheel/rail contact forces

The contact patch shape is simplified as an ellipse. Its axises a and b are calculated according to the theory of Hertz as a function of the normal load N and of the four radii of curvature of the surface of the two contacting bodies.

With the nonlinear expressions for the creepages (see appendix A) we get the tangential forces $T_\xi$ and $T_\eta$ respectively the spin moment $M_\zeta$. The creep forces can either be determined by the linearized relations introducing a force-limitation

$$
\begin{aligned}
T_\xi &= - Gc^2 \, C_{11} \, \nu_\xi \\
T_\eta &= - Gc^2 (C_{22} \, \nu_\eta + C_{23} c \nu_\zeta) \\
M_\zeta &= - Gc^3 (-C_{23} \nu_\eta + C_{33} c \nu_\zeta)
\end{aligned}
\tag{7}
$$

or by the simplified nonlinear theory of Kalker [6,7].

When the forces and the moment within the contact patches are evaluated they have to be transformed into the inertial system (fig. 2) by the aid of the contact angles $\delta_{L,R}$. The resulting forces-refered to the centre of gravity of the wheelset-are designated in the following as $K_i$ and $M_i$ (i = 1,3).

With the resulting creep forces the matrices of equation (1) may be formulated and split up into their linear and nonlinear parts (see appendix B).

Concerning the rotational DOF $\phi_y$ we assume the wheelset moving along the track with an undisturbed constant velocity $v_o$. Pure rolling of the wheelset in its centered position would yield a rotational velocity about the axis of revolution of the wheelset

$$\omega_o = v_o / r_o \tag{8}$$

where $r_o$ is the rolling radius.

Depending on the lateral displacement of the wheelset the rolling radius r of one wheel looks qualitatively as shown in fig. 4:

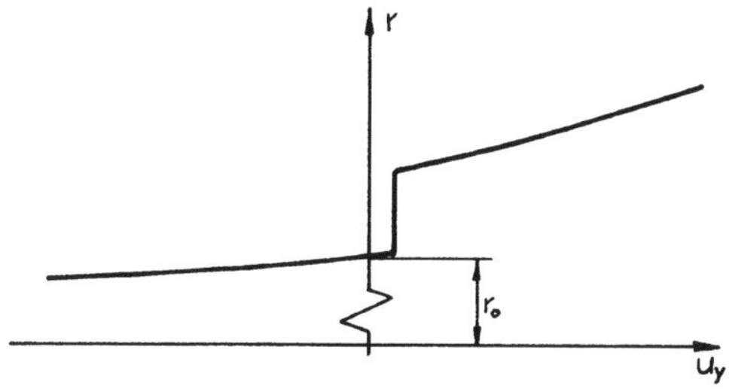


Fig. 4: Rolling radius r for left wheel depending on the lateral displacement $u_y$

We can see that the average $r_m$ between left and right radius of rolling is not equal to the radius $r_o$ for most of the given amplitudes $u_y$. If a wheelset would be forced to rotate with a constant velocity $\omega_o$ according to equation (8), there would arise very high longitudinal creepages in the contact patches which would cause unrealisticly large longitudinal forces.

By admitting a rotational degree of freedom in addition to the constant rotation $\omega_o$

$$\bar{\omega} = \omega_o + (\Delta\omega + \dot{\phi}_y)\,\omega_s \tag{9}$$

this can be avoided.

The undisturbed rotation $\omega_o$ is corrected by a constant term and a periodical part $\dot{\phi}_y$. With this new DOF a procedure may be developed which allows to determine the driving forces necessary to keep the wheelset at constant speed.

## 4. NUMERICAL RESULTS

In the following the latest results of our calculations will be presented. In order to save computing time, the nonlinear functions of the geometric parameters were calculated in advance and stored in

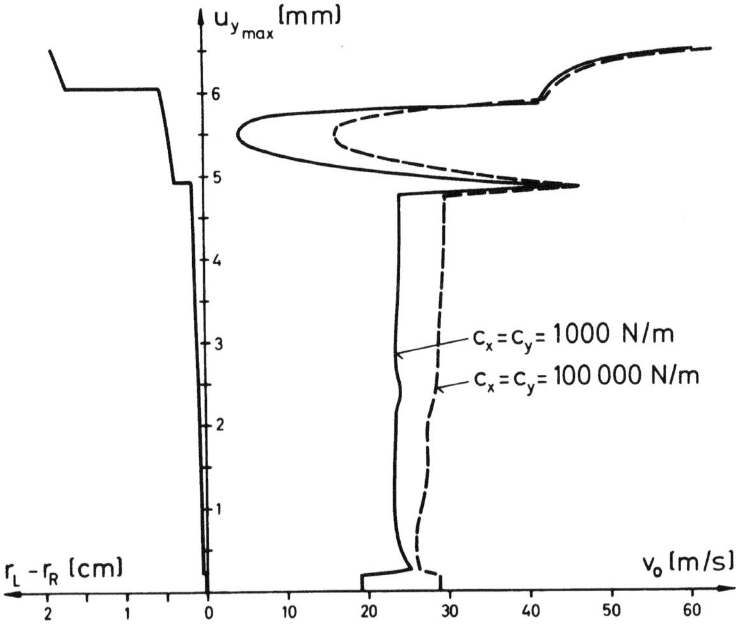


Fig. 5: Limit cycle lateral amplitudes v/s forward speed $v_o$ for two different suspension stiffnesses

arrays. For this procedure, we used a program developed by W. Hauschild [5]. Furthermore the linearized force-creepage relationships were applied and when reaching the limiting value μN, the tangential force was cut off.

The figures 5-7 show on their right sides the amplitudes of lateral deflection $u_{ymax}$ of the limit cycle for a given forward speed $v_o$. On their left sides, the differences of rolling-radii $r_L$-$r_R$ as functions of lateral displacement $u_y$ are to be seen.

Fig. 5 shows the limit cycle amplitudes $u_y$ for a wheelset with a very low suspension stiffness (nearly free) and a more realistic stiffness of $c_x = c_y = 10^5$ N/m. Both curves start from the horizontal axis with the limiting speed of the linearized theory (19 m/s resp. 28 m/s). Each discontinuity in the rolling radii difference (at 0.2 mm; 4.8 mm; 6.1 mm) corresponds to an abrupt change in the limit cycle behaviour. In comparison with the weak suspension, the stiffer one produces a higher forward speed for a given lateral amplitude. The results for amplitudes greater than 6 mm (flange contact) are somewhat questionable. In this range the tables of the geometric functions were to rough.

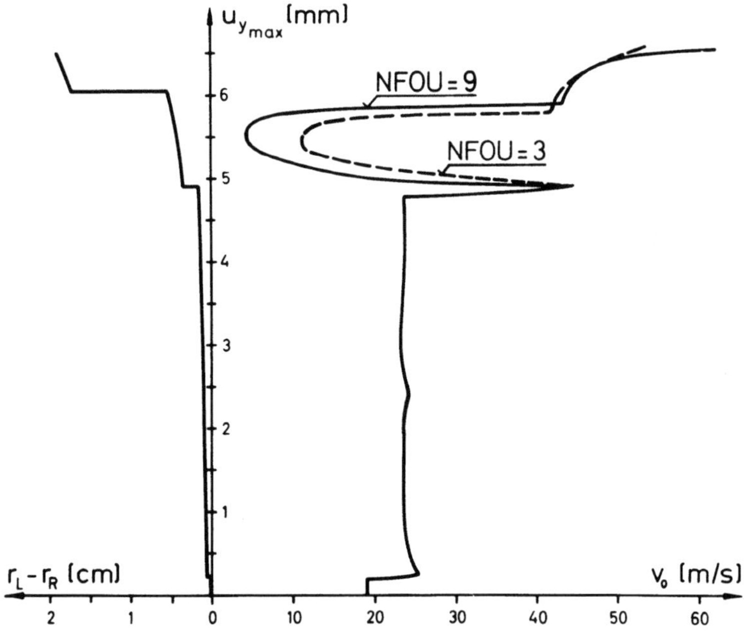


Fig. 6: Limit cycle lateral amplitudes v/s forward speed $v_o$ for different order Fourier approximations (NFOU)

Figure 6 shows limit cycles calculated by taking into account different numbers of Fourier-terms (NFOU). For lower amplitudes (< 4.8 mm) the results do not differ. But as soon as the lateral amplitude becomes larger, the higher order Fourier approach turns out to be more appropriate for investigating the increased nonlinear effects.

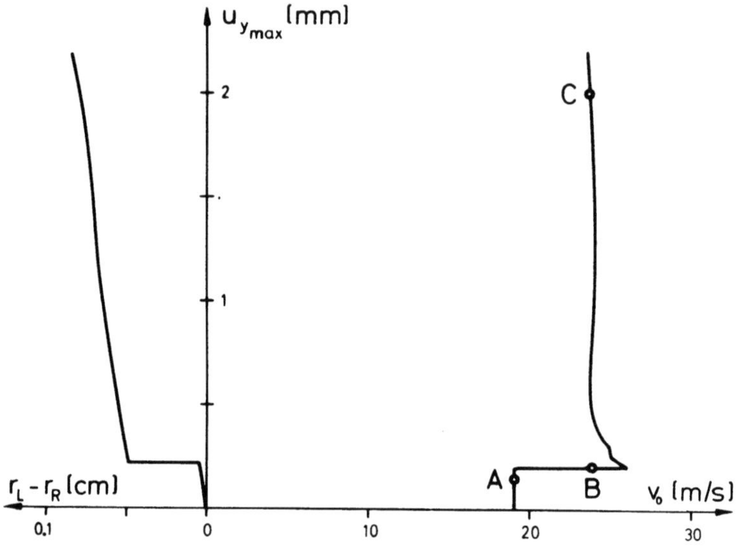


Fig. 7: Limit cycle lateral amplitudes v/s forward speed $v_o$

Figure 7 is a blown up part of fig. 6 for small amplitudes. The combination of the wheel profile ORE S 1002 with the rail profile UIC 60 used for our calculations yields a jump of the wheel-rail contact point for a lateral displacement of $u_y$ = 0.2 mm according to the given flange clearance. This results in an abrupt rise of the difference of rolling radii. Below this amplitude the limit cycle shows the linear critical speed, point A. Point B is located in the discontinuity of the rolling radii difference. Point C is far beyond this area.

The curves in Fig. 8 show the state variables as functions of the dimensionless time $0 \leq t \leq 2\pi$ for the three states of motion A, B, and C. Due to the factor $\omega_s$ by which the time is reduced, the accelerations have the dimension [m]. In the state A, the displacements as well as the accelerations show a harmonic behaviour which indicates that the dynamic system is almost linear. The periodic part of the angular displacement $\phi_y$ about the axis of revolution of the wheelset nearly vanishes, because - if we assume a constant forward speed $v_o$ - it mainly depends on the average radius of rolling

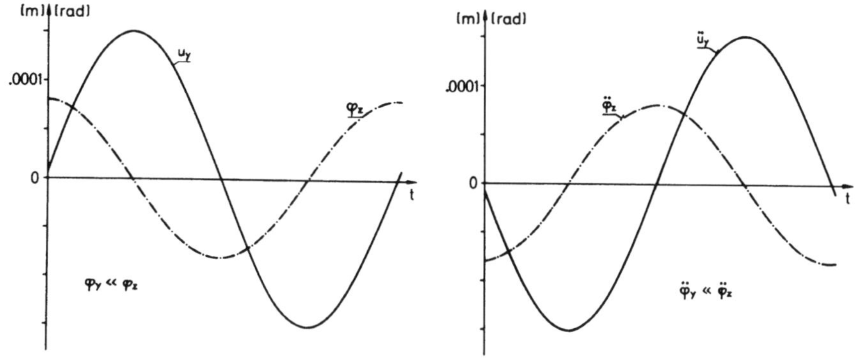


a) Situation A

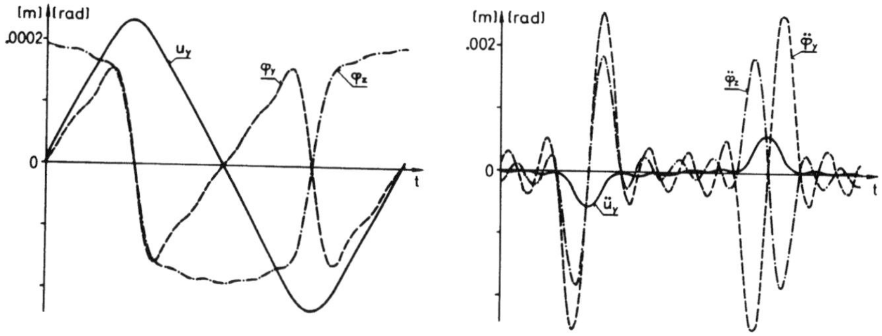


b) Situation B

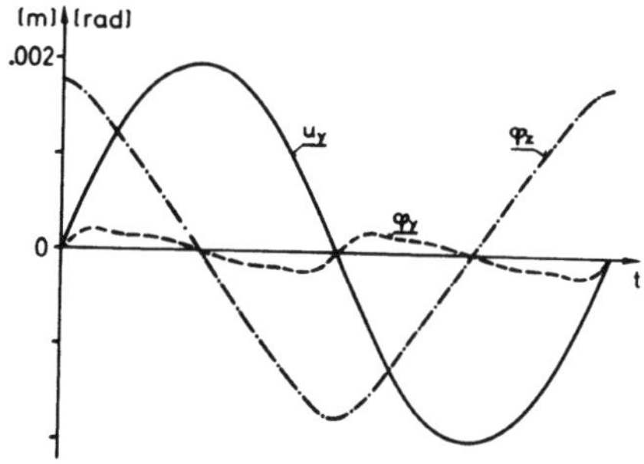


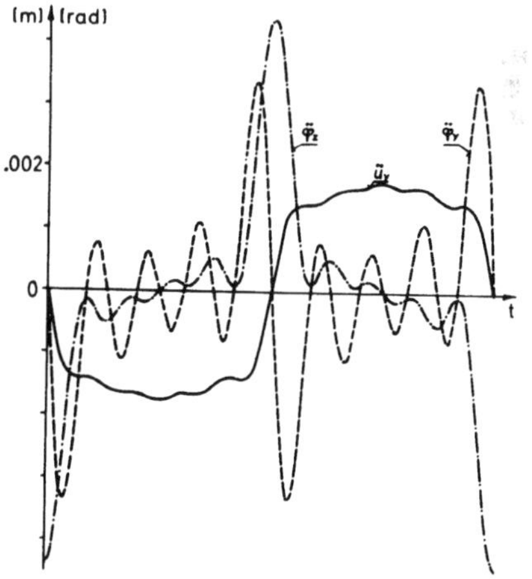


c) Situation C

Fig. 8: Time history of the state variables (Situation A, B, C)

$$r_m = (r_L + r_R) / 2 \qquad (10)$$

which is in this state nearly constant and equal to $r_o$.

In the state B, the time history of the displacements differs strongly from a harmonic motion. Concerning the displacement $\phi_y$, it should be mentioned that its lowest frequency is double as high as the lowest frequencies of the other displacements. This is a consequence of the fact that $\phi_y$ depends on $r_m$ which has its minimal value $r_o$ at a lateral displacement $u_y = 0$ and reaches its maximum when $u_y$ reaches its maximum, too. The peak of the accelerations of all DOFs lie in the section of the lateral amplitude $u_y = 0.2$ mm where the jump of the contact patch takes place.
In the state C shich is characterized by the lateral amplitudes about $u_y = 2$ mm the peak values of acceleration occur about $u_y = 0{,}2$ mm just as in the state B, but the nonlinear influence is obviously less strong than before.

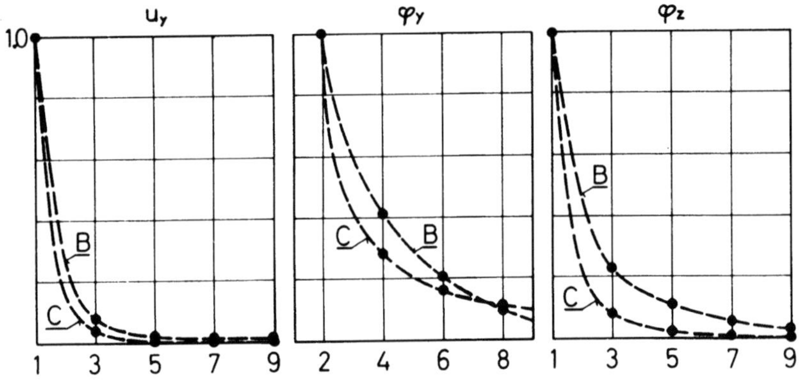


Fig. 9: Amplitudes v/s order of Fourier terms for the situations B and C

Figure 9 shows the relative values of different order Fourier terms used to approximate the three DOFs $u_y$, $\phi_y$, and $\phi_z$ in the states of motion B and C. In the state A, all terms of higher than first order are nearly zero. We can clearly observe that in the state B the higher Fourier terms are more important than in the state C in particular when the yaw motion $\phi_z$ is regarded. According to the fact that time history of the displacements is different from the first harmonic form the forces between wheel and rail are non-harmonic, too.

In fig. 10, the resulting creep force in lateral direction $K_2$ shows a smooth course regarding state A compared to the states B and C where high alternations of this force occur.

We found that the situation described above concerning the states A, B, and C will recur qualitatively in the same way at the next jump of the contact point at $u_y = 4{,}8$ mm.

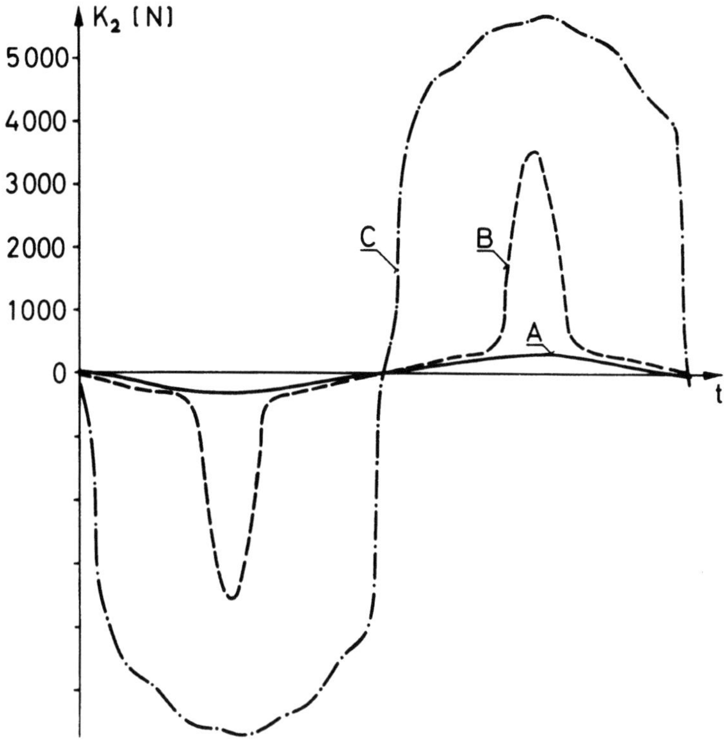


Fig. 10: Time history of lateral creep force for the situation A, B, and C

CONCLUDING REMARKS

The numerical results show that the nonlinear behaviour of a single wheel set-model is affected by the nonlinear geometric functions, especially by the difference of rolling-radii. When the amplitudes of the motion of the wheelset exert values which cause discontinuities of the geometric data, it is important

- to describe the wheelset by a 3-DOF model,
- to choose a high-order Fourier-approximation.

Here with it is possible not only to examine the limit cycle lateral amplitudes as a function of speed but also to evaluate the time history of the motion and of the nonlinear forces with sufficient accuracy. If, however, only the limit cycle amplitude is of interest, it is sufficient to consider only the first-harmonic-approximation as long as flange contact does not occur (here $u_y \leq 4.8$ mm).

The main benefit of the method described above is the fact that arbitrary degree of accuracy may be obtained. With one program cursory investigations with the first harmonic approximation (K.-B., quasi-linearization, describing-function) can be made, as well as the computation of an almost exact periodic solution of the system.

ACKNOWLEDGEMENT

This paper is part of a work supplied by the German Ministry of Research and Technology (FRG).

APPENDIX A:

Nonlinear Creep

$$\nu_{\xi L,R} = \frac{1}{\bar{v}_{L,R}} \left\{ v_o - \bar{\omega}\, r_{L,R} \mp e_{L,R}\, \omega_s\, \dot{\phi}_z \right\}$$

$$\nu_{\eta L,R} = \frac{1}{\bar{v}_{L,R}} \{ -v_o \cos \delta_{L,R}\, \phi_z + \omega_s \Big[ \left( \dot{u}_y + r_{L,R}\, \dot{\phi}_x \right) \cos \delta_{L,R} +$$

$$+ (\dot{u}_z \mp e_{L,R}\, \dot{\phi}_x)(\pm \sin \delta_{L,R} + \phi_x) \Big] \}$$

$$\nu_{\zeta L,R} = \frac{1}{\bar{v}_{L,R}} \left\{ (\bar{\omega}\, \phi_x + \omega_s\, \dot{\phi}_z) \cos \dot{\delta}_{L,R} - \bar{\omega}\, (\pm \sin \delta_{L,R} + \phi_x) \right\}$$

$$\bar{v}_{L,R} = \frac{1}{2} \left( v_o + \bar{\omega} r_{L,R} \mp e_{L,R}\, \omega_s\, \dot{\phi}_z \right)$$

$$\bar{\omega} = \omega_o + \omega_s\, (\Delta\omega + \dot{\phi}_y)$$

$$\dot{u}_z = \frac{\partial u_z}{\partial u_y}\, \dot{u}_y + \frac{\partial u_z}{\partial \phi_z}\, \dot{\phi}_z$$

$$\dot{\phi}_x = \frac{\partial \phi_x}{\partial u_y}\, \dot{u}_y + \frac{\partial \phi_x}{\partial \phi_z}\, \dot{\phi}_z$$

APPENDIX B:

Equations of motion

$$\omega_s^2 \begin{bmatrix} m + mf_{11}^2 + \theta_{x1} f_{21}^2 & o & mf_{12}f_{12} + \theta_x f_{21}f_{22} \\ o & \theta_y & o \\ mf_{11}f_{12} + \theta_x f_{21}f_{22} & o & \theta_z + mf_{12}^2 + \theta_x f_{22}^2 \end{bmatrix} \underline{\ddot{x}}$$

$$+ \omega_s \begin{bmatrix} d_y + d_z(f_{11}^2 + e_{dz}^2 f_{21}^2) & o & d_z(f_{11}f_{12} + e_{dz}^2 f_{21}f_{22}) - \theta_y \bar{\omega} f_{21} \\ o & o & o \\ d_z (f_{11}f_{12} + e_{dz}^2 f_{21}f_{22}) + \theta_y \bar{\omega} f_{21} & o & d_x e_{dx}^2 + d_z(f_{11}^2 + e_{dz}^2 f_{22}^2) \end{bmatrix} \underline{\dot{x}}$$

$$+ \begin{bmatrix} c_y - K_3 f_{21} & o & - K_1 + M_2 f_{21} \\ o & o & - M_1 \\ K_1 - K_3 f_{22} & o & c_x e_{cx}^2 + M_2 f_{22} \end{bmatrix} \underline{x}$$

$$+ \begin{bmatrix} -K_2 - (P - mg + K_3)\, f_{11} - M_1 f_{21} + u_z\, c_z f_{11} + K_2 f_{21} + \phi_x\, K_3 - K_2 f_{11} + c_z e_{cz}^2 f_{21} \\ - M_2 + u_z K_1 + \phi_x M_3 \\ - M_3 - (P - mg + K_3)\, f_{12} - M_1 f_{22} + u_z\, c_z f_{12} + K_2 f_{22} + \phi_x -M_2 - K_2 f_{12} + c_z e_{cz}^2 f_{22} \end{bmatrix} = 0$$

$$\begin{bmatrix} f_{11} & f_{12} \\ f_{21} & f_{22} \end{bmatrix} = \begin{bmatrix} \frac{\partial u_z}{\delta u_y} & \frac{\partial u_z}{\delta \phi_z} \\ \frac{\partial \phi_x}{\delta u_y} & \frac{\partial \phi_x}{\delta \phi_z} \end{bmatrix}$$

$$\left.\begin{matrix} K_1, & K_2, & K_3 \\ M_1, & M_2, & M_3 \end{matrix}\right\} = \text{Creep forces and moments described in the body-fixed coordinate system 2}$$

REFERENCES:


1. Urabe, Galerkin's Procedure for Nonlinear Periodic Systems. Arch. Rat. Mech. Anal., Vol. 20, 1965.

2. Pietruszka, Numerische Bestimmung periodischer Lösungen und deren Stabilität in selbsterregten Systemen. Gastvortrag an der TU Berlin, Febr. 1977.

3. Moelle, Bewegungsgleichungen für den Lauf eines angetriebenen Radsatzes. ILR-Mitteilungen, Nr. 46, TU Berlin, 1978.

4. Hauschild, The Application of Quasilinearization to the Limit Cycle Behaviour of the Nonlinear Wheel-Rail-System, this volume, pp. 146-163.

5. Hauschild, Die Kinematik des Rad-Schiene-Systems. Berichte aus dem 2. Inst. f. Mechanik, TU Berlin, März 1977.

6. Kalker, Simplified Theory of Rolling Contact. Delft Progr. Rep. Series C: Mechanical and Aeronautical Engineering and Shipbuilding, 1 (1973), p. 1-10.

7. Knothe, Moelle, Steinborn: Rolcon - Ein schnelles vielseitiges Digitalprogramm zum rollenden Kontakt. ILR-Mitteilungen, Nr. 55, TU Berlin, 1978.

# COVARIANCE ANALYSIS OF NONLINEAR STOCHASTIC GUIDEWAY-VEHICLE-SYSTEMS

P. C. Müller, K. Popp and W. O. Schiehlen

Technical University Munich and University Stuttgart, F.R.G.

SUMMARY

The dynamic analysis of random vehicle vibrations and the consequences to human comfort requires an integrated investigation of guideway roughness, nonlinear vehicle dynamics and human response to vibration exposure. The covariance analysis represents an efficient integrated technique to determine the stochastic response characteristics such as standard deviations of guideway loading and human ride comfort. Performing covariance analysis the complete system has to be characterized in the time domain. Therefore, (i) a time domain model of guideway irregularities is defined with respect to measurements, (ii) the vehicle dynamics are represented including nonlinear springs and dashpots as well as multi-axle excitation inputs, (iii) the human response to broadband random vibrations is extensively discussed resulting in a perception shape filter. Statistical linearization and well established Lyapunov matrix equation techniques are applied to complete the covariance analysis. Its efficiency is illustrated by the random vibration analysis of a 4-DOF automobile model.

## 1. INTRODUCTION

Rail and road vehicles traveling on irregular guideways are subject to random vibrations affecting the guideway loading and the ride comfort of the passengers. In particular the ride comfort has to meet the International Standard ISO 2631 [1].

The dynamic analysis of random vehicle vibrations and the consequences to human comfort requires an integrated investigation of guideway roughness, nonlinear vehicle dynamics and human response to vibration exposure. Many aspects of these problems are dealt with in numerous publications, e.g. in the classic book by MITSCHKE [2] on dynamics of road vehicles and in more recent papers of HEDRICK, BILLINGTON and DREESBACH [3], KARNOPP [4] and DAHLBERG [5]. To improve the former techniques of analysis and achieving a better design of vehicles with respect to the human perception of random vibrations, this paper illustrates how nonlinear multi-degree-of-freedom models of vehicles on irregular guideways can be systematically analyzed by the covariance analysis using time domain state variable methods. In consequence of this aim, in section 2 the guideway roughness is defined in the time domain with respect to published measurements, in section 3 the non-linear equations of motions are established taking into account multi-axle vehicles, and in section 4 the human response is interpreted for broadband random vibrations by a perception shape filter. Efficient numerical methods for the stochastic analysis of the total system of guideway irregularities, vehicle, and human response are presented in section 5. Finally, in section 6 the covariance method is effectively applied to the analysis of a 4-degree-of-freedom model of a randomly excited automobile.

## 2. GUIDEWAY IRREGULARITY MODELS AND VEHICLE EXCITATION

Research in the field of guideway roughness models is going on for a long time. In recent publications there is a tendency towards standardization of roughness representation. Measurements on road profils, e.g. MITSCHKE [6], BRAUN [7], WENDEBORN [8], and on railway tracks, cf. ORE [9], have shown that irregularities can be described by stationary ergodic Gaussian random variables with zero mean value. The characterisation is nearly always given in the frequency domain using single sided power spectral densities (PSD). A simple but useful road roughness model reads, cf. MITSCHKE [2], VOY [10],

$$\Phi_\zeta(\Omega) = \Phi_o\left(\frac{\Omega_o}{\Omega}\right)^w , \tag{1}$$

where $\Omega$[rad/m] is the (space) angular frequency, $\Phi_\zeta(\Omega)[m^2/\text{rad/m}]$ is the PSD of the random variable $\zeta(x)$ which is a function of distance $x$, and $\Omega_o$, $\Phi_o = \Phi_\zeta(\Omega_o)$ and $w$ are constants describing reference angular frequency, unevenness and waviness, respectively. Usually, the waviness ranges between $1{,}75 \le w \le 2{,}25$. A similar but more sophisticated description is suitable for roads and tracks as well, see DODDS, ROBSON [11], HEDRICK, ANIS [12],

$$\Phi_\zeta(\Omega) = \begin{cases} \Phi_o\left(\frac{\Omega_o}{\Omega}\right)^{w_1} & \text{for } \Omega \le \Omega_o , \\ \Phi_o\left(\frac{\Omega_o}{\Omega}\right)^{w_2} & \phantom{\text{for }} \Omega \ge \Omega_o , \end{cases} \tag{2}$$

where different exponents $w_1$ and $w_2$ are introduced. In case of track irregularities eq. (2) is used to describe equally vertical profile, lateral alignment, gauge and cross-level. Eqs. (1) and (2) are approximations to measured PSD in a distinct frequency range $0<\Omega_1\le\Omega\le\Omega_2$. In either case, the limit $\Omega \to 0$ would yield $\Phi_\zeta(\Omega\to 0) \to \infty$ and thus an infinite variance of the irregularities would follow which is not realistic. To avoid these difficulties two other roughness models are sometimes used, cf. DINCĂ, THEODOSIU [13], FÁBIÁN [14], SUSSMAN [15],

$$\Phi_\zeta(\Omega) = \frac{2\alpha\sigma^2}{\pi}\,\frac{1}{\alpha^2+\Omega^2} , \tag{3a}$$

$$\Phi_\zeta(\Omega) = \frac{2\alpha\sigma^2}{\pi}\,\frac{\Omega^2+\alpha^2+\beta^2}{(\Omega^2-\alpha^2-\beta^2)^2+4\alpha^2\Omega^2} , \tag{4a}$$

where $\alpha,\beta$ and $\sigma^2$ are constants. Since (3) and (4) are valid in the entire frequency range, the autocorrelation function $R_\zeta(\xi)$ which is the equivalent space domain characterization can be calculated by the inverse Fourier transform

$$R_\zeta(\xi) = E\{\zeta(x)\zeta(x-\xi)\} = \int_0^\infty \Phi_\zeta(\Omega)\cos\Omega\xi d\Omega \ , \tag{5}$$

where $E\{\cdot\}$ denotes the expected value. Applied to (3a) and (4a) it yields

$$R_\zeta(\xi) = \sigma^2 e^{-\alpha|\xi|} \ , \tag{3b}$$

$$R_\zeta(\xi) = \sigma^2 e^{-\alpha|\xi|}\cos\beta\xi \ . \tag{4b}$$

Here, $\sigma^2 = R_\zeta(0)$ describes the (finite) variance of the irregularities.

From the roughness representation (1-4) the vehicle excitation models can be obtained in frequency domain and also in time domain using $d\omega = v d\Omega$ , $dx = v dt$ , $d\xi = v d\tau$ , where $v$[m/s] is the vehicle speed and $\omega$[rad/sec] means the (time) angular frequency. Since the variance $R_\zeta(0)$ has to remain unchanged from eq. (5) it follows

$$\Phi_\zeta(\omega)d\omega = \Phi_\zeta(\Omega)d\Omega \ . \tag{6}$$

Using model (1) for example, where $w=2$ is chosen, one gets

$$\Phi_\zeta(\omega) = \frac{1}{v}\Phi_0\left[\frac{v\Omega_0}{\omega}\right]^2 = v\,\Phi_0\left[\frac{\Omega_0}{\omega}\right]^2 \ . \tag{7}$$

Here, a technical approximation should be mentioned, cf. KARNOPP [4]. The PSD $\Phi_{\dot\zeta}(\omega)$ for the derivated process $\dot\zeta(t)$ is $\omega^2\Phi_\zeta(\omega)$ due to classical theory. Applied to (7) it follows a white noise PSD,

$$\Phi_{\dot\zeta}(\omega) = v\,\Phi_0\Omega_0^2 = \text{const} \ , \tag{8a}$$

corresponding to the autocorrelation function

$$R_{\dot\zeta}(\tau) = q_{\dot\zeta}\delta(\tau) \ , \quad q_{\dot\zeta} = \pi v\,\Phi_0\Omega_0^2 \ , \tag{8b}$$

where $q_{\dot\zeta}$ is the intensity and $\delta(\cdot)$ denotes the Dirac function. The white noise approximation results also in an infinite variance which is not realistic. On the other hand, the amount of calculations can be reduced considerably using the simple models (8a), (8b) and $\zeta(t)$ as vehicle input.

A better vehicle excitation model $\zeta(t)$ is given by stationary Gaussian colored noise which can be obtained from a white noise process $w(t)$ by means of a shape filter,

$$\zeta(t) = \underline{h}^T\underline{v}(t) \ , \tag{9}$$

$$\dot{\underline{v}}(t) = \underline{F}\,\underline{v}(t) + \underline{g}\,w(t) \ , \ \mathrm{Re}\,\lambda(\underline{F}) < 0 \ , \ w(t) \sim (0, q_w) \ ,$$

where $\underline{v}(t)$, $\underline{F}$ , $\underline{g}$ and $\underline{h}$ determine the shape filter. The white noise is assumed to have zero mean and intensity $q_w$ . For colored noise characterized by (3a), (4a) or (3b), (4b) the corresponding shape filter quantities read

$$\underline{F} = -\alpha v \ , \ \underline{g} = g \ , \ \underline{h} = 1 \ , \ (g^2 q_w = 2\alpha v\sigma^2) \ , \tag{3c}$$

$$\underline{F} = \begin{bmatrix} 0 & 1 \\ -(\alpha^2+\beta^2)v^2 & -2\alpha v \end{bmatrix} , \ \underline{g} = g\begin{bmatrix} 0 \\ 1 \end{bmatrix} , \ \underline{h} = \begin{bmatrix} v\sqrt{\alpha^2+\beta^2} \\ 1 \end{bmatrix} , (g^2 q_w = 2\alpha v\sigma^2) \ . \tag{4c}$$

All vehicle excitation models up to now are scalar models where only single contact is taken into account. However, real vehicles have multiple contact with the guideway. Thus, the time delay between successive contact points has to be regarded, see section 3.

## 3. VEHICLE DYNAMICS

Vehicles are modeled for dynamic investigations with respect to frequencies less than 30 Hz by multibody systems consisting of rigid bodies, springs, dashpots and bearings, see e.g. SCHIEHLEN and KREUZER [16]. The resulting equations of motion are nonlinear due to springs and dashpots as well as suspension kinematics, cf. SCHIEHLEN [17],

$$\underline{M}(\underline{y})\ddot{\underline{y}}(t) + \underline{k}(\underline{y},\dot{\underline{y}}) = \underline{q}(\underline{y},\dot{\underline{y}},t) \tag{10}$$

where $\underline{M}$ is the symmetric, positive definite inertia matrix, $\underline{y}$ the generalized position vector, $\underline{k}$ the gyro and centrifugal force vector and $\underline{q}$ the vector of generalized damping, stiffness and excitation forces. However, the generalized force vector may also depend on additional force state variables represented by the vector $\underline{z}$ due to serial damper-spring elements in the vehicle:

$$\underline{q} = \underline{q}(\underline{y},\dot{\underline{y}},\underline{z},t) \ , \tag{11}$$

$$\dot{\underline{z}} = \dot{\underline{z}}(\underline{y},\dot{\underline{y}},\underline{z},t) \ . \tag{12}$$

From (10), (11) and (12) the vehicle's state equation is found as

$$\dot{\underline{x}}(t) = \underline{A}\underline{x}(t) + \underline{f}(\underline{x}) + \underline{B}\underline{\xi}(t) \tag{13}$$

where

$$\underline{x}(t) = [\underline{y}^T(t) \ \dot{\underline{y}}^T(t) \ \underline{z}^T(t)]^T \tag{14}$$

is the state vector, $\underline{A}$ the time-invariant state matrix, $\underline{f}(\underline{x})$ a nonlinear vector function, $\underline{B}$ the time-invariant excitation matrix and $\underline{\xi}(t)$ a stochastic vector process due to the guideway irregularities. The state equation (13) applies to vehicles with linear geometry and force characteristics between guideway and wheels. However, geometry and force characteristics of parts within the vehicles may remain nonlinear and, then, they are identified by the nonlinear vector function $\underline{f}(\underline{x})$, which is assumed to be an odd function, $\underline{f}(-\underline{x}) = -\underline{f}(\underline{x})$. The stochastic excitation process $\underline{\xi}(t)$ of multi-axle vehicles is characterized by time delays

$$t_i = \frac{l_i}{v} \ , \quad i = 1,..,r, \tag{15}$$

where $l_i$ is the distance between the front axle and the axle $i$, and $v$ represents the vehicle's velocity. Then, the excitation term in (13) has to be written as

$$\underline{B} \ \underline{\xi}(t) = \sum_{i=1}^{r} \underline{b}_i \ \zeta_i(t) \ ,$$

$$\zeta_i(t) = \zeta(t-t_i) \ , \ 0 = t_1 < t_2 < \ldots < t_r \ , \tag{16}$$

where the scalar stochastic excitation process $\zeta(t)$ is given by the guideway irregularities (9).

## 4. HUMAN RESPONSE

Stochastic guideway irregularities result in random vehicle vibrations acting on the human body. Therefore, the theory of stochastic processes has to be applied not only to the vehicle but also to the human body. The evaluation of the human response on stochastically excited vehicles depends on the state and excitation vector process. Both processes are assumed ergodic, Gaussian and stationary.

The objectively measurable mechanical vibrations acting on the guideway, the vehicle and the human body can be evaluated by scalar vibration variables

$$w_k(t) = \underline{c}_k^T \underline{x}(t) + \underline{d}_k^T \underline{\xi}(t) \ , \ k = 1,2,\dots, \tag{17}$$

where $\underline{c}_k$ and $\underline{d}_k$ are the corresponding weighting vectors. However, the scalar vibration variables $w_k(t)$ are also ergodic, Gaussian, stationary stochastic processes which are uniquely characterized by the mean values

$$m_{wk} = \underline{c}_k^T \underline{m}_x + \underline{d}_k^T \underline{m}_\xi = 0 \tag{18}$$

and the variances

$$\sigma_{wk}^2 = \underline{c}_k^T \underline{P}_x \underline{c}_k + 2\underline{c}_k^T \underline{P}_{x\xi} \underline{d}_k + \underline{d}_k^T \underline{P}_\xi \underline{d}_k \ . \tag{19}$$

Here, $\underline{m}_x$ and $\underline{m}_\xi$ are the vanishing mean values and $\underline{P}_x$, $\underline{P}_\xi$ the nonvanishing covariance matrices of the processes $\underline{x}(t)$ and $\underline{\xi}(t)$. The covariance matrix $\underline{P}_{x\xi}$ characterizes the coupling of the process $\underline{x}(t)$ with respect to the process $\underline{\xi}(t)$. If the excitation follows from white noise guideway irregularities, $\underline{P}_\xi \to \infty$, only vibration variables with $\underline{d}_k \equiv \underline{0}$ are admissible. This may be a restriction for the dynamic load computation with respect to the riding safety.

Mechanical vibrations subjectively perceptible by man differ from the objectively measurable mechanical vibrations. Numerous physiological investigations have shown that the subjective perception by man is proportional to the acceleration and depends on the dynamics of human organs which may be modeled by low order systems. It is possible, however, to characterize the subjectively perceptible vibrations by scalar perception variables, see VDI [18]. The perception variables $\bar{w}_k(t)$ depend on the position of man and on the objective vibration variables $w_k(t)$, $k = 1,2,\dots$ . The relation between perception variables $\bar{w}_k(t)$ and vibration variables $w_k(t)$ can be represented by shape filters given in the time domain using differential equations or by frequency response methods. However, the perception variables $\bar{w}_k(t)$ are also stochastic processes to be characterized by their variances.

In the time domain, the variance of a perception variable $\bar{w}_k(t)$ reads as

$$\sigma_{\bar{w}k}^2 = \alpha^2 \underline{\bar{h}}_k^T \underline{P}_k \underline{\bar{h}}_k \ , \tag{20}$$

where

$$\bar{w}_k(t) = \alpha \underline{\bar{h}}_k^T \underline{\bar{v}}_k(t) \tag{21}$$

and

$$\dot{\underline{\bar{v}}}_k(t) = \underline{\bar{F}}_k \underline{\bar{v}}_k(t) + \underline{\bar{g}}_k w_k(t) \ , \quad \mathrm{Re}\ \lambda(\underline{\bar{F}}_k) < 0 \ . \tag{22}$$

Here, $\underline{\bar{v}}_k(t)$ is an auxiliary state vector, $\underline{\bar{F}}_k$ the time-invariant shape filter matrix, $\underline{\bar{g}}_k$ the input vector, $\underline{\bar{h}}_k$ the output vector and $\alpha$ means a constant following from physiological experiments. Further, $\underline{P}_k$ is the covariance matrix of $\underline{v}_k$.

In the frequency domain, the squared standard deviation of a perception variable $\bar{w}_k(t)$ follows from an improper integral

$$\sigma^2_{\bar{w}k} = \int_{-\infty}^{\infty} S_{\bar{w}k}(\omega)\,d\omega = \int_{0}^{\infty} \Phi_{\bar{w}k}(\omega)\,d\omega \ , \tag{23}$$

where $S_{wk}(\omega)$ is the spectral density or $\Phi_{wk}(\omega) = 2\,S_{wk}(\omega)$ the one-sided spectral density, respectively. The spectral density of the perception variable reads as

$$S_{\bar{w}k}(\omega) = \alpha^2 |f_k(\omega)|^2\, S_{wk}(\omega) \ , \tag{24}$$

where $S_{wk}(\omega)$ means the spectral density of the corresponding vibration variable (17) and $f_k(\omega)$ represents the frequency response of the shape filter,

$$f_k(\omega) = \underline{\bar{h}}_k^T (i\omega\underline{E} - \underline{\bar{F}}_k)^{-1}\underline{\bar{g}}_k$$
$$= \frac{b_o + b_1(i\omega) + \ldots + b_{r-1}(i\omega)^{r-1} + b_r(i\omega)^r}{a_o + a_1(i\omega) + \ldots + a_{s-1}(i\omega)^{s-1} + (i\omega)^s} \ , \ r < s \ . \tag{25}$$

Theoretically, the representation in the time domain and the representation in the frequency domain are equivalent. From the numerical point of view, however, the time domain representation may be more economic since an improper integral has not to be solved.

The ride comfort in vehicles with respect to the longitudinal position of man is specified by the vertical acceleration $a(t)$ at the seat as vibration variable and the shape filter for the evaluation of the corresponding perception variable $\bar{a}(t)$. According to VDI [18] the standard deviation of the perception variable is used to describe the perception,

$$K = \sigma_{\bar{a}} = \alpha \sqrt{\int_0^{\infty} \Phi_a(\omega)\,|f_k(\omega)|^2 d\omega} \ . \tag{26}$$

Then, according to ISO [1] the tolerable exposure time can be found from Fig. 1 for the three main human criteria: preservation of comfort, working efficiency and health, respectively.

The shape filter for the evaluation of the perception variable $\bar{a}(t)$ is given in the standards ISO [1] and VDI [18] by the same frequency response, Fig. 2. However, the given frequency response can only be realized by a high order shape filter. For low order shape filters permissible deviations are also given in the standards and shown in Fig. 2. For a second order shape filter the coefficients in (25) read as

$$b_o = 500\,\frac{1}{s^2} \ , \ b_1 = 50\,\frac{1}{s} \ , \ a_o = 1200\,\frac{1}{s^2} \ , \ a_1 = 50\,\frac{1}{s} \tag{27}$$

and the constant in (24) is given by

$$\alpha = 20\,\frac{s^2}{m} \ . \tag{28}$$

Fig. 2 shows that the second order shape filter fits very well. In the time domain the corresponding shape filter is represented by the vec-

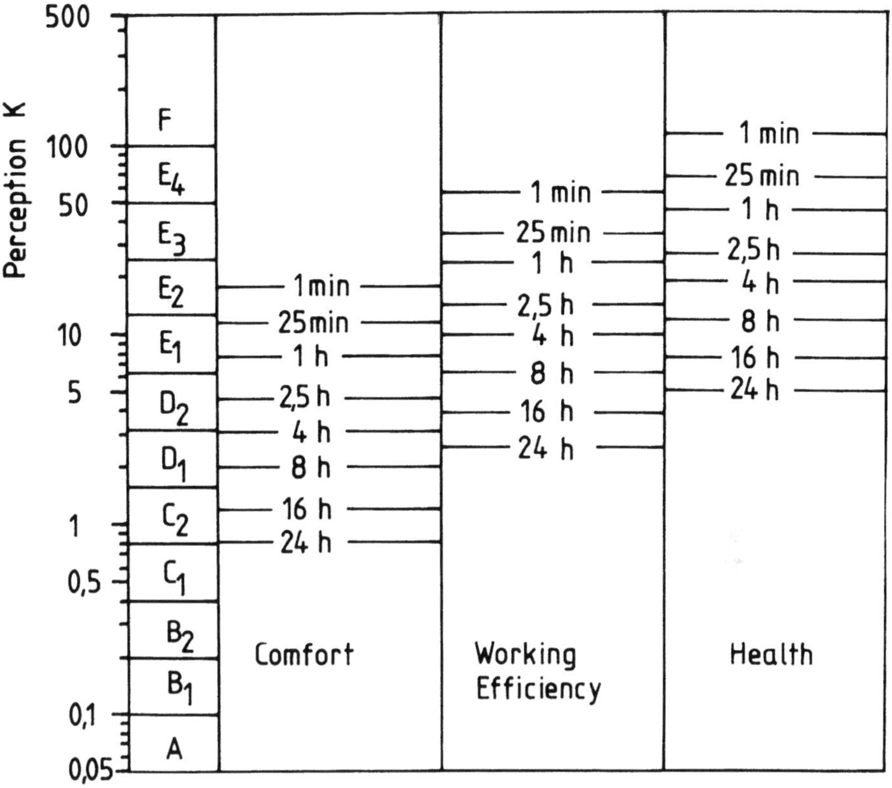


Fig. 1 Human perception and tolerable exposure time to vibration with respect to preservation of comfort, working efficiency and health.

tors

$$\underline{g}_k = \begin{bmatrix} 0 \\ 1 \end{bmatrix} , \quad \underline{h}_k = \begin{bmatrix} b_0 \\ b_1 \end{bmatrix} \tag{29}$$

and the time-invariant matrix

$$\underline{F}_k = \begin{bmatrix} 0 & 1 \\ -a_0 & -a_1 \end{bmatrix} , \tag{30}$$

where the coefficients (27) are also applied.

For the evaluation of ride comfort in vehicles there have also been special experiments executed independent of the more general experiments for the standards, cf. SMITH, McGEHEE and HEALEY [19]. In [19] the ride indicator $R$ is used and it is related immediately to the standard deviation $\sigma_a$ in $m/s^2$ of the vibration variables,

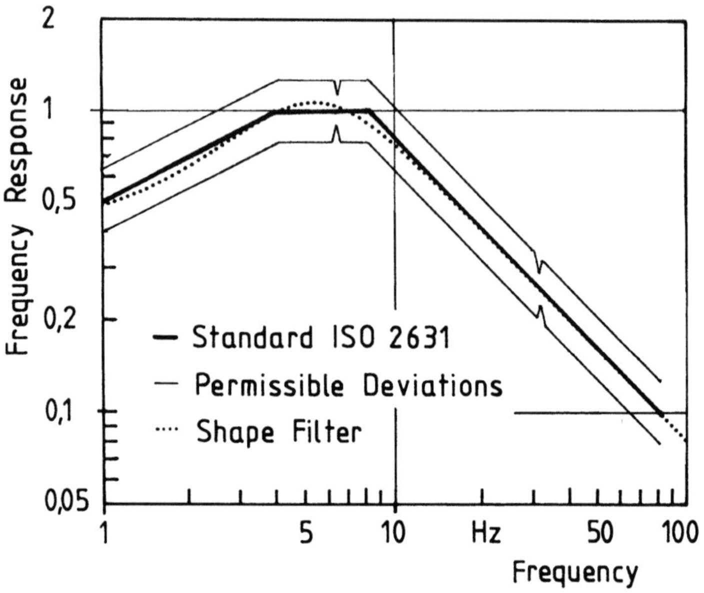


Fig. 2 Frequency response of a shape filter for human exposure to longitudinal vibrations.

$$R = 5{,}43 - 4\,\sigma_a \, . \qquad (31)$$

This means that the dynamical response of the human organs is neglected

$$f_k(\omega) = 1 \, . \qquad (32)$$

Then, from (26), (28), (31) it follows

$$K = 27{,}15 - 5\,R \, . \qquad (33)$$

Thus, special and general experiments on the human response to random vehicle vibrations coincide fairly well and may be used for advanced vehicle design.

## 5. METHODS

To determine the variances of dynamic wheel load, car body acceleration or human ride comfort, frequency domain methods are usually applied, cf. DAHLBERG [5]. In contrast, here the time domain covariance analysis is prefered. However, both methods are primarily related only to linear problems. Therefore, firstly the nonlinear state equation (13) has to be simplified by the statistical linearization technique, cf. HEDRICK, COOPERRIDER and LAW [20], resulting in

$$\dot{\underline{x}}(t) = (\underline{A} + \underline{N}(\underline{P}_x))\underline{x}(t) + \underline{B}\underline{\xi}(t) \, . \qquad (34)$$

In the case of linear models the system matrix $\underline{N}$ vanishes; otherwise it depends on the covariance matrix $\underline{P}_x$ of the system response.

Applying the method of power spectral density (PSD) the variance of the perception variable (21) is determined as follows:

$$\sigma^2_{\underline{wk}} \approx \int_{\omega_1}^{\omega_2} \Phi_{\underline{wk}}(\omega)\,d\omega \tag{35}$$

where

$$\Phi_{\underline{wk}}(\omega) = 2S_{\underline{wk}}(\omega) = 2|f_k(\omega)|^2[\underline{c}_k^T \underline{F}(\omega)\underline{B}+\underline{d}_k^T]\underline{S}_\xi(\omega)[\underline{B}^T\underline{F}^T(-\omega)\underline{c}_k+\underline{d}_k] , \tag{36}$$

$$\underline{F}(\omega) = \underline{F}(\omega;\underline{P}_x) = [i\omega\underline{E} - \underline{A} - \underline{N}(\underline{P}_x)]^{-1} , \tag{37}$$

and $\omega_1 = 2\pi \ s^{-1}$ , $\omega_2 = 80 \cdot 2\pi \ s^{-1}$ according to ISO [1].

If $\underline{N}$ vanishes, then eqs. (35-37) can be directly used. For numerical reasons the values of $S_{\underline{wk}}(\omega)$ - and therefore also those of $\underline{c}_k^T \underline{F}(\omega)\underline{B}$ and $\underline{S}_\xi(\omega)$ - are needed for each point $\omega_1$ of an integration scheme performing (35). Even more laborious the PSD method works for nonlinear problems, $\underline{N} \neq \underline{O}$ . Then, in an intermediate step the nonlinear matrix equation

$$\begin{aligned} \underline{P}_x &= \int_{-\infty}^{+\infty} \underline{S}_x(\omega;\underline{P}_x)\,d\omega \\ &= \int_{-\infty}^{+\infty} \underline{F}(\omega;\underline{P}_x)\underline{B}\ \underline{S}_\xi(\omega)\underline{B}^T\underline{F}^T(-\omega;\underline{P}_x)\,d\omega \end{aligned} \tag{38}$$

has to be solved, e.g. by an iteration method

$$\underline{P}_x^{(i+1)} = \int_{-\infty}^{+\infty} \underline{S}_x(\omega;\underline{P}_x^{(i)})\,d\omega . \tag{39}$$

Sometimes, if only one or two nonlinearities have to be considered, eqs. (38, 39) can be reduced to one or two scalar equations of the same type. Nevertheless, the PSD method is cumbersome, particularly in the case of nonlinear systems.

In contrast, the time domain covariance analysis is more suitable to determine the variances. E.g. for a problem with one contact point the procedure of covariance analysis is performed using the shape filter (9) of vehicle excitation, the linearized state equation (34) of vehicle dynamics, the shape filter (21, 22) of human perception, resulting in the Lyapunov matrix equation

$$\begin{bmatrix} \underline{F} & \underline{O} & \underline{O} \\ \underline{b}_1\underline{h}^T & \underline{A}+\underline{N}(\underline{P}_x) & \underline{O} \\ \bar{\underline{g}}_k d_k \underline{h}^T & \bar{\underline{g}}_k \bar{\underline{c}}_k^T & \bar{\underline{F}}_k \end{bmatrix} \underline{P} + \underline{P} \begin{bmatrix} \underline{F} & \underline{O} & \underline{O} \\ \underline{b}_1\underline{h}^T & \underline{A}+\underline{N}(\underline{P}_x) & \underline{O} \\ \bar{\underline{g}}_k d_k \underline{h}^T & \bar{\underline{g}}_k \bar{\underline{c}}_k^T & \bar{\underline{F}}_k \end{bmatrix}^T$$

$$+ \begin{bmatrix} q_w \underline{g}\,\underline{g}^T & \underline{O} & \underline{O} \\ \underline{O} & \underline{O} & \underline{O} \\ \underline{O} & \underline{O} & \underline{O} \end{bmatrix} = \underline{O} . \tag{40}$$

The matrix $\underline{P}$ includes all covariances of the shape filters as well as of the system response,

$$\underline{P} = \begin{bmatrix} \underline{P}_v & \underline{P}_{xv}^T & \underline{P}_{kv}^T \\ \underline{P}_{xv} & \underline{P}_x & \underline{P}_{kx}^T \\ \underline{P}_{kv} & \underline{P}_{kx} & \underline{P}_k \end{bmatrix} . \tag{41}$$

Hence, the variance of the perception variable follows as (20),

$$\sigma_{wk}^2 = \alpha^2 \, \bar{\underline{h}}_k^T \, \underline{P}_k \, \bar{\underline{h}}_k .$$

The solution of the Lyapunov matrix equation (40) is effectively implemented by the algorithm of SMITH [21]. If $\underline{N}$ vanishes eq. (40) can directly be solved by this algorithm; however if a nonlinear problem is considered, $\underline{N} \neq \underline{O}$, then two subrelations of (40) have to be solved simultaneously by iteration techniques, cf. MÜLLER [22]:

$$[\underline{A} + \underline{N}(\underline{P}_x)] \, \underline{P}_{xv} + \underline{P}_{xv} \, \underline{F}^T + \underline{b}_1 \underline{h}^T \underline{P}_v = \underline{O} , \tag{42}$$

$$[\underline{A} + \underline{N}(\underline{P}_x)] \, \underline{P}_x + \underline{P}_x \, [\underline{A} + \underline{N}(\underline{P}_x)]^T + \\ + \underline{b}_1 \underline{h}^T \underline{P}_{xv}^T + \underline{P}_{xv} \underline{h} \, \underline{b}_1^T = \underline{O} , \tag{43}$$

where $\underline{P}_v$ is known from

$$\underline{F} \, \underline{P}_v + \underline{P}_v \, \underline{F}^T + q_w \underline{g} \, \underline{g}^T = \underline{O} . \tag{44}$$

In the case of successive contact points, eq. (40) has to be modified by some additional terms, cf. MÜLLER and POPP [23]. E.g. for a linear problem with two contact points, the covariance matrix $\underline{P}_x$ of the system response is obtained by the modified Lyapunov equation

$$\begin{bmatrix} \underline{F} & \underline{O} & \underline{O} \\ \underline{O} & \underline{F} & \underline{O} \\ \underline{b}_1 \underline{h}^T & \underline{b}_2 \underline{h}^T & \underline{A} \end{bmatrix} \begin{bmatrix} \underline{P}_v & \underline{P}_{v21}^T & \underline{P}_{xv1}^T \\ \underline{P}_{v21} & \underline{P}_v & \underline{P}_{xv2}^T \\ \underline{P}_{xv1} & \underline{P}_{xv2} & \underline{P}_x \end{bmatrix} + \begin{bmatrix} \underline{P}_v & \underline{P}_{v21}^T & \underline{P}_{xv1}^T \\ \underline{P}_{v21} & \underline{P}_v & \underline{P}_{xv2}^T \\ \underline{P}_{xv1} & \underline{P}_{xv2} & \underline{P}_x \end{bmatrix} \begin{bmatrix} \underline{F} & \underline{O} & \underline{O} \\ \underline{O} & \underline{F} & \underline{O} \\ \underline{b}_1 \underline{h}^T & \underline{b}_2 \underline{h}^T & \underline{A} \end{bmatrix}^T +$$

$$+ q_w \begin{bmatrix} \underline{g} \, \underline{g}^T & e^{\underline{F}(t_2-t_1)} \underline{g} \, \underline{g}^T & \underline{O} \\ \underline{g} \, \underline{g}^T e^{\underline{F}^T(t_2-t_1)} & \underline{g} \, \underline{g}^T & \underline{g} \, \underline{g}^T \underline{S}^T \\ \underline{O} & \underline{S} \, \underline{g} \, \underline{g}^T & \underline{O} \end{bmatrix} = \underline{O} \tag{45}$$

where $\underline{S}$ satisfies

$$\underline{A} \, \underline{S} - \underline{S} \, \underline{F} = e^{\underline{A}(t_2-t_1)} \underline{b}_1 \underline{h}^T - \underline{b}_1 \underline{h}^T e^{\underline{F}(t_2-t_1)} . \tag{46}$$

This procedure generalizes the results of HEDRICK, BILLINGTON, and DREESBACH [3] to general colored noise excitation (instead of white noise excitation).

Although the writing of above equations seems to be extensive, the solution is based on well established methods of linear algebra and easily performed by the algorithm of SMITH [21] or by other methods reported by KOUPAN and MÜLLER [24].

Since the variances are the essential results, the common PSD method goes a detour while the covariance analysis directly determines the desired values. In the following example covariance analysis is effectively applied.

## 6. EXAMPLE

The covariance analysis is demonstrated by the example shown in Fig. 3. A three-mass vehicle with four degrees of freedom is excited by colored noise $\zeta(t)$ according to (7) or white noise (8) for the derivated process $\dot{\zeta}(t)$.

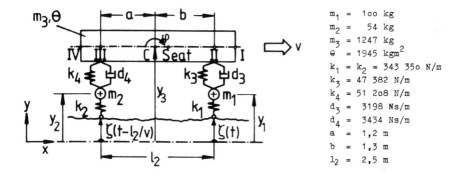

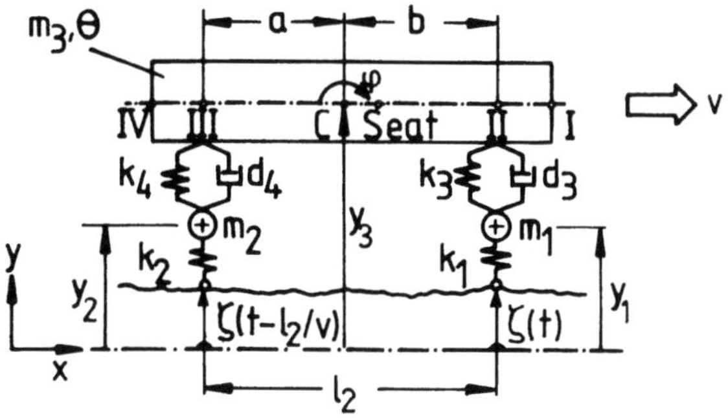


$m_1$ = 100 kg
$m_2$ = 54 kg
$m_3$ = 1247 kg
$\Theta$ = 1945 $kgm^2$
$k_1 = k_2$ = 343 350 N/m
$k_3$ = 47 382 N/m
$k_4$ = 51 208 N/m
$d_3$ = 3198 Ns/m
$d_4$ = 3434 Ns/m
a = 1,2 m
b = 1,3 m
$l_2$ = 2,5 m

Fig. 3 Vehicle model and system parameters.

Small motions are assumed and the time lag $t_2 = l_2/v$ between front and rear contact is taken into account. The generalized coordinates are summarized in the position vector $\underline{y}(t)$, $\underline{y}(t) = [y_1(t), y_2(t), y_3(t), \varphi(t)]^T$. Then, the equation of motion reads

$$\underline{M}\,\ddot{\underline{y}}(t)+\underline{D}\,\dot{\underline{y}}(t)+\underline{K}\,\underline{y}(t) = \underline{s}_1\zeta_1(t)+\underline{s}_2\zeta_2(t), \quad \zeta_1(t)=\zeta(t), \zeta_2(t)=\zeta(t-t_2), \tag{47}$$

where $\underline{M}$, $\underline{D}$, $\underline{K}$ characterize inertia, damping and stiffness matrix, respectively, and $\underline{s}_1$, $\underline{s}_2$ are the excitation input vectors,

$$\underline{M} = \underline{\text{diag}}[m_1, m_2, m_3, \Theta], \quad \underline{s}_1 = [k_1\,0\,0\,0]^T, \quad \underline{s}_2 = [0\,k_2\,0\,0]^T,$$

$$\underline{D} = \begin{bmatrix} d_3 & 0 & -d_3 & bd_3 \\ 0 & d_4 & -d_4 & -ad_4 \\ -d_3 & -d_4 & d_3+d_4 & -bd_3+ad_4 \\ bd_3 & -ad_4 & -bd_3+ad_4 & b^2d_3+a^2d_4 \end{bmatrix}, \tag{48}$$

$$\underline{K} = \begin{bmatrix} k_1+k_3 & 0 & -k_3 & bk_3 \\ 0 & k_2+k_4 & -k_4 & -ak_4 \\ -k_3 & -k_4 & k_3+k_4 & -bk_3+ak_4 \\ bk_3 & -ak_4 & -bk_3+ak_4 & b^2k_3+a^2k_4 \end{bmatrix} . \tag{48}$$

From (47) the vehicle state space representation can be obtained,

$$\dot{\underline{x}}(t) = \underline{A}\,\underline{x}(t) + \underline{b}_1\zeta_1(t) + \underline{b}_2\zeta_2(t) = \underline{A}\,\underline{x}(t) + \underline{B}\underline{\xi} , \tag{49}$$

$$\underline{x}(t) = \begin{bmatrix} \underline{y}(t) \\ \dot{\underline{y}}(t) \end{bmatrix}, \underline{\xi}(t) = \begin{bmatrix} \zeta_1(t) \\ \zeta_2(t) \end{bmatrix}, \underline{A} = \begin{bmatrix} \underline{0} & \underline{E} \\ -\underline{M}^{-1}\underline{K} & -\underline{M}^{-1}\underline{D} \end{bmatrix}, \underline{b}_1 = \begin{bmatrix} \underline{0} \\ \underline{M}^{-1}\underline{s}_1 \end{bmatrix}, \underline{b}_2 = \begin{bmatrix} \underline{0} \\ \underline{M}^{-1}\underline{s}_2 \end{bmatrix},$$

$$\underline{B} = \begin{bmatrix} \underline{b}_1 & \underline{b}_2 \end{bmatrix} .$$

In order to use white noise vehicle excitation, $\dot{\zeta}(t) \equiv w(t)$, eq. (49) has to be differentiated,

$$[\dot{\underline{x}}(t)]^{\bullet} = \underline{A}\,[\dot{\underline{x}}(t)] + \underline{B}\,\underline{w}(t) . \tag{50}$$

Here, the autocorrelation matrix of the white noise vector process $\underline{w}$ reads

$$\underline{R}_w(\tau) = E\{\underline{w}(t)\underline{w}^T(t-\tau)\} = q_\zeta \begin{bmatrix} \delta(\tau) & \delta(\tau+t_2) \\ \delta(\tau-t_2) & \delta(\tau) \end{bmatrix} . \tag{51}$$

Thus, the covariance matrix $\underline{P}_{\dot{x}}$ in steady state follows from the algebraic Lyapunov equation

$$\underline{A}\,\underline{P}_{\dot{x}} + \underline{P}_{\dot{x}}\underline{A}^T + \underline{Q} = \underline{0},\ \underline{Q} = q_{\dot{\zeta}}\left[\underline{b}_1\underline{b}_1^T + \underline{b}_2\underline{b}_2^T + e^{\underline{A}t_2}\underline{b}_1\underline{b}_2^T + \underline{b}_2\underline{b}_1^T e^{\underline{A}^T t_2}\right] , \tag{52}$$

which can be solved e.g. by SMITH's method [21]. From $\underline{P}_{\dot{x}}$ the variances $\sigma_a^2$ of the car body acceleration $\ddot{y}_P(t)$ at any point P with coordinates $y_P(t) = \underline{a}^T\underline{y}(t)$ can immediately be calculated by

$$\sigma_a^2 = [\underline{0}^T\underline{a}^T]\,\underline{P}_{\dot{x}} \begin{bmatrix} \underline{0} \\ \underline{a} \end{bmatrix} . \tag{53}$$

A similar expression holds for the variances $\sigma_F^2$ of the wheel loads $F_i = k_i(\zeta_i - y_i)$ , $i = 1,2$, but here (47) has to be used to eliminate the displacements. Fig. 4 shows some numerical results, where a smooth road ($q_{\dot{\zeta}} = 3{,}14 \cdot 10^{-6} \cdot v\ m^2/s$ - solid line) and a rough road excitation ($q_{\dot{\zeta}} = 24{,}7 \cdot 10^{-6} \cdot v\ m^2/s$ - dashed line) is assumed. In Fig. 4a) the standard deviation $\sigma_F/F$ of the load variation of front and rear wheel is plotted against speed $v$ . It increases with $\sqrt{v}$ since the excitation intensity is proportional to $v$ . In Fig. 4b) the standard deviation $\sigma_a/g$ of the car body acceleration versus vehicle length is given, which clearly shows a minimum for the center of mass position at different speeds. Neglecting the time lag between front and rear excitation (dotted line) can lead to incorrect results especially near the seat position. Covariance analysis has also an important application in optimization problems. Fig. 4c) shows a simple example, where the influence of the nondimensional damping ratio $D = D_i \equiv d_i/(2\sqrt{k_i \bar{m}_i})$, $i = 1,2$, ($\bar{m}_1 = m_3 a/l_1$, $\bar{m}_2 = m_3 b/l_1$) on the standard deviation of seat acceleration, $\sigma_{a_S}/g$ , is studied. Best results are obtained in the

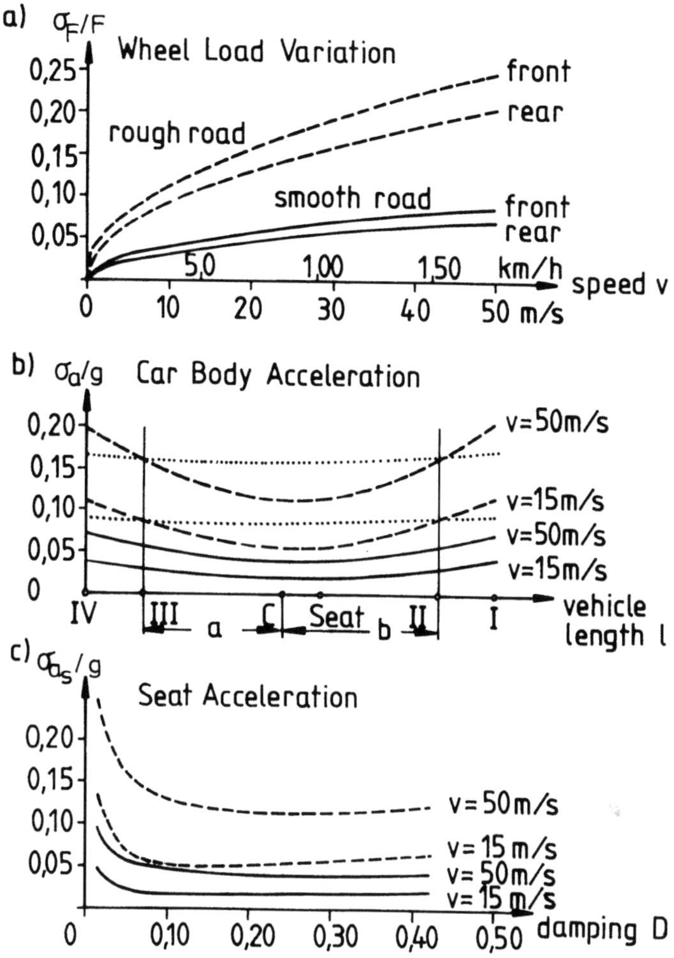


Fig. 4 Random response variances, a) wheel load variation depending on vehicle speed, b) car body acceleration over vehicle length, c) seat acceleration versus damping (---- rough road, ——— smooth road).

range $0.1 \leq D \leq 0.25$ . However, tradeoffs with the wheel load variation call for higher damping. The computation time for all results of Fig. 4 was about 10 seconds at a CYBER 175 machine. Other examples applying the covariance analysis to problems in vehicle dynamics may be found in POPP and SCHIEHLEN [25] and POPP [26].

REFERENCES

1. International Standard ISO 2631: Guide for the Evaluation of Human Exposure to Whole-Body Vibrations. Int. Org. Standardization, 1974.

2. MITSCHKE, M.: Dynamik der Kraftfahrzeuge. Springer-Verlag, Berlin-Heidelberg-New York, 1972.

3. HEDRICK, J.K.; BILLINGTON, G.F.; DREESBACH, D.A.: Analysis, Design and Optimization of High Speed Vehicle Suspensions Using State Variable Techniques. J. Dyn. Syst. Meas. Control, Transact. ASME, Ser. G, 96, pp. 193-203, 1974.

4. KARNOPP, D.C.: Vehicle Response to Stochastic Roadways. Vehicle System Dynamics 7, pp. 97-109, 1978.

5. DAHLBERG, T.: Ride Comfort and Road Holding of a 2-DOF Vehicle Travelling on a Randomly Profiled Road. J. Sound Vibration 58 (2), pp. 179-187, 1978.

6. MITSCHKE, M.: Beitrag zur Untersuchung der Fahrzeugschwingungen (Theorie und Versuch). Deutsche Kraftfahrforschung und Straßenverkehrstechnik, Heft 157, VDI-Verlag, Düsseldorf, 1962.

7. BRAUN, H.: Untersuchungen von Fahrbahnunebenheiten und Anwendung der Ergebnisse. Dissertation, Braunschweig, 1969.

8. WENDEBORN, J.O.: Die Unebenheiten landwirtschaftlicher Fahrbahnen als Schwingungserreger landwirtschaftlicher Fahrzeuge. Grundlagen der Landtechnik 15, Nr. 2, 1965.

9. ORE, Frage 116: Wechselwirkungen zwischen Fahrzeugen und Gleis; Bericht Nr. 1: Spektrale Dichte der Unregelmäßigkeiten in der Gleislage; Teil 1. ORE-Report, C116/RP1/D, Utrecht, Oktober 1971.

10. VOY, C.: Die Simulation vertikaler Fahrzeugschwingungen. Dissertation, Berlin, 1976.

11. DODDS, C.J.; ROBSON, J.D.: The Description of Road Surface Roughness. J. Sound Vibration 31 (2), pp. 175-183, 1973.

12. HEDRICK, J.K.; ANIS, Z.: Proposal for the Characterization of Rail Track Irregularities. MIT, MA 02139, Sept. 1978. Presented at ISO-Meeting, Berlin, Sept. 1978.

13. DINCǍ, F.; THEODOSIU, C.: Nonlinear and Random Vibration. Academic Press, New York-London, 1973.

14. FÁBIÁN, L.: Zufallsschwingungen und ihre Behandlung. Springer-Verlag, Berlin-Heidelberg-New York, 1973.

15. SUSSMAN, N.E.: Statistical Ground Excitation Models for High Speed Vehicle Dynamic Analysis. High Speed Ground Transportation Journal 8 (3), pp. 145-154, 1974.

16. SCHIEHLEN, W.; KREUZER, E.: Symbolic Computerized Derivation of Equations of Motion. In: Dynamics of Multibody Systems, Ed. K. Magnus, Springer-Verlag, Berlin-Heidelberg-New York, 1978.

17. SCHIEHLEN, W.: Dynamical Analysis of Suspension Systems. In: The Dynamics of Vehicles, Ed. A. Slibar and H. Springer, Swets & Zeitlinger, Amsterdam-Lisse, 1978.

18. VDI-Richtlinie 2057, Beurteilung der Einwirkung mechanischer Schwingungen auf den Menschen. Verein Dt. Ing., Düsseldorf, 1975/1979.

19. SMITH, C.C.; McGEHEE, D.Y.; HEALEY, A.J.: The Prediction of Passenger Riding Comfort from Acceleration Data. J. Dyn. Syst. Meas. Control, Transact ASME, Ser. G, 100, pp. 34-41, 1978.

20. HEDRICK, J.K.; COOPERRIDER, N.K.; LAW, E.H.: The Application of Quasi-Linearization Techniques to Rail Vehicle Dynamic Analysis. Final Report for U.S. Department of Transportation, DOT-TSC-902, 1978.

21. SMITH, R.A.: Matrix Equation XA+BX = C. SIAM J. Appl. Math. 16, pp. 198-201, 1968.

22. MÜLLER, P.C.: Kovarianzanalyse von stationären nichtlinearen Zufallsschwingungen. Z. Angew. Math. Mech. 59, pp. T142-T143, 1979.

23. MÜLLER, P.C.; POPP, K.: Kovarianzanalyse von linearen Zufallsschwingungen mit zeitlich verschobenen Erregerprozessen. Z. Angew. Math. Mech. 59, pp. T144-T146, 1979.

24. KOUPAN, A.; MÜLLER, P.C.: Zur numerischen Lösung der Ljapunovschen Matrizengleichung $A^T P + PA = -Q$. Regelungstechnik 24, pp. 167-169, 1976.

25. POPP, K.; SCHIEHLEN, W.: Dynamics of Magnetically Levitated Vehicles on Flexible Guideways. In: The Dynamics of Vehicles on Roads and on Railway Tracks, Ed. H.B. Pacejka, Swets & Zeitlinger, Amsterdam, 1976.

26. POPP, K.: Beiträge zur Dynamik von Magnetschwebefahrzeugen auf geständerten Fahrwegen. Habilitationsschrift TU München 1978. Appeared as Fortschr.-Ber. VDI-Z., Series 12, No. 35, Düsseldorf, 1978.

# VIBRATIONAL CHARACTERISTICS OF ELECTROMAGNETIC LEVITATION VEHICLES – GUIDEWAY SYSTEM

Masao Nagai[1] and Masakazu Iguchi[2]

[1] Department of Mechanical Engineering, Tokyo University of Agriculture and Technology, Tokyo
[2] Department of Mechanical Engineering, University of Tokyo,
Tokyo, Japan

SUMMARY

At first we show how to design preliminarily the suspension of the electromagnetic levitation (EML) vehicle by analyzing statistically a simplified dynamic model in order to satisfy the criteria of the contact frequency and the ride quality. After the preliminary study, we have investigated the dynamic interaction between multiple vehicles and their guideway with the rough surface. We have developed the computer simulation program in which vehicle dynamic equations that consist of 4-degree-of-freedom motions are fully coupled with guideway dynamic equations by distributed suspension forces. Main results of the numerical analysis are: (1) There is a distribution effect of magnetic forces which reduces the dynamic deflection of the guideway and also the vehicle vibration. (2) According to the comparison of this system with conventional railroads, the transient response of the vibration transmitted to guideway supports is of less amplitude and smoother. (3) There is a resonant phenomenon caused by multiple vehicles repeatedly moving over the flexible guideway, and the vertical variation of each vehicle tends to be different from each other, especially in a high speed operation.

## 1. INTRODUCTION

Electromagnetic levitation (EML) vehicles are expected to reduce the burden of the vibration and noise, the cost of the maintenance, and to have a possibility of a high speed operation, because EML vehicles do not contact with the guideway and suspension forces are distributed.

We have investigated the vibrational characteristics of multiple EML vehicles and their guideway in viewpoints of safety and comfortability of this new transportation system. Because of the little clearance between magnets and the guideway surface, it is essential to consider how to suppress a danger of the magnet collision with the guideway as well as to realize a good riding comfort.

At first we show how to design preliminarily this new suspension by analyzing statistically a simplified dynamic model in order to satisfy the criteria of the contact frequency and the ride quality. D.A.Hullender has proposed a criterion of the contact frequency which means the expected number of crossings of a certain level per unit time[1]. We extend this criterion in order to express also the contact velocity if the contact occurs. We also use the UTACV specification for a criterion of the ride quality.

After the preliminary study above, we have investigated the dynamic interaction between multiple EML vehicles and their guideway. Recently many works about the dynamic interaction between vehicles and their guideway have been done, as reviewed by H.H. Richardson et al [2]. The dynamic deflection of the guideway under multiple vehicle loads has been dealt with by H.H.Richardson et al and J.F.Wilson et al [2,3]. But they assume vehicle loads to be constant without considering vehicle dynamics. A.L.Doran et al have dealt with the periodic motion of a single vehicle or multiple vehicles on the flexible guideway for PRT ( personal rapid transit ) system. They show both fully coupled equations and partially coupled equations, but they show only results of a single vehicle case [4].

We have developed the computer simulation program for the numerical analysis of the vibrational characteristics of a train of multiple EML vehicles running over the flexible guideway with the rough surface. Vehicle dynamic equations are fully coupled with guideway dynamic equations by distributed suspension forces. Each vehicle is assumed to be a 4-degree-of-freedom model, which consists of heave and pitch motions of the vehicle body and heave motions of two trucks. We solve a set of these equations by the Runge-Kutta-Gill method with the digital computer HITAC 8700/8800 in the University of Tokyo.

We have also compared the transient response of the vibration transmitted to the guideway support of this system with that of conventional railroads, to make sure that this new transportation system will reduce the burden of the vibration.

## 2. PRELIMINARY STUDY

### *Simplified Dynamic Model*

We show how to design this vehicles-guideway system in order to suppress a danger of the magnet collision with the guideway and also to keep a good riding comfort. For a preliminary design of this new suspension, we use the simplified dynamic model and the criteria of the contact frequency and the ride quality of the UTACV specification. We analyze statistically this model, which is simplified but fundamental, in order to satisfy these criteria. We make several assumptions in this chapter as follows.

(1) A magnetic force is assumed to be concentrated and linearized.

(2) The ratio of the natural frequency of the primary suspension to that of the secondary suspension may be large enough that we assume this suspension system to be semi-coupled as follows. The transfer function of the primary suspension is assumed to be

$$G_1(s)=Y_1(s)/Y_0(s)=\omega_1{}^2/(s^2+2\zeta_1\omega_1 s+\omega_1{}^2) \tag{1}$$

and that of the secondary suspension is assumed to be

$$G_2(s)=Y_2(s)/Y_1(s)=(2\zeta_2\omega_2+\omega_2{}^2)/(s^2+2\zeta_2\omega_2 s+\omega_2{}^2) \tag{2}$$

where $Y_0$, $Y_1$, $Y_2$ denote vertical displacements of the guideway surface, the truck and the vehicle body.

(3) The guideway is assumed to be rigid and the surface roughness is assumed to be a stationary gaussian random process, and the power spectral

density (PSD) of it in spatial frequency has the form

$$S_{y_0}(\phi)=C\phi_1^2/\{(\phi^2+\phi_0^2)(\phi^2+\phi_1^2)\} \quad , \ \phi_0<\phi_1 \tag{3}$$

where $\phi$ denotes the wave number, C the roughness coefficient. As the influence of the low frequency range on the vehicle dynamics is considered to be small, we assume that $\phi_0$ in eq.(3) is zero in this chapter. If the vehicle speed is V, this PSD is transformed to a PSD in temporal frequency $\omega$ (=$\phi$V), such as

$$\Phi_{y_0}(\omega)=(CV/\omega^2)\cdot(\phi_1 V)^2/\{\omega^2+(\phi_1 V)^2\} \tag{4}$$

(4) The contact frequency is assumed to be designed as small as possible, and the transient motion, when the low velocity contact occurs, is assumed to fade within a short guideway section. This assumption may be useful for a preliminary design in a viewpoint of surveying the vehicle dynamics in a long operation time.

*Contact Frequency*

According to S.O.Rice, the expected number of crossings of the level $y=y_c$ with positive slope per unit time is given by

$$f_c=\int_0^\infty \dot{y}\cdot Pr(y_c,\dot{y})d\dot{y} \tag{5}$$

where $Pr(y,\dot{y})$ is the joint probability density of y(t) and $\dot{y}(t)$ [5]. D.A. Hullender applied it to the design of the minimum clearance between the magnet and the guideway surface. Now we are interested in the crossing velocity at the time of crossing the level. Then we use the expected number of crossings of the level $y(t)=y_c$ with $\dot{y}(t)\geqq v_c$ ,

$$f_c=\int_{v_c}^\infty \dot{y}\cdot Pr(y_c,\dot{y})d\dot{y} \tag{6}$$

On the assumption that the guideway surface roughness is a stationary gaussian random process, eq.(6) is reduced to

$$f_c=(\sigma_{\dot{y}}/2\pi\sigma_y)\cdot\exp(-y_c^2/2\sigma_y^2-v_c^2/2\sigma_{\dot{y}}^2) \tag{7}$$

where $\sigma_y$ and $\sigma_{\dot{y}}$ denote standard deviations of y and $\dot{y}$, respectively. From eq.(1) and eq.(4), standard deviations of y and $\dot{y}$ are given by

$$\sigma_y^2=\int_0^\infty |G_1(j\omega)-1|^2\,\Phi_{y_0}(\omega)d\omega$$
$$=[\pi CV\cdot\phi_1 V\cdot\{8\zeta_1^3\omega_1+(1+4\zeta_1^2)\cdot\phi_1 V\}]/[4\zeta_1\omega_1\cdot\{\omega_1^2+2\zeta_1\omega_1\phi_1 V+(\phi_1 V)^2\}] \tag{8}$$

$$\sigma_{\dot{y}}^2=\int_0^\infty \omega^2\cdot|G_1(j\omega)-1|^2\,\Phi_{y_0}(\omega)d\omega$$
$$=[\pi CV\cdot(\phi_1 V)^2\cdot\{(1+4\zeta_1^2)\omega_1+2\zeta_1\phi_1 V\}]/[4\zeta_1\cdot\{\omega_1^2+2\zeta_1\omega_1\phi_1 V+(\phi_1 V)^2\}] \tag{9}$$

If we substitute these values into eq.(7), we easily obtain the extended contact frequency which means the expected number of magnet contacts with the guideway, with the contact velocity larger than a certain positive value $v_c$. If the value of $\phi_1$V is much larger than the natural frequency $\omega_1$ in eqs.(8),(9), the eq.(7) is reduced to eq.(10) approximately.

$$f_c\simeq\left[\frac{2\zeta_1 f_1}{1+4\zeta_1^2}\cdot\frac{\phi_1 V}{2\pi}\right]^{0.5}\cdot\exp\left[-\frac{4\zeta_1 f_1 y_c^2}{(1+4\zeta_1^2)CV}-\frac{v_c^2}{\pi\cdot C\phi_1 V^2}\right] \tag{10}$$

We calculate the extended contact frequency in two typical cases of the guideway roughness, the first of which is as rough as a monorail guideway (C=0.628·10$^{-5}$ rad·m) and the second of which is as smooth as a welded rail (C=0.628·10$^{-6}$ rad·m). Fig.1(a), Fig.1(b) show that the con-

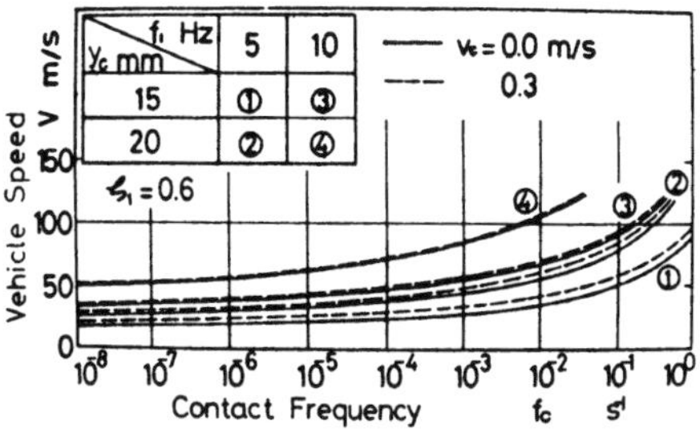


Fig.1(a). Contact frequency vs. vehicle speed ($C=0.628\cdot10^{-5}$ rad·m)

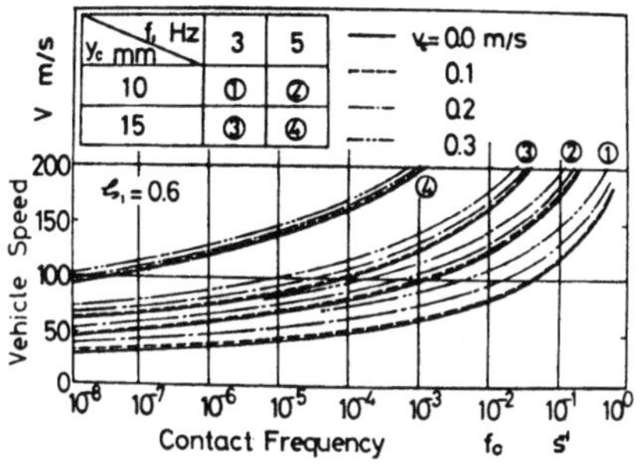


Fig.1(b). Contact frequency vs. vehicle speed ($C=0.628\cdot10^{-6}$ rad·m)

tact frequency $f_c$ is greatly dependent on the roughness coefficient, the vehicle speed and suspension parameters. For example, at a set of parameters such as $f_1$=5Hz, $\zeta_1$=0.6, $y_c$=10mm and $v_c$=0.0m/s, $f_c$ is $1.8\cdot10^{-3}s^{-1}$ over the monorail at a speed 30m/s, but it is only $8.6\cdot10^{-7}s^{-1}$ over the welded rail even at a speed 120m/s. The second case in Fig.1(b) shows also the ratio of $f_c$ with the low contact velocity ($\dot{y}\leq0.3$m/s) is very large.

*Ride Quality*

On the assumptions described previously, the PSD curve of the vertical acceleration of the vehicle body is given by

$$\Phi_{\ddot{y}_2}(\omega)=|(j\omega)^2\cdot G_1(j\omega)\cdot G_2(j\omega)|^2\cdot\Phi_{y_0}(\omega) \tag{11}$$

This curve may be considered to be approximately a piece-wise linear when $\zeta_1$, $\zeta_2$ are nearly $\sqrt{2}/2$, that is, this curve is divided into three parts such as $\omega<\omega_2$ (40db/dec), $\omega_2<\omega<\omega_1$ (0db/dec) and $\omega_1<\omega$ (-80db/dec). Then we get schematically the condition in order that this curve should not go over the PSD curve of the UTACV specification, such as

$$f_1<6\text{Hz}\ ;\ f_1^2\cdot f_2\leq 0.112/\sqrt{CV}\quad:\quad f_1\geq6\text{Hz}\ ;\ f_2\leq 3.11\cdot10^{-3}/\sqrt{CV} \tag{12}$$

These conditions show the desirable natural frequencies of suspensions against the roughness coefficient and the vehicle speed. The fundamental tendency of them is qualitatively reasonable, but it may be necessary to analyze a fully-coupled model more exactly, which is explained in the following chapter.

## 3. VEHICLES - GUIDEWAY MODEL FOR SIMULATION

Fig.2 shows the schematic feature of the multiple vehicles and guideway system for the computer simulation. Our main purpose in this simulation is to investigate the dynamic interaction between a single vehicle or multiple vehicles and their guideway, and to show influences of the guideway flexibility, the roughness, the force distribution length, the truck intervals and the vehicle speed etc., on the vehicle dynamics. We make several simplifications as follows.

(1) Only motions in the vertical plane are dealt with.

(2) Vehicle bodies, trucks and guideway supports are rigid.

(3) The guideway consists of single span beams which are supported simply at uniform intervals.

### *4-DOF Vehicle Model*

As shown in Fig.3, each vehicle model consists of 4-degree-of-freedom motions, such as heave and pitch motions of the vehicle body and heave motions of two trucks. The secondary suspension is assumed to be passive with a spring and a damper. Equations of each vehicle (iv=1 - Nv) are described as follows.

$$\ddot{y}_{1f}=-P_{iv,1}/m_1+R_v\cdot\{\omega_2{}^2(y_{2f}-y_{1f})+2\zeta_2\omega_2(\dot{y}_{2f}-\dot{y}_{1f})\} \tag{13}$$

$$\ddot{y}_{1r}=-P_{iv,2}/m_1+R_v\cdot\{\omega_2{}^2(y_{2r}-y_{1r})+2\zeta_2\omega_2(\dot{y}_{2r}-\dot{y}_{1r})\} \tag{14}$$

$$\ddot{y}_{2f}=-(1+d^2/r^2)/2\cdot\{\omega_2{}^2(y_{2f}-y_{1f})+2\zeta_2\omega_2(\dot{y}_{2f}-\dot{y}_{1f})\} \\ -(1-d^2/r^2)/2\cdot\{\omega_2{}^2(y_{2r}-y_{1r})+2\zeta_2\omega_2(\dot{y}_{2r}-\dot{y}_{1r})\} \tag{15}$$

$$\ddot{y}_{2r}=-(1-d^2/r^2)/2\cdot\{\omega_2{}^2(y_{2f}-y_{1f})+2\zeta_2\omega_2(\dot{y}_{2f}-\dot{y}_{1f})\} \\ -(1+d^2/r^2)/2\cdot\{\omega_2{}^2(y_{2r}-y_{1r})+2\zeta_2\omega_2(\dot{y}_{2r}-\dot{y}_{1r})\} \tag{16}$$

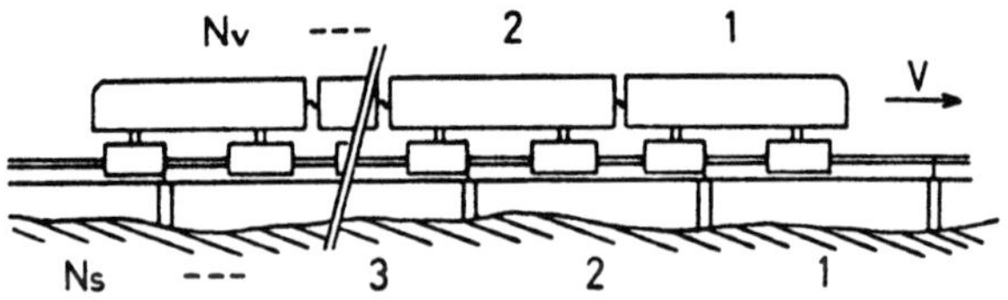


Fig.2. Scheme of a long train of EML vehicles and their guideway

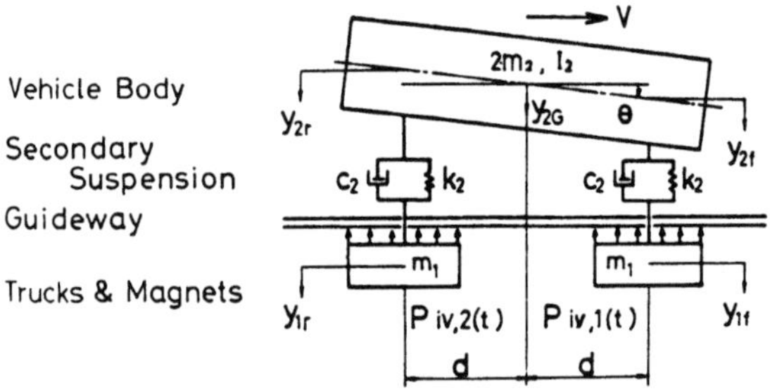


Fig.3. Scheme of a 4-degree-of-freedom model of a EML vehicle

where $y_{1f}$, $y_{1r}$, $y_{2f}$, $y_{2r}$ denote displacements of the front and rear truck and those of the front and rear vehicle body, respectively, and $P_{iv,1}$, $P_{iv,2}$ denote the front and rear magnetic force without the vehicle weight. Parameters in eqs.(13)-(16) correspond to those in Fig.3 as follows.

$$y_{2g}=(y_{2f}+y_{2r})/2,\quad \theta=(y_{2f}-y_{2r})/2d,\quad I_2=2m_2\cdot r^2,\quad m_2/m_1=R_v,$$
$$k_2/m_2=\omega_2{}^2,\quad c_2/m_2=2\zeta_2\omega_2$$

In eqs.(15),(16), the connecting forces between vehicle bodies are neglected on the assumption that each vehicle is very softly connected with others in the vertical direction.

The primary suspension force is assumed to be stabilized by the voltage control of the magnet with the compensation of the gap sensor signal (relative displacement) and the accelerometer signal (truck vertical acceleration). The total magnetic force of each truck is the summation of partial forces distributed from the front edge to the rear edge of the magnet. The relation between the displacement of the truck and that of the guideway surface is written as

$$P_{iv,1}+K_1\cdot\dot{P}_{iv,1}=K_2\cdot(y_{1f}-\bar{y}_{0,iv,1})+K_3\cdot(\dot{y}_{1f}-\dot{\bar{y}}_{0,iv,1})+K_4\cdot\dot{y}_{if}+K_5\cdot\ddot{y}_{if} \quad (17)$$
$$P_{iv,2}+K_1\cdot\dot{P}_{iv,2}=K_2\cdot(y_{1r}-\bar{y}_{0,iv,2})+K_3\cdot(\dot{y}_{1r}-\dot{\bar{y}}_{0,iv,2})+K_4\cdot\dot{y}_{1r}+K_5\cdot\ddot{y}_{1r} \quad (18)$$

where each variable is the deviation from the equilibrium value, and coefficients $K_1$ - $K_5$ are dependent on controller gains. Variables $\bar{y}_{0,iv,1}$ and $\bar{y}_{0,iv,2}$ are mean values of the guideway surface displacements over the length of the trucks (iv,1) and (iv,2), which are the front and rear truck numbers of the iv-th vehicle respectively. The guideway surface displacement is obtained by adding the surface roughness to the guideway deflection which is calculated by the partial differential equation, shown in the next section. The roughness is generated by an approximately gaussian random number in order to realize the PSD of the eq.(3).

*Guideway Model*

When $N_v$ vehicles are moving over $N_s$ spans as shown in Fig.2, equations of $N_s$ spans must be solved simultaneously. The is-th beam (is=1 - $N_s$), over which some trucks are moving, is described by the Bernoulli-Euler beam equation;

$$EI\frac{\partial^4 y_{is}}{\partial x^4}+\rho a\frac{\partial^2 y_{is}}{\partial t^2}+c_b\frac{\partial y_{is}}{\partial t}=F_{is}(x,t) \quad (19)$$

where E denotes the elastic modulus, I the moment of inertia, $\rho$ the mass density, a the cross-sectional area of the beam and $c_b$ the damping coefficient. The guideway deflection $y_{is}(x,t)$ is described as the summation of natural mode solutions of eq.(19) by means of the series expansion method.

$$y_{is}(x,t)\simeq\sum_{n=1}^{N}A_n(x)\cdot\psi_{n,is}(t)\quad;\quad A_n(x)=\sqrt{2/\ell_s}\cdot\sin(n\pi x/\ell_s) \quad (20)$$

where $\psi_{n,is}(t)$ denotes the modal coefficient of the n-th order of the is-th beam and $A_n(x)$ denotes the modal shape function of the n-th order which is normalized and orthogonal over the beam span length $\ell_s$. By substituting eq.(20) into eq.(19), ordinary differential equations are derived for each modal coefficient.

$$\ddot{\psi}_{n,is}(t)+2\zeta_{bn}\omega_{bn}\cdot\dot{\psi}_{n,is}(t)+\omega_{bn}{}^2\cdot\psi_{n,is}(t)=Q_{n,is}(t) \quad (21)$$
$$Q_{n,is}(t)=1/\rho a\cdot\int_0^{\ell_s}F_{is}(x,t)\cdot A_n(x)dx \quad (22)$$

Functions $F_{is}(x,t)$ and $Q_{n,is}(t)$ are divided into 4 cases in general. If some trucks ($i_p$ trucks for example) are moving over the is-th beam at a time t, the patterns of truck loads over the is-th beam are;

(A) the case that there is no load over both supports.

(B) the case that there is no load over the right support.

(C) the case that there is no load over the left support.

(D) the case that there are loads over both supports.

For each case, $F_{is}(x,t)$ and $Q_{n,is}(t)$ are described as follows.

$$\text{(A)}\quad F_{is}=\sum_{i=1}^{ip} P'_{is,i}\cdot D[x,x_1(i_0+2i-1),x_2(i_0+2i-2)] \tag{23}$$

$$Q_{n,is}=e\cdot\sum_{i=1}^{ip} P'_{is,i}\cdot[\cos\{n\pi x_1(i_0+2i-1)/\ell_s\}-\cos\{n\pi x_2(i_0+2i-2)/\ell_s\}]$$

$$\text{(B)}\quad F_{is}=\sum_{i=1}^{ip-1} P'_{is,i}\cdot D[x,x_1(i_0+2i-1),x_2(i_0+2i-2)]+P'_{is,ip}\cdot D[x,0,x_2(i_0+2i_p-2)]$$

$$Q_{n,is}=e\cdot[\sum_{i=1}^{ip-1} P'_{is,i}\cdot[\cos\{n\pi x_1(i_0+2i-1)/\ell_s\}-\cos\{n\pi x_2(i_0+2i-2)/\ell_s\}]$$

$$+P'_{is,ip}\cdot[1-\cos\{n\pi x_2(i_0+2i_p-2)/\ell_s\}]] \tag{24}$$

$$\text{(C)}\quad F_{is}=P'_{is,1}\cdot D[x,x_1(i_0),\ell_s]+\sum_{i=2}^{ip} P'_{is,i}\cdot D[x,x_1(i_0+2i-2),x_2(i_0+2i-3)]$$

$$Q_{n,is}=e\cdot[P'_{is,1}\cdot[\cos\{n\pi x_1(i_0)/\ell_s\}-\cos n\pi] \tag{25}$$

$$+\sum_{i=2}^{ip} P'_{is,i}\cdot[\cos\{n\pi x_1(i_0+2i-2)/\ell_s\}-\cos\{n\pi x_2(i_0+2i-3)/\ell_s\}]]$$

$$\text{(D)}\quad F_{is}=P'_{is,1}\cdot D[x,x_1(i_0),\ell_s]+\sum_{i=2}^{ip-1} P'_{is,i}\cdot D[x,x_1(i_0+2i-2),x_2(i_0+2i-3)]$$

$$+P'_{is,ip}\cdot D[x,0,x_2(i_0+2i_p-3)] \tag{26}$$

$$Q_{n,is}=e\cdot[P'_{is,1}\cdot[\cos\{n\pi x_1(i_0)/\ell_s\}-\cos n\pi]+\sum_{i=2}^{ip-1} P'_{is,i}\cdot[\cos\{n\pi x_1(i_0+2i-2)$$

$$/\ell_s\}-\cos\{n\pi x_2(i_0+2i-3)/\ell_s\}]+P'_{is,ip}\cdot[1-\cos\{n\pi x_2(i_0+2i_p-3)/\ell_s\}]]$$

where $D[x,x_1,x_2]$ denotes the uniform distribution function

$$D[x,x_1,x_2]=1/\ell_m \quad (x_1\leqq x\leqq x_2) \quad ; \quad 0 \quad (x\leqq x_1,\ x\geqq x_2)$$

and the coefficient e is equal to $\sqrt{2\ell_s}/\ell_m\rho a n\pi$, $\ell_m$ denotes the distribution length of the magnetic force. The subscript of $x_1,x_2$ is the serial number of truck edges numbered from the front one. The subscript $i_0$ denotes the minimum edge number in the $i_s$-th span and depends on the beam span length, the truck length, the vehicle length, the interval of vehicles as well as the vehicle speed. It is needless to say that $x_1,x_2$ are functions of time t. The variable $P'_{is,i}$ ($is=1$ - $N_s$, $i=1$ - $i_p$) is the magnetic force of the i-th truck on the $i_s$-th beam. Fig.4 shows the load pattern for the case (A).

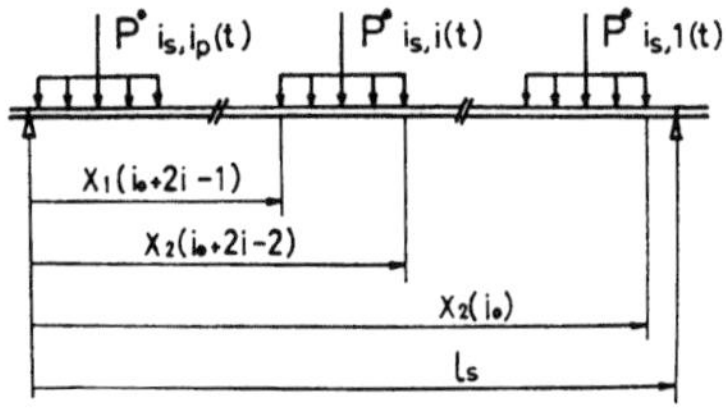


Fig.4. Load pattern of the $i_s$-th beam ; the case (A)

*Algorithm of Computation*

For the computation of the dynamic interaction between $N_V$ vehicles and $N_S$ beams, we need a set of equations of $N_V$ vehicle bodies, $2 \cdot N_V$ trucks and $N_S$ beams. The number $N_S$ is sufficiently a minimum integer to satisfy

$$N_S \geqq (\ell_V/\ell_S) \cdot N_V + 1 \tag{27}$$

where $\ell_S$ and $\ell_V$ denote the beam span length and the vehicle length, respectively. The second term is for one span over which vehicles move. After the front vehicle reaches the guideway support, the same computer algorithm can be repeated for the next set of beams only if numbers corresponding to the new set of beams are updated and conditions of continuity of each variable are satisfied.

It is essential to find how vehicle equations are connected with beam equations at a time $t_i$, because the load pattern on each beam changes with vehicles moving. For the convenience of the general expression, we have described vehicle equations with notations $P_{iv,if}$ and $y_{0,iv,if}$, which mean values of the $i_f$-th ($i_f$=1,front;$i_f$=2,rear) truck of the $i_v$-th vehicle, on the other hand beam equations with notations $P'_{is,i}$ and $y'_{0,is,i}$ which mean values of the i-th truck on the $i_s$-th beam. Therefore we must identify the load pattern of each beam at each time.

The relation between $P_{iv,if}$ and $P'_{is,i}$ or $y_{0,iv,if}$ and $y'_{0,is,i}$ is divided into two cases. In a case when the magnetic force is distributed over one beam, the force $P_{iv,if}$ corresponds to one of $P'_{is,i}$ and also the displacement $y_{0,iv,if}$ to one of $y'_{0,is,i}$. In the other case when the magnetic force is distributed over two beams, e.g. the $i_s$-th beam and the $i_s$+1-th beam, the force $P_{iv,if}$ corresponds to $P'_{is,ip}$ which equals to $P'_{is+1,1}$. In this case the average value $\bar{y}_{0,iv,if}$ in eqs.(17),(18) must be the average of $y'_{0,is,ip}$ and $y'_{0,is+1,1}$ over the magnet length.

After all in the computer simulation program, there are several stages such as, the identification of the load pattern of each beam at a time $t_i$, the connection of $P_{iv,if}$ with $P'_{is,i}$ and $y_{0,iv,if}$ with $y'_{0,is,i}$, respectively, the calculation of equations by the numerical method and the updating of the beam number with the modification of initial values of each variable when the front vehicle crossing the guideway support.

The time required for this simulation is nearly proportional to the number of vehicles, and it is about 30 - 40 times the real travelling time for the case of five vehicles with the computer HITAC 8700/8800.

## 4. RESULTS OF NUMERICAL ANALYSIS

At first we investigate the distribution effect of the magnetic force on the vehicle dynamics and the guideway deflection. And we compare the dynamics of the EML vehicle with that of the conventional wheeled vehicle. In this analysis, we assume that the EML vehicle is a 2-DOF model which consists of a half vehicle body and a truck, and that the wheeled vehicle is a 3-DOF model which consists of a half vehicle body, a truck and a wheelset. The wheelset is assumed to be in point-contact with the rail. Specifications of two types of vehicles and the guideway structure are summarized in Tables 1,2,3 in which values are written for a full model.

*Distribution Effect of Magnetic Forces*

Fig.5(a) shows the PSD of the vertical acceleration of the truck $\ddot{y}_1$. The PSD of the high frequency range greatly decreases compared with the result

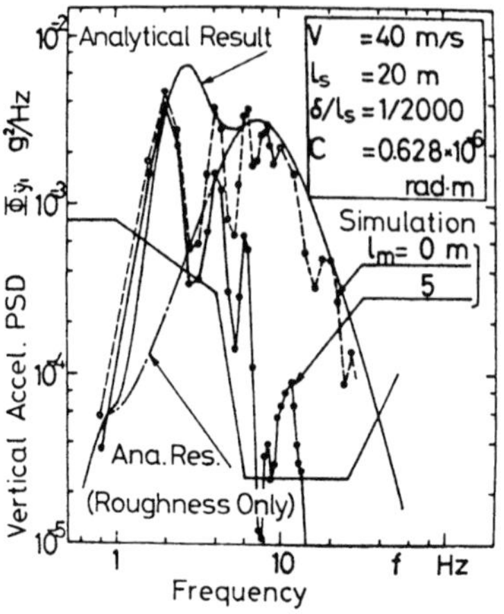


Fig.5(a). Vertical acceleration PSD for a truck; guideway (A)

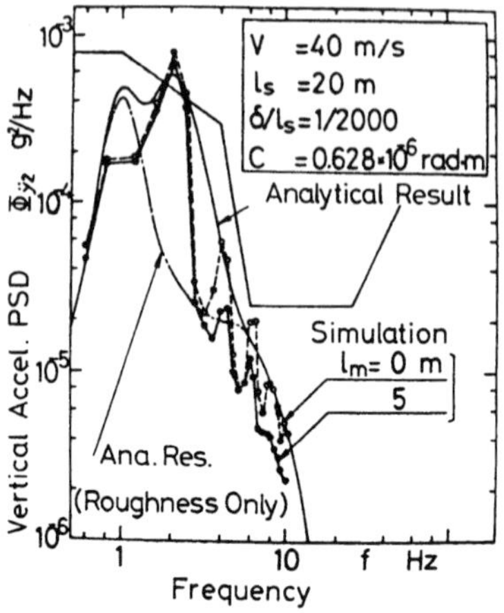


Fig.5(b). Vertical acceleration PSD for a vehicle body; guideway (A)

of the calculation by neglecting the force distribution. The attenuation ratio of the amplitude by the distributed force to that by the concentrated force is described by

$$G(\lambda)=(\lambda/\pi\ell_m)\cdot\left|\sin(\pi\ell_m/\lambda)\right| \tag{28}$$

where $\lambda$, $\ell_m$ denote the wave length of the surface displacement and the magnet length, respectively. According to this equation, the ratio at 4Hz is 0.637 and the ratio at 12Hz is 0.212, when the vehicle speed is 40m/s and the magnet length is 5m. The decrease of PSD is proportional to the second power of this ratio.

Fig.5(b) shows the PSD of the vertical acceleration of the vehicle body $\ddot{y}_2$, and that the distribution effect does not clearly appear as in Fig.5(a). It is because the natural frequency of the secondary suspension is sufficiently smaller than that of the primary suspension. Fig.5(b) also shows approximately analytical results which are obtained by the semi-coupled and concentrated force model. From these results, numerical results are nearly equal to analytical results. The ride quality of this example may be satisfactory according to the UTACV specification.

*Comparison with Conventional Railroads*

Fig.6 shows the transient response of the force transmitted to the guideway support when both vehicles respectively travel over the support at a speed 40m/s. From Fig.6 the maximum value of the force by the EML vehicle is about 75% of that by the wheeled vehicle, and also the response curve by the EML vehicle is smoother than that by the wheeled vehicle.

This tendency becomes large with the ratio of the magnet length to the beam span length.

*Dynamic Deflection of Guideway*

Fig.7 shows the ratio of the dynamic deflection of the beam to the static

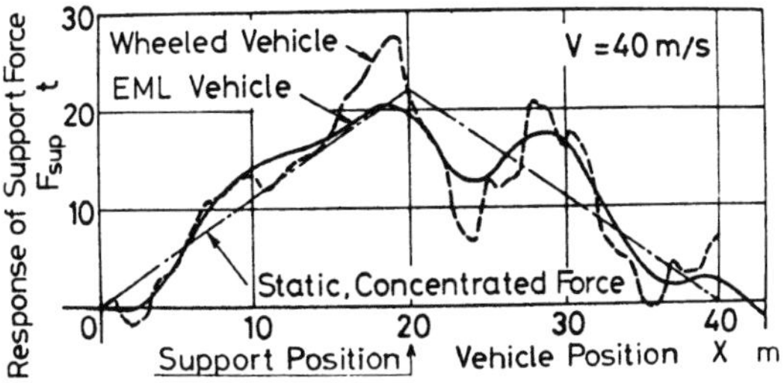


Fig.6. Comparison of the transient response of the vibration transmitted to the guideway support by the EML vehicle with that by the conventional wheeled vehicle; guideway (A)

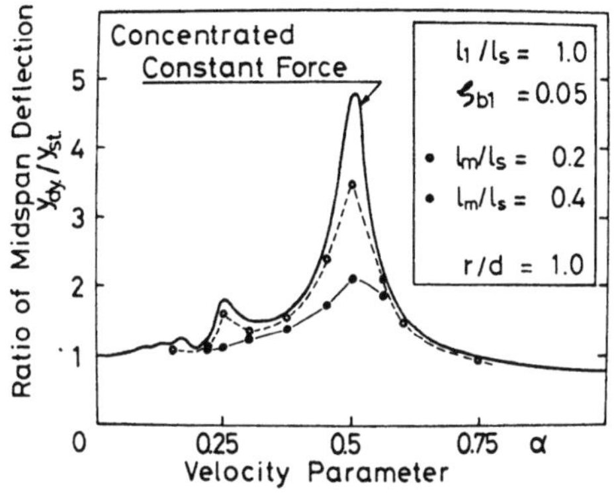


Fig.7. Ratio of the dynamic deflection of the midspan to the static deflection of it when multiple vehicles running at uniform intervals

deflection of it when multiple vehicles are running at uniform intervals. The solid line in Fig.7 is the analytical result when multiple concentrated constant forces are moving, and several peaks are shown on this line. These peaks exist when the ratio of the length of load intervals to the vehicle speed, that is the periodic time of loads, is equal to the natural periodic time of the beam or multiplies of it. This relation is reduced to

$$\alpha=\ell_1/(2m\ell_s) \quad , \quad m=1,2,3,\ldots \quad ; \quad \alpha\neq 1/(2n+1) \quad , \quad n=1,2,3,\ldots \qquad (29)$$

where $\ell_1$, $\ell_s$, $\alpha$ denote the length of load intervals, the beam span length and the velocity parameter, respectively. The second condition above is obtained by the fact that the first mode vibration of the beam does not remain after a constant and concentrated load passes it. In Fig.7 numerical results are shown for two cases that the ratio of the force distribution length to the beam span length, that is $\ell_m/\ell_s$, is 0.2 and 0.4. The first peak (m=1) of numerical results exists clearly almost at the same value $\alpha$=0.5 as the solid line, but the second peak (m=2) is not clearly shown when the ratio $\ell_m/\ell_s$ is 0.4.

It may be concluded that the ratio of the dynamic deflection of the beam to the static deflection of it has several resonant peaks, but the ratio decreases and higher order peaks are extinguished with the force distribution length.

*Response of Each Vehicle*

In following figures, we show vertical accelerations of vehicle bodies with r.m.s. values at front and rear suspended points of vehicle bodies. These points are numbered from the front of the first vehicle to the rear of the last vehicle.

At first Fig.8, Fig.9 show results when vehicles and trucks move at uniform intervals. Fig.8 shows vertical accelerations of vehicle bodies for the guideway (A). There are two peaks at lower speeds than 60m/s in Fig.8. The peak at 20m/s is caused by the resonanse between the secondary suspension and the static deflection of the beam, that is when the secon-

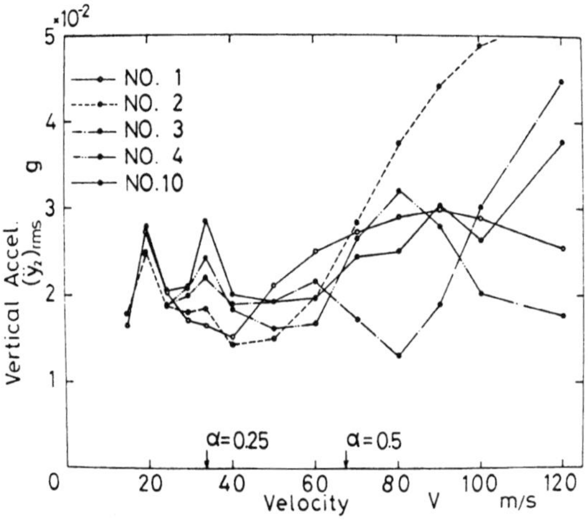


Fig.8. Vertical accelerations of vehicle bodies when trucks running at uniform intervals ; guideway (A)

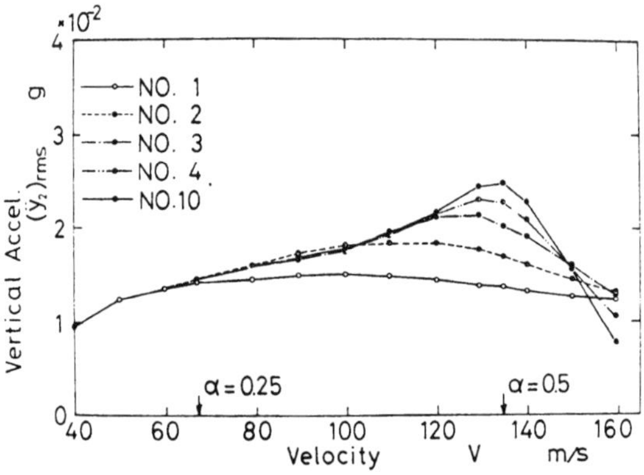


Fig.9. Vertical accelerations of vehicle bodies when trucks running at uniform intervals ; guideway (B)

dary natural frequency is equal to the ratio of the vehicle speed to the beam span length. The second peak at 33.7m/s ($\alpha$=0.25) is caused by the dynamic deflection of the beam under repeatedly moving loads. The vertical acceleration at the front of the first vehicle (No.1) is $0.0165g_{rms}$, and values of following vehicles tend to increase one by one to be 0.0285 $g_{rms}$ at the rear of the last vehicle (No.10) at this resonant speed ($\alpha$= 0.25). At higher speeds than 60m/s, values are shown to become complex and approximately increase with the vehicle speed. It may be because the dynamic deflection of the beam at higher velocity parameters than 0.5 is sensitively dependent on the velocity parameter and the vehicle load pattern, and also because the speed becomes to be near to the resonant speed when the primary natural frequency is equal to the ratio of the vehicle speed to the beam span length.

Fig.9 shows vertical accelerations of vehicle bodies for the guideway (B), which consists of a half length span of the guideway (A), and the phase velocity of which is the twice as large as the guideway (A). The first resonant peak ($\alpha$=0.5) is also clearly shown. At this resonant speed 134.6m/s, the vertical acceleration at the front of the first vehicle ( No.1) is $0.0138g_{rms}$, and that at the rear of the last vehicle (No.10) is $0.0247g_{rms}$.

In Fig.8 and Fig.9 values from No.5 to No.9 are not plotted, because these values converge to nearly the same value at No.10 at a resonant speed and at lower speeds than that. The higher resonant peaks given by eq.(29) do not exist in these figures. There is almost no difference between each vehicle at lower velocity parameters than 0.25, that is, in this lower velocity parameter range the influence of the dynamic deflection of the beam on the vehicle dynamics is small, especially for the guideway (B).

From these figures, vertical accelerations at following vehicles increase to be twice or so as large as the value at the front of the first vehicle at the first resonant speed. It can be said that these results are almost similar to those of the clearance between the magnet and the guideway surface.

Fig.10 and Fig.11 show vertical accelerations of each vehicle body, in which trucks are running at alternately different intervals over the guideway (A) and (B), respectively. The length between the center of the front truck and that of the rear truck is assumed to be 13m, the magnet length 4m and the length of the vehicle body 20m. These figures show the different tendency from Fig.8 and Fig.9. In Fig.10 values at the front of each vehicle (the odd number) are almost larger than those at the rear of each vehicle (the even number). Fig.11 shows nearly the same tendency as Fig.10, but values at the front of each vehicle are almost smaller than those at the rear of each vehicle.

Now we investigate the influence of the pitch and heave natural frequencies of the vehicle body on the vertical acceleration. Fig.12 shows values at each vehicle body moving over the guideway (B) at a speed 80 m/s ($\alpha$=0.297). The parameter r/d is the ratio of the inertia radius to the length between the gravity center and the suspended point, which is nearly equal to the ratio of the pitch natural frequency to the heave natural frequency. From Fig.12 vertical accelerations at each point tend to wholly increase with the ratio r/d, but the relation between the value of the front and that of the rear is almost similar in spite of changing the ratio r/d.

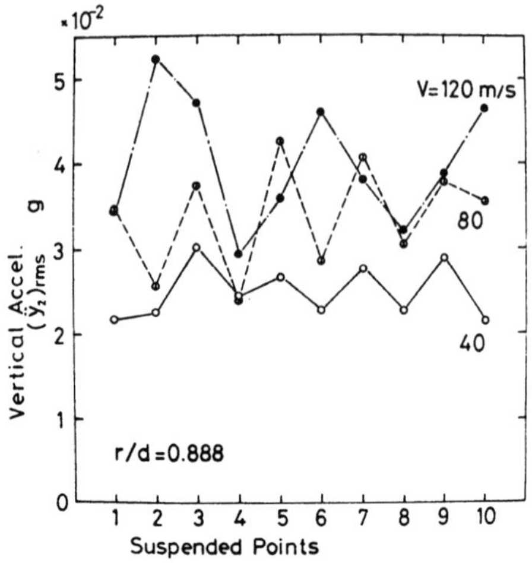


Fig.10. Vertical accelerations of vehicle bodies when trucks running at alternately different intervals ; guideway (A)

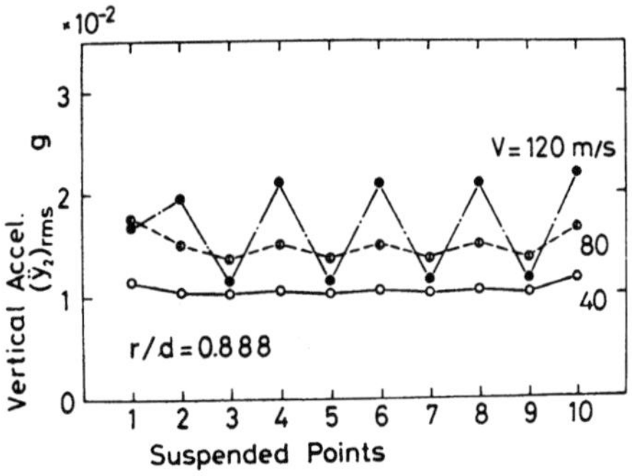


Fig.11. Vertical accelerations of vehicle bodies when trucks running at alternately different intervals ; guideway (B)

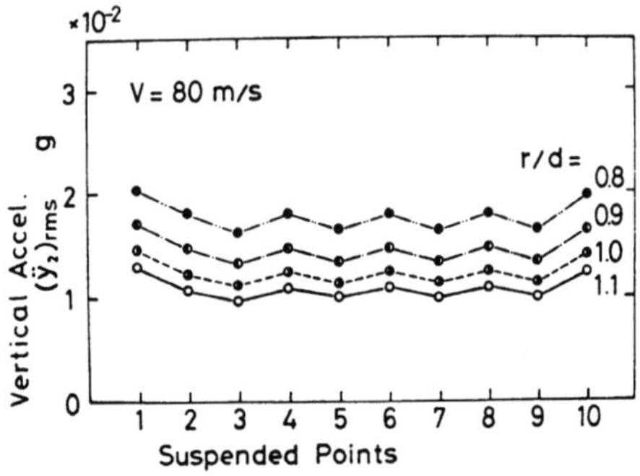


Fig.12. Vertical accelerations of vehicle bodies depending on the heave and pitch natural frequencies ; guideway (B)

Table 1. Standard spec. of EML

| | | |
|---|---|---|
| vehicle weight | | 44.0 t |
| truck weight | | 5.50 t |
| body weight | | 33.0 t |
| vehicle length | $\ell_v$ | 20.0 m |
| inertia radius | r | 5.77 m |
| magnet length | $\ell_m$ | 4.00 m |
| 1-st suspension | $f_1$ | 5.94 Hz |
| | $\zeta_1$ | 0.51 |
| 2-nd suspension | $f_2$ | 1.00 Hz |
| | $\zeta_2$ | 0.20 |

Table 2. Spec. of wheeled vehicle

| | | |
|---|---|---|
| vehicle weight | | 44.0 t |
| wheelset weight | | 2.40 t |
| truck weight | | 2.60 t |
| body weight | | 34.0 t |
| 1-st suspension | $f_1$ | 50.0 Hz |
| | $\zeta_1$ | 0.50 |
| 2-nd suspension | $f_2$ | 5.73 Hz |
| | $\zeta_2$ | 0.00 |
| 3-rd suspension | $f_3$ | 1.77 Hz |
| | $\zeta_3$ | 0.414 |

Table 3. Specifications of guideway structure

| | | | guideway (A) | guideway (B) |
|---|---|---|---|---|
| span length | $\ell_s$ | m | 20.0 | 10.0 |
| bending stiffness | EI | $kg \cdot m^2$ | $0.550 \cdot 10^9$ | $0.138 \cdot 10^9$ |
| weight per unit length | $\rho ag$ | kg/m | $0.734 \cdot 10^4$ | $0.184 \cdot 10^4$ |
| phase velocity | $V_p$ | m/s | 134.6 | 269.2 |
| 1-st natural frequency | $f_{b1}$ | Hz | 3.36 | 13.4 |
| 1-st damping coeff. | $\zeta_{b1}$ | | 0.05 | 0.05 |

## 5. CONCLUSIONS

In the viewpoint of the safety or reliability of the electromagnetic levitation vehicles and guideway system, it is necessary to consider the possibility of the magnet collision with the guideway. The contact frequency used in this paper is one of the criteria for the reliability evaluation of this system, which means the reciprocal of the average time from start to the first collision of the magnet with the guideway. We have analyzed the simplified dynamic model statistically in order to obtain the equation with which we can analytically calculate this value and also we can evaluate the contact velocity when the contact occurs. In a case when the vehicle moves over the roughness as smooth as a welded rail, it may be possible to suppress the contact frequency to be less than $10^{-5}$ $s^{-1}$ ($\simeq 1.0$ $day^{-1}$) and to suppress the contact frequency with the higher contact velocity than o.3m/s to be less than a tenth of the total contact frequency. Although for a strict discussion it must be necessary to analyze a fully-coupled model with the nonlinearity and the distribution of the force, the analysis shown in this paper may be useful for a preliminary design if the probability of the contact is designed to be very small.

After the preliminary design with the statistical study above, we have developed the computer simulation program for the numerical analysis of the dynamic interaction between multiple vehicles and their guideway. We have investigated the influence of the dynamic deflection and roughness of the guideway, that of the force distribution and that of the truck intervals, respectively on the vehicle dynamic characteristics. We have also compared the transient response of the vibration transmitted to the guideway support of this system with that of conventional wheeled railroads. It may be concluded as follows.

(1) There is a distribution effect of the magnetic force which reduces the dynamic deflection of the guideway and absorbs the guideway surface roughness so that the vehicle vibration, especially the high frequency

range of it, tends to decrease.

(2) According to the comparison of this system with conventional wheeled railroads, the transient response of the vibration transmitted to the guideway support is of less amplitude and smoother.

(3) There is a resonant phenomenon when the periodic time of repeatedly moving trucks is equal to the natural periodic time of the guideway or multiplies of it. But the ratio of the dynamic deflection of the guideway to the static deflection of it at resonant conditions decreases greatly with the length of the force distribution.

(4) The vertical acceleration of each vehicle and the clearance between the magnet and the guideway surface tend to be different from each other, especially at a high speed operation, depending on the truck intervals, the beam span length and the velocity parameter. At uniform intervals, variations of following vehicles increase one by one to be twice or so of that of the first vehicle at a resonant speed. Meanwhile at alternately different intervals, variations of the front and rear of each vehicle are almost always alternately different. These dynamic characteristics are caused by the dynamic deflection of the guideway loaded repeatedly by multiple vehicles.

## 6. ACKNOWLEDGEMENTS

The work described in this paper is supported in part by the Japan Industrial Association of Railroad Vehicles, and is a part of the research and developement program administered by the Committee of Low Pollution Railroads. The support is gratefully acknowledged.

## REFERENCES

1. Hullender,D.A., "Minimum Vehicle-Guideway Clearances Based on a Contact Frequency Criterion", Trans. ASME, Series G, June 1974, pp. 213 -217.

2. Richardson,H.H., and Wormley,D.N., "Transportation Vehicle/Beam-Elevated Guideway: Dynamic Interactions - A State of the Art Review", Trans. ASME, Series G, June 1974, pp. 169-179.

3. Wilson,J.F., and Biggers,S.B., "Dynamic Interactions Between Long, High Speed Trains of Air Cushion Vehicles and Their Guideways", Trans. ASME, Series G, March 1971, pp. 16-24.

4. Doran,A.L., and Mingori,D.L., "Periodic Motion of Vehicles on Flexible Guideways", Trans. ASME, Series G, December 1977, pp. 268-273.

5. Rice,S.O., Bell System Tech., 23 (1944), pp. 282-332 ; 24 (1945), pp. 46-156.

# ADAPTION OF MATHEMATICAL VEHICLE MODELS TO EXPERIMENTAL RESULTS

W. Oberdieck, B. Richter and P. Zimmermann

Volkswagenwerk A.G., Wolfsburg, F.R.G.

SUMMARY

An optimization process is shown with which measured vehicle characteristics can be projected with sufficient accuracy by a simple vehicle model.

This means that basically even arbitrarily specified vehicle characteristics can be translated into appropriate model parameters.

This is a welcome aid for driving simulator investigations.

In addition, there appears to be a way of circumventing the problems of side-slip angle measuring with the aid of a simple vehicle model.

## 1. INTRODUCTION AND PROBLEM

The prognostication of vehicle characteristics during the draft design stage offers certain advantages in vehicle development.

This approach basically reduces the number of prototypes, the construction, measuring and modifications of which are expensive and time consuming.

Such prognoses are usually implemented with the aid of mathematical vehicle models. The support of new-vehicle development with model studies of this kind is all the more important the more pronounced the differences are between the new vehicle and its predecessors.

Prognoses can be applied only in those cases where it is known in advance how certain vehicle characteristics can be defined by numerical codes.

Large-scale course-keeping experiments on test subjects can be performed so that correlations may be established between the performance potentials of the driver/vehicle/road system and defined vehicle characteristics.

Driving simulators are particularly useful for this purpose because they permit the simulation of a wide variety of vehicles with different characteristics within a short period of time [1].

This approach frequently requires the use of mathematical models for the simulation of existing vehicles, the characteristics of which were determined by open-loop measurements on the proving ground.

This is difficult in the case of complex vehicle models. If an existing vehicle is to be simulated with a model like that, some of the vehicle components must be physically dismantled so that the characteristics of individual subassemblies can be determined with the aid of measuring apparatus. Furthermore, the agreement between model behavior and real-life vehicle behavior is not always satisfactory despite the substantial outlay for measuring equipment and efforts.

Experience has shown, however, that the study of the interaction between vehicle and driver in the simulator does not require sophisticated vehicle models. Nevertheless, even in regard to simple vehicle models, the question arises of how certain parameters should be determined so that given vehicle characteristics can be obtained.

## 2. VEHICLE MODEL

A suitable model is the well known single-track vehicle model (Bicycle Model), the linear configuration of which is determined by seven parameters:

$a$ distance of c.o.g from front axle

$b$ distance of c.o.g from rear axle

$c_f$ side force coefficient (front axle)

$c_r$ side force coefficient (rear axle)

$i$ steering ratio

$m$ vehicle mass

$J_z$ moment of inertia (z-axis)

The parameters a, b, i, m, and $J_z$ can be measured comparatively easily. The vehicle need not be dismantled.

The determination of the side force coefficients poses problems, though. These coefficients define the correlations between the slip angle (ß) and the cornering force (F) of each axle. The side force coefficients include the tire characteristics, axle kinematics and axle elasticities.

The side force coefficients are constant in the linear configuration of the bicycle model. This means, for instance, that the steering angle input during constant-radius cornering grows linearly along with the lateral acceleration $a_L$. It is known, however, that vehicles display a linear behavior only during small to medium lateral accelerations.

Therefore, a lateral acceleration $a_{Lo}$ is defined, below which the model is to display a linear behavior, i.e.:

$$c_f = c_{fo} \qquad c_r = c_{ro} \qquad \text{for} \quad /a_L/ \leqslant a_{Lo}$$

The side force coefficients should no longer be constants for accelerations beyond $a_{Lo}$ but should be functions of the lateral acceleration. In this approach, polynomes of the third degree were selected in the first step, without a linear portion in order to avoid the occurrence of a salient point at the border between the linear and non-linear range:

$$c_f = c_{fo} + k_{f1}(/a_L/ - a_{Lo})^2 + k_{f2}(/a_L/ - a_{Lo})^3$$

$$c_r = c_{ro} + k_{r1}(/a_L/ - a_{Lo})^2 + k_{r2}(/a_L/ - a_{Lo})^3$$

for $/a_L/ > a_{Lo}$

The following seven parameters are unknown and remain to be determined when the model parameters a, b, i, m and $J_z$ are assumed to be known:

$c_{fo}$, $k_{f1}$, $k_{f2}$, $c_{ro}$, $k_{r1}$, $k_{r2}$, $a_{Lo}$

## 3. VEHICLE BEHAVIOR

Basically, any vehicle variables may be used as functions of time as well as mutual functions in order to identify the unknown model parameters.

Vehicle handling characteristics essentially are defined by steady-state cornering and by the vehicle response to sudden changes in the steering angle input. The time histories of the yaw angle velocity $\dot{\psi}(t)$, lateral acceleration $a_L(t)$ and of the side slip angle $\beta(t)$ are of particular interest in regard to sudden changes of the steering angle input. The measuring of the side slip angle poses special problems. This aspect will be dealt with in greater detail in Section 5.

The vehicle behavior during steady-state cornering is described by the correlation between steering wheel angle $\delta_{SW}$ and lateral acceleration $a_L$.

The following functions were selected for the adaption of the model to proving ground measurements:

1. $\dot{\psi} = f(t)$

2. $\delta_{SW} = f(a_L)$

4. ADAPTATION OF PARAMETERS

The remaining free model parameters are adapted by comparing the measured vehicle responses with the computed model responses. The deviations are computed on the basis of the discrete points given by the measurement. They are then squared and totaled. The resulting values indicate the degree of deviation between real-life vehicle and model.

In this case there are two squared sums of deviations obtained for the functions $\dot{\psi} = f(t)$ and $\delta_{SW} = f(t)$. With the aid of weighting factors it may now be determined which of the two vehicle characteristics is to be preferred in the adaptation process.

The strategy used in the determination of the unknown model parameters was based on the principles of biological evolution [2] :

- mutation of genes
- recombination
- selection
- isolation

The gene should be considered as a parameter. Several parameters form the set of parameters. A mutation means that one or several parameters are randomly changed in regard to quantity and direction. In case of sex-linked traits, the sets of parameters of the two parents can be bisected at randomly selected locations and recombined. The selection is simulated by preferring those parameters for transmission that have the optimum quantity.

The target quantity in this case is a value that is obtained by summing-up the two weighted squared sums of deviation. This value must be reduced to a minimum. Isolation means that no other kinds of sets of parameters are permitted.

The optimization method in accordance with the strategy of evolution is structured in two stages. During the first stage, a population is built on the basis of a starting set of parameters. In this stage parameter variations are performed only to one set of parameters (non sex-linked traits). If the result of these variations is better than the worst one in the population, it is accepted in the population until a given number of parameter sets is reached. If it is worse, it is found unsuitable and is discarded.

Once the maximum number of sets of parameters is reached, the mechanics of sex-linked traits will start. Two sets of parameters are randomly selected from the given population. Preference is given to those parameters that have better target quantities.

These two sets of parameters are the parents of the next generation. Two offspring are generated by parent parameter variation while two additional offspring are generated by an exchange of parts of the parameter set among the former two. This means that each generation procreates four offspring.

Those four parameter sets are ranked by target quantities while the two inferior ones are sorted out. The remaining two are added to the existing population where they are incorporated by quality ranking. The two worst sets of parameters are removed from the expanded population so that the total number of sets within the population remains the same.

This process is repeated until either a given number of computing operations is reached or until the population no longer has any quality differences in target quantity within a given range of precision.

The reason for the two-stage heredity is the fact that non-sex-linked traits permit a broader spectrum of possible variations than do sex-linked traits. In other words, this approach is meant to prevent an excessively fast restriction process at the onset of optimization.

## 5. RESULTS

Figures 1 and 2 are comparisons between the measured vehicle (MV), the adapted simple vehicle model (EV) and a complex vehicle model (SV) that was used for the simulation of the measured vehicle.

It was mentioned earlier in Section 1 that complex lab measurements are required in order to determine the required parameter values of the complex vehicle model.

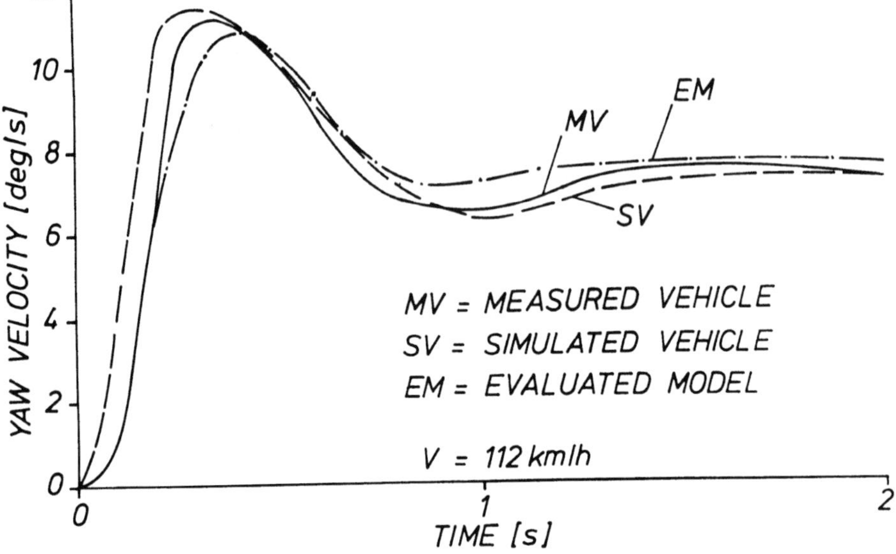


FIG 1: TRANSIENT YAW RESPONSE

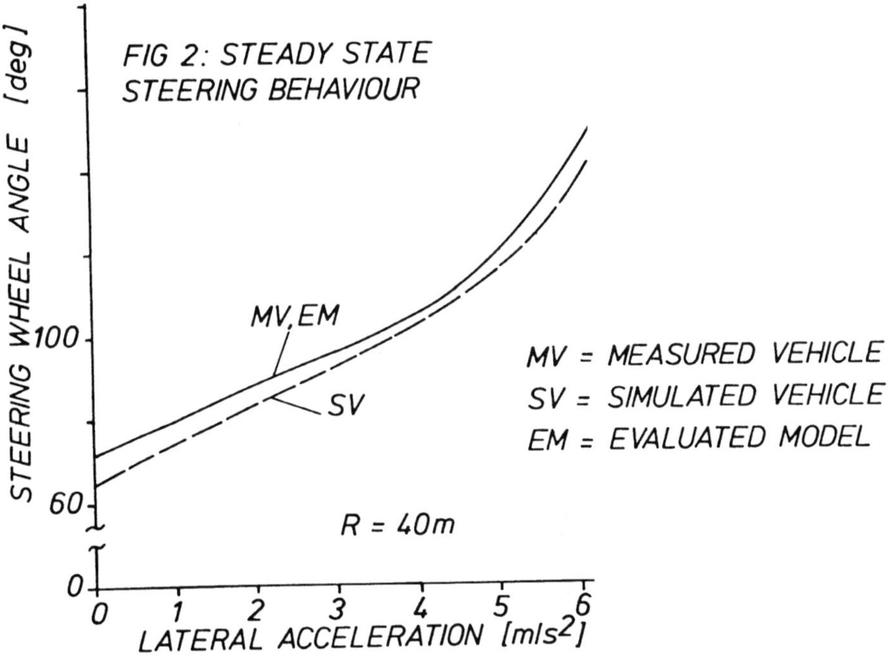


FIG 2: STEADY STATE STEERING BEHAVIOUR

The transient yaw response of the measured vehicle (MV) in Figure 1 lies between that of the simulated vehicle (SV) and that of the adapted simple model (EM). Agreement can be described as satisfactory.

There are virtually no differences between the measured vehicle (MV) and the adapted model (EM) during constant-radius steady-state cornering. The simulated vehicle shows minor deviations.

Figure 3 shows the side slip angle response during transient yaw response for the EM and EV. The differences are in the neighborhood of 0.5° in the maximum range, and at 0.25° for steady-state values. The reproduction of the side-slip angle measurements posed problems (optical sensor). The steady-state values ranged between 1.6° and 3.6°.

In order to be able to better evaluate the adaptability of the simple model, the complex vehicle was defined as an example because there is no problem of measuring reproducibility.

This investigation was performed at two very different load conditions:

Load condition 1: The vehicle is occupied by the driver only.

Load condition 2: The vehicle is occupied by the driver. The trunk is loaded until the permissible rear-axle load is reached.

Figures 4 and 5 show the given vehicle reactions (SV) and the adapted reactions of the simple model (EM) for load condition 1. There appears to be good agreement.

It is important, though, to perform a comparison on the basis of those variables that are not used for the adaptation. This is shown in Figures 6 (lateral acceleration) and 7 (side slip angle response) to transient response. The lateral acceleration deviations are negligible. The difference in the steady-state range is approximately 0.2° for the side-slip angle. This is not very much when considering the unsatisfactory reproducibility of measurements.

Load condition 2 is most unfavorable for a vehicle with rear luggage compartment. This may be seen from the slow yaw response of the vehicle (Figure 8) that is close to the limit of stability. Nevertheless, it may be said that the simple model is sufficiently adaptable. The same applies to constant-radius steady-state cornering (Figure 9).

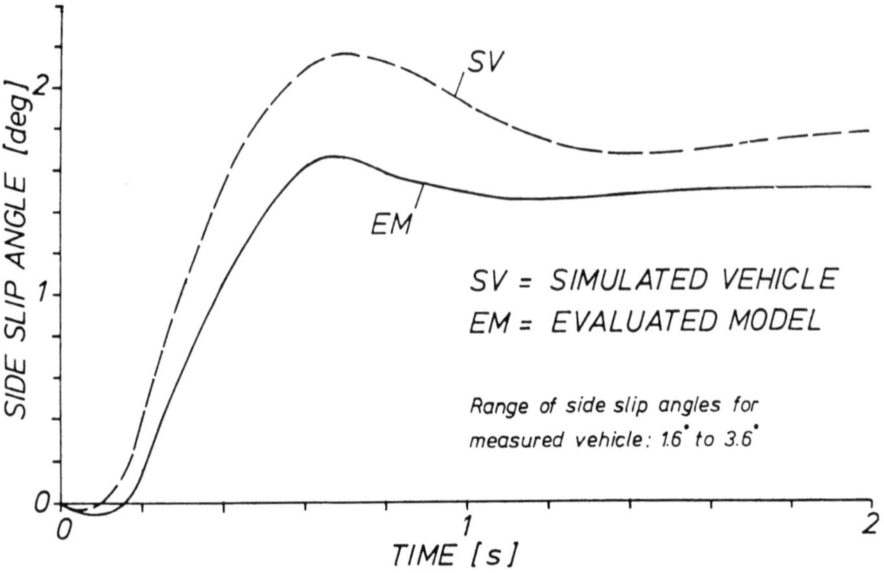


FIG 3: SIDE SLIP ANGLE RESPONSE

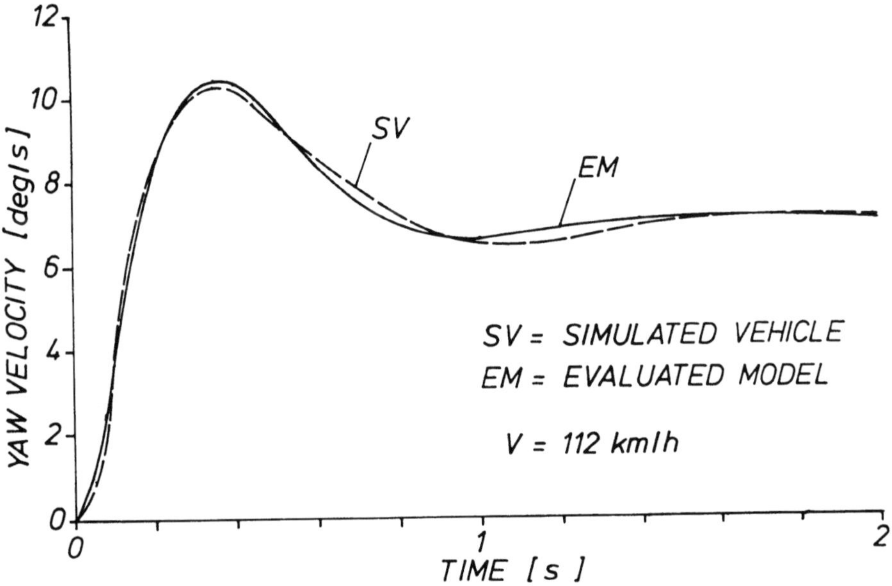


FIG 4: TRANSIENT YAW RESPONSE LOAD 1

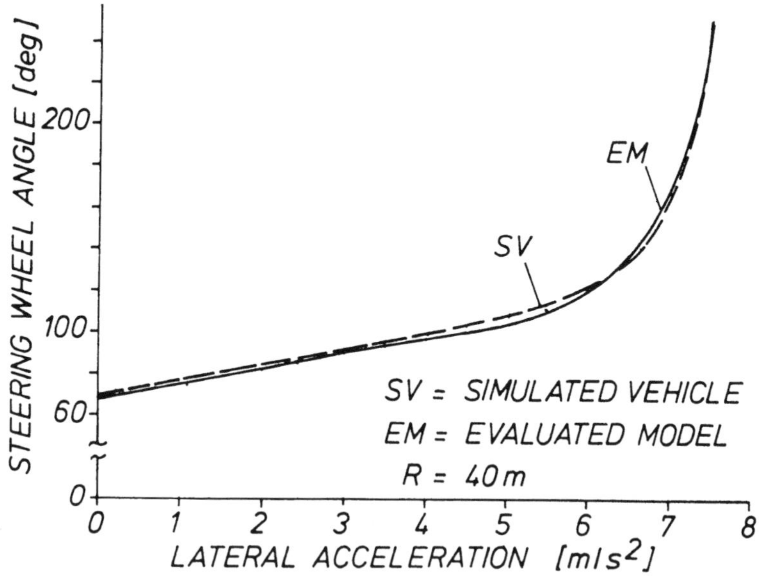


FIG 5: STEADY STATE STEERING BEHAVIOUR LOAD 1

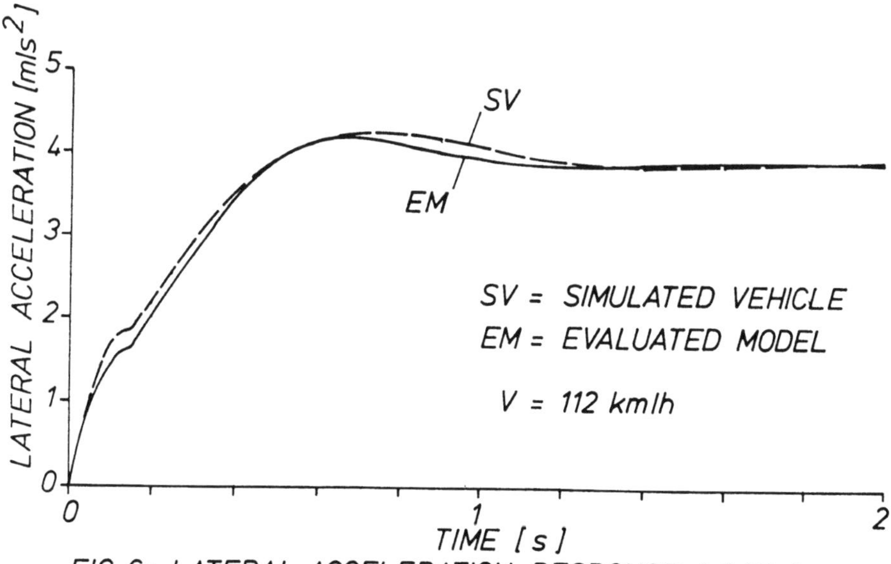


FIG 6: LATERAL ACCELERATION RESPONSE LOAD 1

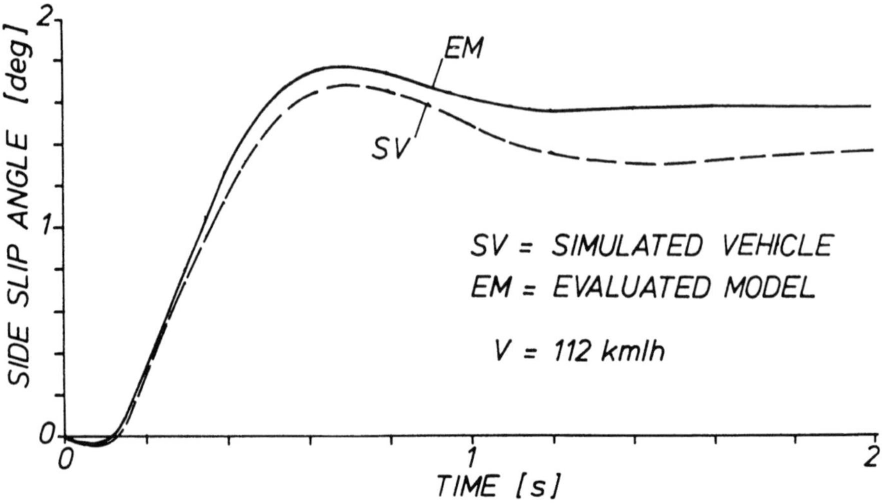


FIG 7: SIDE SLIP ANGLE RESPONSE LOAD 1

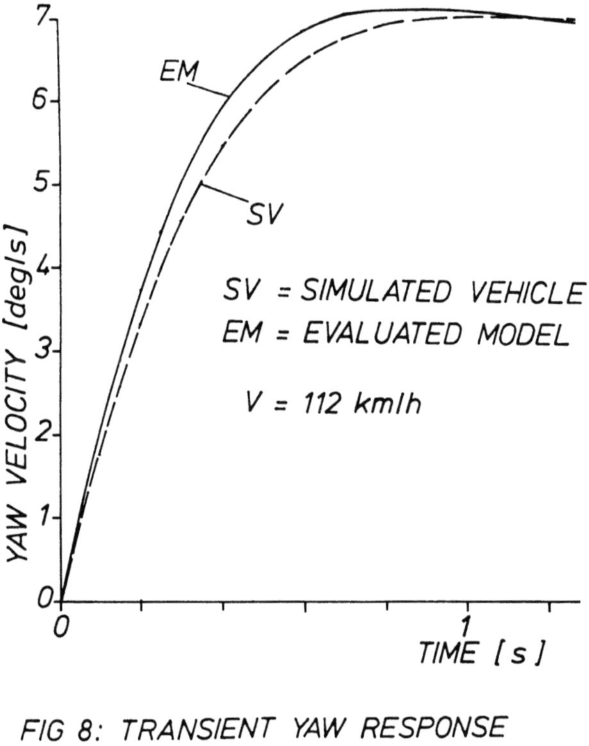


FIG 8: TRANSIENT YAW RESPONSE
LOAD 2

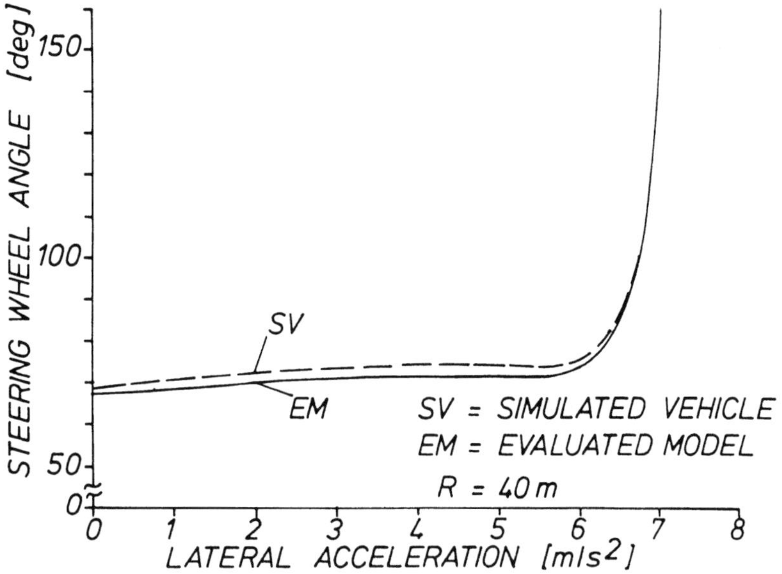


FIG 9: STEADY STATE STEERING BEHAVIOUR LOAD 2

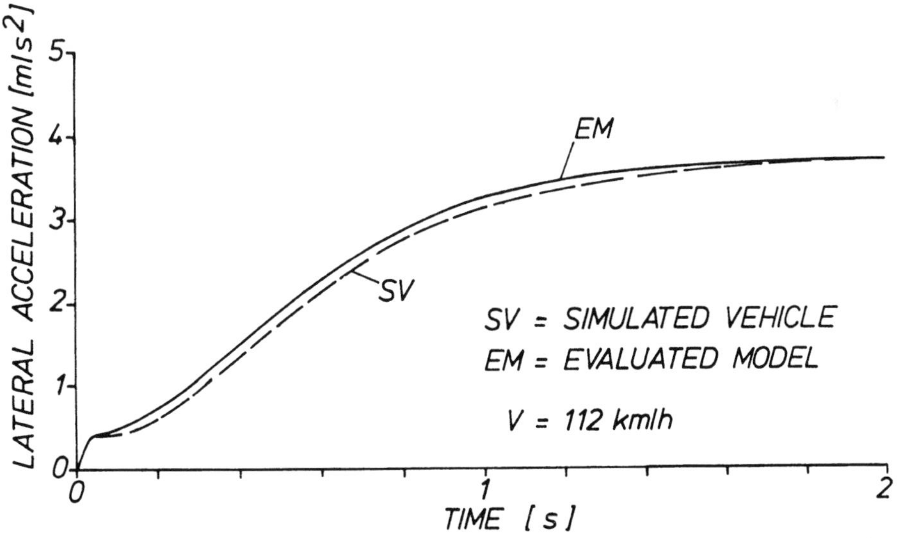


FIG 10: LATERAL ACCELERATION RESPONSE LOAD 2

There is remarkably good agreement during the lateral acceleration that is not used for adaptability (Figure 10).

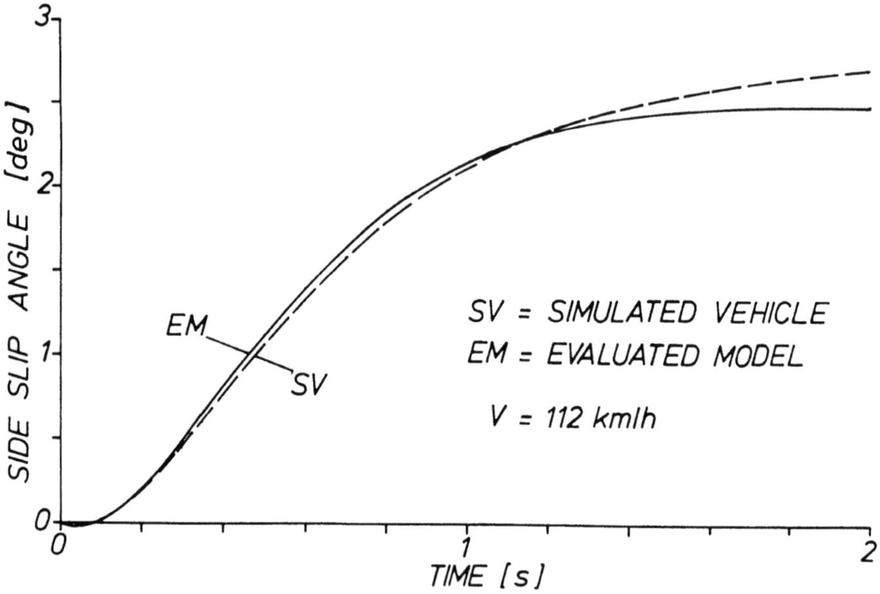


FIG 11: SIDE SLIP ANGLE RESPONSE LOAD 2

The differences for the side slip angle response in the time (Figure 11) range downwards to one second are small. The deviation in the steady-state range is roughly 0.3°. This is acceptable.

6. References


Ref. 1: Richter, Bernd: Driving Simulator Studies - The Influence of Vehicle Parameters on Safety in Critical Situations; SAE - Paper 741105 (1974)

Ref. 2: Rechenberg, Ingo: Evolutionsstrategie; Problemata 15, Verlag Frommann-Holzboog, Stuttgart-Bad Cannstadt (1973)

# VIBRATIONAL MODES OF SINGLE-TRACK VEHICLES IN CURVES

H. B. Pacejka and C. Koenen

Delft University of Technology, Delft, The Netherlands

SUMMARY

The dynamic behaviour of single-track vehicles under cornering conditions differs from the behaviour they exhibit when driving straight ahead. This difference is mainly caused by the coupling between symmetrical and anti-symmetrical degrees of freedom. The coupling becomes stronger when cornering is more pronounced. By means of a mathematical model of a single-track vehicle, capable of cornering at large roll angles, the effects of the above mentioned coupling are studied. In this paper an introduction to the mathematical model is given and some results of the investigations are presented.

## 1. INTRODUCTION

The investigation discussed in this paper forms a part of a more ambitious programme which aims at the development of a sufficiently correct model description and analysis of the motion of single-track vehicles moving in curves and subjected to irregularities of the road surface both regarding friction coefficient and geometry.

A natural first step is the study of the steady-state curving manoevre and the small parasitic free vibration around this motion on a level and smooth road surface. Up to now investigations have been restricted to steady-state cornering [1], straight-line free vibrations [2, 3, 4, 5] also in connection with the in-plane (symmetrical) motion [6] and studies concerning the simulation of certain manoevres [7]. Experimental observations of Jennings [8] and of Weir [9] indicate that the damping of lateral oscillation may become considerably lower when the motor cycle negotiates a curve. Sharp [6] suggests that this may be due to an interaction between the in-plane and the lateral motions of the vehicle.

The present study investigates this phenomenon using a theoretical model of the machine featuring basically four lateral and four in-plane degrees of freedom. Because of the present lack of sufficient information on the riders passive response to motion inputs, the rider has been modelled as a block rigidly attached to the main frame. The steering torque excited by the rider when negotiating a curve remains constant also in the perturbed motion.

The cycle model allows for the following motion components: lateral, yaw, roll and steer and: bounce and pitch and wheel hop, front and rear. Besides, we have the lateral deflections of the tyres. The tyre is modelled to describe two aspects viz the steady-state response to wheel motions and the transient response. The steady-state response functions have been developed using measured data. The differential equations describing the transient response have been assumed considering the existing general knowledge in this area.

The steady-state cornering condition is described by a set of non-linear algebraic equations. The perturbed motion is governed by a set of linear differential equations.

In Fig. 1 the root locus curves have been shown for the free vibrational modes of a vehicle moving in the neighbourhood of a straight path. At the left hand side the course of the roots of the characteristic equation for the lateral motion has been depicted with the speed of travel u as parameter. The equation is of the eighth degree. It has been found that two or four roots may become real. The wobble mode is of a relatively high frequency with a pronounced steering vibration. The weave mode is a low frequency combined lateral-yaw-roll oscillation with some steer. The other well-damped mode may be referred to as rear wobble considering the mode shape with large amplitudes of the rear tyre side force. At the right-hand side the four complex roots for the in-plane modes have been shown. They appear as dots since they are independent of speed in the model considered.

For motions almost along a straight path, the average roll-angle is almost zero and coupling between small lateral and in-plane motions can be neglected. Changes in the location of the in-plane roots do not affect the out-of-plane modes.

This will not remain true, however, when the average roll angle becomes larger. Notably in sharp fast curves appreciable coupling between the modes is expected to occur. The nature of interaction may be illustrated by the examples given in Fig. 2. A coupling from lateral accelerations to in-plane accelerations and vice-versa obviously occurs when the frame is inclined. In addition to this dynamic coupling a one-sided coupling occurs which is due to the load dependent properties of the tyres. In-plane motions give rise to changes in the normal load acting on the tyre. This in turn causes changes in various properties of the tyre for instance in the cornering stiffness. As an effect of the latter the side slip of the tyre alters and thus the lateral motion of the vehicle. The third example shows the geometrical coupling between steer angle and vertical motion which is due to the finite magnitude of castor length and of wheel radius. These examples represent only a few of the numerous coupling terms which appear in the equations discussed briefly in the sequel.

## 2. DEVELOPMENT OF THE MODEL

In Fig. 3 the model of the single-track vehicle has been shown. Both the motion variables and the geometrical parameters have been indicated. As to the motion variables we distinguish:

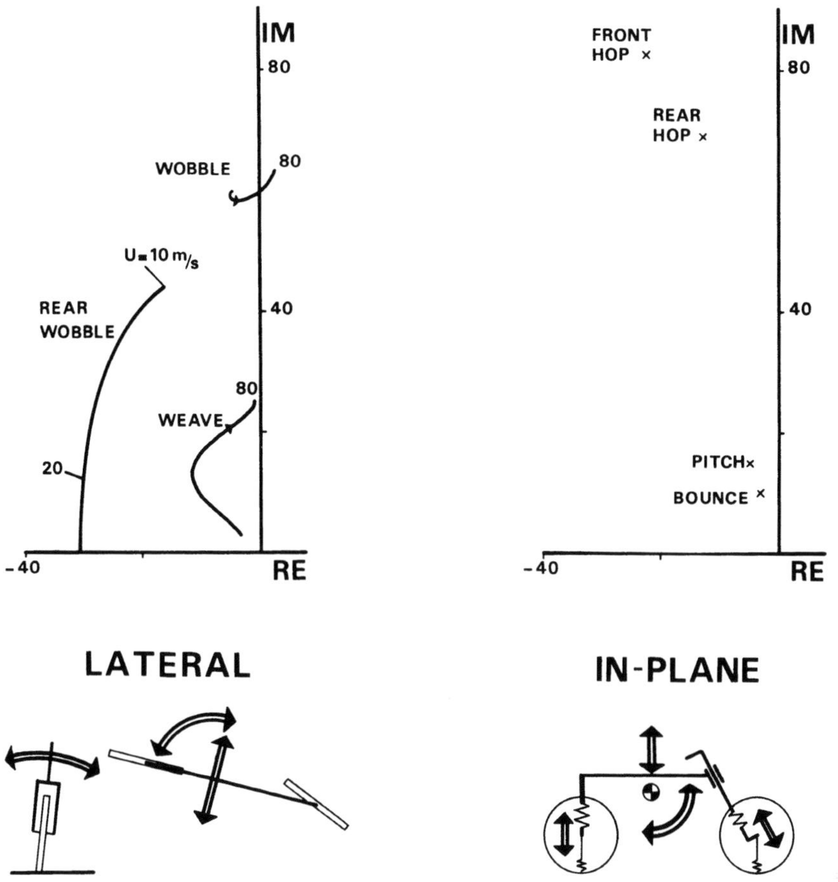


Fig. 1. Root loci for two groups of uncoupled modes of vibration.

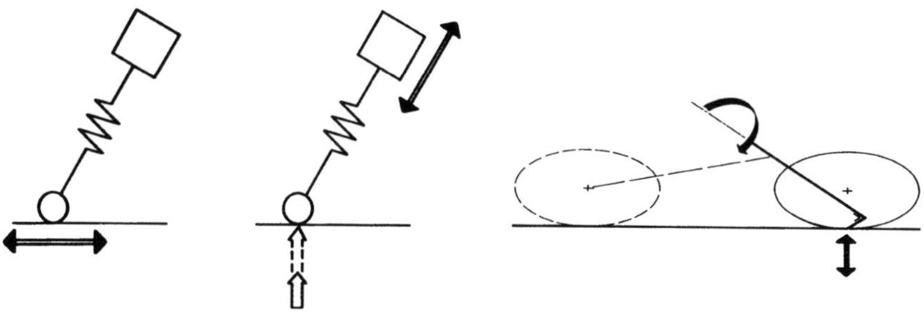

Fig. 2. Examples showing nature of interaction between in-plane (∿ vertical) and out-of-plane (∿ lateral) motions.

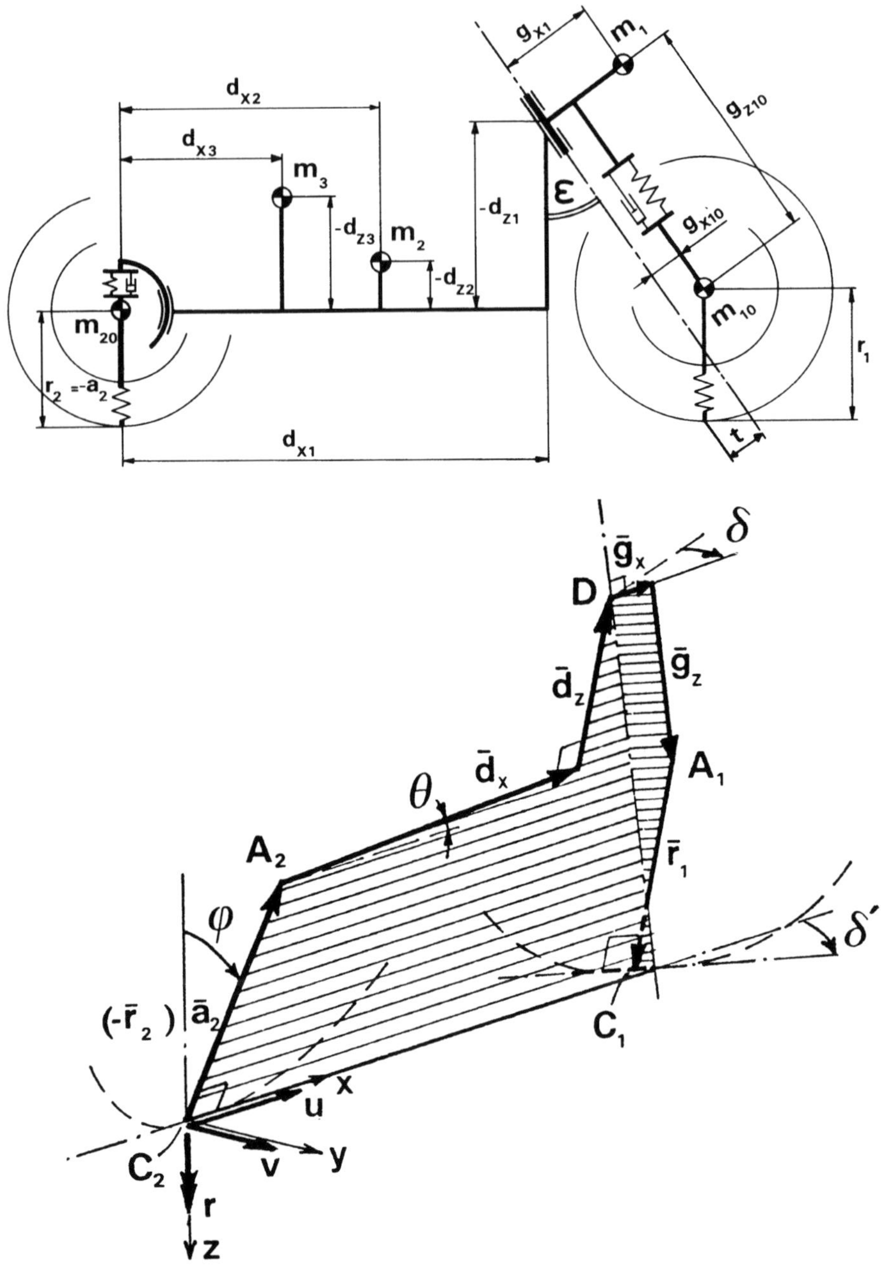


Fig. 3. Geometry of the model and vectors defining the location of points of interest with respect to the moving triad $(C_2, x, y, z)$.

| *in-plane variables* | | *out-of-plane variables* | |
|---|---|---|---|
| ($u$ | longitudinal speed (constant)) | $v$ | lateral velocity |
| $\theta$ | pitch angle | $r$ | yaw velocity |
| $\lambda_1$ | front spring compression | $\phi$ | roll angle |
| $\lambda_2$ | rear spring compression | $\delta$ | steer angle |
| ($\lambda_{t1}$ | front tyre compr. (dependent)) | $\eta_1$ | front lateral tyre deflection |
| $\lambda_{t2}$ | rear tyre compression | $\eta_2$ | rear lateral tyre deflection |

In the model we distinguish the following parts:
rear sprung mass (main or rear frame plus rigid rider)
rear unsprung mass (wheel, axle etc.)
front sprung mass (front frame, handle-bar) hinged to the mainframe, rake angle $\varepsilon$
front unsprung mass (wheel, axle etc.).
The following assumptions and restrictions hold:
- the longitudinal velocity of the rear contact centre remains constant;
- both front and rear frames are considered to be rigid;
- the rider body is regarded as a rigid part of the main frame; controlling actions of the rider are not regarded;
- the tyre crown radius is set equal to zero ($M_x = 0$);
- the self-aligning torque $M_z$ is neglected;
- to achieve a steady-state the desired corresponding steer torque is applied; this torque remains unchanged in the perturbed motion;
- the rotation energy of parts of the engine and further propulsion mechanism is not taken into account in the case of a crankshaft parallel to the x-axis (parameter set A);
- aerodynamic forces and their effects are not regarded;
- the road surface is flat and even;
- the transient responses of the tyre side force to variations of side slip, turn slip, camber angle and vertical load of the wheel are assumed to be similar.

The derivation of the equations of motion turns out to become extremely complex and voluminous. It was necessary to restrict the effort. This can be done by introducing assumptions regarding the magnitude of certain motion variables.

A variable quantity, in general denoted with $q$ consists of a steady-state constant part $q_s$ and a variable part $\tilde{q}$ describing the perturbed motion with respect to the steady-state condition:

$$q = q_s + \tilde{q} \tag{1}$$

Those steady-state quantities $q_s$ which would also exist in case we were dealing with a rigid vehicle system (in-plane deflections and steer angle equal to zero) are considered as large quantities. These are (beside speed $u$): $v_s$, $r_s$, $\phi_s$ and the lateral forces $F_{y1s}$ and $F_{y2s}$. The quantities $\delta_s$, $\theta_s$ and the spring and tyre deflections $\lambda_s$, $\eta_s$ are considered small. Products of these small quantities are neglected in the final differential equations. Also products of these small steady-state quantities and the perturbation variables $\tilde{q}$ are neglected in these equations. The same holds for products of the latter quantities themselves. The result is a linear set of differential equations in $\tilde{q}$.

## 3. MODEL OF THE CYCLE

The method of Lagrange is used to establish the equations of motion. The Lagrangian equations are modified to suit the use of quasi-coordinates in addition to true coordinates (cf. [10]). The quasi-coordinates x and y are definded as the non-existing integrals of velocities u and v. No true coordinates exist which by differentiation yield u and v. The following set of modified Lagrangian equations are used:

$$
\begin{aligned}
&\frac{d}{dt}\frac{\partial T}{\partial u} - r\frac{\partial T}{\partial v} = Q_x \\
&\frac{d}{dt}\frac{\partial T}{\partial v} + r\frac{\partial T}{\partial u} = Q_y \\
&\frac{d}{dt}\frac{\partial T}{\partial r} + u\frac{\partial T}{\partial v} - v\frac{\partial T}{\partial u} = Q_\psi \\
&\frac{d}{dt}\frac{\partial T}{\partial \dot{\phi}} - \frac{\partial T}{\partial \phi} + \frac{\partial D}{\partial \dot{\phi}} + \frac{\partial U}{\partial \phi} = Q_\phi
\end{aligned}
\qquad (2)
$$

etc.
with T denoting the kinetic energy, D the dissipation function, U the potential energy and Q the generalized forces. The use of these equations immediately yields equations of motion in terms of u, v and r which is convenient when the motions to be investigated are not restricted to the neighbourhood of a straight path.

The velocities u, v and r describe the motion of a moving axis system ($C_2$,x,y,z). Its origin is defined as the contact centre $C_2$ of the rear wheel with the ground. The x-axis coincides with the line of intersection of the plane of symmetry of the rear frame and the road plane. The z-axis points downwards.

With respect to this system of axes the position and attitude of the various vehicle parts are defined. Having derived expressions for the location of front wheel centre $A_1$, the attitude and the linear and angular velocities of the wheel and from these the corresponding terms on the left hand side of the Lagrangian equations it is easy, by simple modifications, to determine the terms corresponding to other parts of the vehicle.

In vectorial form the location of $A_1$ with respect to the moving axis system reads (cf.Fig. 3):

$$\bar{a}_1 = \bar{A}^\phi\cdot\bar{a}_2 + \bar{A}^\phi\cdot\bar{A}^\theta\cdot(\bar{d}_x + \bar{d}_z) + \bar{A}^\phi\cdot\bar{A}^\theta\cdot\bar{A}^\varepsilon\cdot\bar{A}^\delta\cdot(\bar{g}_x + \bar{g}_z) \qquad (3)$$

Vectors $\bar{a}_2$, $\bar{d}_{x,y}$ and $\bar{g}_{x,y}$ are seen with respect to triads which are rotated with respect to each other over angles $\phi$, $\theta$, $\varepsilon$ and $\delta$. Matrices $\bar{A}$ represent rotation or coordinate transformation matrices.

Vector $\bar{r}_1 = \overline{A_1C_1}$ is needed to determine the radial tyre deflection and to establish the location of the front contact point needed to acquire slip velocities of the front wheel with respect to the road. The vector can be found with the aid of the following three conditions:

$$\bar{r}_1 \cdot \bar{j}_1 = 0$$
$$(\bar{r}_1 \times \bar{j}_1) \cdot \bar{k} = 0 \tag{4}$$
$$\bar{r}_1 \cdot \bar{k} = -\bar{a}_1 \cdot \bar{k}$$

where the unit vector $\bar{j}_1$ represents the normal to the wheel plane and $\bar{k}$ the normal to the road plane.

The angular velocity of the front axle is composed of the velocity r and the relative angular speed with respect to the moving axis system $(C_2, x, y, z)$. Seen with respect to the axis system attached to the front axle we obtain

$$\bar{\omega}_1 = r.\bar{A}^{-1}\,\bar{k} + \bar{\omega}_{1r} \qquad \text{with} \tag{5}$$

$$\bar{A} = \bar{A}^{\phi}\,\bar{A}^{\theta}\,\bar{A}^{\varepsilon}\,\bar{A}^{\delta} \tag{6}$$

The relative angular velocity $\bar{\omega}_{1r}$ is obtained from

$$\begin{pmatrix} 0 & -\omega_{1rz} & \omega_{1ry} \\ \omega_{1rz} & 0 & -\omega_{1rx} \\ -\omega_{1ry} & \omega_{1rz} & 0 \end{pmatrix} = \bar{A}.\dot{\bar{A}}^{-1} \tag{7}$$

The attitude of the front wheel plane with respect to $(C_2, x, y, z)$ is determined by unit vector $\bar{j}_1$

$$\bar{j}_1 = \bar{A}.\bar{j} \tag{8}$$

The camber angle is obtained from

$$\sin \gamma_1 = j_{1z} \tag{9}$$

and steer angle at road level

$$\tan \delta' = -j_{1x}/j_{1y} \tag{10}$$

The location of the front contact centre $C_1$ with respect to $(C_2, x, y, z)$ becomes

$$\bar{c}_1 = \bar{a}_1 + \bar{r}_1 \tag{11}$$

The vector of the slip velocity of a material point $C_1^*$ of the wheel at the instant considered located in $C_1$ reads

$$\dot{\bar{s}}_1 = \begin{pmatrix} u \\ v \\ 0 \end{pmatrix} + \begin{pmatrix} 0 \\ 0 \\ r \end{pmatrix} \times \bar{c}_1 + \dot{\bar{c}}_1 - \Omega_1 r_{e1} \bar{A}^{\delta'} \begin{pmatrix} 1 \\ 0 \\ 0 \end{pmatrix} \tag{12}$$

with $\Omega_1$ denoting the angular speed of rolling of the front wheel and $r_{e1}$ the effective rolling radius. Matrix $\bar{A}^{\delta'}$ accomplishes the transformation of the components of the rolling speed to those directed along the x and y axes. The angular speed of rolling consists of a constant part $\Omega_{10}$ and a variable part $\dot{\chi}_1$

$$\Omega_1 = \Omega_{10} - \dot{\chi}_1 \tag{13}$$

with $\chi_1$ denoting an additional coordinate indicating the variation of the angle of rolling of the wheel with respect to the average value $\Omega_{10} = u/r_{e1}$.
The slip speed vector seen with respect to an axis system rotated over the angle $\delta'$ becomes

$$\dot{\underline{s}}'_1 = \underline{\underline{A}}^{-\delta'} . \dot{\underline{s}}_1 \tag{14}$$

By assuming that the longitudinal wheel slip $\dot{s}'_{1x}$ vanishes, a constraint equation for $\chi_1$ is obtained. This enables us to eliminate $\chi_1$ (and $\chi_2$) from the final set of equations. The slip angle $\alpha_1$ can be obtained from

$$\tan \alpha_1 = - \dot{s}'_{1y} / u_1 \tag{15}$$

with the forward speed of the contact centre $u_1 = \Omega_1 r_{e1}$. The turn slip of the wheel is defined as the yaw rate of the line of intersection of the wheel plane and the road divided by the forward speed of the contact centre (at steady turning: one over the turn radius):

$$\rho_1 = (r + \dot{\delta}')/u_1 \tag{16}$$

The tyre force $F_{z1}$ (normal to the ground and positive downwards) is obtained from the radial tyre force $F_{r1}$ and the lateral tyre force $F_{y1}$

$$F_{z1} \cos \gamma_1 = F_{r1} + F_{y1} \sin \gamma_1 \tag{17}$$

$F_{r1}$ is directly obtainable from the variation in tyre radius and the tyre radial stiffness. $F_{y1}$ follows from the lateral force generating properties of the tyre discussed here after. Space does not permit us to write down the complete equations of motion of the vehicle. For the interested reader we refer to the original report [11] and to [12].

## 4. MODEL OF THE TYRE

The tyre input and output quantities have been shown in Fig. 4. The motion variables $\alpha$, $\gamma$ and $\rho$ and the vertical force $F_z$ are considered here as the input into the tyre block. The force $F_y$ and the moments $M_x$ and $M_z$ are the output quantities. The force $F_x$ and the moment $M_y$ have been disregarded. The moments $M_x$ and $M_z$ will be neglected in the present study.

We distinguish the steady-state component of the lateral force $F_{ys}$ and the variable part $\tilde{F}_y$.

### *Steady-state model*

In Fig. 5 a typical example of a measured side force versus slip-angle plot has been shown. The measurements have been conducted on dry asphalt, at 40 km/h. The approximate linear relationship with slip angle $\alpha$ has been indicated. In the small range of $\alpha$ which usually is not exceeded in practice this relation is acceptable also in quantitative respect for not too large camber angles $\gamma$. Qualitatively, this approximation seems reasonable also for large camber angles.

The following approximate mathematical represention of the steady-

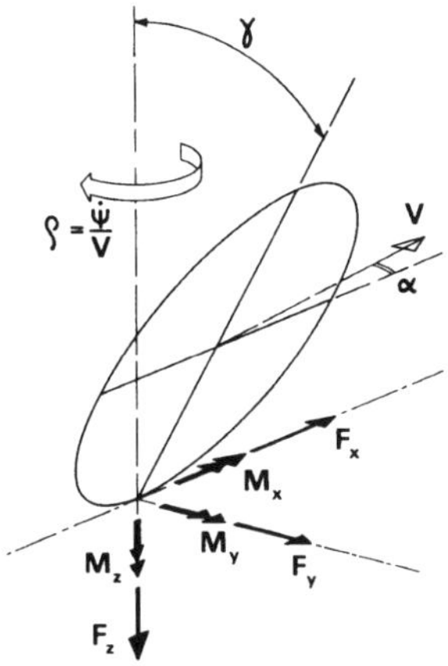


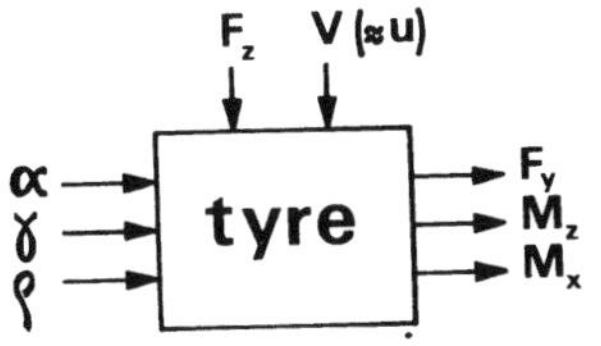


Fig. 4. Tyre input and output quantities.

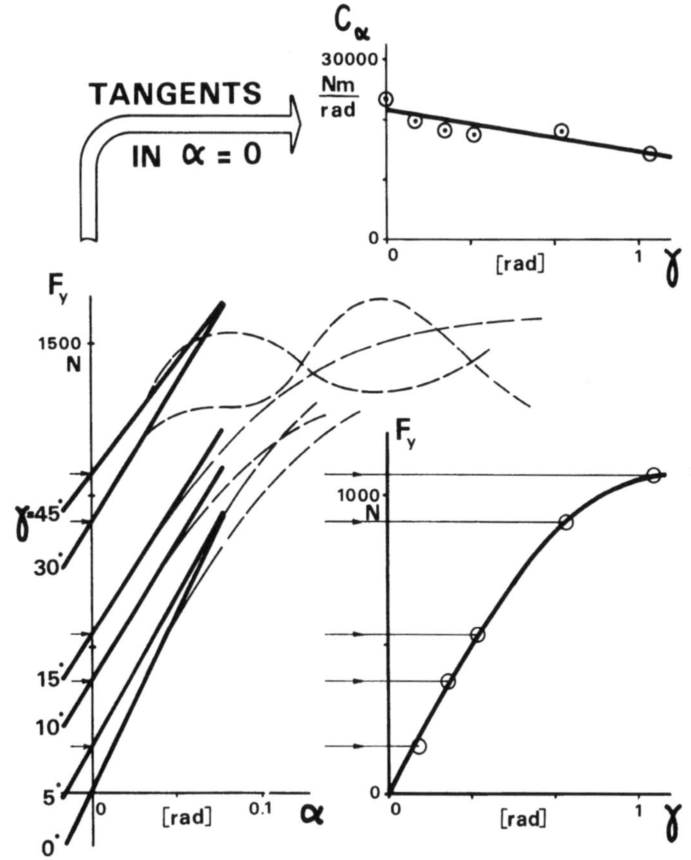


Fig. 5. Measured tyre side force data and model representation of $C_\alpha(\gamma)$ and $F_y(\gamma)_{\alpha=0}$

state response has been used (formally, all variable quantities should be provided with subscript s):

$$F_{ys}(\alpha,\gamma,\rho,F_z) = C_\alpha(\gamma,F_z).\alpha + F_\gamma(\gamma,F_z) - C_\rho(F_z).\rho$$

$$= (1 + A.\Delta F_z)C_{\alpha o}(\gamma).\alpha + (1 + B.\Delta F_z)\{F_{\gamma o}(\gamma) - C_{\rho o}.\rho\} \qquad (18)$$

Subscript o refers to the situation at nominal load $F_{zo}$. The linear approximate relation with the increment $\Delta F_z$ is used in the transient theory of the tyre side force response. In the steady-state representation the indicated load dependency might be used when the load transfer due to aerodynamic drag is taken into account or when relatively small charges in the loading situation are regarded. In Fig. 5 an example is given of $\gamma$ - dependencies derived from measurements on the road. The mathematical approximations employed read:

$$C_{\alpha o}(\gamma) = C_{\alpha oo}(1 - c_{\alpha\gamma}\gamma)$$

$$F_{\gamma o}(\gamma) = C_{\gamma o}\gamma(1 - c_{\gamma\gamma}\gamma) \qquad (19)$$

$$C_{\rho o} = r_e C_{\gamma o}$$

The latter assumption stems from considerations of Ref. [13,p.782] and is in good quantitative agreement with the lateral response theory based on the string-type tyre model. Note that at steady-state conditions $\rho = \rho_s = r_s/u$. Its influence is very small and might be neglected. In the vibratory state, however, the influence of turn slip variations is not negligible. The final function for the steady-state side force generation now reads (with subscript s properly introduced):

$$F_{ys} = (C_{\alpha oo} - \eta_\alpha \Delta F_{zs})(1 - c_{\alpha\gamma}\gamma_s)\alpha_s + (C_{\gamma o} - \eta_\gamma \Delta F_{zs})\{\gamma_s(1 - c_{\gamma\gamma}\gamma_s) - r_e\rho_s\} \qquad (20)$$

Values of the parameters occurring in this equation are listed in the Appendix. It turns out that $C_\gamma$ is approximately proportional to $-F_z$ so-that $\eta_\gamma \approx C_{\gamma o}/(-F_{zo})$.

*Transient state tyre model*

The description of the transient or non-stationary tyre behaviour is based on the string-type tyre model. This holds for all input quantities (cf. [13] p.810,831) except for camber angle $\tilde{\gamma}$. As yet, insufficient data is available to model the response to camber change. In Ref. [14] information is given on the response to a step input of the camber angle. It seems that two different mechanisms play a role in generating the camber thrust. For this limited study it has been assumed that the response time for $\tilde{F}_y$ is equal in magnitude for all input quantities $\tilde{\alpha}$, $\tilde{\rho}$, $\tilde{\gamma}$ and $\tilde{F}_z$. Hence, the following approximate differential equation with relaxation length $\sigma$ is assumed to hold.

$$\frac{\sigma}{u}\dot{\tilde{F}}_y + \tilde{F}_y = \frac{\partial F_{ys}}{\partial \alpha_s}\tilde{\alpha} + \frac{\partial F_{ys}}{\partial \gamma_s}\tilde{\gamma} + \frac{\partial F_{ys}}{\partial \rho_s}\tilde{\rho} + \frac{\partial F_{ys}}{\partial F_{zs}}\tilde{F}_z \qquad (21)$$

The partial derivatives are obtained from Eq.(20). For instance (for $\Delta F_{zs} = 0$):

$$\frac{\partial F_{ys}}{\partial \gamma_s} = - C_{\alpha oo} c_{\alpha\gamma} \alpha_s + C_{\gamma o}(1 - 2c_{\gamma\gamma}\gamma_s) \quad (22)$$

Again we refer to [11] and [12] for the complete expressions.

5. METHODS OF SOLUTION

The steady-state condition is governed by a set of ten non-linear algebraic equations.

$$\bar{f}(\bar{q}_s) = \bar{0} \qquad \text{with} \quad (23)$$

$$\bar{q}_s = (v_s, r_s, M_{\delta s}, \delta_s, F_{y1s}, F_{y2s}, \theta_s, \lambda_{1s}, \lambda_{2s}, \lambda_{t2s})^T \quad (24)$$

For convenience, the roll angle $\phi_s$ is used as a given quantity while instead the steer torque $M_{\delta s}$ is regarded as an unknown quantity. The iteration method of Newton has been employed to find the solution describing the steady-state motion.

The free perturbed motion about the calculated steady-state circular motion is described by a set of 10 linear differential equations with the variables

$$\tilde{q} = (\tilde{v}, \tilde{r}, \tilde{\phi}, \tilde{\delta}, \tilde{F}_{y1}, \tilde{F}_{y2}, \tilde{\theta}, \tilde{\lambda}_1, \tilde{\lambda}_2, \tilde{\lambda}_{t2})^T. \quad (25)$$

The variables $\tilde{v}, \tilde{r}, \tilde{F}_{y1}$ and $\tilde{F}_{y2}$ appear in the equations up to their first order time derivatives. The remaining variables appear up to their second-order derivatives. Hence, the total order of the system amounts to 16. For the determination of eigenvalues and eigenvectors this set of 10 equations is transformed to a set of 16 linear first-order differential equations

$$\bar{A}\dot{\tilde{p}} = - \bar{B}\tilde{p} \quad (26)$$

The matrices $\bar{A}$ and $\bar{B}$ contain elements which depend on the steady-state condition established in the preceding stage of the computing process. The complete equations are given in [11] and [12].

DISCUSSION OF RESULTS

For two different sets of parameter values given in the Appendix the steady-state motion and the free vibrations in curves have been investigated. One of the parameter sets (A) have been experimentally determined in the Vehicle Research Laboratory Delft. The parameters of vehicle B are largely based on data given in the literature.

Figure 6 presents the steady-state relationship of a number of quantities with the centripetal acceleration $a_y$ (expressed in g's). Obviously at higher speeds $\tan\phi$ changes approximately linearly with $a_y$. Furthermore, both the slip angles $\alpha_1$ and $\alpha_2$ don't show a visible variation with speed at a given $a_y$. The rear slip angle is directly related to the lateral speed v of the rear contact centre. The large dif-

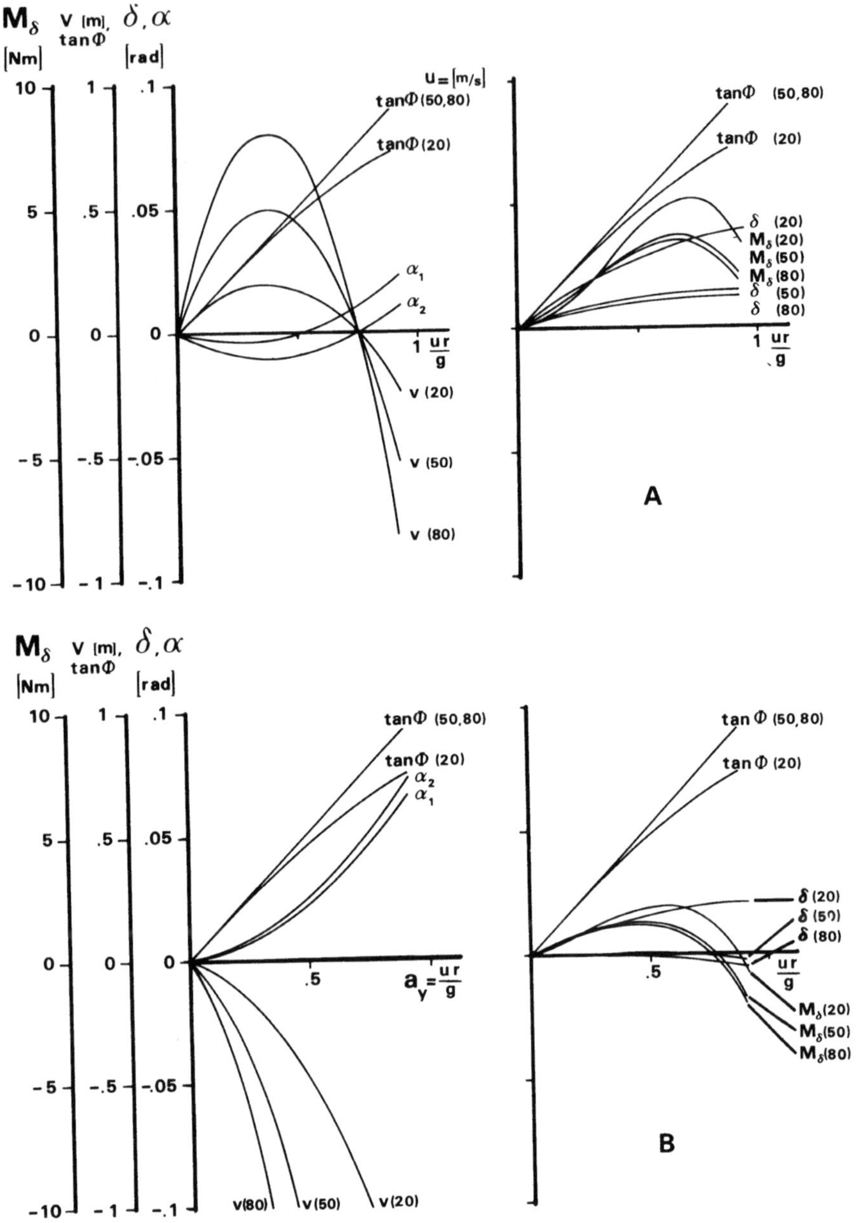


Fig. 6. Steady-state relationships with the centripetal acceleration $a_y$ = ur/g in [g] at different values of the speed u in [m/s].

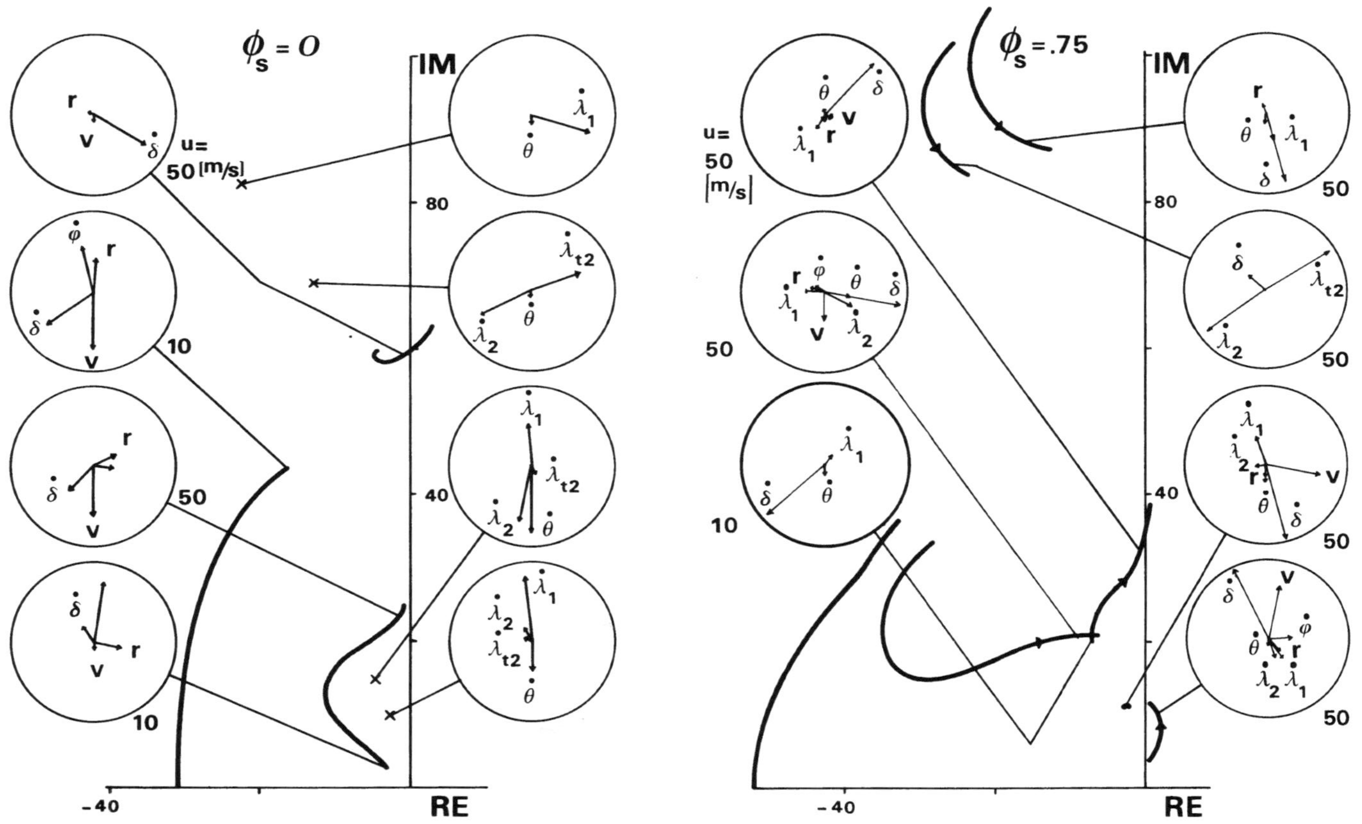


Fig. 8. Mode shapes belonging to plots of Fig. 7a and 7e.

ference between the courses of the slip angles for the two machines is due to the much larger camber stiffnesses of vehicle A.

The vibrational behaviour of the machines is depicted in Figs. 7 to 9 . For each of the five steady-state roll angles $\phi_s$ considered a separate plot of the root-loci belonging to the various modes is given. The root-locus curves indicate the course of the eigenvalues as a function of the speed of travel u that varies along the curves (starting at 10 m/s and ending at 80 m/s). Arrows indicate the direction in which the speed u increases. At the location of the arrows the speed u = 30 m/s. For two typical cases the eigenvectors belonging to certain points of the root locus curves have been given.

For the motorcycle with parameter set A Fig. 7a shows the original root locus plot occurring at straight-line rolling with $\phi_s = 0$. The plots of Figs. 7 b,c,d and e show the changes which occur at increasing $\phi_s$ in the modes originally referred to as pitch, bounce, front and rear wheel hop, weave, front and rear wobble. An interesting interaction of the weave mode and the bounce and the pitch mode turns out to occur. First with the pitch mode (Fig. 7b) and then with the bounce mode as well (Fig. 7c). In the latter figure the three modes seem to blend partly together. At $\phi_s = 0.5$ the weave-bounce mode curve emerges at low speed from the real axis and ends at high speeds near the original bounce eigenvalue. Another branch starts at or near this point and ends near the original pitch root. From or near that point the third branch representing a combined weave pitch mode emerges. At larger roll angles the mode curves appear to separate again. One of the branches crosses the imaginary axis and causes the weave-bounce mode to become unstable. At $\phi_s = 0.75$ rad the mode appears to become unstable already at very low speed where the frequency is a little below 1 Hz.

The figure indicates furthermore that both wheel hop modes do not show any coupling with the other modes. On the contrary, the hop modes and the wobble mode appear to separate. The wobble frequency decreases considerably. For this particular vehicle an interaction with the weave-pitch mode is established. The two remaining real roots one of which appears to become positive beyond a certain speed (capsize) are not shown.

In Figs. 8a and b the eigenvectors have been given for a number of cases in the complex plane. The complex amplitude of the angular velocities (rad/s) and linear velocities (m/s) have been shown. The phasors rotate anti-clockwise.

Figure 9a presents the original root loci for the vehicle with parameter set B. The bounce mode eigenvalue almost coincides with the weave mode eigenvalue at a certain speed of travel. Like in the preceding case, the weave mode interacts with the low-frequency in-plane modes. At $\phi_s = 0.5$ rad a situation arises similar to the one occurring with vehicle A at $\phi_s = 0.75$ rad (Fig. 7e). At further increase of the roll angle a further separation of the curves occurs. The weave-bounce mode becomes more unstable. Also, the original wobble mode (now interacting with pitch and weave) moves to the right giving rise to a wobble type instability also at low speeds.

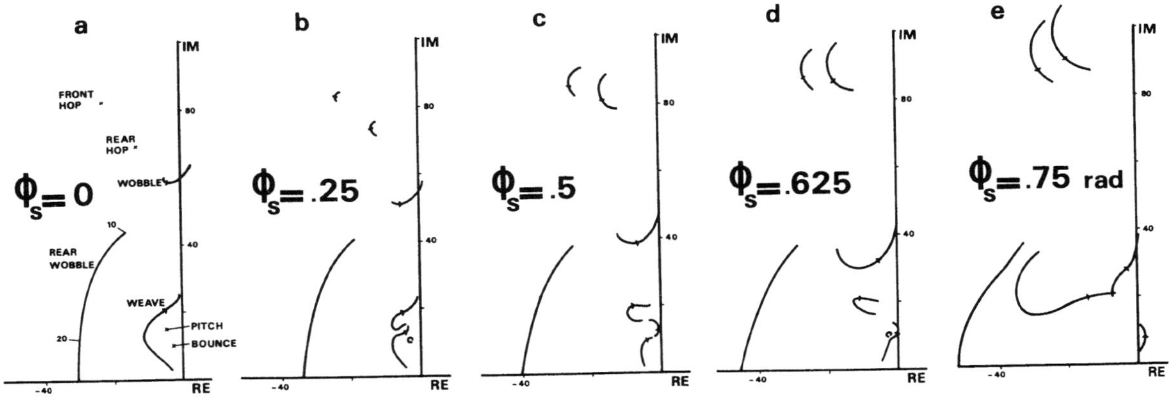


Fig. 7. Root-loci at increasing values of roll angle $\phi$. Speed u varies from 10 to 80 m/s along curves (parameterset A).

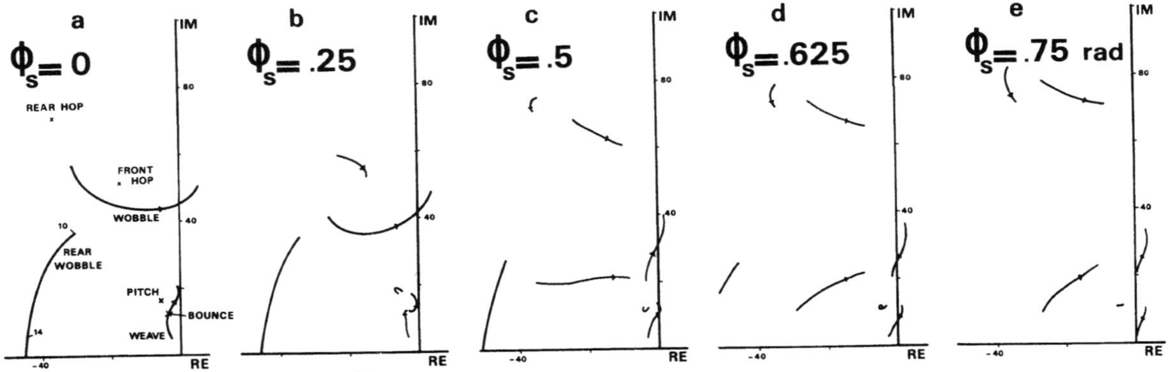


Fig. 9. Root-loci for parameterset B.

## CONCLUSIONS

Based on the present investigation that has been restricted to the examination of only two vehicle configurations including a limited parameter study the following conclusions may be drawn.

In general it has been found that at increasing roll angle the frequency of the wobble mode decreases. The weave and the pitch mode appear to form a combined mode especially in the higher speed range. Interaction with the wobble mode may occur and a considerable decrease in stability of the motion in particular in the lower speed range shows up.

Interaction of the weave mode with the bounce mode may also occur which is followed by a destabilization of the weave-bounce mode. The frequency of this mode is relatively low.

A parameter study where the vertical tyre stiffnesses have been changed revealed that the wheel hop modes do not appear to interact with other modes. Only an increase in wheel hop frequency with increasing roll angle has been observed in most cases.

The reduction in damping of certain modes which occurs at increasing roll angle may be attributable to the coupling of the originally uncoupled modes. No doubt, part of these changes will also be caused by the changes in parameter values due to the roll angle.

## LIST OF SYMBOLS

- $\bar{a}$ position vector (cf. Fig. 3)
- $\bar{A}$ rotation matrix/mass matrix in (26)
- $c_{\alpha\gamma}$ coefficient in tyre model
- $c_{\gamma\gamma}$ coefficient in tyre model
- $C$ spring stiffness
- $C_t$ tyre spring stiffness
- $\bar{d}$ position vector (cf. Fig. 3)
- $F_y$ tyre side force
- $F_z$ tyre normal load (negative)
- $\bar{g}$ position vector (cf. Fig. 3)
- $J$ moment of inertia
- $J_{yw}$ polar moment of inertia of wheel
- $K$ suspension damping coefficient
- $K_\delta$ steer damping coefficient
- $M_x$ overturning couple
- $M_z$ aligning torque
- $M_\delta$ steer torque
- $m$ mass
- $r$ yaw rate, wheel radius
- $r_e$ effective rolling radius
- $\bar{r}$ position vector (cf. Fig. 3)
- $\dot{\bar{s}}$ vector of slip velocity
- $t$ mechanical trail = $r_{1o} \sin\varepsilon - g_{x1o}$
- $u$ longitudinal velocity (cf. Fig. 3)
- $v$ lateral velocity (cf. Fig. 3)
- $\alpha$ sideslip angle
- $\gamma$ camber angle
- $\delta$ steer angle
- $\delta'$ steer angle in the road plane
- $\varepsilon$ rake angle
- $\theta$ pitch angle
- $\phi$ roll angle
- $\dot{\chi}$ variable part of $\Omega$
- $\lambda$ spring compression
- $\lambda_t$ tyre spring compression
- $\rho$ turn slip
- $\sigma$ relaxation length
- $\Omega$ angular speed of rolling of the wheel
- $\Omega_o$ constant part of $\Omega$

<u>Subscripts</u>

- 1 concerning front assembly
- 2 concerning rear assembly
- 3 concerning rider body
- o unsprung part/nominal condition
- s concerning the steady state

<u>Superscripts</u>

- ~ variable part

REFERENCES


1. Hayhoe, G.F., An Introduction to the Dynamics of Single Track Vehicles. Doctoral Dissertation. Cranfield Institute of Technology, 1973.
2. Sharp, R.S. , The Stability and Control of Motorcycles. J. Mech. Engng. Sci., 1971,13 (No. 5), 316-329.
3. Eaton, D.J. , Man-machine Dynamics in the Stabilisation of Single Track Vehicles. Doctoral Dissertation, Highway Safety Research Institute, The University of Michigan, Ann. Arbor, 1973.
4. Weir, D.H. , Motorcycle Handling Dynamics and Rider Control and the Effect of Design Configuration on Response and Performance. Doctoral Dissertation, University of California, Los Angeles, 1972.
5. Singh, D.V. and Goel, V.K., Stability of Single Track Vehicles. IUTAM Symposium on the Dynamics of Vehicles, Delft University of Technology, 1975, 187-196.
6. Sharp, R.S. , The Influence of the Suspension System on Motorcycle Weave-mode Oscillations. Vehicle System Dynamics 5 (1976), 147-154.
7. Roland, R.D., Computer Simulation of Bicycle Dynamics. ASME meeting Mechanics and Sport,Nov. 1973,Detroit 35-83.
8. Jennings, G., A Study of Motorcycle Suspension Damping Characteristics. SAE 740628 (1974), 10 p.
9. Weir, D.H. and Zellner, J.W., Experimental Investigation of the Transient Behavior of Motorcycles. SAE 790266 (1979), 16 p.
10. Pacejka, H.B.,Principles of Plane Motions of Automobiles. IUTAM Symposium on the Dynamics of Vehicles, Delft University of Technology, 1975, 33-59.
11. Koenen, C., The Equations of Motion of a Single Track Vehicle moving along a Circular Path. Veh. Res. Lab. Rep.No. 0134, Delft 1979 (in preparation).
12. Koenen, C. and Pacejka, H.B., Vibrational Modes of Motorcycles in Curves. Int.Motorcycle Safety Conference. May 1980, Washington D.C.
13. Pacejka, H.B.,The Tire as a Vehicle Component. Chap. 7.4. of Mechanics of Pneumatic Tires (Ed. S.K. Clark). Wash.D.C. 1971, N.B.S. Monograph 122.
14. Segel, L. and Wilson, R., Requirements for Describing the Mechanics of Tires used on Single Track Vehicles, IUTAM Symp. on The Dynamics of Vehicles, Delft Univ. of Techn., 1975, 173-186.

APPENDIX

*Values of parametersets A and B.*

| | | A | B |
|---|---|---|---|
| $m_{10}$ | (kg) | 17.5 | 20.0 |
| $m_1$ | | 13.1 | 18.0 |
| $m_{20}$ | | 25.6 | 22.0 |
| $m_2$ | | 173.7 | 228.0 |
| $m_3$ | | 72.5 | 0 |
| $J_{xx1}$ | ($kgm^2$) | 0 | 0 |
| $J_{xx2}$ | | 12.4 | 28.6 |
| $J_{xx3}$ | | 10.0 | 0 |
| $J_{yy1}$ | | 1.2 | 1.3 |
| $J_{yy2}$ | | 27.1 | 29.2 |
| $J_{yy3}$ | | 12.0 | 0 |
| $J_{zz1}$ | | .5 | .67 |
| $J_{zz2}$ | | 17.5 | 16.9 |
| $J_{zz3}$ | | 2.0 | 0 |
| $J_{xz2}$ | | -3.3 | 2.0 |
| $J_{yw1}$ | | .58 | .58 |
| $J_{yw2}$ | | .74 | 1.06 |
| $C_{t1}$ | (kN/m) | 115 | 73 |
| $C_{t2}$ | | 170 | 117 |
| $C_1$ | | 9 | 12.8 |
| $C_2$ | | 25.7 | 35.1 |
| $K_1$ | (Ns/m) | 550 | 782 |
| $K_2$ | | 1100 | 1502 |
| $K_\delta$ | (Nms/rad) | 7.4 | 1.36 |
| $a_2$ | (m) | -.321 | -.305 |
| $d_{x1}$ | | 1.184 | 1.060 |
| $d_{z1}$ | | -.485 | -.636 |
| $d_{x2}$ | | .718 | .527 |
| $d_{z2}$ | | -.171 | -.330 |
| $d_{x3}$ | | .550 | 0 |
| $d_{z3}$ | | -.643 | 0 |
| $g_{x10}$ | | .066 | .054 |
| $g_{x1}$ | | .015 | .056 |
| $g_{z10}$ | | .600 | .710 |
| $r_{10}$ | | .319 | .336 |
| $r_{20}$ | | .321 | .305 |
| $t$ | | .093 | .101 |
| $\varepsilon$ | (rad) | .52 | .48 |
| $C_{\alpha 001}$ | (kN/rad) | 18.500 | 11.710 |
| $C_{\alpha 002}$ | | 25.800 | 18.900 |
| $C_{\gamma 01}$ | | 1.710 | .935 |
| $C_{\gamma 02}$ | | 2.800 | 1.290 |
| $\eta_{\gamma 1}$ | (-) | 1.33 | .806 |
| $\eta_{\gamma 2}$ | | 1.68 | .776 |
| $\eta_{\alpha 1}$ | | 5.4 | 3.81 |
| $\eta_{\alpha 2}$ | | 5.8 | 4.27 |
| $c_{\gamma\gamma 1}$ | (-) | .36 | |
| $c_{\alpha\gamma 1}$ | | .46 | |
| $c_{\gamma\gamma 2}$ | | .50 | |
| $c_{\alpha\gamma 2}$ | | .40 | |
| $\sigma_1$ | (m) | .25 | .092 |
| $\sigma_2$ | | .25 | .152 |

# THE EXACT THEORY OF THE MOTION OF A SINGLE WHEELSET MOVING ON A PERFECTLY STRAIGHT TRACK

A. D. de Pater

Delft University of Technology, Delft, The Netherlands

SUMMARY

Paper deals with an important problem in the field of railway dynamics: the motion of a single wheelset on a perfectly straight and rigid track for any arbitrary profiles of rail and tyre. The exact equations of motion are shown and their derivation is discussed. Moreover, the reduction of the equations in the case that the translational coordinates and the changes of location of the contact points are small with respect to the track gauge and the rotatory coordinates are small with respect to unity. In such a way a consistent first order non-linear theory for the wheelset with slight parasitic movements can be established, from which afterwards a likewise consistent completely linear theory can be derived.

## 1. INTRODUCTION

For a wheelset moving on a purely straight track, the profiles of the cross-sections of the rails and the wheels being arbitrary, it is possible to derive the equations of motion which hold for all values of the coordinates. Such equations are highly non-linear.

In the framework of a symposium paper it is impossible to discuss the derivation of the equations and to enunciate the underlying theory in full detail. We shall restrict ourselves to a global description of the most important results (Sec.2). The complete derivation is exposed in [1].

For practical purposes the equations of motion are too complicated. The motion can best be determined by replacing the equations by simpler ones in such a way that the most important non-linearities are maintained, as we shall show in Sec.3. The reduced equations are suitable as a starting point for deriving a theory for the motion of a complete railway vehicle.

For the sake of completeness we shall also discuss briefly the case of very small displacements, in which the equations of motion become completely linear.

## 2. THE EXACT THEORY

We consider a purely straight and rigid railway track and a single wheelset moving on it. In the central position of the wheelset the system is completely symmetrical with respect to a vertical plane through the centre line of the track. The profiles of the rails are arbitrary and so are the profiles of the tyres.

The system has four degrees of freedom. We indicate the position of the mass centre of the wheelset by the three coordinates $s_i$ (longitudinal), $v_i$ (lateral) and $w_i$ (vertical) and by the three angles $\phi_i$, $\theta_i$, $\psi_i$: $\psi_i$ about a vertical axis, $\theta_i$ about the axis of revolution of the wheelset and $\phi_i$ about a horizontal axis which is perpendicular to the two first-mentioned axes. Because each tyre contacts a rail (normally in one point), there must be two constraint relations between the six coordinates. The best way to derive them is to prescribe $v_i$ and $\psi_i$ and to consider $w_i$ and $\phi_i$ as functions of $v_i$ and $\psi_i$; it is easily seen that they are independent of $s_i$ and $\theta_i$.

In normal railway practice, the coordinates $v_i$ and $w_i$ are always small with respect to the track gauge 2b, whereas $\phi_i$ and $\psi_i$ are small with respect to unity; $s_i$ and $\theta_i$ can have any given values. However, we can show that it is possible to derive the exact equations of motion for any given values of $v_i$, $w_i$, $\phi_i$ and $\psi_i$ as well; afterwards, approximations can be found to any degree wanted. For this purpose we consider $\psi_i$, $\phi_i$ and $\theta_i$ as Euler angles.

Moreover, we introduce (Fig.1 and Fig.2)

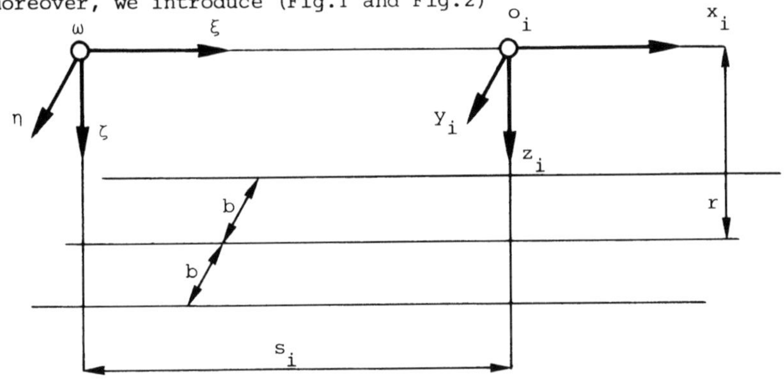


Fig. 1. Geometrical description of the fundamental motion of the wheelset.

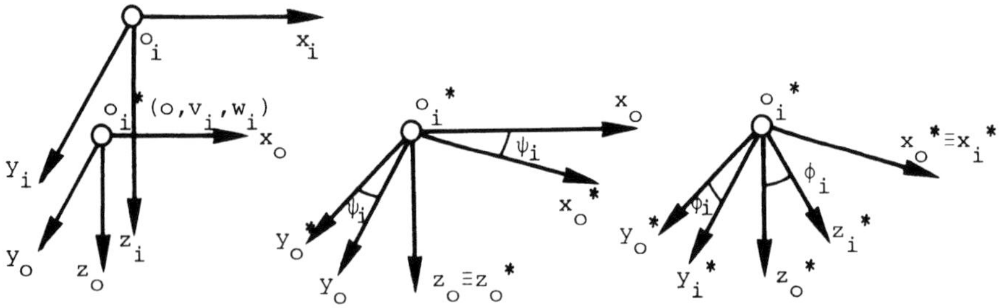


Fig. 2. Geometrical description of the parasitic motion of the wheelset.

two coordinate systems, viz. $(o_i, x_i, y_i, z_i)$ and $(o_i^*, x_i^*, y_i^*, z_i^*)$. The first one moves along the track in a purely translational way which is such that the origin $o_i^*$ always is situated in the coordinate plane $y_i\, o_i\, z_i$. The origin of the second one, $o_i^*$, coincides with the centre of gravity of the wheelset; the axis $o_i^*\, y_i^*$ with its axis of revolution; the axis $o_i^*\, x_i^*$ is horizontal. Thus the matrix which describes the rotation from the second system to the first one, is equal to

$$\bar{G}_i = \begin{pmatrix} \cos\psi_i & -\cos\phi_i \sin\psi_i & \sin\phi_i \sin\psi_i \\ \sin\psi_i & \cos\phi_i \cos\psi_i & -\sin\phi_i \cos\psi_i \\ 0 & \sin\phi_i & \cos\phi_i \end{pmatrix}. \qquad (2\text{-}1)$$

In a contact point we can define (Fig.3) the

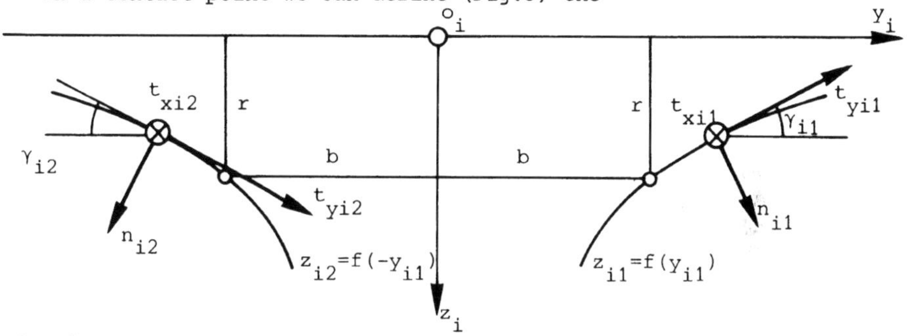


Fig. 3. A cross-section of the track.

normal direction $n_{ij}$ (which is normal to the surfaces of the rail and the tyre), a longitudinal tangential direction $t_{xij}$ and a lateral tangential direction $t_{yij}$. The coordinates $(x_{ij}, y_{ij}, z_{ij})$ with respect to the coordinate system $(o_i, x_i, y_i, z_i)$ and the coordinates $(x_{ij}^*, y_{ij}^*, z_{ij}^*)$ with respect to the system $(o_i^*, x_i^*, y_i^*, z_i^*)$ can be considered as unknowns, and, in addition, the conicity angles $\gamma_{ij}$, $\gamma_{ij}^*$ and the coordinates $w_i$ and $\phi_i$ (note that j is the index for the rail, whereas an asterisk indicates that a quantity is related to the wheelset rather than to the rail). By means of the relations which express that a contact point is situated both on the rail and on the tyre surface and by means of the relations expressing that the normal directions on both surfaces coincide, we find a sufficient number of (rather complicated) equations to calculate the coordinates $w_i$, $\phi_i$, the coordinates of the contact points and the angles $\gamma_{ij}$, $\gamma_{ij}^*$ as functions of $v_i$ and $\psi_i$.

The virtual displacements of the origin $o_i^*$, the virtual rotations of the wheelset (see Fig.4) and the virtual displacement of a tyre in its momentary contact point can be found by varying the coordinates. The components of the last-mentioned displacement can be determined in the normal and the tangential directions: that in the normal direction must be zero. In this way the partial derivatives $\partial w_i/\partial v_i$, $\partial w_i/\partial \psi_i$, $\partial \phi_i/\partial v_i$ and $\partial \phi_i/\partial \psi_i$ can be found.

The translational and angular velocities of the wheelset and, in particular, its velocities in a contact point can easily be determined from the virtual displacements, which allow us, moreover, to find the work coefficients of the normal and the tangential forces in a contact

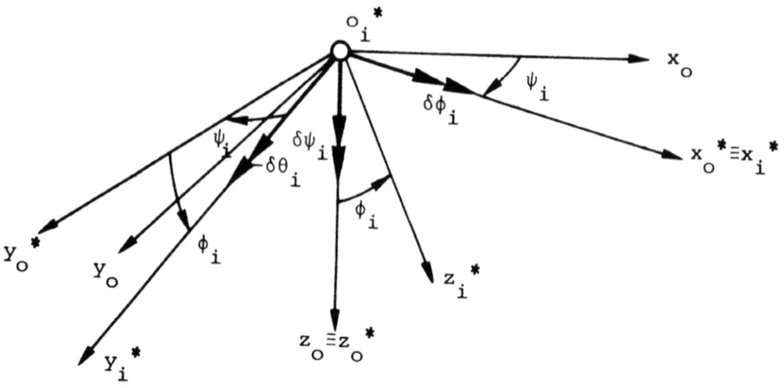


Fig. 4. The translational and rotatory virtual displacements of the wheelset.

point (the normal forces being absent when the constraint relations hold). Besides the contact point forces, a vertical force $G_i$ (gravity together with spring forces) applies on the wheelset in the centre of gravity. We shall give here the expressions for the components of the velocity in a contact point in the directions $n_{ij}$, $t_{xij}$, $t_{yij}$; they read:

$$
\begin{aligned}
W_{nij} &= \pm\dot{v}_i \sin\gamma_{ij} + \dot{w}_i \cos\gamma_{ij} \\
&+ \dot{\phi}_i\{(y_{ij}{}^{*}\cos\phi_i - z_{ij}{}^{*}\sin\phi_i)\cos\gamma_{ij} \mp (y_{ij}{}^{*}\sin\phi_i + z_{ij}{}^{*}\cos\phi_i)\sin\gamma_{ij}\cos\psi_i\} \\
&\pm \dot{\psi}_i(x_{ij}{}^{*}\cos\psi_i - y_{ij}{}^{*}\cos\phi_i\sin\psi_i + z_{ij}{}^{*}\sin\phi_i\sin\psi_i)\sin\gamma_{ij},
\end{aligned}
\tag{2-1}
$$

$$
\begin{aligned}
W_{txij} &= \dot{s}_i + \dot{\phi}_i(y_{ij}{}^{*}\sin\phi_i + z_{ij}{}^{*}\cos\phi_i)\sin\psi_i - \dot{\theta}_i(z_{ij}{}^{*}\cos\psi_i - x_{ij}{}^{*}\sin\phi_i\sin\psi_i) \\
&- \dot{\psi}_i(x_{ij}{}^{*}\sin\psi_i + y_{ij}{}^{*}\cos\phi_i\cos\psi_i - z_{ij}{}^{*}\sin\phi_i\cos\psi_i),
\end{aligned}
\tag{2-2a}
$$

$$
\begin{aligned}
W_{tyij} &= \dot{v}_i\cos\gamma_{ij} \mp \dot{w}_i\sin\gamma_{ij} \\
&- \dot{\phi}_i\{(y_{ij}{}^{*}\sin\phi_i + z_{ij}{}^{*}\cos\phi_i)\cos\gamma_{ij}\cos\psi_i \pm (y_{ij}{}^{*}\cos\phi_i - z_{ij}{}^{*}\sin\phi_i)\sin\gamma_{ij}\} \\
&- \dot{\theta}_i\{x_{ij}{}^{*}(\cos\gamma_{ij}\sin\phi_i\cos\psi_i \pm \sin\gamma_{ij}\cos\phi_i) + z_{ij}{}^{*}\cos\gamma_{ij}\sin\psi_i\} \\
&+ \dot{\psi}_i\{x_{ij}{}^{*}\cos\psi_i - (y_{ij}{}^{*}\cos\phi_i - z_{ij}{}^{*}\sin\phi_i)\sin\psi_i\}\cos\gamma_{ij}
\end{aligned}
\tag{2-2b}
$$

Note that in a combination of signs ($\pm$ or $\mp$) the upper sign refers to the right-hand side (j=1) and the lower sign to the left-hand side (j=2). Because of the constraint relations the various quantities must have such values that the normal velocities $W_{nij}$ are always zero.

The tangential velocities are important for the determination of the tangential forces and so are the components of the angular velocity $\omega_{nij}$ of the wheel around the direction of the normal in the contact point $n_{ij}$:

$$\omega_{nij}=\pm\dot{\phi}_i\sin\gamma_{ij}\sin\psi_i-\dot{\theta}_i(\pm\sin\gamma_{ij}\cos\phi_i\cos\psi_i+\cos\gamma_{ij}\sin\phi_i)+\dot{\psi}_i\cos\gamma_{ij}. \qquad (2\text{-}3)$$

The expression for the potential energy simply reads

$$U=-G_i w_i. \qquad (2\text{-}4)$$

The expression for the kinetic energy can be found from the tranlational and angular velocities of the wheelset:

$$T=\tfrac{1}{2}m_i(\dot{s}_i^2+\dot{v}_i^2+\dot{w}_i^2)+\tfrac{1}{2}J_i(\dot{\phi}_i^2+\dot{\psi}_i^2\cos^2\phi_i)+\tfrac{1}{2}J_{yi}(\dot{\theta}_i-\dot{\psi}_i\sin\phi_i)^2, \qquad (2\text{-}5)$$

where $m_i$ is the mass of the wheelset, $J_i$ its moment of inertia both with respect to the axis $o_i^* x_i^*$ and to the axis $o_i^* z_i^*$, whereas $J_{yi}$ is the moment of inertia with respect to the axis $o_i^* y_i^*$.

Altogether, we are able to find the equations of motion as Lagrange's equations. They read:

$$m_i\ddot{s}_i=Q_{si}, \quad m_i\ddot{v}_i=Q_{vi}, \quad m_i\ddot{w}_i=G_i+Q_{wi}, \qquad (2\text{-}6)$$

$$J_i\ddot{\phi}_i+(J_i-J_{yi})\dot{\psi}_i^2\cos\phi_i\sin\phi_i+J_{yi}\dot{\theta}_i\dot{\psi}_i\cos\phi_i=Q_{\phi i}, \qquad (2\text{-}7a)$$

$$J_{yi}(\ddot{\theta}_i-\ddot{\psi}_i\sin\phi_i-\dot{\phi}_i\dot{\psi}_i\cos\phi_i)=Q_{\theta i}, \qquad (2\text{-}7b)$$

$$-J_{yi}\ddot{\theta}_i\sin\phi_i+(J_i\cos^2\phi_i+J_{yi}\sin^2\phi_i)\ddot{\psi}_i-J_{yi}\dot{\phi}_i\dot{\theta}_i\cos\phi_i$$

$$-2(J_i-J_{yi})\dot{\phi}_i\dot{\psi}_i\cos\phi_i\sin\phi_i=Q_{\psi i}. \qquad (2\text{-}7c)$$

In the relations (2a) and (2c) the gyroscopic terms

$$J_{yi}\dot{\theta}_i\dot{\psi}_i\cos\phi_i \text{ and } -J_{yi}\dot{\theta}_i\dot{\psi}_i\cos\phi_i$$

are conspicuous.

For the work coefficients which appear in the right-hand members of the Eqs. (1)-(2c), we have

$$Q_{si}=-(T_{xi1}+T_{xi2}), \qquad (2\text{-}8a)$$

$$Q_{vi}=-(N_{i1}\sin\gamma_{i1}-N_{i2}\sin\gamma_{i2})-(T_{yi1}\cos\gamma_{i1}+T_{yi2}\cos\gamma_{i2}), \qquad (2\text{-}8b)$$

$$Q_{wi}=-(N_{i1}\cos\gamma_{i1}+N_{i2}\cos\gamma_{i2})+(T_{yi1}\sin\gamma_{i1}-T_{yi2}\sin\gamma_{i2}), \qquad (2\text{-}8c)$$

$$Q_{\phi i}=-(N_{i1}y_{i1}^*\cos\gamma_{i1}+N_{i2}y_{i2}^*\cos\gamma_{i2})\cos\phi_i$$

$$+(N_{i1}z_{i1}^*\cos\gamma_{i1}+N_{i2}z_{i2}^*\cos\gamma_{i2})\sin\phi_i$$

$$+(N_{i1}y_{i1}^*\sin\gamma_{i1}-N_{i2}y_{i2}^*\sin\gamma_{i2})\sin\phi_i\cos\psi_i$$

$$+(N_{i1}z_{i1}^*\sin\gamma_{i1}-N_{i2}z_{i2}^*\sin\gamma_{i2})\cos\phi_i\cos\psi_i$$

$$-(T_{xi1}y_{i1}^*+T_{xi2}y_{i2}^*)\sin\phi_i\sin\psi_i$$

$$-(T_{xi1}z_{i1}^*+T_{xi2}z_{i2}^*)\cos\phi_i\sin\psi_i$$

$$+(T_{yi1}y_{i1}^{*}\cos\gamma_{i1}+T_{yi2}y_{i2}^{*}\cos\gamma_{i2})\sin\phi_i\cos\psi_i$$

$$+(T_{yi1}z_{i1}^{*}\cos\gamma_{i1}+T_{yi2}z_{i2}^{*}\cos\gamma_{i2})\cos\phi_i\cos\psi_i$$

$$+(T_{yi1}y_{i1}^{*}\sin\gamma_{i1}-T_{yi2}y_{i2}^{*}\sin\gamma_{i2})\cos\phi_i$$

$$-(T_{yi1}z_{i1}^{*}\sin\gamma_{i1}-T_{yi2}z_{i2}^{*}\sin\gamma_{i2})\sin\phi_i, \tag{2-9a}$$

$$Q_{\theta i}=(T_{xi1}z_{i1}^{*}+T_{xi2}z_{i2}^{*})\cos\psi_i$$

$$-(T_{xi1}x_{i1}^{*}+T_{xi2}x_{i2}^{*})\sin\phi_i\sin\psi_i$$

$$+(T_{yi1}x_{i1}^{*}\cos\gamma_{i1}+T_{yi2}x_{i2}^{*}\cos\gamma_{i2})\sin\phi_i\cos\psi_i$$

$$+(T_{yi1}x_{i1}^{*}\sin\gamma_{i1}-T_{yi2}x_{i2}^{*}\sin\gamma_{i2})\cos\phi_i$$

$$+(T_{yi1}z_{i1}^{*}\cos\gamma_{i1}+T_{yi2}z_{i2}^{*}\cos\gamma_{i2})\sin\psi_i, \tag{2-9b}$$

$$Q_{\psi i}=-(N_{i1}x_{i1}^{*}\sin\gamma_{i1}-N_{i1}x_{i2}^{*}\sin\gamma_{i2})\cos\psi_i$$

$$+(N_{i1}y_{i1}^{*}\sin\gamma_{i1}-N_{i2}y_{i2}^{*}\sin\gamma_{i2})\cos\phi_i\sin\psi_i$$

$$-(N_{i1}z_{i1}^{*}\sin\gamma_{i1}-N_{i2}z_{i2}^{*}\sin\gamma_{i2})\sin\phi_i\sin\psi_i$$

$$+(T_{xi1}x_{i1}^{*}+T_{xi2}x_{i2}^{*})\sin\psi_i$$

$$+(T_{xi1}y_{i1}^{*}+T_{xi2}y_{i2}^{*})\cos\phi_i\cos\psi_i$$

$$-(T_{xi1}z_{i1}^{*}+T_{xi2}z_{i2}^{*})\sin\phi_i\cos\psi_i$$

$$-(T_{yi1}x_{i1}^{*}\cos\gamma_{i1}+T_{yi2}x_{i2}^{*}\cos\gamma_{i2})\cos\psi_i$$

$$+(T_{yi1}y_{i1}^{*}\cos\gamma_{i1}+T_{yi2}y_{i2}^{*}\cos\gamma_{i2})\cos\phi_i\sin\psi_i$$

$$-(T_{yi1}z_{i1}^{*}\cos\gamma_{i1}+T_{yi2}z_{i2}^{*}\cos\gamma_{i2})\sin\phi_i\sin\psi_i. \tag{2-9c}$$

Multiplying the first Eq. (6) with $\dot{s}_i$, the second one with $\dot{v}_i$, etc., and adding the results, we obtain the energy balance

$$\frac{d}{dt}(T+U)=P \tag{2-10}$$

with

$$P=-\sum_{j=1}^{2}(N_{ij}W_{nij}+T_{xij}W_{txij}+T_{yij}W_{tyij}), \tag{2-11}$$

which can be verified by means of the relations (4), (5), (1), (2a) and (2b).

The equations of motion (6)-(7c) can also be found directly by applying Newton's rules on the motion of the mass centre of the wheelset and by writing down Euler's equations for the wheelset. The results agree completely with the six equations (6)-(7c).

Besides these equations we dispose on the two constraint relations for the quantities $w_i$ and $\phi_i$ and on four relations between the tangential forces $T_{xij}$, $T_{yij}$, the normal forces $N_{ij}$ and the creep and spin quantities

$$\upsilon_{xij}=W_{txij}/V, \quad \upsilon_{yij}=W_{tyij}/V, \quad \phi_{ij}=\omega_{nij}/V. \tag{2-12}$$

These twelve equations determine completely the twelve unknowns of the problem, viz. the coordinates $s_i$, $v_i$, $w_i$, $\phi_i$, $\theta_i$, $\psi_i$; $N_{ij}$, $T_{xij}$, $T_{yij}$.

3. THE FIRST ORDER THEORY

The equations of motion, shown in Sec. 2, are rather complicated and it is desirable to simplify them. This can be done in a systematical way by replacing the coordinates $x_{ij}$, ... $z_{ij}^*$ by

$$\left.\begin{aligned} &\xi_{ij}=x_{ij}, \quad \eta_{ij}=b\mp y_{ij}, \quad \zeta_{ij}=z_{ij}-r, \\ &\xi_{ij}^*=x_{ij}^*, \quad \eta_{ij}^*=b\mp y_{ij}^*, \quad \zeta_{ij}^*=z_{ij}^*-r \end{aligned}\right\} \tag{3-1}$$

and by assuming

$$\left.\begin{aligned} &|v_i| \ll b, \quad |w_i| \ll b, \quad |\phi_i| \ll 1, \quad |\psi_1| \ll 1, \\ &|\xi_{ij}| \ll b, \quad |\eta_{ij}| \ll b, \quad |\zeta_{ij}| \ll b, \\ &|\xi_{ij}^*| \ll b, \quad |\eta_{ij}^*| \ll b, \quad |\zeta_{ij}^*| \ll b. \end{aligned}\right\} \tag{3-2}$$

Here b is the half-distance between the contact points and r the wheel radius, both in the central position of the wheelset.

The simplified equations are not given here; they even yet are essentially non-linear because of the conicity angles $\gamma_{ij}$, $\gamma_{ij}^*$, which still can have any given values, and because the creep and spin quantities often are so large that the relations for the tangential forces have to be considered as non-linear.

In normal railway practice a wheelset is incorporated in a vehicle frame. We consider the particular case of a frame which translates with a constant speed $\dot{s}_i=V$ in a direction, parallel to the track, linear springs connecting the wheelset with the frame. We introduce the displacement $u_i$ of the wheelset with respect to the frame. Moreover, we replace the angle $\theta_i$ by the angle $\chi_i$:

$$\theta_i=s_i/r-\chi_i; \tag{3-3}$$

then in normal cases the angular velocity $|\dot{\chi}_i|$ will be small as compared with V/r.

We now again obtain six equations of motion. They contain not only the tangential but also the normal forces. The latter ones can be eliminated by making use of the constraint relations; then we find four equations of motion for $u_i$, $v_i$, $\chi_i$ and $\psi_i$, containing only the tangential forces:

$$m_i\ddot{u}_i+c_{xi}u_i=-(T_{xi1}+T_{xi2}), \tag{3-4a}$$

$$J_{yi}\ddot{\chi}_i=-(T_{xi1}+T_{xi2})r, \tag{3-4b}$$

$$m_i(1+w_i'^2)\ddot{v}_i+J_i\phi_i'^2\ddot{v}_i+J_{yi}V\phi_i'\dot{\psi}_i/r+c_{yi}v_i+c_{zi}w_i'w_i+c_{zi}b_i^2\phi_i'\phi_i-G_iw_i'$$

$$=-\frac{2b}{2b-r(tg\gamma_{i1}+tg\gamma_{i2})}\left(\frac{T_{yi1}}{\cos\gamma_{i1}}+\frac{T_{yi2}}{\cos\gamma_{i2}}\right), \tag{3-5a}$$

$$J_i\ddot{\psi}_i-J_{yi}V\phi_i'\dot{v}_i/r+c_{xi}b_i^2\psi_i=(T_{xi1}-T_{xi2})b. \tag{3-5b}$$

Here we have put

$$w_i'=\partial w_i/\partial v_i, \quad \phi_i'=\partial\phi_i/\partial v_i, \tag{3-6}$$

$w_i$ and $\phi_i$ now being independent from $\psi_i$, whereas $c_{xi}$, $c_{yi}$ and $c_{zi}$ represent the spring rigidities.

The expressions for the potential and the kinetic energies now read

$$U^*=-G_iw_i+\tfrac{1}{2}c_{xi}(u_i^2+b_i^2\psi_i^2)+\tfrac{1}{2}c_{yi}v_i^2+\tfrac{1}{2}c_{zi}(w_i^2+b_i^2\phi_i^2), \tag{3-7}$$

$$T^*=\tfrac{1}{2}m_i(V+\dot{u}_i)^2+\tfrac{1}{2}m_i(1+w_i'^2)\dot{v}_i^2+\tfrac{1}{2}J_i(\phi_i'^2\dot{v}_i^2+\dot{\psi}_i^2)$$

$$+\tfrac{1}{2}J_{yi}\{(V/r-\dot{\chi}_i)^2-2V\dot{\psi}_i\phi_i/r\}. \tag{3-8}$$

For the energy balance we find

$$\frac{d}{dt}(T^*+U^*)=P_{ex}^*+P_{in}^*, \tag{3-9}$$

with

$$P_{ex}^*=\dot{L}_i^*V, \quad P_{in}^*=-\sum_{j=1}^{2}(T_{xij}W_{txij}+T_{yij}W_{tyij}), \tag{3-10}$$

$$L_i^*=m_i\ddot{u}_i-J_{yi}\ddot{\chi}_i/r-J_{yi}(\dot{\psi}_i\phi_i+\dot{\phi}_i\dot{\psi}_i)/r-(T_{xi1}\zeta_{i1}^*+T_{xi2}\zeta_{i2}^*)/r$$

$$-(T_{yi1}/\cos\gamma_{i1}+T_{yi2}/\cos\gamma_{i2})\psi_i, \tag{3-11}$$

$$W_{txij}=\dot{u}_i+r\dot{\chi}_i\mp b\dot{\psi}_i-V\zeta_{ij}^*/r, \tag{3-12a}$$

$$W_{tyij}=\left\{\frac{2b\dot{v}_i}{2b-r(tg\gamma_{i1}+tg\gamma_{i2})}-V\psi_i\right\}\cos^{-1}\gamma_{ij}, \tag{3-12b}$$

$L_i^*$ being the tractive effort, applying on the frame, which maintains the velocity V constant. The asterisk indicates that the constraint relations for $w_i$ and $\phi_i$ hold.

For the sake of completeness we give the expressions for the tangential forces in the linear case (for vanishing creep and speed):

$$T_{xij}=K_{xij}\upsilon_{xij}, \quad T_{yij}=K_{yij}\upsilon_{yij}+K_{zij}\phi_{ij}. \tag{3-13}$$

The coefficients of proportionality are equal to

$$K_{xij}=Gc^2C_{11},\quad K_{yij}=Gc^2C_{22},\quad K_{zij}=Gc^3C_{23}, \tag{3-14}$$

G being the shear modulus, $C_{11}$, $C_{22}$, $C_{23}$ three functions of the axis ratio a/b and Poisson's ratio $\nu$, given by [2 p. 90], and c the square root of the product of the contact ellipse axes:

$$c=\sqrt{ab}. \tag{3-15}$$

For the creep and spin quantities the relations (2-12) hold again, with the velocities (12a-b) and the angular velocity

$$\omega_{nij}=\pm\dot{\chi}_i\sin\gamma_{ij}+\dot{\psi}_i\cos\gamma_{ij}\mp Vr^{-1}\sin\gamma_{ij}-V\phi_i r^{-1}\cos\gamma_{ij} \tag{3-16}$$

The quantities a, b and c can be calculated by means of Hertz' theory, once the normal force $N_{ij}$ and the radii of curvature in the contact point are known. For $N_{ij}$ we need the equations of motion for $w_i$ and $\phi_i$ which hold when the constraint relations are left out of consideration:

$$m_i\ddot{w}_i+c_{zi}w_i=G_i-(N_{i1}\cos\gamma_{i1}+N_{i2}\cos\gamma_{i2})+(T_{yi1}\sin\gamma_{i1}+T_{yi2}\sin\gamma_{i2}), \tag{3-17a}$$

$$J_i\ddot{\phi}_i+J_{yi}V\dot{\psi}_i/r+c_{zi}b_i^2\phi_i=-(N_{i1}\cos\gamma_{i1}-N_{i2}\cos\gamma_{i2})b+(N_{i1}\sin\gamma_{i1}+N_{i2}\sin\gamma_{i2})r$$
$$+(T_{yi1}\cos\gamma_{i1}+T_{yi2}\cos\gamma_{i2})r+(T_{yi1}\sin\gamma_{i1}+T_{yi2}\sin\gamma_{i2})b. \tag{3-17b}$$

The mechanical meaning of the first order theory is that in deriving the equations of motion by Newton's rules one assumes that the forces apply on the wheelset when it is in the central position ($v_i=w_i=\phi_i=\psi_i=0$). Thus, in the equations of this theory no products of forces and displacements appear, with the exception of equation (11) for the tractive effort, in which it is essential that also second order terms are taken into account.

The equations of motion of the first order theory can be solved by means of harmonic balance methods. This has been done by author: the analytical results are shown in [3]; numerical results (which show that under the critical speed, predicted by the linear theory, a limit cycle often still is possible) will be published soon; moreover, at the moment the results are compared with those obtained by a direct numerical integration of the equations of motion. There is every appearance that when the non-linearity ensues from the rail and tyre profiles, the investigation of the non-linear behaviour of a complete vehicle by means of harmonic balance methods is not very much more complicated than that of the linear behaviour.

For solving the equations of the first order theory it is useful to introduce reduced (non-dimensional) parameters and coordinates. We shall not enter into the details of this part of the calculations.

## 4. SECOND ORDER THEORY

From the exact theory also a second order theory can be derived by retaining the terms with the squares of the coordinates as well. The mechanical meaning of this method is that in deriving the equations of motion by means of Newton's rules one takes into account products of the forces and the coordinates $v_i$, $w_i$, $\phi_i$, $\psi_i$. The merits of the first order theory can be judged by comparing its results with those of the (more exact) second order theory; it will be interesting to perform such an investigation.

## 5. THE COMPLETELY LINEAR CASE

The equations of the first order theory can be linearised completely for the case in which the variation of the conicity remains very small. In this case the equations of motion split up into two separate sets of equations, one for the symmetrical motion (which always is stable) and one for the lateral motion (which often is unstable).

The method of replacing the parameters and coordinates by reduced ones, mentioned in Sec.3, can also be applied in the completely linear case. But in this case a second reduction method mostly is more advantageous because it gives a smaller number of reduced parameters.

## 6. CONCLUDING REMARKS

For the more complicated case of a circular track (a railway curve with a constant radius) it is also possible to draw up an exact theory. On the contrary, when the track (straight or curved) present small irregularities, it is appropriate to start with a first order theory from the beginning.

The purpose of the present paper is only to supply building-stones for forthcoming investigations. As such we may mention:

a. the shape of the track: straight, curved with constant radius and cant; curved with varying radius and cant; straight or curved with small irregularities; deformable;
b. the construction of the vehicle: one single wheelset; a single frame with various wheelsets; a vehicle with a main frame, bogie frames and various wheelsets;
c. the nature of the equations: completely linear; first order; second order; completely exact;
d. the nature of the solution: by linear means, by harmonic balance methods; by numerical integration.

With regard to the last category we observe that in using harmonic balance methods various assumptions are possible. The simplest way is to suppose that the linear creep and spin laws (3-13) hold and that the normal forces $N_{ij}$ are constant. The first refinement is to take into account the variations of $N_{ij}$ and the variations of the dimension of each contact ellipse. Next, one may also take into account the variations of the shape of the contact ellipse. Most complicated is to investigate a model in which the non-linearities of the creep and spin laws are taken into account completely.

At last mention should be made of the very thorough study performed during the last years by Aldington, who investigated the behaviour of a wheelset running through a curve with constant radius, starting from the exact equations of motion. In this investigation Aldington assumed that the surface of the inner rail is not purely toroidal but presents a lateral deviation which can be represented by a harmonic function of the position coordinate ($s_i$). Aldington used the exact theory wherever possible and succeeded to integrate the equations of motion numerically. In order to obtain an as realistic as possible model Aldington also takes into account the motion of that half of the bogie frame which is placed over the wheelset, and also the motion of a quarter part of the vehicle body (main frame). When the proof of the present paper was being corrected, author became aware of the title of Aldington's publication [4].

REFERENCES


1. Pater, A.D. de, The exact theory of the motion of a single wheelset moving on a purely straight track, Delft University of Technology, Department of Mechanical Engineering, Laboratory of Engineering Mechanics Report No. 648, to be published.
2. Kalker, J.J., On the rolling contact of two elastic bodies in the presence of dry friction, Thesis Delft (1967), 7 + 155 pp.
3. Pater, A.D. de, A nonlinear model of a single wheelset moving with constant speed along a purely straight track, Materiały na II konferencję naukową "Optymalizacja środków technicznych organizacyjnych i ekonomicznych w transporcie", III, Techniczne środki transportu, Warszawa (1978) p. 170-179.
4. Aldington, T., The stability of a single wheelset on existing track-an exact quantitative analysis, to be published.

# COUPLING AND ORDER-REDUCTION OF COMBINED LINEAR STATE-SPACE AND MULTIBODY SYSTEMS WITH APPLICATION TO THE DYNAMICS OF A WHEEL-RAIL VEHICLE

R. Richter and A. Jaschinski

DFVLR, Oberpaffenhofen, The Netherlands

SUMMARY

The dynamic modelling of vehicles usually deals with complex systems. Therefore, in the modelling process the linear vehicle system often is split into a main system and subsystems, which are modelled separately.

After transforming the subsystems assumed to be linear state-space systems into a modal form they are coupled with the main system assumed to be a linear multibody system. As a result one obtains overall system equations, which can be of very high order. In the case of complex subsystems the reduction of the total system order can be achieved by reducing the subsystem order. A procedure based on eigenmode-discarding of the subsystems is proposed.

To treat large order systems a computer program for the coupling and order reduction procedure was developed.

The method is applied to a wheel-rail vehicle with a 15 degree-of-freedom wheelset model. The objective is to approximate the wheelset subsystem by a drastically reduced-order model and still represent the stability of the hunting modes accurately.

## 1. INTRODUCTION

The dynamic modelling of railway vehicles often leads to linear differential equations of high order. For these systems there is a need for order-reduction to save computer time,avoid numerical difficulties and simplify design considerations.

If the linear vehicle system is partitioned into a main system (e. g. bogie frame) and subsystems (e. g. wheelsets) a procedure for coupling the main system with the subsystems is presented. The equations of the main system are generated by a multibody formalism whereas the subsystems are described by state-space equations, e. g. first-order differential equations. After transforming these equations into a modal form the subsystems are coupled with the main system. This modal representation allows prior reduction of the order of the subsystems by systematically discarding eigenmodes, with a consequent reduction of the overall system order.

The order-reduction is applied to the wheelset model described in [1], [2] having 15 degrees of freedom. The corresponding bogie model with two wheelsets is described by 74 first-order differential equations, the complete vehicle model with two bogies leads to 168 equations. A schematic of the wheelset model within the bogie is shown in figure 1.

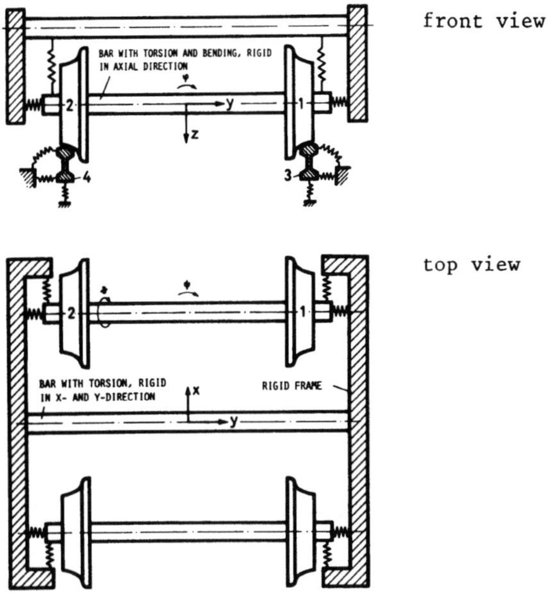


Fig. 1: Schematic of wheelset and bogie model

The objective is to obtain a reduced-order model for the stability of the hunting modes.

In this paper, after deriving the system equations the order-reduction method is discussed and applied to the bogie model mentioned above.

## 2. SYSTEM EQUATION SYNTHESIS

In the following the equations for coupling the main system with the subsystems are presented. The resulting total system equations are the basis for the order reduction. In deriving the equations of motion for a complex system it may be useful to partition it into a main system and subsystems, which have to be coupled.

For each system the equations of motion may be set up independently as follows:

- second order differential equations generated by a multibody algorithm describe the main system

- first order differential equations describe the subsystems behavior, which may be modelled separately

The system synthesis is performed by introducing the coupling forces between the main system and the subsystems.

With respect to our present application the assumptions for the coupling elements are considered as follows:

- they consist of linear springs and dampers in parallel
- the displacements of the attachment points of the subsystems are given as a function of the state-vector of these systems.

The equations of motion of the multibody system as generated by the formalism described in [3] are

$$M\ddot{\underline{p}} + D\dot{\underline{p}} + K\underline{p} = R\underline{f} \tag{1}$$

where M, D, K, $\underline{p}$ are mass-, damping-, stiffness-matrices and the coordinate-vector, respectively. $\underline{f}$ is the vector of the coupling forces between the main system and the subsystems and matrix R distributes the forces to the appropriate coordinates.

For reasons to be discussed later, we need two concepts for subsystem models:

- a "free" subsystem without connection to any other system
- a "constrained" subsystem with a connection to an inertial frame.

For a "free" or "constrained" subsystem the first-order differential equations have the form

$$\dot{\underline{x}} = A\underline{x} + B\underline{f} \tag{2}$$

where A and B are the system matrices. The state-vector $\underline{x}$ consists of the position- and velocity coordinates $\underline{q}$, $\dot{\underline{q}}$:

$$\underline{x} = \begin{bmatrix} \underline{q} \\ \dot{\underline{q}} \end{bmatrix} \tag{3}$$

For calculating the forces $\underline{f}$ we have to distinguish between a "free" and a "constrained" subsystem.

Under the assumptions mentioned above the forces $\underline{f}$ for a "free" subsystem are a function of the multibody state-vector components $\underline{p}$, $\dot{\underline{p}}$ and the subsystem state vector $\underline{x}$:

$$\underline{f} = \underline{f}^{o}\ (\underline{p}, \dot{\underline{p}}) + \underline{f}^{1}\ (\underline{x}) \tag{4}$$

For a "constrained" subsystem the contribution $\underline{f}^{1}$ $(\underline{x})$ is already incorporated in the matrix A and therefore we have

$$\underline{f} = \underline{f}^{o}\ (\underline{p}, \dot{\underline{p}}) \tag{5}$$

By transforming eq. (2) into a modal form using the transformation with the eigenvector matrix X

$$\underline{x} = X\ \underline{y} \tag{6}$$

one obtains

$$\underline{\dot{y}} = \Lambda\, \underline{y} + Q\, \underline{f} \tag{7}$$

where $\Lambda$ and Q are given by

$$\Lambda = X^{-1} A\, X, \quad Q = X^{-1} B \tag{8}$$

If the coupling elements between the main system and the subsystems are consisting of linear springs and dampers in parallel one can find an expression for the coupling forces $\underline{f}$ by calculating the functions $\underline{f}^o\,(\underline{p},\, \underline{\dot{p}})$ and $\underline{f}^1(\underline{x})$ corresponding to eq. (4). In this case the coupling forces depend on the relative displacements $\underline{s}$ and the relative velocities $\underline{\dot{s}}$ of the attachment points:

$$\underline{f} = \mathrm{diag}\,[k_i]\,\underline{s} + \mathrm{diag}\,[d_i]\,\underline{\dot{s}} \tag{9}$$

$k_i$ are the spring constants, $d_i$ the damper constants. Introducing the coordinates we get [3]

$$\begin{aligned} \underline{s} &= \overline{S}_j\, \underline{p} + S_j\, \underline{q}_j \\ \underline{\dot{s}} &= \overline{S}_j\, \underline{\dot{p}} + S_j\, \underline{\dot{q}}_j \end{aligned} \tag{10}$$

This relation holds for the coupling of each subsystem j with the main system. The matrix $\overline{S}_j$ is constructed by the multibody algorithm, the matrices $S_j$ describing the displacements of the attachment points of the subsystems are given on assumption.

Expressing the physical coordinates $\underline{q}_j$, $\underline{\dot{q}}_j$ by the modal coordinates $\underline{y}_j$ we get

$$\begin{aligned} S_j\, \underline{q}_j &= [S_j \mid 0] \begin{bmatrix} \underline{q}_j \\ \underline{\dot{q}}_j \end{bmatrix} = [S_j \mid 0]\ \underline{x}_j \\ S_j\, \underline{\dot{q}}_j &= [0 \mid S_j] \begin{bmatrix} \underline{q}_j \\ \underline{\dot{q}}_j \end{bmatrix} = [0 \mid S_j]\ \underline{x}_j \end{aligned} \tag{11}$$

$$\begin{bmatrix} S_j\, \underline{q}_j \\ S_j\, \underline{\dot{q}}_j \end{bmatrix} = \begin{bmatrix} S_j & 0 \\ 0 & S_j \end{bmatrix} \underline{x}_j = \begin{bmatrix} S_j & 0 \\ 0 & S_j \end{bmatrix} \begin{bmatrix} X^j_{11} & X^j_{12} \\ X^j_{21} & X^j_{22} \end{bmatrix} \underline{y}_j \tag{12}$$

The eigenvector-matrix $X^j$ has been partitioned here. Using equations (10), (12) we find

$$\underline{f}^j = \mathrm{diag}\,[k_i]\,(\overline{S}_j\, \underline{p} + T^j_1\, \underline{y}_j) + \mathrm{diag}\,[d_i]\,(\overline{S}_j\, \underline{\dot{p}} + T^j_2\, \underline{y}_j) \quad \text{with} \tag{13}$$

$$\begin{aligned} T^j_1 &= S_j\, [X^j_{11} \mid X^j_{12}] \\ T^j_2 &= S_j\, [X^j_{21} \mid X^j_{22}] \end{aligned} \tag{14}$$

So the functions $\underline{f}^o(\underline{p}, \dot{\underline{p}})$ and $\underline{f}^1(\underline{x})$ of eq. (4) are given by

$$\begin{aligned} \underline{f}_j^o(\underline{p}, \dot{\underline{p}}) &= \text{diag}\,[k_i]\,\overline{S}_j\,\underline{p} + \text{diag}\,[d_i]\,\overline{S}_j\,\dot{\underline{p}} \\ \underline{f}_j^1(\underline{y}) &= \text{diag}\,[k_i]T_1^j\,\underline{y}_j + \text{diag}\,[d_i]T_2^j\,\underline{y}_j \end{aligned} \tag{15}$$

where $\underline{x}_j$ is replaced by $X^j \cdot \underline{y}_j$.

Let the main system be connected with m subsystems. The total vector $\underline{f}$ of all coupling forces of eq. (1) then is

$$\underline{f} = [\underline{f}_1^T\ \underline{f}_2^T \ldots \underline{f}_m^T]^T \tag{16}$$

Consequently, the matrix R of eq. (1) can be partitioned

$$R = [R_1 | \ldots | R_j | \ldots | R_m] \tag{17}$$

i. e. a submatrix $R_j$ distributes the forces $\underline{f}_j$ of subsystem j on the coordinates of the main system. Using equations (1), (7), (13), (16), (17) the equation synthesis results in the following set of differential equations:

$$\begin{aligned} M\,\ddot{\underline{p}} + D^*\dot{\underline{p}} + K^*\underline{p} &= \sum_{j=1}^{m} U_j\,\underline{y}_j \\ \dot{\underline{y}}_j &= \Lambda_j^*\,\underline{y}_j + V_j\,\underline{p} + W_j\,\dot{\underline{p}} \quad (j = 1, 2, \ldots m) \end{aligned} \tag{18}$$

where the new matrices are given by:

$$\begin{aligned} D^* &= D && \text{for "free" subsystems} \\ D^* &= D - R\,\text{diag}\,[d_i]\,\overline{S} && \text{for "constrained" subsystems} \end{aligned} \tag{19a}$$

$$\begin{aligned} K^* &= K && \text{for "free" subsystems} \\ K^* &= K - R\,\text{diag}\,[k_i]\,\overline{S} && \text{for "constrained" subsystems} \end{aligned} \tag{19b}$$

$$\overline{S} = \begin{bmatrix} \overline{S}_1 \\ \vdots \\ \overline{S}_m \end{bmatrix} \tag{20}$$

$$\text{diag}\,[d_i] = \text{diag}\begin{bmatrix} \text{diag}\,[d_i]_1 & & \\ & \ddots & \\ & & \text{diag}\,[d_i]_m \end{bmatrix} \tag{21a}$$

$$\text{diag}\,[k_i] = \text{diag}\begin{bmatrix} \text{diag}\,[k_i]_1 & & \\ & \ddots & \\ & & \text{diag}\,[k_i]_m \end{bmatrix} \tag{21b}$$

$$\Lambda_j^* = \Lambda_j + Q_j \left(\text{diag}[k_i]_j \, T_1^j + \text{diag}[d_i]_j \, T_2^j\right)$$
$$\Lambda_j^* = \Lambda_j \qquad (22)$$

for "free" and "constrained" subsystems respectively,

$$U_j = R_j \left(\text{diag}\,[k_i]_j \, T_1^j + \text{diag}\,[d_i]_j \, T_2^j\right) \qquad (23)$$

$$V_j = Q_j \, \text{diag}\,[k_i]_j \, \overline{S} \qquad (24a)$$

$$W_j = Q_j \, \text{diag}\,[d_i]_j \, \overline{S} \qquad (24b)$$

## 3. ORDER-REDUCTION

Complex subsystems (e. g. wheelsets) lead to a high order of the system equations (18). The objective is to reduce the order of the subsystem for a certain investigation. The method proposed here is based on the representation by eigenvalues and eigenvectors [5], [6].

Therefore, the system of equations (18) is transformed into a system of first order differential equations:

$$\frac{d}{dt}\begin{bmatrix} \underline{p} \\ \dot{\underline{p}} \\ \underline{y}_1 \\ \underline{y}_2 \\ \vdots \\ \underline{y}_m \end{bmatrix} = \begin{bmatrix} 0 & I & 0 & 0 & \dots & 0 \\ -M^{-1} K^* & -M^{-1} D^* & M^{-1} U_1 & M^{-1} U_2 & \dots & M^{-1} U_m \\ V_1 & W_1 & \Lambda_1^* & 0 & \dots & 0 \\ V_2 & W_2 & 0 & \Lambda_2^* & \dots & 0 \\ \dots & \dots & \dots & \dots & \dots & \dots \\ V_m & W_m & 0 & 0 & \dots & \Lambda_m^* \end{bmatrix} \begin{bmatrix} \underline{p} \\ \dot{\underline{p}} \\ \underline{y}_1 \\ \underline{y}_2 \\ \vdots \\ \underline{y}_m \end{bmatrix} \qquad (25)$$

One begins the order-reduction by discarding eigenmodes of the "free" or "constrained" subsystem. If the eigenmodes of the "constrained" subsystem are already a good approximation to the behavior within the total system (strong coupling) it is recommended that order-reduction starts with the "constrained" subsystem. On the other hand in the case of weak coupling the eigenmodes of the "free" subsystem can be used.

The process of discarding eigenmodes is performed in several steps by systematic trials:

Those subsystem-eigenmodes, which have presumably minor influence are discarded step by step. After each step the reduced subsystem is coupled with the main system (see chap. 2) and the eigenvalues of the reduced system have to be compared with those of the complete system. In the decision whether a particular eigenmode should be discarded, also the eigenvector can be helpful, because its components indicate a possible contribution to the motion of interest. Therefore, to judge the accuracy of the reduced-order system, the resulting eigenvalues and eigenvectors are compared with those of the complete system.

To apply the presented method of coupling and order-reduction to large order systems a computer-program basing on [3] was developed.

## 4. APPLICATION EXAMPLE

The order-reduction procedure is applied to the bogie, consisting of a bogie frame (the main system) and two wheelsets (the subsystems) as described in [1].

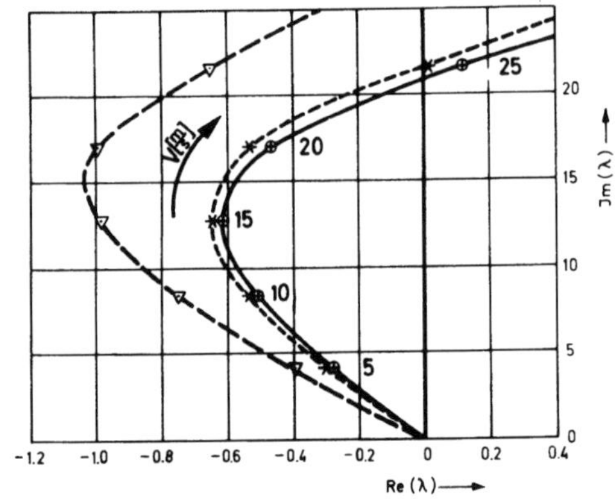


Fig. 2: Eigenvalues of reduced-order models

- o complete model
- \+ reduced model with hunting, torsion and lateral eigenmodes (coinciding with o)
- * reduced model with hunting and torsion eigenmodes
- ∇ reduced model with hunting eigenmode

Since the coupling of a wheelset with the bogie frame is rather strong the "constrained" wheelset-eigenmodes for order-reduction have been used. The objective is to obtain a reduced-order model for the stability of the hunting modes.

The bogie hunting mode that becomes unstable first with increasing velocity represents the in-phase lateral motion of the wheelsets. The second hunting mode goes unstable at a higher speed and represents the out-of-phase lateral motion.

Figure 2 shows the eigenvalues of the in-phase hunting motion for the complete and three reduced bogie models as a function of the bogie-velocity. The complete bogie model is represented by 74 first-order differential equations. The curve in fig. 1 coinciding with the curve of the complete model represents a reduced model with 26 first-order differential equations. For this reduced model only three complex conjugate eigenforms (six differential equations) of each wheelset-subsystem have been used. They are associated with the wheelset

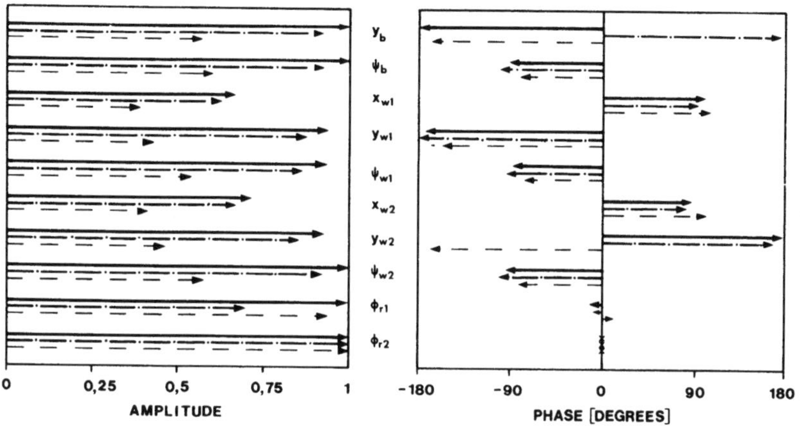


Fig. 3: Relevant coordinates of hunting eigenvector (V = 25 m/s) for different reduced order models

— COMPLETE MODEL —·— REDUCED MODEL (HUNTING, TORSION, LATERAL MOTION) – – REDUCED MODEL (HUNTING, TORSION)

hunting, torsion of the wheelset and wheelset lateral motion (HTL-model). The curve next to the one just discussed represents a reduced bogie model of order 22. Here each wheelset is represented by the two eigenforms associated with wheelset hunting and torsion of the wheelset (HT-model). This model still approximates the eigenvalues of the complete model very well. The last curve of fig. 1 shows the eigenvalues of a reduced model, taking into consideration only the hunting eigenform of each wheelset.

Figure 3 shows relevant components of the eigenvector of the in-phase hunting motion for the complete bogie model and two reduced models for velocity V = 25 m/s. The eigenvectors are of interest, because a better approximation of the eigenvector of the complete model usually results in improved eigenvalues. To compare the eigenvectors of different models the biggest component of each eigenvector is normalized to one and has the phase angle zero. The modal coordinates have been transformed back to the physical coordinates by eq.(6).

The eigenvector of the HTL-model closely approximates the eigenvector of the complete model. Only the rail bending $\phi_{rl}$ is not represented well, because the corresponding wheelset eigenmode has been discarded. Furthermore, one can see that the eigenvectors of the reduced HT-model and the complete model do not agree well, although the eigenvalues differed only slightly.

The eigenvalues and eigenvectors of the second hunting mode, representing the out-of-phase lateral motion of the wheelsets are approximated accurately by the most reduced model, which takes into consideration only the hunting eigenform of each ("constrained") wheelset. The reason for the excellent agreement of the complete model with the most reduced model is that the bogie frame almost is at rest except from a small yaw motion. Therefore, the hunting eigenform of the "constrained" wheelset (i. e. coupled to inertial space) is almost not changed.

The investigation has also been performed for the complete vehicle, consisting of a carbody, two bogie frames and four wheelsets. It has been shown that the stability results for this vehicle agree well with the results obtained for one bogie, [4]. Further investigations have also shown that the reduced wheelset model of 6th order is valid in a rather wide range of parameter variations (e. g. stiffness of wheelset axle, spring constants, conicity, coefficient of friction), [4].

## CONCLUSIONS

The order-reduction method described in this paper has been applied successfully to a wheel-rail vehicle. The objective of obtaining a reduced bogie model for the stability of the hunting modes has resulted in the reduced model with 6 first-order differential equations per wheelset instead of 30.

One should emphasize that different objectives will lead to different reduced order models, for example a reduced order model for the hunting stability may be quite different than a model for the vertical ride comfort.

The reduced wheelset model can be combined with different bogie- and carbody-models to analyse the stability of the system. The obtained reduced-order system can also be used for parameter design studies of the main system, because all physical parameters of the main system (e. g. carbody with bogie-frames) remain unchanged.

## 5. ACKNOWLEDGEMENT

The model of the wheelset on the rails and the thereon based models for bogie and vehicle have been made available by Fried. Krupp GmbH, Krupp Industrie- und Stahlbau (KIS). The investigations have been performed at DFVLR under subcontract of KIS. In particular we would like to express our gratitude for valuable discussions to Dipl.-Math. H. Schutzbach and Dipl.-Ing. W. Michels.

REFERENCES

1. Michels, W., Problemkreis Zusammenwirken von Fahrzeug und Fahrweg Übersicht über den Problemkreis.
Vortrag 3, Status-Seminar V, Spurgeführter Fernverkehr Rad/Schiene-Technik, Willingen, April 1978
2. Michels, W., Schutzbach, H., Contribution to Investigations of the Dynamics of Railway Vehicles.
6th IAVSD-IUTAM-Symposium Berlin 1979
3. Duffek, W., Federl, U., Kortüm, W., Lehner, M., Richter, R., Wallrapp, O., FADYNA ein Programmsystem zur Rechnersimulation der Fahrzeug-Fahrweg-Dynamik spurgeführter Verkehrssysteme.
DFVLR, A 552-78/14
4. Richter, R., Jaschinski, A., Modellsynthese und Parametervariation für die lineare Dynamik von Rad-Schiene-Fahrzeugen.
DFVLR, A 552-79/1
5. Davison , E. J., A Method for Simplifying Linear Dynamic Systems.
IEEE Transactions on Automatic Control
Vo. AC 11, No. 1, January 1966
6. Kortüm, W., Lehner, M., Richter, R., Multibody Systems Containing Active Elements: Generation of Linearized System Equations, System Analysis and Order-Reduction
Dynamics of Multibody Systems.
Symposium Munich/Germany, August 29 - Sept. 3, 1977,
Springer-Verlag, Berlin, Heidelberg, New York 1978

# RANDOM PROCESSES AND TRANSFER FUNCTION OF RAILWAY VEHICLE

Ladislav Rus

ČKD Praha, Záv. Lokomotivka, Research Institute of Diesel Locomotive, Praha, CSSR

SUMMARY

A railway vehicle could be represented as a dynamic system, excited both kinematically (by the rail irregularities) and dynamically (by the unbalanced masses in the motor). The excitations are correlated and noncorrelated, stationary and ergodic. The finite element method is used for solution of the dynamic behaviour of the vehicle. This method is based on the principle the balance of the whole structure, which is composed of basic elements. System excitation in form of power spectral density was applied to all wheel - sets both in vertical and in horizontal direction. The resulting diagrams of transfer functions and power spectral densities of the displacements and stresses are shown in the paper. They serve for calculations of ride quality and characteristics of the dynamic load and stress in the vehicle. Further calculations of structure life expectation or location of dynamically most exposed point in the structure are also possible.

## 1. INTRODUCTION

The finite element method has been in earlier times predominantely used for static calculations of the beam structures. The most frequently applied calculations were for bridges, cranes, oil rigs, antenna towers etc. In recent years the application of this method also entered into the field of dynamic calculations of various mechanical assemblies, since the development of digital computers made the solution of lager systems of equations possible, which is vitally important for dynamic calculations.

In Research Institute for Diesel Locomotives - CKD Praha, a program was written for ICL 1905 computer. This program serves for the solution of dynamic problems in bogies and locomotive main frames. Basically, it is possible to solve general dynamic problems in a space arrangement, for systems having both displacement and torsional vibrations. The dynamic systems can be build up of masses, bogies, beams, springs and dampers. Excitation can be introduced into all nodal points of the system, either in the force or kinematic form. The program is designed so as to provide for a solution of the forced vibrations of dynamic systems with excitation of a stochastic nature. The scope of utilisation of this program for the solution of general dynamic systems is shown in Fig. 1.

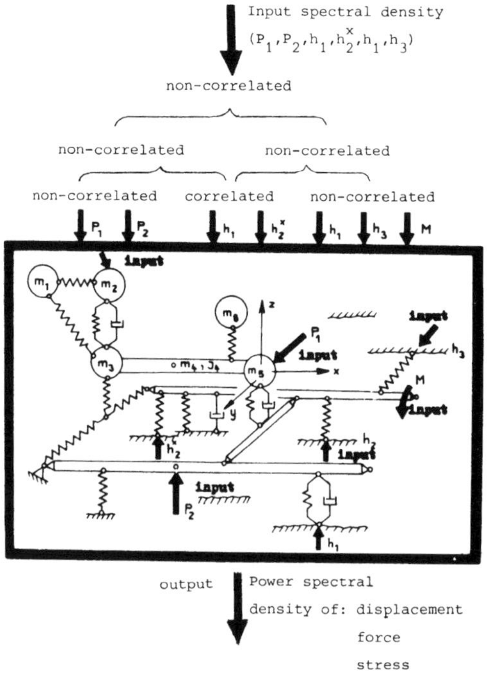


Fig. 1. Possibilities of the solved dynamic system, input exciting variables and the form of the computer output.

## 2. FINITE ELEMENT METHOD

Solution of the dynamic systems using the finite element method is based on the principle of solution the balance of the whole structure which is composed of basic elements. Suitably written equations of motion are transformed to the system of algebraic equations. Since the number of the equations is very high, the notation is carried out in matrix form, and the problem is solved using digital computer.

In matrix form, the column matrices of forces and displacements are referred to as P, p, U, u (main and local coordinate system).

The whole structure will be described by coordinates system within the main coordinate system (MCS). For the actual computation it is necessary to transform the variables from the main coordinate system to the local coordinate system. For this purpose transformation matrix T serves

$$p = T . P \quad ; \quad u = T . U \tag{1}$$

Since matrix T is a square and orthogonal $T^t = T^{-1}$

$$P = T^t . p \; ; \quad U = T^t . u \tag{2}$$

If there is a linear dependence between the displacement matrix u and matrix of forces p

$$p = k \, . \, u \tag{3}$$

then, by means of relations (1), (2), equation (3) can be rewritten to the form

$$TP = kTU \rightarrow P = KU \quad ; \quad K = T^t \, . \, k \, . \, T \, . \tag{4}$$

Characteristic values for individual elements of the structure are input in local coordinate system. From equation (4) is obvious, that one of the basic input parameters is the element stiffness matrix, which could be expressed in the form:

$$k_N = \begin{bmatrix} k_{ii} & k_{ij} \\ k_{ji} & k_{jj} \end{bmatrix} \tag{5}$$

For the space arrangement the matrix is of order (12, 12). Considering the notation in (5), the transformation matrix is given in the form:

$$T = \left[\begin{array}{c|c} \lambda_\lambda & 0 \\ \hline 0 & \lambda_\lambda \end{array}\right] = \left[\begin{array}{c|c} \lambda_N & 0 \\ \hline 0 & \lambda_N \end{array}\right] \tag{6}$$

where $\lambda_N$ is the matrix of direction cosines of order (6, 6).

If the individual connections of the basic elements within the whole structure are denotes by numbers 1, 2, 3, ... n, in accordance

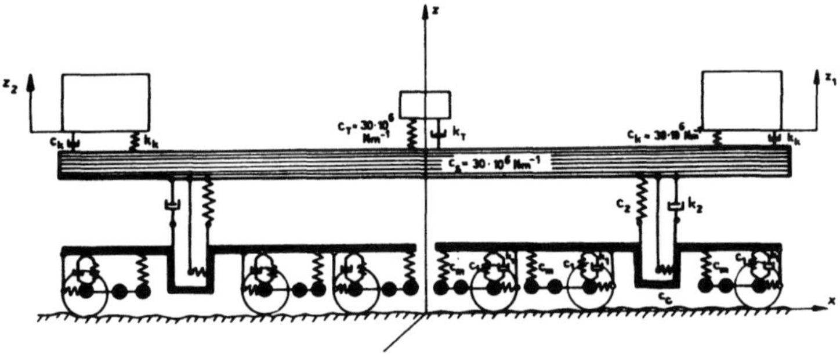

Fig. 2. Dynamic model of the locomotive with considering the bending vibration of the locomotive body.

with relation (4), the relation between the displacement matrix U and the matrix of forces P can be expressed, using the stiffness matrix K of the whole structure like follows:

$$\begin{bmatrix} P_1 \\ P_2 \\ P_3 \\ P_4 \\ \cdot \\ \cdot \\ \cdot \\ P_n \end{bmatrix} = \begin{bmatrix} K_{11} & K_{12} & K_{13} & \cdots & K_{1n} \\ K_{21} & K_{22} & K_{23} & \cdots & K_{2n} \\ K_{31} & K_{32} & K_{33} & \cdots & K_{3n} \\ K_{41} & K_{42} & K_{43} & \cdots & K_{4n} \\ \cdot & \cdot & \cdot & & \cdot \\ \cdot & \cdot & \cdot & & \cdot \\ \cdot & \cdot & \cdot & & \cdot \\ K_{n1} & K_{n2} & K_{n3} & \cdots & K_{nn} \end{bmatrix} \cdot \begin{bmatrix} U_1 \\ U_2 \\ U_3 \\ U_4 \\ \cdot \\ \cdot \\ \cdot \\ U_n \end{bmatrix} \qquad (7)$$

$$P = K \cdot U \qquad (8)$$

If the system (8) is divided into the matrices of known nodal load P with unknown displacements U and nodal load $P_p$ with known displacements $U_p$, we obtain:

$$\begin{bmatrix} P \\ P_p \end{bmatrix} = \begin{bmatrix} K_A & K_B \\ K_B^t & K_p \end{bmatrix} \cdot \begin{bmatrix} U \\ U_p \end{bmatrix} \qquad (9)$$

After modification of equation (9) we obtain the relation for computation of unknown displacements.

$$U = K_A^{-1} \cdot P - K_A^{-1} K_B U_p , \qquad (10)$$

or the relation for computation of forces $P_p$

$$P_p = K_B^t \cdot K_A^{-1} P - K_B^t K_A^{-1} K_B U_p + K_p U_p . \qquad (11)$$

The aim of the solution is to find general method for computing of dynamic models, which consist of beams, masses, springs and dampers, with the possibility of excitation both dynamically in all points of the structure. In matrix notation, any structure can be described by an equation of the form:

$$M_{r1} \cdot \ddot{u} + C_{r1} \dot{u} + K_{r1} u = P - C_{r2} \dot{h} - K_{r2} \cdot h \qquad (12)$$

where $C_{r2}$ and $K_{r2}$ are reduced damping matrices, respectively stiffness matrices of the whole structure C.

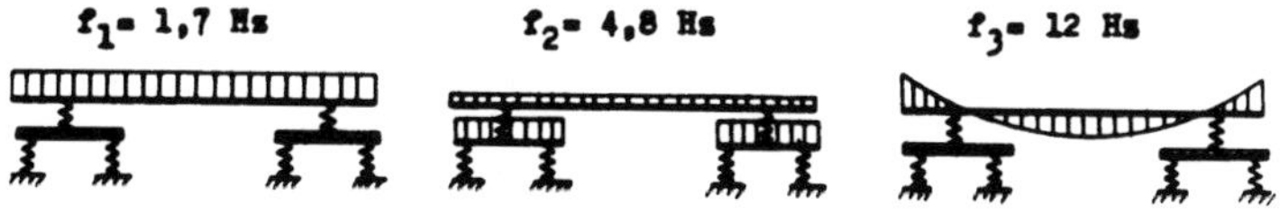


Fig. 3. The forms of vibrations in the resonance areas.

The exciting variables P and h are of random nature. For railway vehicles they consist of excitation from the motor P and excitation from the rail irregularities h and $\dot{h}$. After Fourrier transformation of equation (12) we obtain:

$$(K_{r1} - \omega^2 M_{r1} + i\omega\, C_{r1})\bar{U} = \bar{P}_o - (K_{r2} + i\omega\, C_{r2})\bar{H}_o \quad , \tag{13}$$

while $\bar{U}$, $\bar{P}_o$, $\bar{H}_o$ are Fourrier transforms of variables u, P, h expressed in complex form (the bar over the variable). For further processing, equation (13) can be rewritten into the form:

$$\bar{D} \cdot \bar{U} = \bar{P}_o - \bar{B}\bar{H}_o \tag{14}$$

Equation (14) is actually a system of equations with complex coefficients. To make the subsequent solution of the system easier, system (14) is transformed, using the comparison of real and imaginary parts, to the system with the number of equations doubled.

$$\begin{bmatrix} Re\bar{D} & -Jm\bar{D} \\ Jm\bar{D} & Re\bar{D} \end{bmatrix} \cdot \begin{bmatrix} Re\bar{U} \\ Jm\bar{U} \end{bmatrix} = \begin{bmatrix} Re\bar{P}_o \\ Jm\bar{P}_o \end{bmatrix} + \begin{bmatrix} Re\bar{B} & -Jm\bar{B} \\ Jm\bar{B} & Re\bar{B} \end{bmatrix} \cdot \begin{bmatrix} Re\bar{H}_o \\ Jm\bar{H}_o \end{bmatrix} \tag{15}$$

Using relations (13), (14), equation (15) can be expressed in the form:

$$\begin{bmatrix} K_{r1} - \omega^2 M_{r1} & ; -\omega C_{r1} \\ \omega C_{r1} & ; K_{r1} - \omega^2 M_{r1} \end{bmatrix} \cdot \begin{bmatrix} Re\bar{U} \\ Jm\bar{U} \end{bmatrix} = \begin{bmatrix} Re\bar{P}_o \\ Jm\bar{P}_o \end{bmatrix} - \begin{bmatrix} K_{r2} & ; -\omega C_{r2} \\ \omega C_{r2} & ; K_{r2} \end{bmatrix} \cdot \begin{bmatrix} Re\bar{H}_o \\ Jm\bar{H}_o \end{bmatrix} \tag{16}$$

or, after rewriting

$$D_1 \cdot \bar{U}_1 = \bar{P}_{o1} - B_1 \cdot \bar{H}_{o1} \tag{17}$$

System (17) is the final equation system, from which is possible, for given unoity exċitation $\bar{P}_{o1}$ and $\bar{H}_{o1}$ to compute the required transfer function of the structure for the correlated inputs, or the matrix of transfers for non-correlated inputs.

$$\bar{U}_1 = \begin{bmatrix} Re\bar{U} \\ Jm\bar{U} \end{bmatrix} = D_1^{-1} (\bar{P}_{o1} - B_1\bar{H}_{o1}) \quad . \tag{18}$$

The transfer vector of the system, or the square of the transfer absolute value, is then defined by the following expression:

$$H\ (i\omega) = Re\bar{U} + iJm\bar{U} \tag{19}$$

$$/H(i\omega)/^2 = /Re\bar{U}/^2 + /Jm\bar{U}/^2$$

Stiffness matrix K, damping matrix C and massa matrix M of the whole structure, whose reduced forms can be found in equation (12), are described in [1]. The principle used for deriving their coefficients is based upon developing unity displacement and rotation of i-th and j-th node of an element and finding out of forces and torques, that were for it necessary.

According to above mentioned theoretical basis, a programme for ICL 1905 computer was written. The results of the computation are used for computation of dynamic system transfer absolute value squared $/H(i\omega)/^2$.

For the computation of the output power spectrum the following relation defining the dependence of the input and output power spectral density are available.

Correlated inputs

$$Gy(f) = H(f) \cdot G_{xx}(f) \cdot H^{+T}(f) \qquad (20)$$

where

$Gy(f)$ - power spectral density at the output of the system
$H(f)$ - matrix of individual system transfers
$G_{xx}(f)$ - matrix of the cross-spectral densities of the system inputs
$H^{+T}$ - conjugate transposed matrix of the system

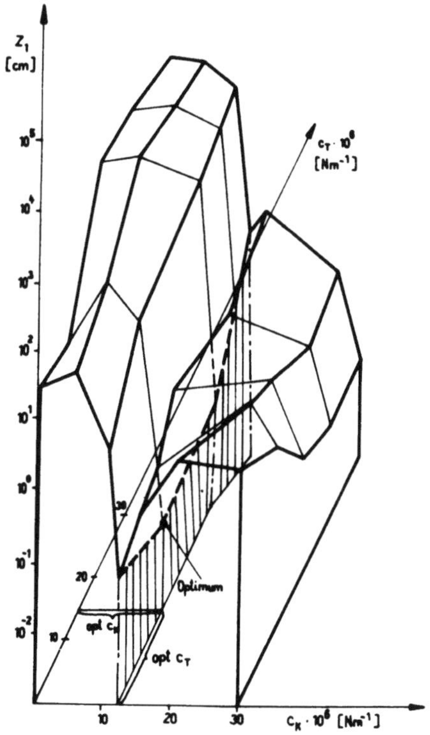


Fig. 4. Influence of the engine driver's cabin seating stiffness and of the motor bedding stiffness to the locomotive transfer function course.

Non-correlated inputs

$$G\dot{y}(f) = \sum_{i=j}^{N} /H_i(f)/^2 \cdot G_{ii}(f) \qquad (21)$$

where

$/H_i(f)/^2$ - absolute value of the ith system transfer squared

$G_{ii}(f)$ - power spectral density of the ith input to the system

## 3. APPLICATION OF THE PROGRAM - COMPUTING THE LOCOMOTIVE CAB IN VIBRATION

The dynamic calculation for the system illustrated in Fig. 2 was performed by means of the finite element method. This locomotive has two bogies and is provided with double spring suspension and damping. The suspended traction motors were considered as well. Both the bogie frame and the locomotive body frame are elastic. The engine driver's cabin and motor are elastically connected to both ends, respectively in the center, to the body frame. In case of an electric locomotive, there is a transformer. The system is excited by the vertical rail irregularities having random character. The vehicle is excited at all wheel - sets and is considered mutually fully correlated.

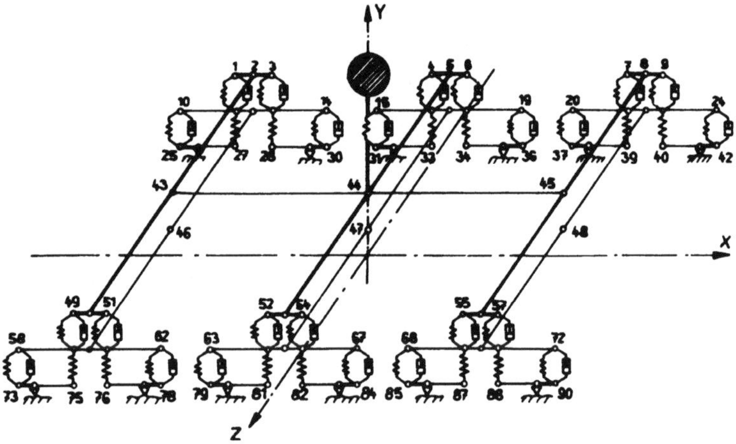


Fig. 5. Model of a vehicle - Finite Element Method.

The objective is to determine the influence of individual vehicle parameters upon the intensity of locomotive body bending vibration and further to find such a combination, that would minimize the bending vibrations. The examined parameter was the vibration of engine driver's cabin. The vibration amplitude $Z_1$ at this point was defined as a measure for assessing the optimality of the proposed locomotive parameters.

After a series of computations, it was found, that from a range of parameters, that can be changed without drastical changes in the locomotive design, the greatest influence upon the engine driver's cabin

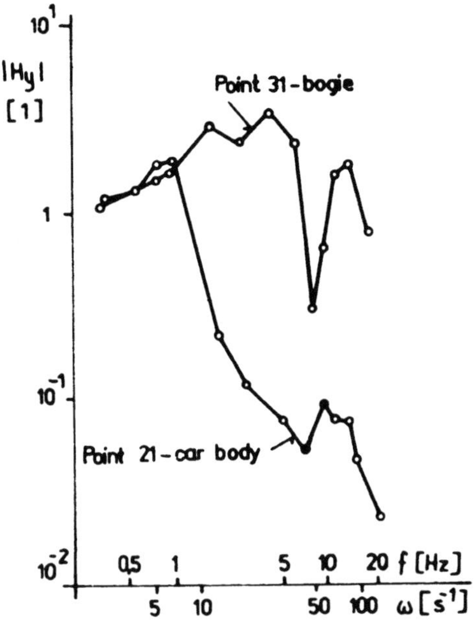


Fig. 6. Transfer function for displacement of points 23 and 31.

vibration have the stiffness of its seating, the stiffness of the seating of the motor upon the locomotive frame, the stiffness of the spring in the locomotive bogie pivot and the combinations of the previously mentioned parameters. In the course of the transfer function of the driver's cabin occur resonance peaks, corresponding to the locomotive body hunting movement, bogie hunting movement and locomotive body bending vibrations. The vibration forms for individual resonance peaks are shown in Fig. 3. The three dimensional dependence of the resonance peak magnitude for various mutual combinations of the motor seating stiffness $c_t$ and of the cabin seating stiffness $c_k$ is illustrated in Fig. 4. From the diagram it is possible to find the optimum values of $c_T$ and $c_K$, which provide the minimum amplitude of the resonance bending vibrations and facilitate the decrease in the influence of the vibrations upon the locomotive crew to the minimum value.

From the performed computations and from previously mentioned diagrams, following conclusions can be drawn:

1. The decrease of locomotive body bending vibration can be achieved by means of proper changing of the motor seating stiffness $C_T$.
2. To the decrease of the amplitude of the locomotive body bending vibrations, the ratio of motor seating stiffness $C_T$ and engine driver's cabin seating stiffness $C_K$ are also of some influence.
3. From the transfer function, the decrease of the resonance peak for the suitable combination of motor seating stiffness $C_T$ and $C_C$ can be observed, ($C_C$ - locomotive bogie pivot spring stiffness).
4. If the attention is concentrated to the influence of ratio of the engine driver's cabin seating stiffness $C_K$ and the bogie pivot spring

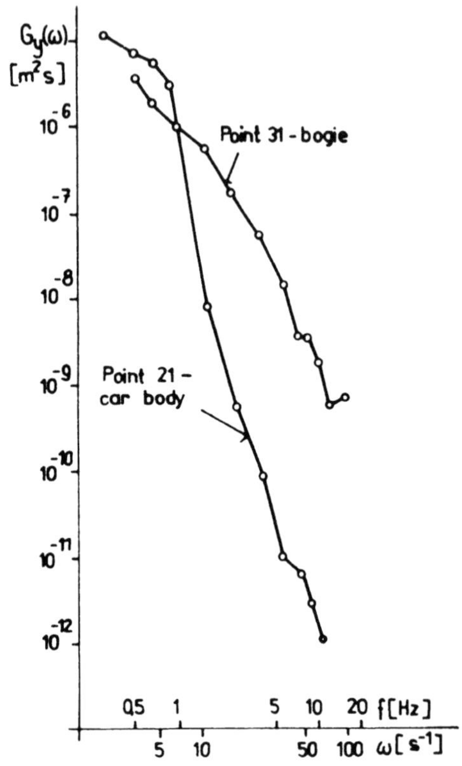


Fig. 7. Power spectral density of displacement of points 21 and 31.

stiffness $C_c$ to the magnitude of the bending vibration resonance peak amplitude, the three dimensional dependence shown in Fig. 4 can be determined. From the diagram, the optimal values of $C_C$ and $C_K$ can be obtained, for which the value of the bending vibration resonance peak is at the minimum level.

## 4. RESULTS OF THE COMPUTATIONS

Above mentioned theory was applied to the solution of a locomotive, dynamic chart of which is shown in Fig. 5. This model was used for the vertical and horizontal excitation.

System excitation in form of power spectral density was applied to all wheel - sets both in vertical and in horizontal direction. The resulting diagrams of transfer functions and power spectral densities of the displacements and stresses are shown in Fig. 6, 7, 8. These diagrams can be computed for all points of the structure. They serve for calculations of ride quality and characteristics of the dynamic load and stress in the vehicle (Fig. 8). Further calculations of structure life expectation or location of dynamically most exposed point in the structure is also possible.

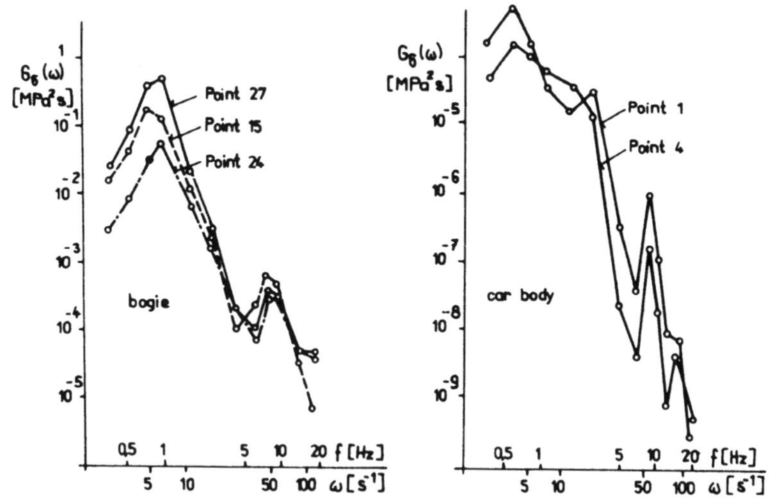


Fig. 8. Power spectral stress density of points 1, 27, 15, 24, 4.

5. REFERENCES


1. Rus, L., Finite Element Method, Used in Dynamics System Computation, Report VUML 645.37.
2. Rus, L., Hunting motion of bogies of railway vehicles. Communications of VSD Zilina, Vol. 12, 1971, p. 155 - 168.
3. Rus, L., Running stability and railway vehicle transfer function, solved by method of statistical linearisation. The dynamics of vehicles, Vienna 1977.
4. Sixta, P., Programme for Calculation Dynamical respones, Finite Element Method, CKD Praha, VUML.

# AN ADAPTIVE CONTROL FOR VEHICLE SUSPENSIONS

H. K. Sachs

Department of Mechanical Engineering, Wayne State University,
Detroit, MI, U.S.A.

SUMMARY

This paper discusses the simulation of an automotive suspension control system. Statistical data is converted into Fourier series. A hypothetical vehicle is assumed to pass over the terrain at a given speed yielding measurable acceleration values at the axle and vehicle body for the purpose of reconstructing the terrain configuration. The terrain information together with sprung mass, unsprung mass and tire spring rate allows determination of optimal spring rates and shock absorber rates as a function of speed. This data is to be stored on the on-board microprocessor together with the reconstructed terrain configuration as computed on the vehicle. By means of pattern recognition of the respective terrains the suspension rates are set to the optimal values.

INTRODUCTION

A pilot study entitled "Development of Computerized Vehicle Suspension" [1] laid the foundation for the work presented in this manuscript. The notion of automatic suspension appeared first in connection with a complete design in a paper by Federspiel-Labrosse, [2]. A similar study was carried out by Westinghouse in 1965, [3]. The necessity of improving suspension characteristics was soon put on a more scientific basis. Works by Bender, Paul, Fenoglio, Karnopp (at Massachusetts Institute of Technology at the time [4,5,6,7]) considered feed-back automatic control systems for vehicles. It was next shown that a transfer function can be synthesized (Wiener [8]). Also Thompson [9] considered the optimal active suspension for bounce, pitch and roll control of passenger cars. None of the above publications suggested the use of an adaptive control employing on-board minicomputer circuits. The reason for this is that computer technology was not as advanced at that time, as it is today. Neither size nor price form obstacles for the adoption of such control methods now, and will be much less so in the future.

The terms "optimal" control and "adaptive" control in connection with active vehicle suspension designs require further explanation. In a strictly literal sense there does not exist a "best" or "optimal" set of suspension parameters that is realizable, because the "best" type of suspension is one which completely isolates the sprung mass from any form of road shock. Any and all suspensions, no matter how soft, are force transducers and the softer the suspension the greater must be the excursions of the unsprung masses, which are limited. But, obviously, within given design criteria there are better performing suspension systems than others and the object of this investigation is to find means to select the best possible set of suspension parameters that meet the design criteria.

The term adaptive control is commonly used by control engineers to indicate that the control process is adapted to the source of the disturbance or perturbation. Another alternative is the feedback control wherein the response of the system is compared with a desired output and the controller acts to minimize that difference between the actual and desired response and in some instances the two control strategies are combined.

In this investigation we employ an adaptive control algorithm that allows pre-sampling of the source of disturbance, namely the terrain roughness, from statistical data, [10,11]. From the infinitely wide range of terrain configurations that exist a large and substantially representative class of terrains, referred to herein as model terrains, can be assembled and their power spectral density (p.s.d.) functions can be placed side-by-side. It can be shown that in most cases the distribution of the roughness follows similar patterns which can be mathematically expressed by Fourier transforms.

This implies, of course, the existence of a maximal wavelength that contains all irregularities of the terrain that are representative for that sample of the terrain. There is some arbitrariness in the choice of that wave length, and based on the available statistical data [10,11] a sample length 130 m is chosen.

For this investigation a total of 12 such model terrain statistics are made available and one is selected for the purpose of analysis. The computer algorithm dedicated to the on-board microprocessor is simulated and combined with the off-board terrain and optimal suspension parameter selection process, so as to close the loop of the process. The following paragraphs will explain in detail the logic of the program and its operational features.

## THE SUSPENSION MODEL AND TRANSFER FUNCTION

In Fig. 1 is shown a single wheel suspension model. For the purpose of determining the optimal suspension parameters we shall assume linearity of all suspension components, even though one could use describing function techniques to obtain the linear equivalent of a nonlinear element, such as air-springs. The tire spring rate K is assumed to be invariant and the forward speed u known or measured. The equations of motion for the $n^{th}$ harmonic perturbation are:

$$m_b \ddot{x}_b + c(\dot{x}_b - \dot{x}_a) + k(x_b - x_a) = 0 \;; \quad j = \sqrt{-1} \qquad (1)$$

$$m_a \ddot{x}_a - c(\dot{x}_b - \dot{x}_a) - k(x_b - x_a) + Kx_a = Kh_n \exp(jn\omega t) \tag{1}$$

$$x_b - x_a = y$$

Let $x_b = X_b^n \exp(j\omega nt)$  $y = Y^n \exp(jn\omega t)$

(1) can be conveniently written in matrix form

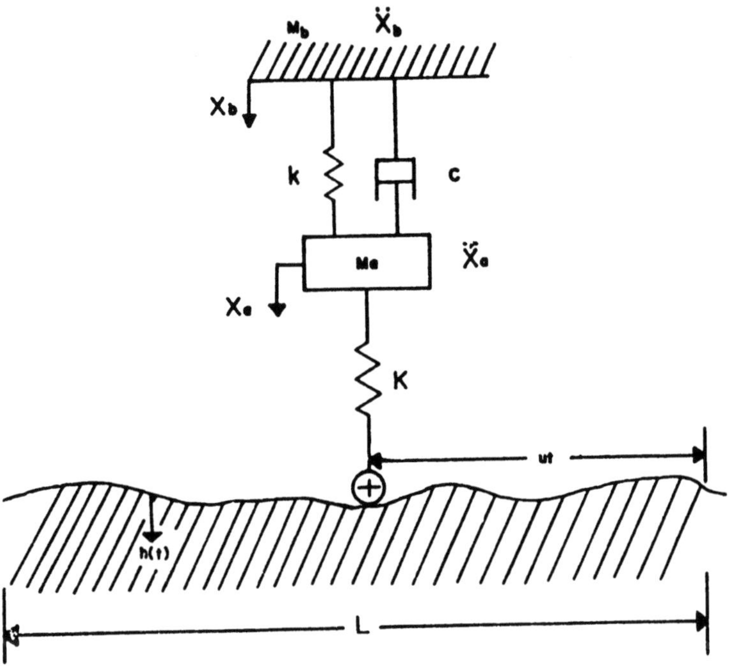


FIGURE 1 SCHEMATIC VIEW OF SUSPENSION SYSTEM

$$\begin{bmatrix} \left(-\omega^2 n^2 + \frac{k+K}{m_a} + \frac{k}{m_b} + c((m_a+m_b)/m_a m_b) j\omega n\right) & -K/m_a \\ \left(\frac{c}{m_b} j\omega n + \frac{k}{m_b}\right) & -\omega^2 n^2 \end{bmatrix} \begin{Bmatrix} Y^n \exp j\phi \\ X_b^n \exp j\psi \end{Bmatrix} = \begin{Bmatrix} Kh_n/m_a \\ 0 \end{Bmatrix} \tag{2}$$

where in (2) $X_b^n$ and $Y^n$ are complex (solution vectors).
Letting

$$a_{11} = -\omega^2 n^2 + ((k+K)/m_a) + (k/m_b) + c((m_a + m_b)/m_a m_b) j\omega n$$

$$a_{12} = -K/m_a \; ;$$

$$a_{21} = (jc\omega n/m_b) + k/m_b \; ;$$

$a_{22} = -\omega^2 n^2$ ; $a_{10} = Kh_n/m_a$ ; $a_{20} = 0$ ; and the determinant of (2) is $(a_{11}a_{22} - a_{12}\, a_{21}) = \bar{D}$ then (2) has the solution

$$|Y|^n = \begin{vmatrix} a_{10} & a_{12} \\ 0 & a_{22} \end{vmatrix} / \bar{D}$$

$$|X_b|^n = \begin{vmatrix} a_{11} & a_{10} \\ a_{21} & 0 \end{vmatrix} / \bar{D} \qquad (3)$$

The expression (3) are outputs which one obtains for this two degree of freedom system. Division of (3) by the input vector of length $h_n$ yields the two transfer functions relating the $n^{th}$ component of the input to the $n^{th}$ component of the relative and absolute displacement series. The corresponding transmissibility functions are obtained by taking the root of the sum of the squares of real and imaginary parts and the tangents of the corresponding phase angles by computing the ratio of imaginary and real parts of (3).

Our objective is to find sets of spring rate values $k_r$, and damping coefficients $c_r$ such that the transmitted acceleration values for a random terrain that can be represented by Fourier series is a minimum without exceeding the design axle clearance more often than 3 times in 1000 working cycles. Since the disturbances are of a random nature it is necessary to define the terrain functions first.

TERRAIN ANALYSIS AND SYNTHESIS

The significance of terrain configurations is that once their pattern is established it is possible to reconstruct them without a substantial loss of accuracy relative to the response characteristics of vehicles passing over such terrains. From terrain measurements (approximately 130 for each terrain sampled over a length of circa 130 meters) one can obtain equivalent Fourier series which are considered representative for each terrain configuration. Conveniently the "Fast Fourier Transform" [12] can be employed to obtain the coefficients of each sine and cosine term. The reconstruction of the terrain from 128 terms considered in the series shows an accuracy of between 97% and 99% relative to the distribution of the measured data.

It is well known that power spectral density (psd) data varies for different terrains, but when plotted on log-log paper the totality of all points appear to lie in proximity of a straight line of negative slope (Fig. 2). This pattern repeats itself for terrains of quite different roughness configuration so that it is possible to assemble and classify terrains by their psd characteristics. For the purposes

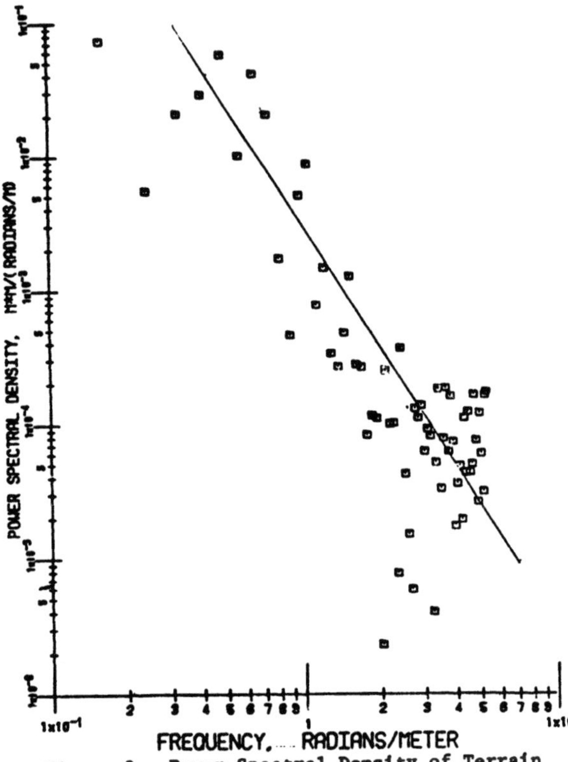


**Figure 2. Power Spectral Density of Terrain**

of this investigation twelve different terrain samples were made available. Obviously, if a continuous record of roadway, say from Detroit to Los Angeles (approximately 3800 km) were available, this data together with optimal suspension parameter data could be stored on the vehicle and upon terrain reconstruction en route and its association with the precomputed optimal parameters one could affect optimal suspension control on the basis of a continuous previewing concept. The disadvantage of this idea is obviously the storage of an enormous data bank requiring a huge memory of the on-board computer. Besides, it may not necessarily produce better results than the previewing of a limited sample size of representative terrain - or road profiles.

It is assumed that the wheels of the vehicle passing over the terrain will stay in contact with the terrain profile at all times. Also, because of the substantially larger sample length in comparison to wheelbase the phasing between front and rear wheel sets is here neglected. One can, of course, provide suspension control for each wheel independently.

The vehicle is to be equipped with low frequency sensitive accelerometers on the superstructure (body) and high frequency accelerometers on the axle (unsprung mass). With $m_b$ the pro-rated sprung mass, $m_a$ the pro-rated unsprung mass, k the suspension stiffness rate, K the tire stiffness rate and c the shock absorber constant we can write the general equations of motion in the form:

$$m_b \ddot{x}_b + k x_b + c\dot{x}_b - kx_a - c\dot{x}_a = 0$$
$$m_a \ddot{x}_a + (K+k)x_a + c\dot{x}_a - kx_b - c\dot{x}_b = Kh(t) \qquad (1.a)$$

The displacements in (1) are $x_b$ (body) and $x_a$ (axle). h(t) is the wheel lift and drop due to terrain roughness. Adding the two equations (1) we obtain

$$m_b\ddot{x}_b + m_a\ddot{x}_a + Kx_a = Kh(t) \qquad (4)$$

$$x_a = \int_\tau (\int_\tau \ddot{x}_a dt)dt \qquad (5)$$

From (4) we obtain the terrain profile h(t), namely

$$h(t) = \ddot{x}_b(\frac{m_b}{K}) + \ddot{x}_a(\frac{m_a}{K}) + x_a \qquad (4a)$$

If $\ddot{x}_b$ and $\ddot{x}_a$ are measured and converted into electronic impulses then (5) can be obtained by integrating twice and inverting the sign on summing amplifiers. Alternately, the accelerometer outputs could be digitized and the integration done numerically on the microprocessor. In either case the result of (4a) yields a reading of the reconstructed terrain profile. It is noted that suspension rate k and damping c do not appear in (4) and (5) and therefore the computation algorithm for the reconstruction of the terrain profile is not affected by the characteristic of the restoring force and dissipative force functions.

In order to demonstrate the accuracy of the reconstruction process the spatial Fourier series of the terrain are converted into a time series based on constant speed motion of the vehicle. Let the $n^{th}$ term of the series be sin $(nax_f)$ where n is an integer, a is the wave parameter $(a=2\pi/\ell)$, $\ell$ is wave length in meters and $x_f$ is the forward motion displacement. Then

$$x_f = u.t \qquad (6)$$

where u is speed and t is time, and

$$nax_f = n\ (2\pi u/\ell).t = n.\omega.t \qquad (7)$$

In (7) $\omega = 2\pi u/\ell$ is the circular frequency of the harmonic of "$\ell$" meter wave length. Referring back to (1) we can compute $\ddot{x}_a, \ddot{x}_b$ relative to each and every term contained in the Fourier series representing h(t). These are, in fact, the signals sensed by the on-board accelerometers. $\ddot{x}_b$ and $\ddot{x}_a$ is then reintroduced into (4a) to obtain h(t) and the output is compared with the input. The correlation proves to be on the order of a fraction of a percent.

AVERAGING h(t), POWER SPECTRAL DENSITY

The terrain elevation may be expressed in terms of discrete (measured) values or as a continous function of distance or time. The $n^{th}$ term of a Fourier series expansion of the terrain function may have the form

$$h(t)_n = h_{nc} \cos n\omega t + h_{ns} \sin n\omega t \tag{8}$$

where $h_{nc}$ and $h_{ns}$ are the coefficients associated with the sine and cosine terms of the $n^{th}$ order. Hence

$$h^2(t)_n = h^2_{nc} \cos^2 n\omega t + h^2_{ns} \sin^2 n\omega t + h_{nc} h_{ns} \sin 2n\omega t \tag{9}$$

Averaging the above with respect to one period of length $\tau$ we obtain

$$\bar{h}^2_n = \frac{h^2_{nc}}{\tau} \int_\tau \cos^2 n\, t\, dt + \frac{h^2_{ns}}{\tau} \int_\tau \sin^2 n\, t\omega dt + \frac{h_{nc}}{\tau} h_{ns} \int_\tau \sin 2\, n\omega t dt \tag{10}$$

Now the last integral on the left side of (10) integrated over the full period $\tau$ vanishes and (10) becomes

$$\bar{h}^2_n = \left(\frac{h^2_{nc}}{2} + \frac{h^2_{ns}}{2}\right) = \frac{(h^2_{nc} + h^2_{ns})}{2} \tag{11}$$

Each of the squared amplitude values $\bar{h}^2_n$ belongs to a discrete frequency of order n. Then, the average of all such values over the frequency spectum considered herein is

$$\bar{h}^2 = \frac{1}{m+1} \sum_1^m \bar{h}^2_n \qquad n = 1,2,3, \ldots.m \tag{12}$$

and its root mean square value is

$$\bar{h} = \left(\frac{1}{m+1} \sum_1^m \bar{h}^2_n\right)^{1/2} \tag{12a}$$

For each model terrain we can obtain its meean square amplitude (12) or its root (12a) respectively. The discrete power spectral density curve of the terrain (see Fig. 2) is obtained dividing (11) by the wave parameter $a_n$ and subsequent plotting of the results versus the wave parameter $a_n$. If the respective values are plotted on log-log paper, the dense distribution of points can be seen to lie in proximity of a straight line given by

$$_{10}\log(\bar{h}_1^2/a_1) - {}_{10}\log(\bar{h}_n^2/a_n) = ({}_{10}\log a_1 - {}_{10}\log a_n) f \qquad (13)$$

or $$f = ({}_{10}\log[\bar{h}_n^2\, a_n/\bar{h}_n^2\, a_1])/{}_{10}\log(a_1/a_n) \qquad (13a)$$

where f is the slope of the line. Bender [13] has shown that the power spectral density function (13) can be expressed in terms of frequency rather than wave parameter yielding a straight line image expressed by

$$\phi(n\omega) = \frac{\bar{h}u}{(n\omega)^2} \qquad (14)$$

where

$$\phi(n\omega) = \frac{\bar{h}_n^2}{n\omega} = \text{discrete power spectral density.} \qquad (15)$$

OPTIMALITY

The axle clearance is usually determined by several design criteria other than softness of ride. Therefore, it is necessary to avoid the incidence of the axle striking the axle stops. The probability that the maximal displacement of the axle relative to the body Y will lie between $\pm \alpha\bar{y} = \pm H$ where $\alpha$ is a number, H is the available axle clearance and $\bar{y}$ is the expected suspension deflection, is

$$\text{Prob}\,[-\alpha\bar{y} \leq y(t) \leq \alpha\bar{y}] = \int_{-\alpha\bar{y}}^{\alpha\bar{y}} e^{(-(y/\bar{y})^2/2)} dy/\bar{y}(2\pi)^{1/2} \qquad (16)$$

and for $\alpha = 1,2,3$ this probability is [14]

$$\left.\begin{array}{ll} \alpha = 1 & \text{Prob.} = 68.3\% \\ \alpha = 2 & \text{Prob.} = 95.4\% \\ \alpha = 3 & \text{Prob.} = 99.7\% \end{array}\right\} \quad \text{Gaussian or normal distribution}$$

This means for $\alpha = 3$, in only 3 out of 1000 working cycles would Y exceed the allowable axle clearance H and therefore strike the axle stops.

In seeking an optimal set of suspension parameters, design constraints imposed on the system limit the average maximal displacement across the suspension $\bar{y}$ to $(1/\alpha)H$. For each terrain profile there exists a set of suspension parameters k and c that minimize the maximal body accelerations $|\ddot{x}_b|$ at a certain average forward speed u and at the same time satisfy the design constraint, namely, that the displacement across the suspension should be within the limit H 99.7% of the time. At other speeds, u, there are sets of suspension parameters which are not optimal, yet satisfy the design constraint. These parameters are optimal to the extent that the maximal displacement across the suspension elements is utilized.

SYSTEMS RESPONSE TO RANDOM INPUTS

The transmissibility functions relative to the transfer functions (3) give:

$$\left(\frac{Y_n}{h_n}\right)^2 = \frac{\eta_n^4}{[\eta_n^4\,\omega_b^2/\Omega_a^2 - \eta_n^2(1+\frac{k}{K}+\frac{\omega_b^2}{\Omega_a^2})+1]^2 + 4\zeta^2\eta_n^2[\eta_n^2(\frac{k}{K}+\frac{\omega_b^2}{\Omega_a^2})-1]^2} \tag{17}$$

$$\left(\frac{X_n}{h_n}\right)^2 = \frac{1+4\zeta^2\eta_n^2}{[\eta_n^4\frac{\omega_b^2}{\Omega_a^2} - \eta_n^2\,(1+\frac{k}{K}+\frac{\omega_b^2}{\Omega_a^2})+1]^2+4\zeta^2\eta_n^2[\eta_n^2(\frac{k}{K}+\frac{\omega_b^2}{\Omega_a^2})-1]^2} \tag{18}$$

Where in (17), (18) $\eta = \omega n/\omega_b$ , $k/m_b = \omega_b^2$ , $K/m_a = \Omega_a^2$ ,

$\zeta = c/c_c$ , $c_c = 2(k\ m_b)^{1/2}$ ;

Let it be required that according to the constraint (16)

$$\alpha\ \bar{y} < H \quad \text{or } \bar{y} = H/\alpha = H/3 \quad (\alpha=3) \tag{19}$$

Then with the help of (17), (18), (19) we obtain the mean square response

$$\bar{Y}^2 = \frac{H^2}{9} = \sum_1^m Y_n^2 = \frac{1}{2}\sum_1^m h_n^2\ M_{yn}^2 \ \ / \ (m+1) \tag{20}$$

$$\bar{X}^2 = \frac{1}{2}(\sum_1^m h_n^2\ M_{xn}^2)/(m+1) \tag{21}$$

where in (20), (21) the squared terms of the transmissibilities are:

$M_{yn}^2 = \eta_n^4/D_1^2$ ; $\qquad M_{xn}^2 = (1+4\zeta^2\eta^2 n)/D_1^2$

$D_1^2$ is the expression of the denominators of (17) and (18).

Since the acceleration transmissibility for the n[th] frequency is

$$|a_c|_n = n^2\omega^2 |X_b^n| \tag{22}$$

or in nondimensional form

$$|\bar{a}_c|_n = n^2\omega^2 |X_b^n|/g \quad \text{(g is gravitation constant)}, \tag{22a}$$

we infer that for minimizing the acceleration transmissibility

$$\min(\bar{X}^2) = \frac{1}{2}\sum_1^m h_n^2 (\min(M^2_{xn})/(m+1)) \tag{23}$$

Because of the phase shift between each output term, (20), (21),(23) hold only for small damping parameters $\zeta_i$ . The optimization is predicated on minimizing (23) while satisfying the constraint equation (20) for each selected model terrain of a specified (psd) image versus in-put frequency ($n\omega$) as a function of forward speed u. This data is to be stored on the vehicle minicomputer. A block diagram (Fig.3) illustrates the algorithm with the off-board pre-analysis processor framed by the dashed lines. The solution of (20), (23) is accomplished on a large scale computer by first assigning a damping parameter $\zeta_i$ (trial value) and searching for $k_b$ such that (20) is satisfied and by further iteration on (23) to find the optimal damping parameter $\zeta$ and finally introducing $\zeta$, $k_b$ into (20) until both, (20), (23) are satisfied. In Table 1 we show the results of this operation for one terrain and three different vehicle axle clearances. The data in table (1) is predicated on a vehicle of which the data is presented in table (2).

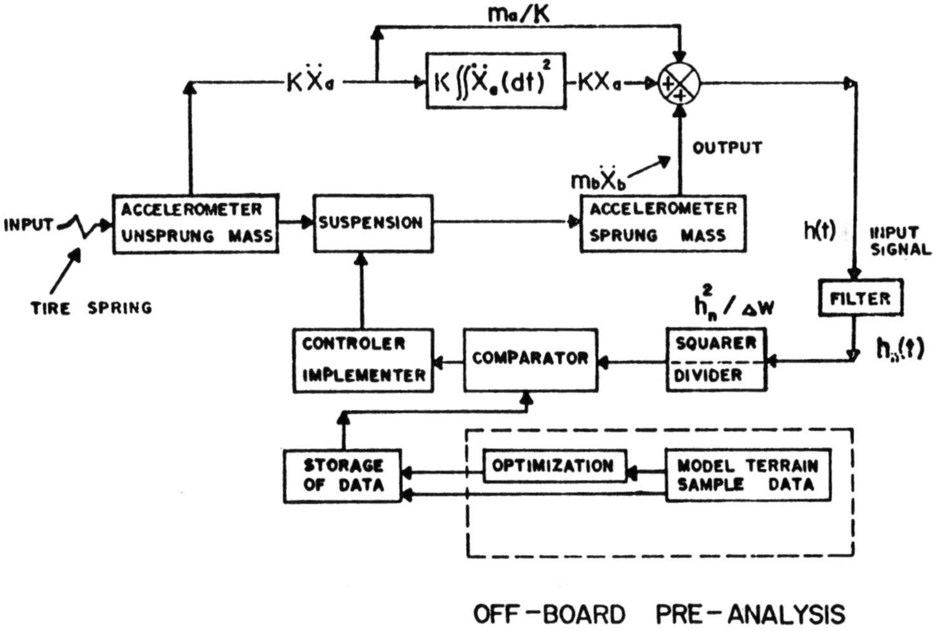


FIGURE 3 BLOCK DIAGRAM OF ADAPTIVE CONTROL

Table I

| H = .15m | | | H = .21m | | | H = .24m | | |
|---|---|---|---|---|---|---|---|---|
| u | $k=\frac{dp_A}{ds}$ | c | u | $k=\frac{dp_A}{ds}$ | c | u | $k=\frac{dp_A}{ds}$ | c |
| 8 | 1000 | 1132 | 8 | 2000 | 267 | 8 | 2000 | 537 |
| 12 | 3000 | 980 | 12 | 4000 | 755 | 12 | 5000 | 422 |
| 16 | 6000 | 462 | 16 | 8000 | 537 | 16 | 9000 | 566 |
| 20 | 9000 | 1132 | 20 | 12000 | 654 | 20 | 14000 | 706 |
| 24 | 14000 | 706 | 24 | 17000 | 778 | 24 | 11000 | 3129 |
| 28 | 18000 | 800 | 28 | 24000 | 924 | 28 | 28000 | 998 |
| m/sec | N/m | N sec/m | m/sec | N/m | $\frac{N\ sec}{m}$ | m/sec | N/m | $\frac{N\ sec}{m}$ |

At axle clearance H = .15m  u = 16 m/sec,
k = 3000 (N/m),  c = 1132 $\frac{N\ sec}{m}$

are optimal values. So are
u = 8m/sec,  k = 2000 N/m
c = 267 $\frac{N\text{-}sec}{m}$

for H = .21m
and u = 12m/sec,  k = 5000 $\frac{N}{m}$
c = 422 $\frac{N\ sec}{m}$

for H = .24m .

Table 2

## Technical Data

**Tandem Axle**

Sprung Mass M: $4.087 \times 10^3$ kg per wheel set

Unsprung Mass m: $3.043 \times 10^2$ kg per wheel set

Sprung Mass

Natural frequency 60 cpm = 1 Hz; $\omega_n^2 = 39.48/sec^2$

Spring rate k = 161,354 kg/sec$^2$

Spring rate K = 1,201,376 kg/sec$^2$

Damping Constant c = 17,976 kg/sec

Damping Parameter $\xi = c/2\ (kM)^{1/2} = .7$

Wave Length L = 100m (m = meter)

Speed u = 25 m/sec (maximum), 20, 15, 10, 5 m/sec

Double amplitude of fundamental wave of length 100m: .1 m (10cm)

Axle clearance:   (Static-Source)

RESULTS

In Figs. 4,5,6, we show the acceleration in units of 'g' of the vehicle body versus speed for three differently chosen axle clearances H = .15m, .21m, .24m. All of these represent optimal data, as explained above, relative to variable suspension spring rates and shock absorber constants. Fig. 7 shows a comparison of acceleration transmissibility between two systems, one with fixed suspension parameters and one with optimized suspension parameters. The improvement is remarkable.

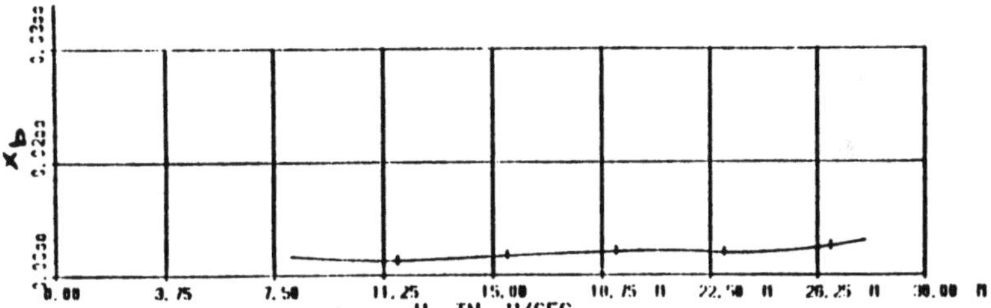


$x_b$ VS U (H= 0.24M)   U IN M/SEC   FIGURE 6

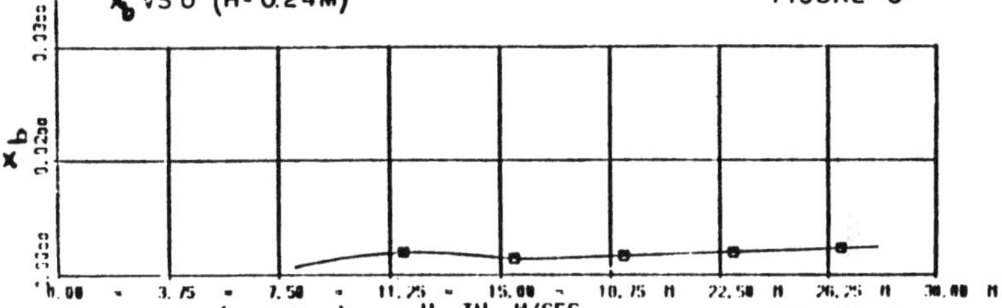


$x_b$ VS U (H=0.21M)   U IN M/SEC   FIGURE 5

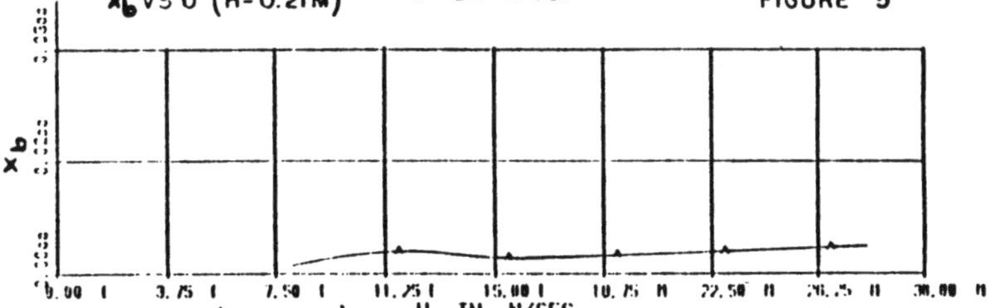


$x_b$ VS U (H=0.15 M)   U IN M/SEC   FIGURE 4

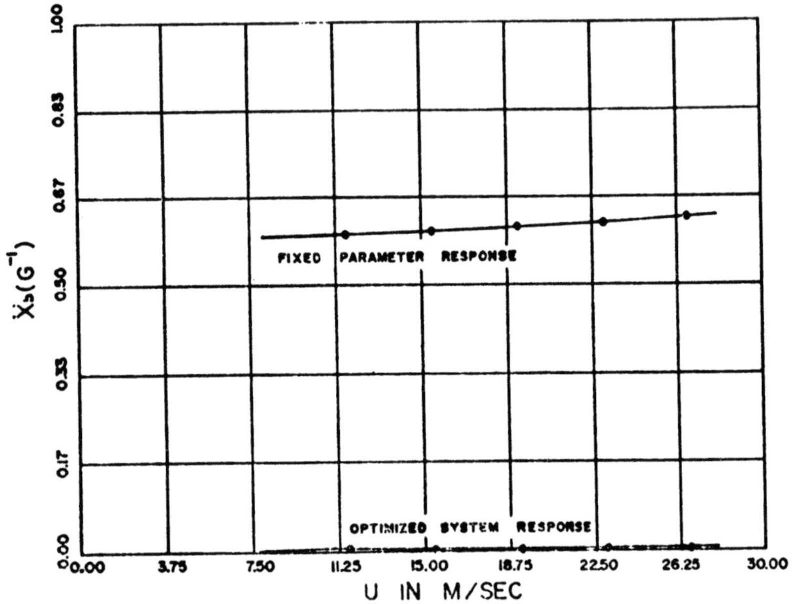


FIGURE 7 **COMPARISON $x_b$ VS. U (H=0.21M) CONSTANT AND OPTIMIZED PARAMETER**

CONCLUSION

In summarizing the results of this investigation it is found that the proposed strategy for the solution of the suspension control problem is most suitable for multi-axle vehicles operating on terrains having power spectral densities ranging from $10^{-1}$ to $10^{-4} m^2$/rad/m within the frequency range of 2 to 6 rad/m. The envisaged control system offers the options of automatic control, manual control (based on the perceptions of the operator, control based on static deflection only due to payload (dynamically fixed control) on smooth terrains (beaches) and paved roads.

Among the advantages of dynamic suspension control are: likelihood of increased fatigue resistance of components due to lowering of the number of operating cycles (however, at higher stress levels due to near limit deflections) and better utilization of dampers (shock absorbers), increased comfort to passengers and better protection of fragile loads.

The particular advantage of this adaptive vs. feedback suspension control lies in its capability of limited previewing of terrains without the need of carrying expensive radar or sonar scope equipment on the vehicle. It has been determined that sampling lengths of 100m to 120m are quite adequate for their description. At 90km/h average vehicle speed (25m/sec) this amounts to about 4 sec. of dwell time (minimum). Updating of parameter settings occur at intervals of sample length.

ACKNOWLEDGEMENT

This work was sponsored by the U.S. Army Tank Automotive Command in Warren, Michigan [1,15]. The author is indebted to Dr. Mahendra Singh, who assumed the responsibility for the computer programs and computer graphics and Mr. R. Siorek from the sponsoring agency, who provided the terrain statistical data and the design data.

REFERENCES

1. H.K. Sachs, Development of Computerized, Active Vehicle Suspension, U.S. Army Tank Automotive Command, Warren, MI. Report No. 12126, March 1976, U.S. Dept. of Commerce Documentation Center, Springfield, VA, Reference Ad-A028390.
2. G.M. Federspiel-Labrosse, "Contribution à l'Etude et au Perfectionement de la Suspension des Véhicules", J. de la SIA-1954, F.I.S.I. T.A., pp. 427-436.
3. W.O. Osbon and L. R. Allen, "Active Suspension Systems for Automotive Military Vehicles", Scientific Paper 65-IDI-HYDRA-PI 1965, Westinghouse Res. Lab., Pittsburgh, PA.
4. I.L. Paul and E.K. Bender, "Active Vibration Isolation and Active Vehicle Suspension", PB.173648, Report DSR-76109-1,-2 (Bibliography, 1966),U.S. Clearinghouse, Springfield, VA.
5. E.K. Bender, D.C. Karnopp and I.L. Paul, "On the Optimization of Vehicle Suspensions Using Random Process Theory", ASME Publication 67-Tran-12.
6. E.K. Bender and I.L. Paul, "Analysis of Optimum and Preview Control of Active Vehicle Suspensions", Report DSR-76109-6, PB 176137, 1967, U.S. Clearinghouse, Springfield, VA.
7. I.L. Paul and B.F. Fenoglio, "Design and Computer Simulation of a Near Optimum Active Vibration Isolation System", Report DSR-76109-8, 1968, U.S. Document Center, Springfield, VA.
8. N. Wiener, "Extrapolation, Interpolation and Smoothing of Stationary Time Series", Technology Press, Cambridge, MA,1949.
9. A.G. Thompson, "Optimum Damping in a Randomly Excited Non-Linear Suspension", Proc. Inst. Mech. Engrs., 1969-1970, Vol. 1841, Pt. 2A.
10. S. Heal and C. Cicillini, "Micro Terrain Measurements", U.S. Army Tank-Automotive Command, Report No. 45866L-REC-9,Warren, MI.
11. F. Kozin, L.S. Cote, and S.L. Bogdanoff, "Statistical Studies of Stable Ground Roughness", U.S. Army Tank-Automotive Center, Report No. 8391, November, 1963.

12. E.O. Brigham, "The Fast Fourier Transform", Prentice Hall, 1977, pp. 148-163.
13. E.K. Bender, "Optimization of the Random Vibration Characteristics of Vehicle Suspensions", Ph.D. Thesis, MIT, June 1967.
14. W.T. Thomson, Theory of Vibrations With Applications, Prentice Hall, Englewood Cliffs, N.H., 1972, Chapters, 3,5,10.
15. H.K. Sachs, "Automotive Suspension Control", U.S. Army Tank Automotive Research and Development Command Report No. 12377, U.S. Dept. of Comm., National Techn. Information Service, Springfield, VA 22151, AD-A068405, June 1978.

# ON THE STABILITY AND FOLLOWABILITY OF THE SEMI-TRAILER VEHICLE WITH AN AUTOMATIC STEERING SYSTEM

Yasushi Saito

Faculty of Engineering, Tokyo University of Agriculture and Technology, Tokyo, Japan

SUMMARY

A theoretical study, with simplified mathematical models, is made of the effects of an automatic steerig system of rear wheels of the semi-trailer, which may influence the directional stability and followability ( off-tracking characteristics ) of the semi-trailer vehicles. The steering angle of rear wheels of the semi-trailer, in this system, is a function of yaw angle of the semi-trailer relative to the tractor and also depends proportionally upon the forward speed of the vehicle combination.

Results of a series of sample calculations are presented, to indicate the manner in which the followability and dynamic stability are influenced by various steering ratios between the relative yaw angle and steering angle of rear wheels of the semi-trailer. The calculations also show that the steering system could be useful to solve the conflicting requirements existing between the stability and followability of the vehicle combination.

INTRODUCTION

Lateral oscillations and off-tracking have been regarded as major drawbacks to articulated vehicles operation and furthermore, geometrical factors or devices to improve the followability have, in general, adverse effects on the stability of these vehicles.

As well known, numerous papers hitherto have been published on the subject of directional stability and control of articulated vehicles. More than ten years ago, above all, a series of the most fundamental and systematic studies of the directional behavior of these vehicles was published by F. Jindra [3][4]. His works are so essential that there is not much, especially in the scope of linear analysis, that can be added to the theoretical studies of the handling characteristics of articulated vehicles. In general, these studies have been related to the mechanical behavior of a combination of a towing vehicle and a few trailers, and a little information has been available on the problem of conflicting requirements existing between the stability and followability of these vehicles.

At first, in the treatment presented here, the followability in a sudden quarter circular turn is investigated for a system consisting of a tractor and semi-trailer. Although this combination is comparatively simple, it possess not a few degrees of freedom, and a lot of parameters influencing the trajectories or paths of the system in such a turn. These circumstances make it necessary to idealize the given system to a simpler one amenable to an analytical treatment. Thus, as an input to the vehicle combination, a sudden quarter circular turn is applied and furthermore, all inertia effects of the vehicle combination in turning are neglected, in other words, all motions of the vehicles are assumed to be kinematic.

Secondly, the directional stability in a straight run at a constant forward speed is investigated for the tractor semi-trailer combination with an automatic steering system, which throuth the steering of rear wheels of the semi-trailer could improve the stability at higher speeds and the followability at lower speeds.

As often assumed in a conventional analysis, the study is restricted to small oscillations, and the typical linear method of stability analysis is adapted. The results of a series of sample calculations then show the general characteristics and effects of the automatic steering system on the stability and followability of the tractor semi-trailer combination.

## FOLLOWABILITY

In the general tracking behavior of a tractor semi-trailer combination, as well known, the trajectories or transitional paths of rear wheels of the tractor and of the semi-trailer do not coincide with those of front wheels of the tractor.

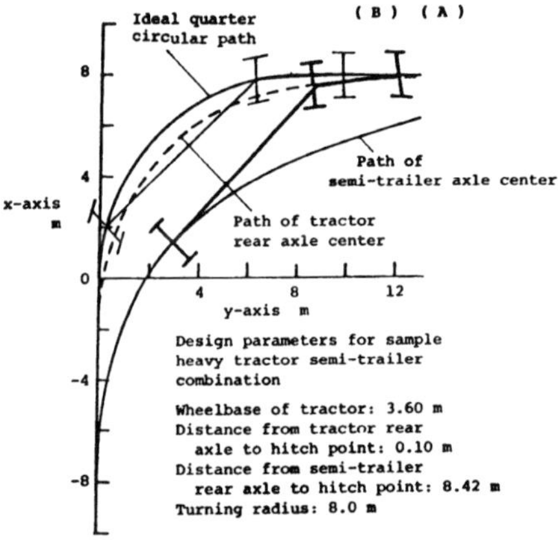


Fig. 1 Tractor semi-trailer excuting a quarter circular turn; ( A ) normal turn, ( B ) special turn, where all centers of three axles are completely on geometrical path.

Consider now, as one of the most typical and practically important cases, the tracking behavior of the tractor and semi-trailer combination excuting a sudden quarter circular turn, after having proceeded previously in a straight line, and straightening out subsequently. For the purpose of simplification, all inertia effects are ignored here, in other words, the study of followability of the vehicle combination is confined to the geometry and kinematics pertaining to tracking.

To provide a better insight into the general characteristics and magnitudes of the off-tracking, a series of sample calculations is made for a heavy tractor semi-trailer combination.

Shown in Fig.1 are two cases, ( A ) and ( B ), of the vehicle combination in such a turn mentioned above, that is, the case ( A ) indicates an ordinary turning behavior and the case ( B ) describes a special one, in which all wheels of the vehicle combination, including the rear wheels of the tractor and of the semi-trailer, are so steered respectively that the centers of three axles of the vehicle combination completely coincide with the ideal quarter circular path. Apparently, in the case ( A ), the deviation of the rear axle center of the semi-trailer from the geometry is considerably large compared with that of the tractor. Therefore, from the practical standpoint, the steering of rear wheels of the semi-trailer is exclusively considered here to improve its followability.

In Fig.2, all the steered angles of the vehicle combination, corresponding to the case ( B ) in Fig.1, are shown as a function of proceeding displacement of the front axle center of the tractor. In these circumstances, there can be considered two inputs to steer the rear wheels of the semi-trailer; one is the steering angle of the tractor and the other is the relative yaw angle of the semi-trailer to the tractor. From the results shown in Fig.2, as the input, the

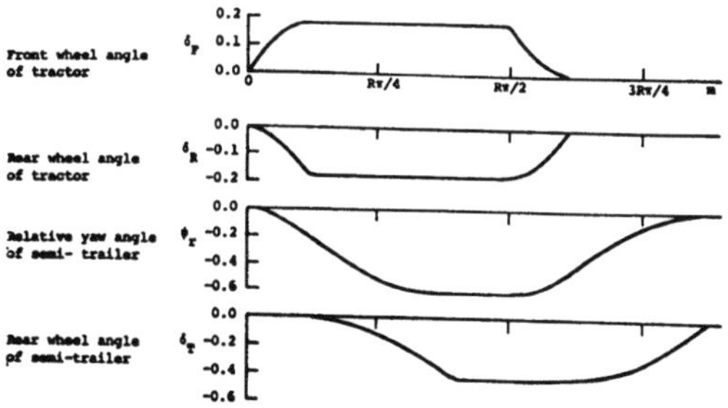


Fig.2 Three steering angles and one relative yaw angle in case ( B ) of Fig.1, R = 8.0 m

relative yaw angle of the semi-trailer seems to be more suited than the steering angle of the tractor, since the phase difference between the relative yaw angle and the steering angle of the semi-trailer is, as a whole, smaller than that between the steering angle of the tractor and the steering angle of the semi-trailer. In addition, as shown later, the steering of rear wheels of the semi-trailer proportionally controlled by the relative yaw angle plays a major role in the improvement of dynamic stability of the vehicle combination. On the contrary, when controlled by the steering angle of the tractor, such a improvement can not be expected.

Subsequently, Fig.3 clearly demonstrates that a higher performance of followability is attained by increasing the steering ratio

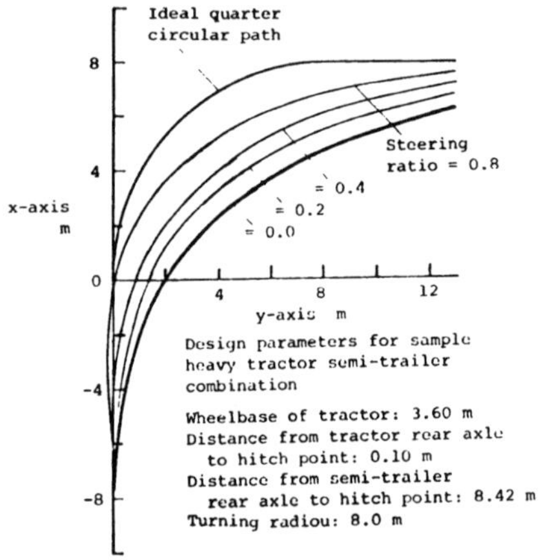


Fig.3 Effect of steering ratio on path of rear axle center of semi-trailer

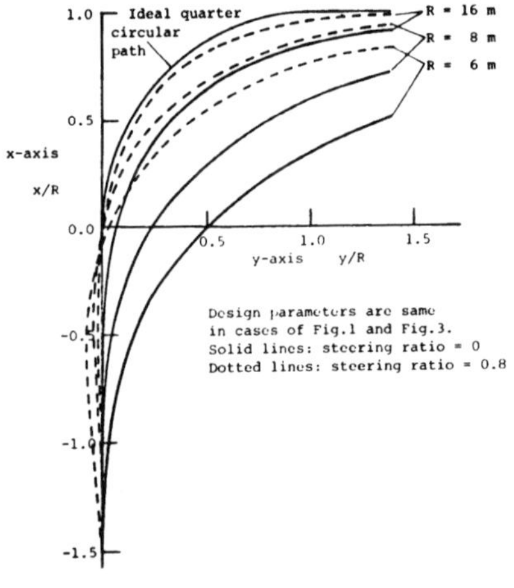


Fig.4 Effect of turning radius on path of rear axle center of semi-trailer and results improved by steering system

between the relative yaw angle and the steering angle of the semi-trailer. However, when the steering ratio reaches to a certain level, 0.8 for instance, a slight but undesirable negative deviation —— on the left of x-axis —— begins to appear at the first part of the transient kinematic path of the rear axle center of the semi-trailer. Therefore, there must be an optimum value of the steering ratio in the quarter circular turn.

In general, as well known, deviations of the rear axle center of the semi-trailer from the geometry are significantly influenced by the length of the turning radius. In order to show the effect of the radius specifically, a few sample calculations are carried out and those results are shown in Fig.4. In the diagram, the solid lines indicate the deviations of the rear axle center of the semi-trailer with no steering system, whereas the dotted lines describe the results improved by the steering system.

## STABILITY

In the study of lateral oscillations of a tractor and semi-trailer combination with an automatic steering system, the standard method of stability analysis is considered to be applicable. Consequently, the equations of motion derived from the simplified analysis are substantially same with the equations written in an ordinary text of vehicle dynamics [1],except a few additional terms involved, which are concerned with steering the rear wheels of the semi-trailer.

To emphasize the effects of the automatic steering system, sample calculations are made for a light tractor semi-trailer comination, since it is likely to be inherently less stable at high speeds.

### *Equations of Motion*

In order to write the equations, a schematic sketch is shown in Fig.5, together with some symbols used in the following equations.

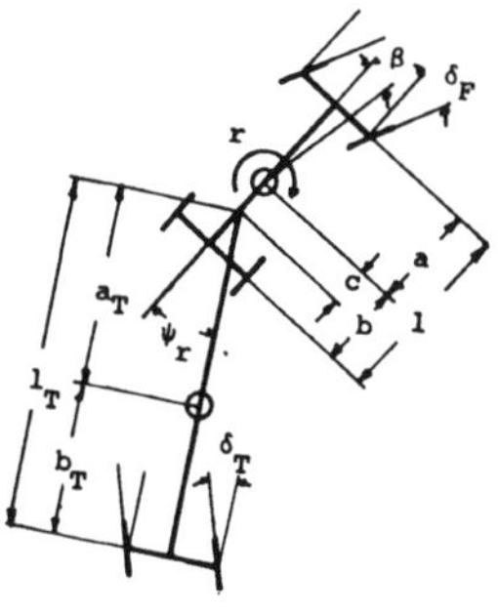


Fig.5 Schematic sketch of tractor semi-trailer combination

The linearized equations of motion for the tractor semi-trailer combination can be derived from the summations of lateral forces and yaw moments about a vertical axis through the hitch point as

$$[(M + M_T)uD + C_A + C_B + C_{TB}]\beta$$
$$+ [(c + a_T)M_T D - (M + M_T)u - \{C_A a - C_B b - C_{TB}(c + l_T)\}/u]r$$
$$+ [M_T a_T D^2 + C_{TB}(l_T D/u + 1 + \alpha)]\psi_r + C_A \delta = 0 \quad (1)$$

$$[McuD + C_A(c + a) + C_B(c - b)]\beta$$
$$- [ID + Mcu + C_A(c + a)a/u - C_B(c - b)b/u]r$$
$$+ C_A(a + c)\delta = 0 \quad (2)$$

$$[M_T a_T uD + C_{TB} l_T]\beta$$
$$+ [I_T D + M_T a_T(c + a_T)D - M_T a_T u + C_{TB}(c + l_T)l_T/u]r$$
$$+ [I_T D^2 + M_T a_T^2 D^2 + C_{TB} l_T^2 D/u + C_{TB}(1 + \alpha)l_T]\psi_r = 0 \quad (3)$$

where, u = forward speed and D = d/dt. Most of symbols used above are listed below and the rest of them are shown in Fig.5.

| Tractor | | Semi-trailer | |
|---|---|---|---|
| $C_A$ | = Cornering stiffness of front wheels | $C_{TB}$ | = Cornering stiffness of rear wheels |
| $C_B$ | = Cornering stiffness of rear wheels | $I_T$ | = Yaw moment of inertia |
| I | = Yaw moment of inertia | $M_T$ | = Mass |
| M | = Mass | $\alpha$ | = Steering ratio |

Note; Both C and $\beta$, in this paper, have opposite sign to the ordinery papers written in English.

In the above equations, the symbol $\alpha$ is the steering ratio between the relative yaw angle $\psi_r$ and the steering angle $\delta_T$, namely, $\psi_r = \alpha\delta_T$. When in a system the steering ratio can decrease —— from positive to negative —— proportionally to increasing of the forward speed or vice versa, this system called here " automatic steering system ". To investigate the effects of the system, a series of numerical calculations was carried out.

*Sample Calculation*

In this calculation, the design parameters of a baseline light tractor semi-trailer combination are taken as follows :

- M = Tractor mass, 288.3 kgf-sec²/m
- $M_T$ = Semi-trailer mass, 974.5 kgf-sec²/m
- I = Tracter yaw inertia, 700 kgf-m-sec
- $I_T$ = Semi-trailer yaw inertia, 3850 kgf-m-sec
- a = Distance from tractor c.g. to front axle, 1.37 m
- b = Distance from tractor c.g. to rear axle, 1.83 m
- c = Distance from tractor c.g. to hitch point, 1.78 m
- l = Wheelbase of tractor, 3.20 m
- $a_T$ = Distance from semi-trailer c.g. to hitch point, 2.29 m
- $b_T$ = Distance from semi-trailer c.g. to rear axle, 1.96 m

$l_T$ = Wheelbase of semi-trailer, 4.25 m
$C_A$ = Tractor front wheels cornering stiffness, 10,800 kgf/rad
$C_B$ = Tractor rear wheels cornering stiffness, 30,000 kgf/rad
$C_{TB}$ = Semi-trailer rear wheels cornering stiffness, 20,000 kgf/rad

Before investigating the effects of steering of rear wheels of the semi-trailer on its dynamic stability, it is of interest to examine how the stability and followability are affected by various steering ratios in a steady state circular turn. The results of sample calculations indicate that the tractor towing the semi-trailer is considered to have oversteer characteristics and also that the yaw velocity and sideslip angle of the tractor are not influenced by changing the value of the steering ratio. In sharp contrast to this, the steering ratio exert a strong influence on the relative yaw angle in a steady turn, as shown in Fig.6. Therefore, the variation of the

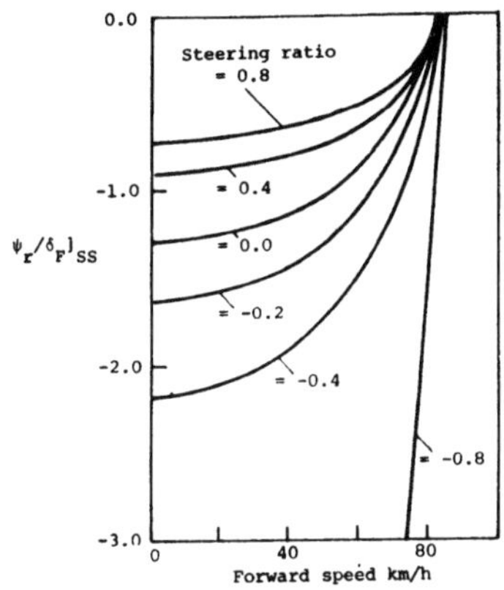


Fig.6 Effect of steering ratio on relative yaw angle of semi-trailer in steady state

steering ratio has a significant bearing on the vehicle steering characteristics, especially on the followability.

Subsequently, in order to examine the effects of the steering ratio on transient behaviors of the vehicle combination, the responses of the relative yaw angle of the semi-trailer, resulted from application of a step input, are shown in Fig.7. From the step responses calculated of the relative yaw angle, the positive ratio lowers the dynamic stability of the semi-trailer and the negative ratio improves oscillatory characteristics.

To provide a more complete information regarding the effects of the steering ratio on the stability of the vehicle combination, the real parts of the roots from the characteristic equation are plotted in Fig.8. It is of interest to note that a pair of the two real roots closely related to the characteristic equation of the single tractor are hardly influenced by changing the value of the steering ratio and that in contrast to the above a conjugate complex pair are considerably influenced by various steering ratios.

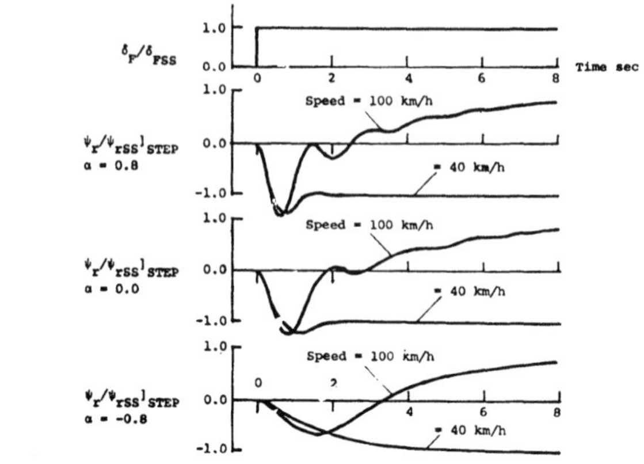


Fig.7 Responses of relative yaw angle $\psi_r$ of semi-trailer to step input of front wheel angle of tractor
( $\psi_{rSS}$ : relative yaw angle in steady state )

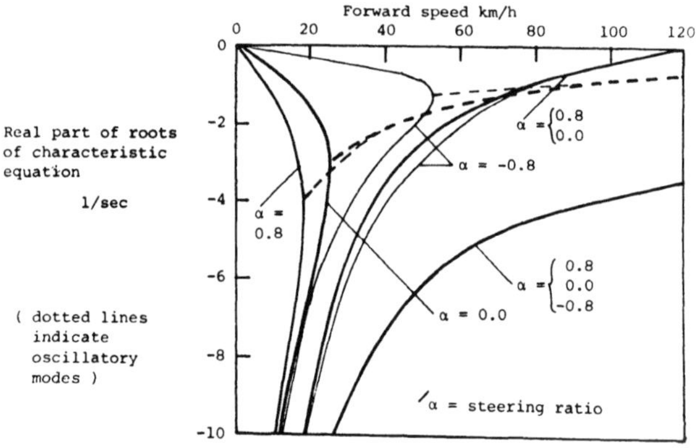


Fig.8 Real part of solution of characteristic equation for sample light tractor semi-trailer combination with steering system

The stability characteristics displayed in Fig.8 are further clarified in Fig.9, where the damping ratio and natural frequency of the oscillatory motion are plotted as a function of the forward speed respectively. In this figure, it is readily seen that the negative steering ratio could be a strong stabilizing factor for the vehicle combination at higher speeds. On the other hand, as mentioned in the foregoing paragraph, the positive value of the steering ratio could improve the followability of the vehicles at lower speeds.

During the design process of the automatic steering system, therefore, a compromise must be made between providing minimum off-tracking at lower speeds and achieving satisfactory dynamic stability

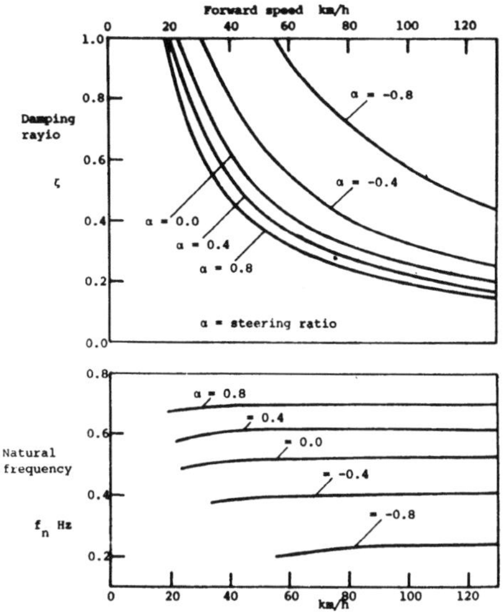


Fig.9 Effect of steering ratio on damping ratio and natural frequency for sample light tractor semi-trailer combination

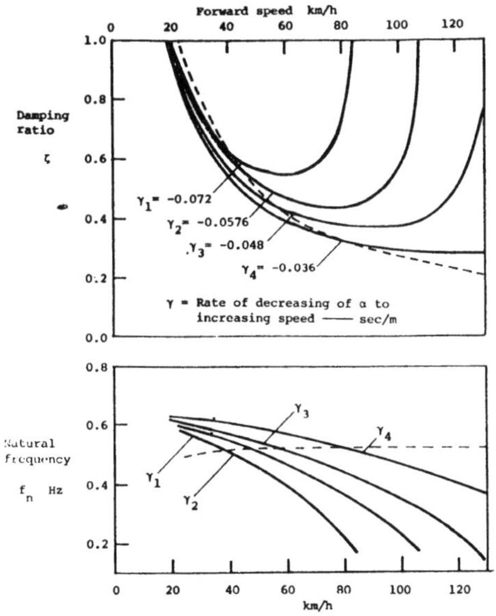


Fig.10 Effect of automatic steering system on damping ratio and natural frequency for sample light tractor semi-trailer combination ( dotted lines indicate $\zeta$ and $f_n$ of baseline vehicle combination )

at higher speeds. A series of plotted curves shown in Fig.10 is the results of sample calculations of the damping ratio and natural frequency for the tractor semi-trailer combination with the automatic steering system, in which the steering ratio is set to be directly proportional to the forward speed with various rates.

Thus, without any modification of the vehicle design parameters —— for example, wheelbase of semi-trailer, position of c.g. of semi-trailer, tire characteristics and so on —— the vehicle combination with the automatic steering system could be expected to have a better handling characteristics throughout the range of normal operating speeds.

## CONCLUSION

The primary objective of this study is to make a presentation or proposal of an automatic steering system, which is a device to solve the conflicting requirements —— stability and followability —— of achieving more satisfactory handling, rather than a refined quantitative analysis of the tractor semi-trailer combination with it. The latter is left to future work.

As far as concerned with the parameters of this system, further study toward establishing an optimum steering ratio not only from the vehicle responses but also from the extensive human factors evaluation would be desirable.

## REFERENCES


1. Ellis, J.R., Vehicle Dynamics, 1969, Business Books Limited, London.
2. Jindra, F., Off-tracking of Tractor-Trailer Combinations. Automobile Engineers, March 1963, p. 96-101.
3. Jindra, F., Tractor and Semi-Trailer handling. Automobile Engineers, October 1963, p. 438-446.
4. Jindra, F., Lateral Stability of a Three-Unit Tractor Train. Ing-Arch XXVII/4, 1963, p. 209-226.

# INVESTIGATION OF THE INFLUENCE OF VARIOUS BRAKING REGULATIONS ON ACCIDENT-AVOIDANCE PERFORMANCE

Michael Sayers and Leonard Segel

Highway Safety Research Institute, University of Michigan, Ann Arbor, MI, U.S.A.

SUMMARY

Passenger car braking regulatings promulgated by the United States (FMVSS 105-75) and various European countries are studied in terms of incompatibilities in brake proportioning that derive from the requirements in the regulations. Analysis shows that the European regulations require a larger forward bias in proportioning than the U.S. regulation. Computer simulations were used to predict differences in performance that result from proportioning two example vehicles (one produced in the U.S. and the other in Europe) to comply with the different regulations. The simulations included two states of vehicle loading and conditions of straight-line braking, braking in a turn, and three surface friction levels. Both vehicles were usually capable of achieving shorter stopping distances when proportioned to meet 105-75 than when proportioned to comply with the European regulations, but the differences were small and could be considered trivial when a car is driven by an ordinary driver. When proportioned to meet 105-75, the rear wheels of the two study cars sometimes locked up first (a condition that the European regulations are designed to prevent), but the corresponding deceleration levels were higher than the ranges covered by any of the regulations.

INTRODUCTION

The proliferation of vehicle performance regulations within the international community has resulted in a variety of braking performance requirements among North American and European countries. These regulations are influenced by different opinions on good braking and handling performance so that they can be incompatible with each other. As a result, some vehicle manufacturers might have to modify brake proportioning on vehicles intended for export.

The various braking regulations that are in force around the world have been previously reviewed and compared extensively [1,2,3,4]. Accordingly, the study reported herein focuses on just two questions: "How do the various brake regulations restrict front/rear proportioning selection on typical passenger cars?" and "Which regulations lead to passenger cars with better accident avoidance capabilities?"

The first question is addressed by first presenting an overview of the technical sections of the various braking regulations that place requirements on the non-failed systems and might affect proportioning selection. Next, by quasi-static analysis, we calculate the range of proportioning permitted by each regulation for two "representative" passenger cars—one of U.S. manufacture, the other of European manufacture. (This study is limited to constant proportioning, thus excluding more sophisticated systems such as load sensitive valves, antilock, etc.)

We address the second question by employing a comprehensive computer model of the passenger car to calculate the braking performances of the two representative vehicles over a wide range of operating conditions that are not necessarily covered in the regulations. Each condition is simulated with proportionings selected, on the basis of quasi-static analysis, to best comply with the U.S. and European regulations. The simulated conditions cover extreme option selection (that is, customer options such as automatic transmission, air conditioning, etc.) and load combinations, straight-line braking, braking in a constant radius turn (two radii), and three surface conditions exhibiting high, medium, and low frictional properties.

Tire and vehicle parameter data are included in an appendix.

## AN OVERVIEW OF EXISTING BRAKING REGULATIONS

In a broad review of existing brake standards, Oppenheimer [1] categorizes the existing regulations into the braking standards of the United States (FMVSS 105-75), Western Europe (ECE R.13 and 71/320/EEC), and Sweden (F-18). The regulations reflect different philosophies towards what constitutes "good braking" and utilize different means of achieving their goals.

Each of the four braking standards contains stopping distance or deceleration limits which are verified by testing a candidate vehicle. The requirements of these standards are summarized in Table 1. This table also shows 90%* target stopping distances and "equivalent decelerations" for each regulation, calculated after assuming that the vehicle deceleration starts at zero, increases linearly to a time $\Delta t$, then remains constant for the duration of the stop. The "equivalent deceleration" is the constant value reached after time $\Delta t$, defined here to be 0.5 second.

Although the European regulations require less deceleration capability, they contain additional provisions aimed at preserving the stability of the vehicle. The different regulations are like "apples and oranges" and, thus, cannot be compared in a generalized manner; however, quasi-static analysis can be used to calculate constraints on brake proportioning applicable to a particular vehicle that derive from the regulations. This analysis involves only the fore/aft load transfer occurring during braking and yields adhesion utilization values at the front and rear axles for a given deceleration level and a given proportioning. The adhesion utilization values, $K_F$ and

---

*When designing for a regulation involving physical testing, a manufacturer will typically allow for different sorts of imprecisions; in this study, a tolerance of 10% is used repeatedly.

Table 1. Target Stopping Distance and Deceleration Capabilities from the Collected Standards.

| Standard | Test | Loading | Velocity (mph) | Stopping Distance (ft) | 90% of Stopping Distance (ft) | Equivalent Deceleration (g's) |
|---|---|---|---|---|---|---|
| USA 105-75 | 2nd Effectiveness | GVWR | 30<br>60<br>80 | 54<br>204<br>383 | 48.6<br>183.6<br>344.7 | .795<br>.744<br>.678 |
| | 3rd Effectiveness | Lightly Laden | 60 | 194 | 174.6 | .787 |
| | 1st & 4th Effectiveness | GVWR | 30<br>60<br>80<br>100 | 57<br>216<br>405<br>673 | 51.3<br>194.4<br>362.7<br>605.7 | .746<br>.697<br>.641<br>.587 |
| 71/320/EEC & ECE R.13 | Type '0' | Lightly Laden & GVWR | 49.7 | 166.2 | 149.6 | .628 |
| 71/320/EEC, ECE R.13, Sweden's F-18 | Retardation | Lightly Laden & GVWR | 49.7 | .592 g's specified, + 10% = | | .651 |

$K_R$, at the front and rear axles, represent the braking forces between the tire and the road, normalized by the vertical loads on the tires. The front and rear adhesion utilizations, respectively, are given by:

$$K_F = \frac{p \cdot A_x}{(1 - a/\ell + h/\ell \cdot A_x} \qquad (1)$$

$$K_R = \frac{(1-p) \cdot A_x}{(a/\ell - h/\ell \cdot A_x} \qquad (2)$$

where p is proportioning (the ratio of the brake torque generated at the front axle to the brake torque generated at both axles), $A_x$ is deceleration (in g's), $a/\ell$ is the longitudinal distance between the c.g. (center of gravity) of the vehicle and the front axle divided by the wheelbase, and $h/\ell$ is the ratio of the height of the c.g. of the vehicle divided by the wheelbase of the vehicle. In deriving the above, it is assumed that none of the wheels are locked and that the rolling radii of all tires are equal. If the brakes on the vehicle are sufficiently powerful and fade resistant to lock up any wheel on dry pavement with a reasonable pedal force application, the only design variable which can be set to keep $K_F$ and $K_R$ within limits defined by the various braking standards is the front/rear proportioning, p. When $K_F$ and $K_R$ are not directly specified, an implied constraint is that K must be less than the available traction in order to prevent lockup.

Accordingly, the four braking regulations of interest are summarized below in terms that make use of Equations (1) and (2) and as displayed in figures which show proportioning constraints imposed by separate regulations on an example vehicle.

*United States: FMVSS 105-75*

The U.S. regulation [5] requires stopping distance tests to be performed on a "roadway having a skid number of 81...without lockup of any wheel at speeds greater than 10 mph." All of the stopping distance requirements for non-failed brakes are listed in Table 1, along with the "equivalent deceleration" levels calculated for 90% stopping distances and a .5-second build-up time for the deceleration.

The front and rear axle adhesion utilizations, $K_F$ and $K_R$, are shown as functions of deceleration and proportioning in Figure 1 for a 1978 Chevrolet Monte Carlo in a "lightly-loaded" condition. The

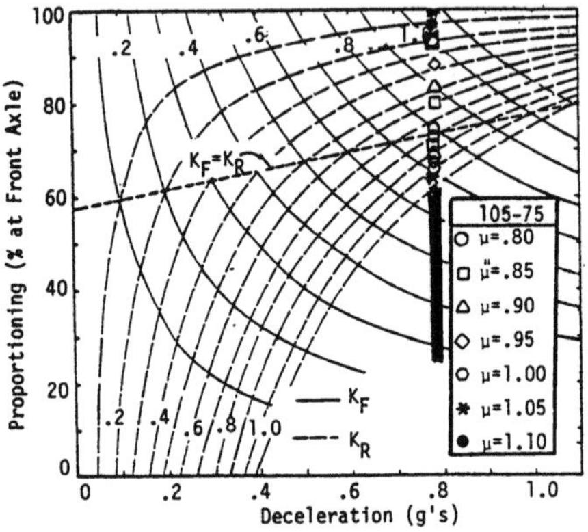


Fig. 1. Lines of constant $K_F$ and $K_R$, and the "proportioning window" imposed by 105-75 on the lightly-loaded Monte Carlo.

curves of constant K can be used to locate proportioning boundaries for various possible levels of tire-road frictional capabilities, as indicated by the friction coefficient, $\mu$. For example, consider a deceleration of .79 g (see Table 1) with no lockup. If p = 70%, we can see that $K_F$ (the adhesion utilization of the front axle) would be .75 and $K_R$ would be .90. If $\mu$ = 0.8, the rear axle would try to generate more longitudinal force than is available and the wheels would lock up. If, however, $\mu$ = .95, the 70% proportioning would be permissible. With $\mu$ = .95, the quasi-static analysis defines a "proportioning window," extending from p = 68% to p = 88%, such that any proportioning within this "window" will allow a .79 g deceleration with $K_F$ and $K_R$ both less than $\mu$, and thus no wheel lockup.

Tire traction limits are usually sensitive to vertical load and therefore the friction coefficient at the front axle might differ from the friction coefficient at the rear axle. For example, dry-asphalt traction data* compiled for the tires installed on the Monte Carlo (see appendix) show values of $\mu_F$ = 1.03 and $\mu_R$ = 1.11 at the normal loads produced by this car in a deceleration of 0.79 g, and thus define a proportioning "window" applicable for dry asphalt, as indicated in Figure 1 by the heavy black lines. This window, as shown, is for one option package and loading condition. But, since the standard is applied to all possible option packages at both GVWR and light loading, a proportioning window must be considered for each combination. The overall window for the car model is defined by the overlap of all of the windows applicable to specific option and

*This data is typical for a surface with an ASTM dry skid number of 81.

loading combinations.

*Western Europe: ECE R.13 and 71/320/EEC**

The European regulations specify stopping distances, given that "the road shall possess a surface having good adhesion," and that the stop must be made without the wheels locking.

A proportioning window determined in the same manner as the window for 105-75 but based on a target deceleration level of 0.65 g's (see Table 1) is shown in Figure 2 for the same vehicle, loading condition,

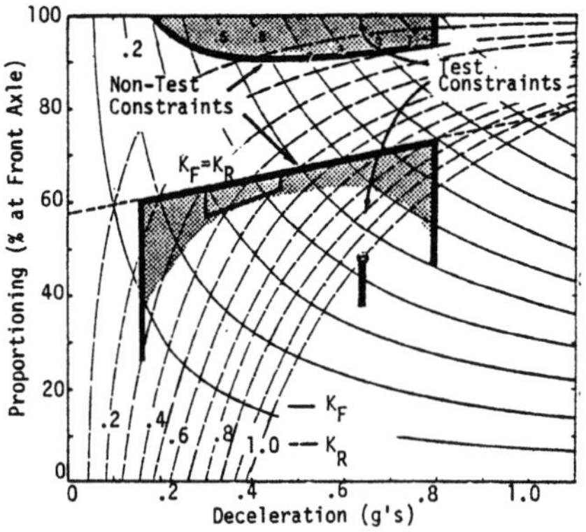


Fig. 2. Proportioning constraints imposed by 71/320/EEC on the lightly-loaded Monte Carlo.

and tire data that were considered in Figure 1. The bounds on proportioning are clearly much wider than is the case for the U.S. requirement, due to the lower deceleration requirement called out in 71/320/EEC.

71/320/EEC also requires the vehicle manufacturer to submit engineering data. These data consist of plots of adhesion utilization (K) versus deceleration ($A_x$) which are calculated according to Equations (1) and (2). The regulation requires that these curves meet the following two conditions:

1) $A_x \geq 0.1 + .85(K_F - .2)$

2) $K_F > K_R$

Combining these inequalities with Equations (1) and (2) yields the following relationships:

*The Common Market regulation 71/320 [7] is nearly identical to the United Nations ECE R.13 [6] in the technical aspects, and unless noted otherwise, terms and phrases cited in this section are excerpted from 71/320/EEC.

$$p > 1 - a/\ell + h/\ell \cdot A_x \quad (3)$$

$$p < \left(\frac{A_x - 1}{.85} + .2\right) \cdot (1 - a/\ell + h/\ell \cdot A_x)/A_x \quad (4)$$

The proportioning constraints imposed by Equations (3) and (4) are shown in Figure 2 and we see that these "nontest" constraints restrict the allowable proportioning much more than the stopping distance criterion. It should be noted, however, that in order to provide curves of K vs. $A_x$, per the regulation, the manufacturer must assume a value for the braking force/line pressure gain at each axle. But this gain depends on the brake lining friction, which varies under different operating conditions. Thus differing curves could be prepared for one vehicle by making different (yet valid within proper contexts) assumptions regarding the lining friction. (This study treats the braking force/line pressure gain of a particular axle as a constant.) A small discrepancy between ECE R.13 and 71/320/EEC is that the latter permits $K_R > K_F$ over a brief interval, as shown in the figure by the notch in the lower boundary. However, this does not affect proportioning selection if the proportioning is a constant.

*Sweden: F-18*

The Swedish regulation F-18 [8] simply requires a deceleration level of .592 g's (Table 1) and that "no wheel will lock at retardations lower than those prescribed...when braking on a carriageway having a coefficient of friction of 0.8...there is also the condition that at a retardation between 5.8 and 8.0 m/sec$^2$ the rear wheels shall not lock before the front wheels."

A departure of F-18 from other regulations is the direct specification of a road coefficient of friction, which is not a theoretical assumption, but an actual test condition. The procedure involves wetting the track, measuring the tire-road friction coefficient, $\mu$, at optimal slip, and, if necessary, changing the water depth until the $\mu$ readings are within .05 of the specified value of .8. (The reference tire used to establish the specified test condition is the ASTM E249-14 tire which is no longer the standard tire in the U.S.) On one hand, available friction is measured rather than sliding friction (per 105-75), but on the other hand, the correlation between measurements made with the ASTM tire and those made with a normal passenger car tire may not be very good. Lacking any data, we assume below that $\mu = .8$ for the tires on the vehicle under test, and that the tire-road friction coefficient is not load or velocity sensitive. (Clearly, this is not the case, but the nature of the friction sensitivity to load depends on the tires and operating conditions and, although tire traction on a wet surface is sensitive to velocity, the friction is lowest at the initial velocity and will only go up as the vehicle slows. Thus, the friction coefficient which exists at the initial velocity should be adequate to predict lockup and to determine the brake torque levels which define the "limit" condition.)

Another important condition in F-18 is that the order of wheel lockup is a criterion for certification and is found through road testing. Therefore we consider manufacturing tolerances.

Brake proportioning can be thought of as a design proportioning subject to error as a result of actual torques being different from the desired torques, i.e.,

$$p(1+\varepsilon_p) = \frac{T_F(1+\varepsilon_F)}{T_F(1+\varepsilon_F) + T_R(1+\varepsilon_R)} \qquad (5)$$

where $T_F$, $T_R$, and p are the design values of front torque, rear torque, and proportioning, and $\varepsilon_F$, $\varepsilon_R$, and $\varepsilon_p$ are the corresponding relative errors. By rewriting Equation (5), the proportioning error can be expressed as:

$$\varepsilon_p = \frac{(\varepsilon_F-\varepsilon_R)(1/p - 1)}{\varepsilon_F - \varepsilon_R + (1+\varepsilon_R)/p} \qquad (6)$$

The proportioning boundaries imposed by F-18 on the example Monte Carlo are sketched in Figure 3. The error tolerances for the front

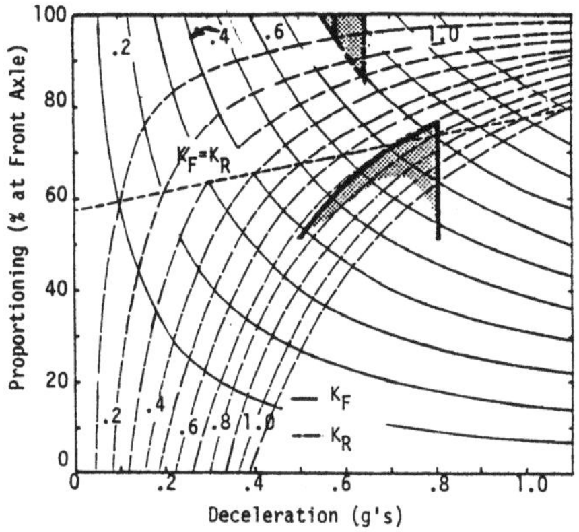


Fig. 3. Proportioning constraints imposed by F-18 on the lightly-loaded Monte Carlo.

and rear brake torque gains were set ± 10% with the signs set to the worst case for each boundary. (For example, if braking performance is limited by impending lockup of the front axle, $\varepsilon_F$ is assumed to be +10%, while $\varepsilon_R$ is assumed to be -10%. The actual proportioning is then greater than the design proportioning, and accordingly, the design proportioning is set lower to compensate for the assumed error.) The upper boundary follows the $K_F$ = .8 curve, with an adjustment for the proportioning tolerance, up to the deceleration which is specified for no wheel lockup. The solid line is for the limit $A_x$ = .59— the value given in the standard—and the dashed line is for $A_x$ = .65— the target value that includes a 10% margin of safety, as per Table 1. The requirement that the rear wheels not lock before the front wheels at decelerations between 5.8 and 8.0 m/sec$^2$ extends the lower proportioning boundary to $A_x$ = .82 g's. (If, however, the available friction is only $\mu$ = .80, as we have assumed, this lower boundary need only be extended to $A_x$ = 0.8.)

A comparison of Figures 1, 2, and 3 reveals the proportioning constraints imposed by F-18 to be more restrictive than those imposed by 105-75 or 71/320/EEC (ECE R.13), but it must be noted that more assumptions lie behind Figure 3 than behind the other two figures.

PROPORTIONING CONSTRAINTS WHICH ARE IMPOSED BY THE VARIOUS REGULATIONS ON TWO REPRESENTATIVE VEHICLES

A typical U.S. passenger car and a typical European car are used to evaluate the limits on those proportioning values which satisfy the various regulations.

The representative U.S. vehicle is the 1978 Chevrolet Monte Carlo and the representative European vehicle is the Volkswagen Golf.

For each of the two vehicles, there are two combinations of options and loading for which the braking efficiency is the most compromised. These are:

1) A lightly-loaded car equipped with many options that add weight to the front end, resulting in a forward center-of-gravity location.

2) A heavily-loaded car equipped such that the gross axle weight rating (GAWR) for the rear axle is highest relative to that of the front axle.

Overall vehicle weight is also a factor in deceleration performance because tire traction decreases with increasing load.

Parameter values for the two cars are listed in the appendix. This appendix also lists the "worst case" option selection and contains tire traction data.

Figure 4 presents the proportioning boundaries imposed by each of the regulations (discussed above) on the Monte Carlo. The lower

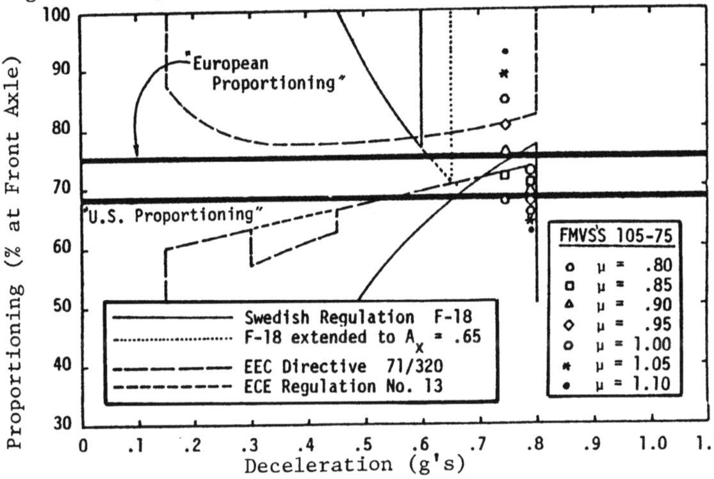


Fig. 4. Proportioning boundaries imposed by various braking regulations on the Monte Carlo.

boundaries all apply to one "worst case" option/load combination, while the upper boundaries are for the other "worst case." A comparison of Figure 4 with Figures 1 through 3 reveals the extent to which the "proportioning windows" tighten when worst cases are considered.

The "window" defined by FMVSS 105-75 depends on the available traction—a function of the tire and road surface properties. Figure 4 shows limits for various μ values, so that tire data can be used to determine the proportioning value which would do the best job of insuring that the car can pass the road tests required by 105-75.

A proportioning value of p = 68% is best, given that the tire data indicate the friction coefficient at the rear axle (under light loading) is 0.12 higher than that of the front axle (under GVWR loading). Thus 68% is selected as a so-called "U.S. proportioning."

71/320/EEC (ECE R.13) is best satisfied by a "European proportioning" value of p = 75%. The boundaries due to the F-18 overlap, as shown in the figure, with p = 75% being a good value for minimizing the degree to which the proportioning boundaries are violated. Thus F-18 and 71/320/EEC, while implemented differently, are best satisfied by the same proportioning.

Figure 5 illustrates the proportioning constraints applicable to the Golf. In this case, the "U.S. proportioning" is 77% and the "European proportioning" is 83%.

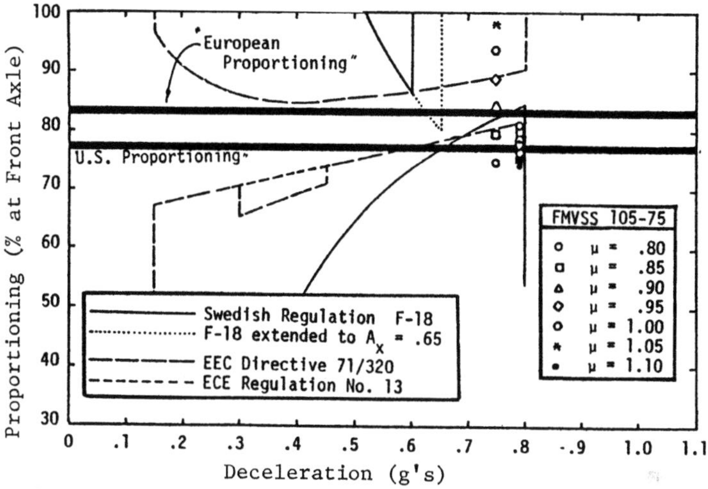


Fig. 5. Proportioning boundaries imposed by various braking standards on the Golf.

It is clear from Figures 4 and 5 that the regulations of the U.S. and Europe impose similar constraints on proportioning selection, although they are implemented differently and reflect different concepts of "good braking." Both result in a high braking efficiency capability on high friction surfaces, and reduced efficiencies on surfaces with lower friction levels. The European regulations require a slightly higher proportioning, therefore braking efficiency will be higher for the vehicle proportioned to meet 105-75 on all but high-friction surfaces, where the European proportioning results in high braking efficiency and a bit more stability than would be the case for a U.S. proportioned vehicle.

## BRAKING PERFORMANCE UNDER CONDITIONS NOT ADDRESSED IN EXISTING REGULATIONS

Having evaluated the incompatibility between the European and U.S. standards, we address the question, "Which regulatory philosophy leads to overall better accident-avoidance capability?"

To obtain appropriate data, a number of computer simulations were conducted with the Golf and Monte Carlo, each with the so-called "U.S. proportioning," and "European proportioning" and at both of the extreme loading/option package combinations identified in the appendix; thus a total of 8 vehicle conditions were examined (2 baseline vehicles x 2 load/option conditions x 2 proportionings). The braking maneuvers included straight-line braking and braking in a constant-radius turn (initial lateral acceleration = 0.4 g and also 0.2 g). These maneuvers were performed on three surfaces, with high, medium, and low friction levels, with the one exception of the severe turn-low friction surface combination, because the surface was not capable of providing the necessary cornering forces.

(The computer simulation program was developed at HSRI during an earlier study of vehicle behavior [9]. It is based on a 14-degree-of-freedom model of the motor car and includes a pneumatic tire model sufficiently complex to predict shear forces generated during combined cornering and braking conditions.)

An automatic steering controller was added to the vehicle model in order to simulate the braking-in-a-constant-radius-turn maneuver. This controller employs a feedback signal that represents the deviation from the intended path of a point lying on the longitudinal axis of the vehicle, a small distance in front of the c.g. A proportional-derivative control scheme is used such that $\delta$, the intended steer angle (not including deflection-steer or compliance-steer effects) at the road, is defined as:

$$\delta = C_p \cdot \Delta x + C_d \cdot \Delta\dot{x} + C_i \quad .$$

where $\Delta x$ is the path deviation and $C_p$ and $C_d$ are feedback gains. Compared to a human driver, this controller is crude and simple, although these drawbacks are mitigated by a response that is quicker than a human controller. On noting that the braking-in-a-turn simulations are not to establish precise magnitudes of vehicle performance indices, but rather to compare the performances of two differently proportioned vehicles, we can be reasonably assured that the controller is adequate to maintain desired vehicle trajectories within the context of this study.

Limit conditions were found by incrementing the brake-line pressure until an axle lockup occurred during straight-line braking, or until the automatic controller was no longer able to maintain a constant radius path during the braking-in-a-turn maneuver.

A more detailed discussion of the vehicle model is presented in [10], which also lists all of the vehicle and tire parameter data needed for simulation.

Tables 2, 3, and 4 summarize the deceleration and stopping distance performances predicted by the simulations. The longitudinal deceleration levels are the steady-state values found for each limit condition. The stopping distances applicable to an initial velocity of 40 mph were obtained directly from the simulation, whereas the stopping distances from 60 mph were calculated by noting that each braking maneuver contains a transient response to the brake input, followed by a more-or-less steady-state behavior (constant $A_x$). The transient response always had decayed by the time t=1.5 sec., thus the distance required to reach zero velocity from other initial velocities can be calculated from the following relationship, viz.:

Table 2. A Summary of Limit Stopping Distances Achieved on a High Friction Surface.

| Vehicle | Loading Condition | Initial $A_y$ (g's) | Proportioning | $A_{xs.s.}$ (g's) | S.D. (ft) $V_o$=40 mph | S.D. (ft) $V_o$=60 mph | ΔV (mph) | ΔS.D. (ft) |
|---|---|---|---|---|---|---|---|---|
| Monte Carlo | Light, Many Options | 0 | U.S. | .91 | 64.6 | 140 | -28.1 | -28.8 |
| | | | European | 1.00 | 60.7 | 131 | -30.1 | -30.7 |
| | | .2 | U.S. | 1.02 | 58.7 | 129 | -30.7 | -31.4 |
| | | | European | 1.00 | 60.9 | 133 | -29.5 | -30.0 |
| | | .4 | U.S. | .98 | 64.0 | 137 | -28.5 | -29.1 |
| | | | European | .88 | 67.8 | 146 | -26.9 | -27.4 |
| | GVWR, Base Vehicle | 0 | U.S. | .98 | 62.0 | 134 | -29.4 | -29.9 |
| | | | European | .86 | 69.0 | 150 | -26.0 | -26.6 |
| | | .2 | U.S. | .92 | 66.9 | 142 | -27.6 | -28.2 |
| | | | European | .84 | 71.9 | 154 | -25.4 | -26.0 |
| | | .4 | U.S. | .96 | 65.4 | 140 | -27.9 | -28.2 |
| | | | European | .80 | 74.5 | 161 | -24.1 | -24.5 |
| Golf | Light, Front-End Options | 0 | U.S. | .86 | 67.4 | 148 | -26.8 | -27.4 |
| | | | European | .81 | 71.6 | 157 | -25.0 | -25.7 |
| | | .2 | U.S. | .87 | 69.4 | 149 | -26.3 | -27.0 |
| | | | European | .81 | 72.4 | 156 | -25.2 | -25.9 |
| | | .4 | U.S. | .84 | 71.8 | 157 | -24.7 | -25.1 |
| | | | European | .80 | 74.4 | 163 | -23.7 | -24.1 |
| | GVWR | 0 | U.S. | .72 | 80.0 | 176 | -22.1 | -22.7 |
| | | | European | .66 | 86.6 | 191 | -20.3 | -20.8 |
| | | .2 | U.S. | .73 | 78.8 | 175 | -22.2 | -22.9 |
| | | | European | .67 | 84.4 | 188 | -20.6 | -21.2 |
| | | .4 | U.S. | .69 | 85.0 | 186 | -20.5 | -21.1 |
| | | | European | .63 | 91.1 | 201 | -19.1 | -19.6 |

Table 3. A Summary of Limit Stopping Distances Achieved on a Medium Friction Surface.

| Vehicle | Loading Condition | Initial $A_y$ (g's) | Proportioning | $A_{xs.s.}$ (g's) | S.D. (ft) $V_o$=40 mph | S.D. (ft) $V_o$=60 mph | ΔV (mph) | ΔS.D. (ft) |
|---|---|---|---|---|---|---|---|---|
| Monte Carlo | Light, Many Options | 0 | U.S. | .62 | 91.5 | 202 | -19.3 | -19.7 |
| | | | European | .55 | 102.2 | 226 | -17.1 | -17.6 |
| | | .2 | U.S. | .66 | 89.6 | 193 | -19.8 | -20.4 |
| | | | European | .61 | 94.5 | 205 | -18.8 | -19.9 |
| | | .4 | U.S. | .66 | 88.1 | 193 | -19.9 | -20.6 |
| | | | European | .59 | 92.9 | 207 | -19.1 | -19.8 |
| | GVWR, Base Vehicle | 0 | U.S. | .53 | 106.1 | 234 | -16.5 | -16.9 |
| | | | European | .48 | 115.7 | 257 | -15.1 | -15.4 |
| | | .2 | U.S. | .57 | 95.7 | 217 | -18.0 | -18.5 |
| | | | European | .49 | 109.4 | 250 | -15.7 | -16.6 |
| | | .4 | U.S. | .57 | 82.7 | 216 | -18.2 | -18.9 |
| | | | European | .51 | 92.2 | 241 | -16.3 | -16.9 |
| Golf | Light, Front-End Options | 0 | U.S. | .56 | 98.7 | 219 | -17.9 | -18.5 |
| | | | European | .51 | 109.4 | 241 | -16.1 | -16.7 |
| | | .2 | U.S. | .58 | 96.1 | 216 | -17.9 | -18.7 |
| | | | European | .51 | 102.7 | 231 | -17.6 | -18.4 |
| | | .4 | U.S. | .53 | 100.7 | 231 | -16.9 | -18.2 |
| | | | European | .49 | 107.5 | 247 | -16.1 | -17.2 |
| | GVWR | 0 | U.S. | .47 | 117.9 | 261 | -14.9 | -15.4 |
| | | | European | .44 | 126.2 | 278 | -14.0 | -14.4 |
| | | .2 | U.S. | .51 | 108.2 | 244 | -15.8 | -16.5 |
| | | | European | .46 | 118.3 | 268 | -14.5 | -15.2 |
| | | .4 | U.S. | .42 | 124.5 | 285 | -14.1 | -15.0 |
| | | | European | .41 | 129.9 | 295 | -13.4 | -14.0 |

Table 4. A Summary of Limit Stopping Distances Achieved on a Low Friction Surface.

| Vehicle | Loading Condition | Initial $A_y$ (g's) | Proportioning | $A_{xs.s.}$ (g's) | S.D. (ft) $V_o$=40 mph | S.D. (ft) $V_o$=60 mph | ΔV (mph) | ΔS.D. (ft) |
|---|---|---|---|---|---|---|---|---|
| Monte Carlo | Light, Many Options | 0 | U.S. | .39 | 142.1 | 315 | -12.2 | -12.5 |
| | | | European | .35 | 154.9 | 348 | -11.2 | -11.5 |
| | | .2 | U.S. | .34 | 159.5 | 357 | -11.0 | -11.3 |
| | | | European | .29 | 182.3 | 415 | - 9.6 | -10.0 |
| | GVWR, Base Vehicle | 0 | U.S. | .34 | 160.9 | 360 | -10.8 | -11.0 |
| | | | European | .30 | 182.7 | 407 | - 9.5 | - 9.7 |
| | | .2 | U.S. | .26 | 210.2 | 465 | - 8.4 | - 8.7 |
| | | | European | .24 | 228.2 | 504 | - 7.8 | - 8.1 |
| Golf | Light, Front-End Options | 0 | U.S. | .39 | 140.5 | 314 | -12.4 | 12.8 |
| | | | European | .35 | 160.1 | 349 | -11.1 | -11.4 |
| | | .2 | U.S. | .33 | 168.4 | 368 | -10.7 | 11.0 |
| | | | European | .29 | 191.4 | 418 | - 9.4 | - 9.8 |
| | GVWR | 0 | U.S. | .32 | 172.4 | 380 | -10.2 | -10.5 |
| | | | European | .30 | 183.5 | 405 | - 9.6 | - 9.9 |
| | | .2 | U.S. | .27 | 206.5 | 448 | - 8.8 | - 9.1 |
| | | | European | .26 | 220.6 | 467 | - 8.3 | - 8.5 |

$$S.D.(final) = S.D.(t=1.5) + \frac{V^2(t=1.5)}{2 \cdot g \cdot A_{xs.s.}} , \quad (7)$$

where

$$S.D.(t=1.5) = V(t=0) \cdot 1.5 + \int_0^{1.5} dt \int_0^{1.5} -A_x(t)dt = V(t=0) \cdot 1.5 + \Delta S.D.$$

$$V(t=1.5) = V(t=0) + \int_0^{1.5} -A_x(t)dt = V(t=0) + \Delta V \quad ,$$

and the values of ΔV and ΔS.D. are included in Tables 2-4. (V should be converted to units that are compatible with S.D. (such as ft/sec) before using Eq. (7).)

The data presented in Tables 2-4 lead to some general observations on the effects that the different proportioning values have on braking capacity, namely:

1) For nearly every condition, the "U.S. proportioning" leads to better limit deceleration capability than does the "European proportioning."

2) Tire properties influence limit stopping performance much more than proportioning (given realistic variations). The tire data available for the two representative vehicles (see appendix and [10]) indicate that greater adhesion is exhibited on high and medium friction surfaces by the tires mounted on the U.S. car. For all of the different maneuvers simulated on these surfaces, the difference in achievable deceleration between similarly proportioned, but differently tired, vehicles is greater than for differently proportioned, but otherwise identical, vehicles.

3) Differences in deceleration capability are largest on the high friction surface, where the differences average .1 g, and smallest on the low friction surface, where they average about .03 g. When differences in limit deceleration are normalized, the percentage change is always about 10%.

4) Limit deceleration for the braking-in-a-turn condition is sometimes greater, sometimes less, than that found in straight-line braking. Only on the low friction surface is a trend evident—the limit deceleration in a turn is always less than that achievable during a straight-line stop.

The outcome of a situation in which an accident is avoided, or reduced in severity, by braking depends on three general factors, namely:

1) the overall reaction time of the driver, which is the delay between the time that the obstacle to be avoided interrupts the driver's line of vision and the time when the driver's foot hits the brake pedal,

2) the ability of the driver to apply the correct brake pedal force needed to achieve limit deceleration, and

3) the limit deceleration capability of the vehicle.

Thus the increased "safety" of a vehicle can be measured by the reduced demands placed on the driver (factors (1) and (2)).

Table 5 was prepared by taking the U.S. proportioning as a baseline and calculating at Δt value for each condition, where Δt is the

Table 5. Extra Reaction Time Deriving from Proportioning the Two Vehicles to Meet 105-75 Instead of 71/320/EEC

| Surface Condition | Vehicle Type | Loading Condition | Initial Lateral Acceleration for Constant Radius Turn | Δt, "Extra Reaction Time" from U.S. Proportioning | |
|---|---|---|---|---|---|
| | | | | $V_0$=40 mph | $V_0$=60 mph |
| High Friction (Dry Asphalt) | Monte Carlo | Light (Many Options) | 0 (g's) | -.07 (sec) | -.10 (sec) |
| | | | .2 | .04 | .05 |
| | | | .4 | .06 | .10 |
| | | GVWR (Base - No Options) | 0 | .12 | .18 |
| | | | .2 | .09 | .14 |
| | | | .4 | .16 | .24 |
| | Golf | Light (Some Front-End Options) | 0 | .07 | .10 |
| | | | .2 | .05 | .08 |
| | | | .4 | .04 | .07 |
| | | GVWR | 0 | .11 | .17 |
| | | | .2 | .10 | .15 |
| | | | .4 | .10 | .17 |
| Medium Friction (Wet Asphalt) | Monte Carlo | Light | 0 | .18 | .27 |
| | | | .2 | .08 | .14 |
| | | | .4 | .08 | .16 |
| | | GVWR | 0 | .16 | .26 |
| | | | .2 | .23 | .38 |
| | | | .4 | .16 | .28 |
| | Golf | Light | 0 | .18 | .25 |
| | | | .2 | .11 | .17 |
| | | | .4 | .12 | .18 |
| | | GVWR | 0 | .14 | .19 |
| | | | .2 | .17 | .27 |
| | | | .4 | .09 | .11 |
| Low Friction (Polished Wet Surface) | Monte Carlo | Light | 0 | .22 | .38 |
| | | | .2 | .39 | .66 |
| | | GVWR | 0 | .37 | .53 |
| | | | .2 | .31 | .44 |
| | Golf | Light | 0 | .33 | .40 |
| | | | .2 | .39 | .57 |
| | | GVWR | 0 | .19 | .28 |
| | | | .2 | .24 | .22 |

extra time available to the driver to react to an emergency situation that would not be available if the vehicle were proportioned to comply with the European standards. Suppose that at 40 mph, on the medium friction surface, a stop must be made in 300 feet by the domestic passenger car navigating a .2 g turn under GVWR loading. From Table 4, the limit stopping distance capability is 109.4 feet with European proportioning, which requires a reaction time of 3.25 sec. With the U.S. proportioning, the limit stopping distance capability is 95.7 feet, leaving 3.48 sec. for the driver to react. The difference is .23 sec., as shown in Table 5.

The values of Δt are small for all of the conditions involving the high friction surface. These values are larger for the low friction surface, but here the driver must be able to come close enough to the correct limit pedal force that differences of several hundredths of a g in deceleration capability between the differently proportioned vehicles are significant (see Table 4).

The driver who over-brakes and locks the front axle, the driver who under-brakes, and the driver who "pumps" the brake pedal, of course, realize negligible benefit from a slight difference in proportioning on the low friction surface.

In light of the general European regulatory philosophy that the adhesion utilization of the front axle should be higher than the utilization of the rear, we also look at the simulated adhesion utilization at each axle—a dynamic variable during the transient portion of a braking run. Table 6 lists the steady-state and peak values (designated "S.S." and "Max.," respectively) of the adhesion utilization at

Table 6. Simulated Adhesion Utilization at the Two Axles on the Medium and High Friction Surfaces.

| Surface Friction | Vehicle | Loading Condition | Proportioning | $A_{xs.s.}$ | $K_F$ S.S. | $K_F$ Max. | $K_R$ S.S. | $K_R$ Max. | First Axle to Lock |
|---|---|---|---|---|---|---|---|---|---|
| High (Dry Asphalt) | Monte Carlo | Light | U.S. | .91 | .84 | 1.03 | 1.09 | 1.12 | Rear |
| | | | European | 1.00 | 1.00 | 1.05 | .95 | 1.11 | Front |
| | | GVWR | U.S. | .98 | 1.00 | 1.04 | .90 | 1.11 | Front |
| | | | European | .86 | 1.01 | 1.05 | .60 | 1.11 | Front |
| | Golf | Light | U.S. | .86 | .83 | .87 | .98 | .99 | Rear |
| | | | European | .81 | .85 | .87 | .61 | .83 | Front |
| | | GVWR | U.S. | .72 | .83 | .85 | .47 | .54 | Front |
| | | | European | .66 | .82 | .86 | .30 | .34 | Front |
| Medium (Wet Asphalt) | Monte Carlo | Light | U.S. | .62 | .62 | .67 | .61 | .68 | Front |
| | | | European | .55 | .62 | .67 | .41 | .48 | Front |
| | | GVWR | U.S. | .53 | .62 | .67 | .40 | .43 | Front |
| | | | European | .48 | .63 | .67 | .28 | .30 | Front |
| | Golf | Light | U.S. | .56 | .58 | .60 | .47 | .53 | Front |
| | | | European | .51 | .57 | .59 | .30 | .32 | Front |
| | | GVWR | U.S. | .47 | .58 | .60 | .27 | .30 | Front |
| | | | European | .44 | .59 | .60 | .19 | .20 | Front |

both axles on the medium and high friction surfaces during straight-line braking, and also notes the axle that locks first. From Table 6 we observe that:

1) The U.S.-proportioned vehicles lock the rear wheels first on the high coefficient surface when the loading is such that the c.g. is in an extreme forward position. The deceleration achieved prior to rear-wheel lockup is very high for both the Monte Carlo and the Golf; in fact, higher than the deceleration capability required by any of the standards.

2) The adhesion utilization of the rear axle is often much greater than that of the front on the high friction surface for both proportionings. Due to the load sensitive behavior of pneumatic tires, however, there is more traction available at the relatively lightly-loaded rear axle than at the heavily-loaded front axle. When the rear axle locks before the front, the difference between the steady-state values of $K_F$ and $K_R$ is .15 for the Golf and .25 for the Monte Carlo.

## SUMMARY AND CONCLUSIONS

Nearly all of the braking regulations in use throughout the Western world are based on one of two prototypes: the U.S. standard,

FMVSS 105-75, and the United Nations' ECE R-13 (often implemented in a nearly identical regulation adopted by the Common Market, Directive 71/320/EEC). FMVSS 105-75 requires short stopping distances on high friction surfaces, whereas ECE R.13 attempts to eliminate rear-axle lockup and the resultant loss of directional stability. Both regulatory philosophies lead to fore/aft brake proportionings that are best for stops on high friction surfaces, however, ECE R.13 leads to somewhat higher proportioning values than does FMVSS 105-75. A third, independent standard, Sweden's F-18, leads to the same proportioning as R.13.

A study employing dynamic simulation showed that the proportioning selection deriving from 105-75 results in a marginally better overall limit deceleration capability than does a proportioning selection deriving from ECE R.13. However, the difference in deceleration capability would not be significance to overall vehicle safety unless extremely skilled drivers are assumed to be in control. Also, the difference in proportioning is not as significant as the choice of tires with which the vehicle is equipped. When a so-called "U.S. proportioning" was assumed, the rear wheels locked before the front when a vehicle in the unloaded condition was braked on a high friction surface. However, the decelerations achieved before lock up occurred were about .9 g's—a deceleration much higher than regulation requires.

ACKNOWLEDGEMENTS

The work reported in this paper was funded by the Motor Vehicle Manufacturers Association, and most of the vehicle data needed for the computer simulations were provided by General Motors and Volkswagen.

REFERENCES

1. Oppenheimer, Paul. "Braking Regulations in Europe." SAE Paper No. 740313, Feb-Mar. 1974.
2. Oppenheimer, Paul. "Braking Regulations for Passenger Cars." SAE Paper No. 770182, Feb.-Mar. 1977.
3. Seiffert, U.W., Marks, H.-G., and Ziwica, K.-H. "Hydraulic Brake System U.S. Versus Common Market." SAE Paper No. 760219, Feb. 1976.
4. Aoki, Kazuhiko. "A Comparison of World Braking Standards with Reference to the Development of Japanese Braking Standards." SAE Paper No. 720030, Jan. 1972.
5. Braking Standard #105-75. Federal Register, Vol. 41, No. 139, Monday, July 19, 1976.
6. United Nations. Agreement Concerning the Adoption of Uniform Conditions of Approval and Reciprocal Recognition of Approval for Motor Vehicle Equipment and Parts. Addendum 12: Regulation No. 13, Uniform Provisions Concerning the Approval of Vehicles with Regard to Braking. Geneva, March 20, 1958.
7. European Communities Council Directive. On the Approximation of the laws of Member States Relating to the Braking of Certain Motor Vehicles and Their Trailers (71/320/EEC), Sept. 6, 1971.

8. Regulations for Braking Systems on Motor Vehicles and Trailers Which are Coupled to Vehicles (F-18). Sweden, Dec. 29, 1971.
9. Bernard, J.E., et al. "Vehicle-In-Use Limit Performance and Tire Factors: The Tire In Use." Final Report, Contract No. DOT-HS-031-3-693, Highway Safety Res. Inst., Univ. of Michigan, March 1975.
10. Sayers, M. and Segel, L. "Investigation of the Influence of Various Braking Regulations on Accident-Avoidance Performance." Final Technical Report, MVMA Project #4.31, Highway Safety Res. Inst., Univ. of Michigan, Rept. No. UM-HSRI-78-51, Nov. 1978.

## APPENDIX

Table A.1. Parameter Values for Two Representative Passenger Cars.

| Car Model | 1978 Chevrolet Monte Carlo | | | | Volkswagen Golf | | | |
|---|---|---|---|---|---|---|---|---|
| Options | None, Base Vehicle | | 5.0 V8, A/C, Auto, Power Windows, Console, AM/FM/TP, Power Seat, etc. | | 4 dr., Sun Roof, R. Window Washer | | Auto, Passive Seat Belts | |
| Loading | 300 lb | GVWR | 300 lb | GVWR | 300 lb | GVWR | 300 lb | GVWR |
| Static Front Load (lb) | 1894 | 2069 | 2143 | 2344 | 1333 | 1454 | 1285 | 1454 |
| Static Rear Load (lb) | 1547 | 2249 | 1597 | 2245 | 752 | 1302 | 830 | 1302 |
| c.g. height/wheelbase ($h/\ell$) | .20 | .20 | .20 | .20 | .22 | .23 | .22 | .23 |
| Normalized Front Axle Load ($1-a/\ell$) | .55 | .48 | .57 | .51 | .64 | .53 | .61 | .53 |
| "Worst Case" | No | Yes | Yes | No | No | Yes | Yes | Yes |

Table A.2. Friction Coefficients on Dry Asphalt for Tires Installed on the Two Representative Vehicles.

| Vehicle | Tire Size | Load (lbs) | $\mu_p$ |
|---|---|---|---|
| Monte Carlo | P205/70R-14 | 800 | 1.11 |
| | | 1368 | 1.03 |
| | | 1716 | .99 |
| | | 2060 | .89 |
| Golf | 155SR-13 | 384 | 1.00 |
| | | 588 | .94 |
| | | 793 | .88 |
| | | 994 | .83 |
| | | 1195 | .81 |

# THE STEERING RESPONSES OF DOUBLES

R. S. Sharp

Department of Mechanical Engineering, University of Leeds, Leeds, United Kingdom

SUMMARY

The performance of fully laden doubles with respect to lateral stability, frequency responses, and lane change behaviour is calculated using a conventional constant forward speed, yaw and sideslip freedom, linear vehicle model. A novel type of hitch connection, joining the trailer to the tractor-semitrailer is proposed and the effects of varying its geometry are determined. It is concluded that its use can lead to some stability improvement and to dramatic response improvement, and the physical principles underlying the response improvement are discussed.

Deterioration of the low speed off-tracking behaviour is seen to be an inevitable consequence of the response improvement, and adjustment of the hitch geometry for low speed running is proposed as desirable and practicable.

1. INTRODUCTION

In this paper a "Double" means a tractor-semitrailer towing a full trailer, the full trailer having two parts, a dolly, and a pup trailer. In commercial use, the pup trailer and the semitrailer are often similar units and are interchangeable, the pup being attached to the dolly by a fifth wheel arrangement. Conventionally, a pin at the rear of the semitrailer, the pintle pin, acts as an attachment for the pintle hook at the front of the tongue, or drawbar, which guides the dolly. Fig. 1 is a sketch of the arrangement.

Notable studies of the lateral motion characteristics of such vehicles have been completed by Jindra [1], Hazemoto [2], Hales [3], and Mallikarjunarao and Fancher [4], and the results of Nordmark and Nordström [5] concerning straight trucks towing full trailers of the type described are relevant.

From these studies, the following conclusions can be drawn:
(a) the dynamic coupling between towing vehicle and towed vehicle is

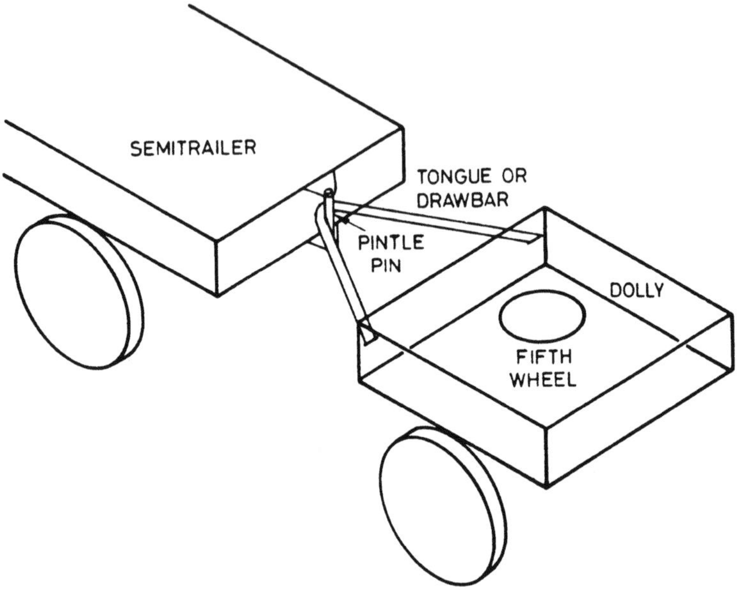


Fig. 1 Conventional hitch representation and terminology

very weak, since the lateral forces at the pintle pin are very small in comparison with the tyre lateral forces. Thus the motions of the tractor-semitrailer are influenced very little by the presence or condition of the full trailer; (b) trailers of the type considered possess one or two lightly damped oscillatory modes of motion typically having natural frequencies of 0.3 to 0.8 Hz. The stability of these modes deteriorates as forward speed increases, as loading increases, as the pup centre of mass moves rearwards, as the pup yaw inertia increases, as the pup wheelbase is shortened, and as the trailer tyre cornering stiffnesses are reduced; (c) in rapid manoeuvres, typically of an obstacle avoidance type, at high road speeds, the lateral motion of the tractor can be considerably amplified at the pup trailer. With the rather low roll stability of many laden commercial vehicles, this amplification can result in the trailer rolling over when the tractor is nowhere near to performing a limit manoeuvre. Because of the small influence of the trailer on the tractor-semitrailer, the driver of the vehicle is likely to be unable to anticipate the mishap. When the vehicle is for petroleum delivery and weighs in excess of 664000 N, the result can be very destructive [4]. The amplification is associated with excitation of the lightly damped modes of the trailer, and modifications to the trailer which increase the damping of these modes tend to decrease the amplification. Moving the pintle pin forwards in the towing vehicle is also suggested as being helpful, and although both Hazemoto [2], and Nordmark and Nordström [5] suggest some advantage in using a short tongue, this is disputed in the present work and by Jindra [1]; (d) there is to some extent a fundamental conflict between good high speed lateral behaviour and good low speed off-tracking behaviour; (e) for relatively low lateral accelerations, a reasonably accurate representation of the vehicle motions can be obtained from a constant forward speed linear model with yaw and sideslip freedoms of each component.

Nordmark and Nordstrom [5] also mentioned the value of pup trailer roll steer but appear to have considered roll steer coefficients an order of magnitude greater than those which are practicable.

The current investigation started with the independent development of a set of equations of motion for a vehicle model similar to that of Mallikarjunarao and Fancher [4], except that lateral flexibility of the pintle pin attachment to the semitrailer was allowed. For very high stiffness of this attachment, it was confirmed that this analysis gave results identical to those of [4]. The analysis was also extended to include a roll degree of freedom for the pup trailer, the main purpose being to examine the effectiveness of practicable amounts of roll steer, and it was confirmed that, for very high roll stiffness, results from this analysis agreed with those from the simpler one. The effects of pintle pin flexibility, yaw stiffness and damping at the pintle pin, yaw stiffness and damping at the dolly turntable, and roll steer on vehicle frequency responses were reported in the extensive summary of this paper [6], and since none of them are particularly helpful, they are not reported again here.

In [6], an alternative hitch geometry which is capable of dramatically improving the frequency response behaviour of the Michigan tanker is described, and subsequent work has been aimed at fully understanding its effectiveness, showing it to be capable of reducing amplification in realistic emergency manoeuvreing, and showing it to be effective when applied to other doubles designs.

Following the restriction of further study to the consequences of modifying the hitch geometry, pintle pin flexibility and pup trailer roll freedom were omitted from the vehicle model employed, which then became identical with that of reference [4]. The effect of the modified hitch, which is shown diagrammatically in Fig. 2, is to move the pintle pin forwards in the semitrailer and to lengthen the tongue by an equal amount, without restriction. These changes are represented by very simple modifications to the parameter values describing a particular vehicle.

## 2. VEHICLE MODEL

As in [4], the following assumptions are implicit in the equations of motion which have been employed: (a) each separate part of the

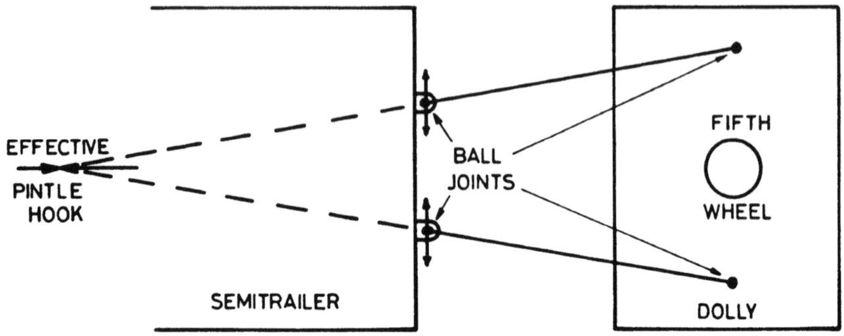


Fig. 2 Diagrammatic representation of modified hitch

vehicle (e.g. the tractor) is a rigid body and the various bodies are constrained by frictionless pin joints relative to one another; (b) the roadway consists of a horizontal plane; (c) only lateral and yaw freedoms exist for each part of the vehicle; (d) the input to the vehicle is road wheel steer angle, (i.e. fixed control is assumed); (e) the tyres respond only to slip angle, and they respond instantaneously producing forces and moments which are linear functions of slip angle, (with double wheels in fact, small moments arising from longitudinal tyre tread and carcase deformations are included in the analysis. These are considered linearly dependent on yaw rate); (f) the motions include only small perturbations from straight running, so that inertial effects may be linearised.

Also the effect of the modified hitch is considered equivalent to that of a conventional hitch having the pintle hook at the instantaneous centre of the modified hitch in its central symmetrical position. The validity of this treatment is examined in Appendix 1.

## 3. RESULTS

The equations of motion have been used to obtain eigenvalues, frequency responses, and responses to one complete sine wave of steer angle (representing a lane change manoeuvre), for each of four vehicles in its fully laden condition. The fully laden condition is not the least stable possible, but it is the least stable which must be tolerated if part loads are distributed sensibly. Vehicle speed has been varied to ensure that the results follow the pattern of previous ones, but most of the results included here are for a speed of 88km/hr. For each vehicle, the hitch geometry has been varied over a wide range, and in the lane change manoeuvre, sine wave periods of 1, 2 and 3 seconds have been employed. The vehicles considered are:

A, the Michigan petroleum delivery tanker, [4]; B, Hazemoto's 24 ft. double; C, Hazemoto's 20 ft. double; and D, Hazemoto's 40 ft. double, [2].

Qualitatively, the effects of speed and hitch geometry on all four vehicles are the same. Also vehicles B and C are sufficiently similar in their parameter values and behaviour properties not to require separate documentation. Vehicle B is most closely representative of doubles in use in the U.S.A., and probably in Japan and Sweden too, so that the results for this vehicle are considered in most detail. Results for vehicles A and D are included where their features are of particular interest.

In the frequency and lane change response calculations, the lateral accelerations of the mass centres of the tractor, semitrailer, dolly, and pup trailer, and of the pintle pin have been calculated. This is because the lateral accelerations of the mass centres relate closely to the likelihood of rollover or other loss of driver control, while the acceleration of the pintle pin relates to the disturbance which is applied to the trailer and, other things being equal, is best minimised. The articulation angle of the dolly relative to the semitrailer has also been determined, since this relates to the degree of change to the hitch geometry which occurs during a manoeuvre.

In the frequency response calculations, a knowledge of the magnitudes of the lateral accelerations of semitrailer mass centre and pintle pin, and their phase relationship, allows the determination of the point on the centreline of the semitrailer which oscillates least, and the extent of this oscillation relative to that of the mass centre (see Appendix 2). Again, other things being equal, locating the pintle pin at this point will minimise the input to the trailer, and thereby minimise the amplification of the tractor lateral motions which occur. Note, however, that other things turn out not to be equal. The weak coupling between the trailer and the tractor-semitrailer allow one to deduce that relocating the pintle pin will not cause much change to the tractor-semitrailer motions, so that the calculations leading to the relocation are not invalidated.

Some results relating to the low speed off-tracking behaviour of the vehicles with modified hitches have also been obtained.

Eigenvalues obtained for vehicles A, B, and D at 88 km/hr forward speed in each of five conditions are given in Table 1. The five conditions are defined as follows: (1) standard; (2) pintle pin moved forwards 3.05 m; (3) pintle pin moved forwards 6.1 m; (4) pintle pin moved forwards 9.15 m; and (5) pintle pin moved forwards 12.2 m. The dolly tongue is of course lengthened as the pintle pin is moved forwards.

The variations of the eigenvalues of two of the standard vehicles with forward speed are included in [4] (vehicle A), and [2] (vehicle B).

| Vehicle type | Condition (88 km/hr in each case) | | | | |
|---|---|---|---|---|---|
| | (1) | (2) | (3) | (4) | (5) |
| A | -3.73±2.47<br>-1.75±3.52<br>-2.24±4.76<br>-0.77±4.73 | -3.71±2.47<br>-1.85±3.47<br>-2.15±3.57<br>-0.69±3.80 | -3.71±2.47<br>-1.86±3.48<br>-1.67±2.86<br>-1.15±3.62 | -3.71±2.46<br>-1.86±3.47<br>-1.53±2.21<br>-1.28±3.77 | -3.71±2.46<br>-1.87±3.46<br>-1.50±1.79<br>-1.30±3.83 |
| B | -1.57±1.02<br>-1.31±3.25<br>-1.23±5.36<br>-0.88±2.61 | -1.56±0.94<br>-1.37±3.29<br>-1.30±3.69<br>-0.75±2.31 | -1.56±0.93<br>-1.36±3.33<br>-1.31±3.32<br>-0.75±1.94 | -1.56±0.92<br>-1.40±3.33<br>-1.25±3.20<br>-0.77±1.65 | -1.56±0.92<br>-1.40±3.34<br>-1.23±3.15<br>-0.78±1.44 |
| D | -1.85±2.94<br>-1.36±0.92<br>-1.35±5.62<br>-1.05±2.02 | -1.88±2.97<br>-1.38±0.88<br>-1.56±3.64<br>-0.89±1.90 | -1.88±2.97<br>-1.38±0.87<br>-1.64±3.06<br>-0.80±1.70 | -1.88±2.97<br>-1.38±0.87<br>-1.67±2.81<br>-0.77±1.51 | -1.89±2.98<br>-1.38±0.86<br>-1.67±2.68<br>-0.77±1.35 |

Table 1. Eigenvalues for vehicles A, B, and D in each of five conditions (from each pair of numbers, the first is the real part, representing the modal damping, the second the imaginary part, representing the circular frequency).

Frequency response gains for vehicle B at 88 km/hr are shown in Figs. 3 and 4. Fig. 3 relates to the standard vehicle, and shows the lateral accelerations of semitrailer, dolly, and pup trailer mass centres, and that of the pintle pin, relative to that of the tractor mass centre. Fig. 4 shows the effect of altering the hitch geometry on the ratio of pup trailer mass centre lateral acceleration amplitude to that of the tractor mass centre.

Figs. 5 and 6 also refer to frequency response calculation results, the former for vehicle A, the latter for vehicle D. In each of these two figures, lateral acceleration amplitudes relative to that of the tractor mass centre for the semitrailer mass centre, the pintle pin, the dolly mass centre, and the pup trailer mass centre of the standard vehicle, and the pup trailer mass centre with hitch condition (5) are shown.

In Fig. 7 calculated lateral acceleration time histories, in response to a single sine wave steering input of 2 sec period, for vehicle B are presented. The parameters plotted are the same as in Figs. 5 and 6. Other lane change simulation results, not included in detail, for the other vehicles and input sine wave periods of 1 and 3 secs, are qualitatively similar to those of Fig. 7, and confirm that the tractor-semitrailer motions are hardly affected by changes to the hitch geometry. Hitch conditions (4) and (5) give very similar results, and the pup trailer mass centre acceleration changes regularly from condition (1), the standard, through condition (4). The main features of the simulation results are summarised in Table 2, in which $R_1$, $R_2$, and $R_3$ are the ratios of maximum lateral acceleration of the pup trailer and tractor mass centres, and $\Gamma_{2M1}$, $\Gamma_{2M2}$, and $\Gamma_{2M3}$ are the maximum dolly/semitrailer articulation angles for a maximum lateral

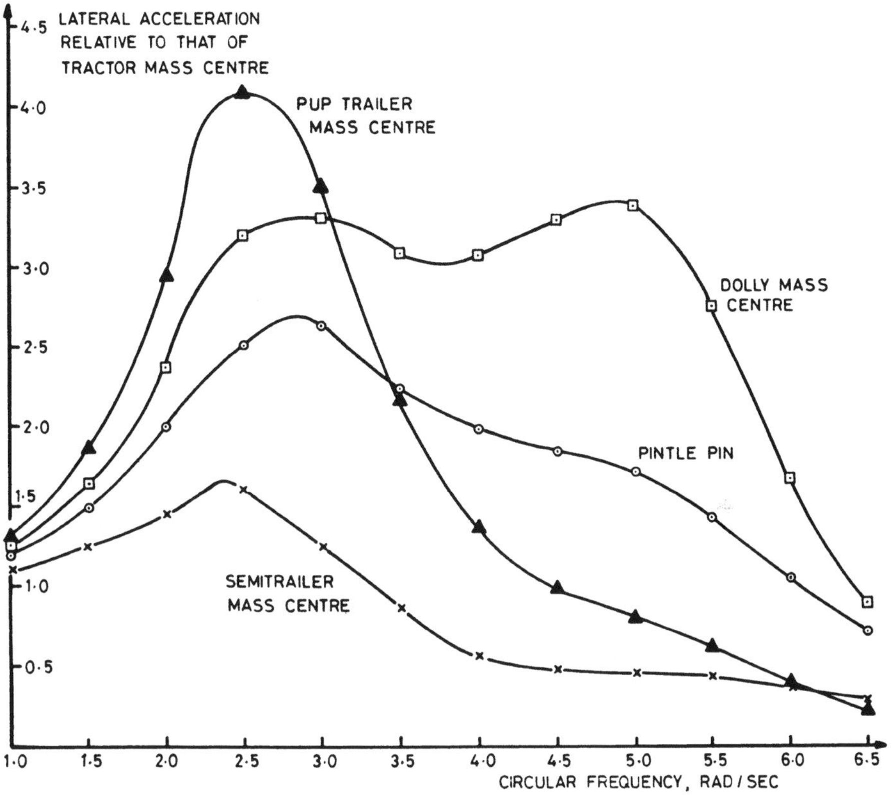


Fig. 3 Lateral acceleration gains for vehicle B (1) at 88 km/hr as functions of steering input frequency

acceleration of either the semitrailer or pup trailer mass centres of 0.3 g, for 1, 2, and 3 sec. sine wave input periods respectively.

| vehicle | condition | $R_1$ | $R_2$ | $R_3$ | $\Gamma_{2M1}^{(deg)}$ | $\Gamma_{2M2}^{(deg)}$ | $\Gamma_{2M3}^{(deg)}$ |
|---|---|---|---|---|---|---|---|
| A | (1) | 2.39 | 2.65 | 1.76 | 2.16 | 1.46 | 1.14 |
|  | (5) | 0.85 | 1.25 | 1.29 | 1.45 | 1.01 | 1.14 |
| B | (1) | 1.50 | 2.22 | 2.18 | 4.27 | 3.00 | 2.18 |
|  | (5) | 0.68 | 1.05 | 1.15 | 2.75 | 2.32 | 2.85 |
| D | (1) | 0.91 | 1.30 | 1.38 | 4.86 | 3.85 | 3.20 |
|  | (5) | 0.42 | 0.70 | 0.90 | 2.74 | 2.59 | 3.46 |

Table 2. 88 km/hr lane change simulation parameters

The low speed off-tracking behaviour of vehicle B in condition (5) turning a 15 m radius curve is sketched in Fig. 8. It is apparent that the hitch modification considerably worsens the off-tracking.

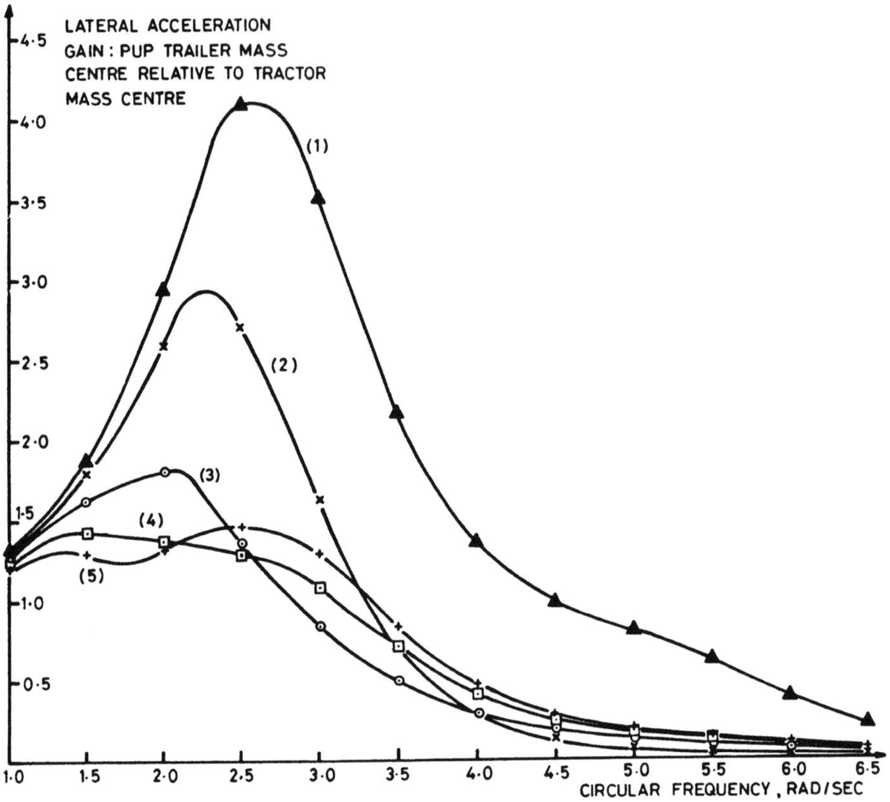


Fig. 4 Lateral acceleration gains as functions of steering input frequency for vehicle B at 88 km/hr with different hitch geometries

## 4. DISCUSSION OF RESULTS

As expected, two of the four eigenvalues in each set (the first two in Table 1) are clearly associated with the tractor-semitrailer motions primarily, and are little affected by changes to the hitch geometry. Moving the pintle pin forwards reduces the natural frequencies of the trailer modes and increases the damping as a proportion of critical damping, even though the magnitude of the (negative) eigenvalue real part is sometimes reduced. These results relate physically to the trailer motions following a small general disturbance from straight running, in which the pintle pin is virtually constrained to travel in a straight line, and take no account of how effective tractor-semi-trailer motions are at exciting pup trailer motions. They do however indicate an improvement of the trailers' stability characteristics.

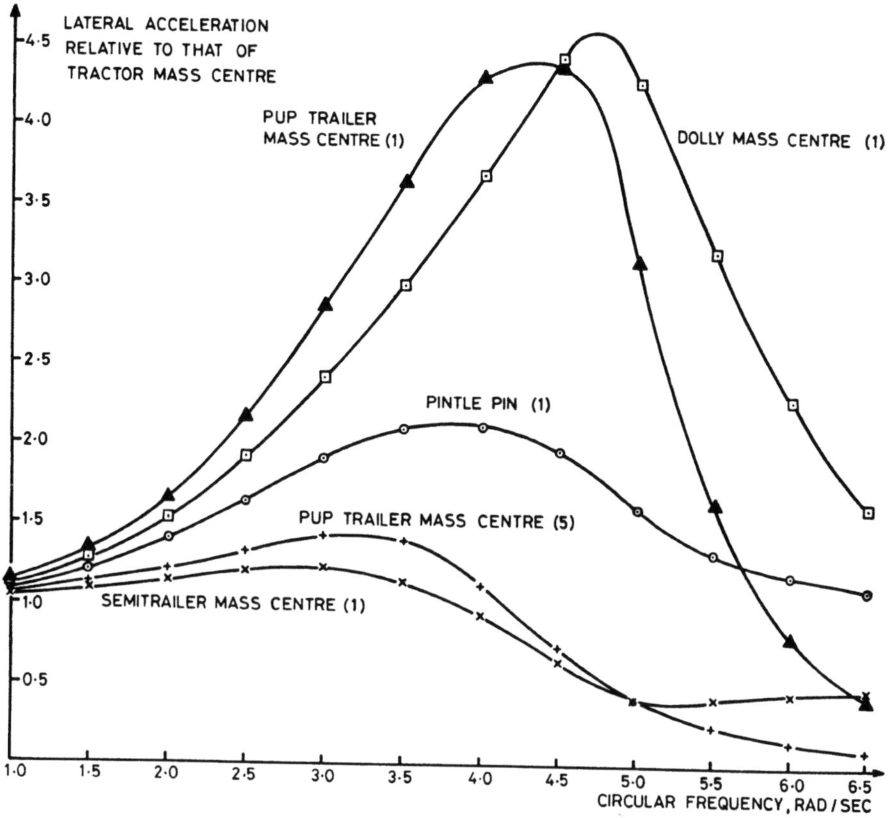


Fig. 5 Lateral acceleration gains as functions of steering input frequency for vehicle A at 88 km/hr

The frequency response results, Figs. 3, 5, and 6, show that the standard pintle pin on each vehicle is positioned such that its lateral motion is considerably greater than that of the semitrailer mass centre, and since the trailer can be viewed as responding to the disturbance of the pintle pin, (because of the weak coupling), relocating the pintle pin at some point in the semitrailer which oscillates less is obviously desirable, see Appendix II. Also desirable is the reduction of the steering input to the dolly wheels resulting from a given lateral displacement of the pintle hook, since this steering input is the main source of the lateral forces which disturb the trailer. Lengthening the dolly tongue achieves this reduction, and this is normally the main mechanism by which the trailer motions are reduced.

For vehicles A and B, peak magnification factors over 4 occur, while vehicle D has a peak of about 2 for the pup trailer mass centre to tractor mass centre acceleration amplitudes. These peaks occur at frequencies near to the natural frequencies of the least damped modes,

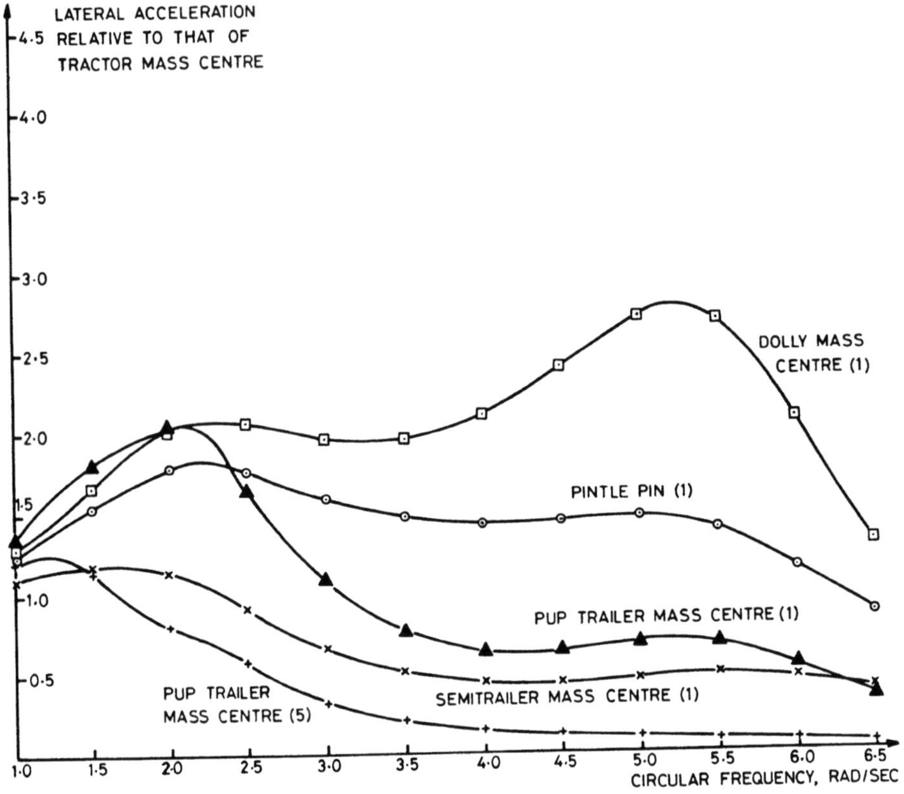


Fig. 6 Lateral acceleration gains as functions of steering input frequency for vehicle D at 88 km/hr

given in Table 1. Fig. 4 shows the successive reductions of the peak magnification factor for vehicle B as the hitch geometry is modified, with diminishing returns becoming apparent when the pintle pin is about 9 m forwards of its standard position, and Figs. 5 and 6 show the effectiveness with vehicles A and D respectively of only the extreme hitch geometry studied.

The lane change simulation results of Fig. 7 and Table 2 indicate that for manoeuvres reasonably representative of real life accident avoidance situations, the modified hitch geometry is very effective in reducing amplification of the tractor lateral acceleration, the largest magnification factor with the best modified hitch being 1.29. Ideally, when this ratio remains greater than unity, the pup trailer should possess somewhat greater roll stability than the semitrailer, so that it is no more likely to overturn.

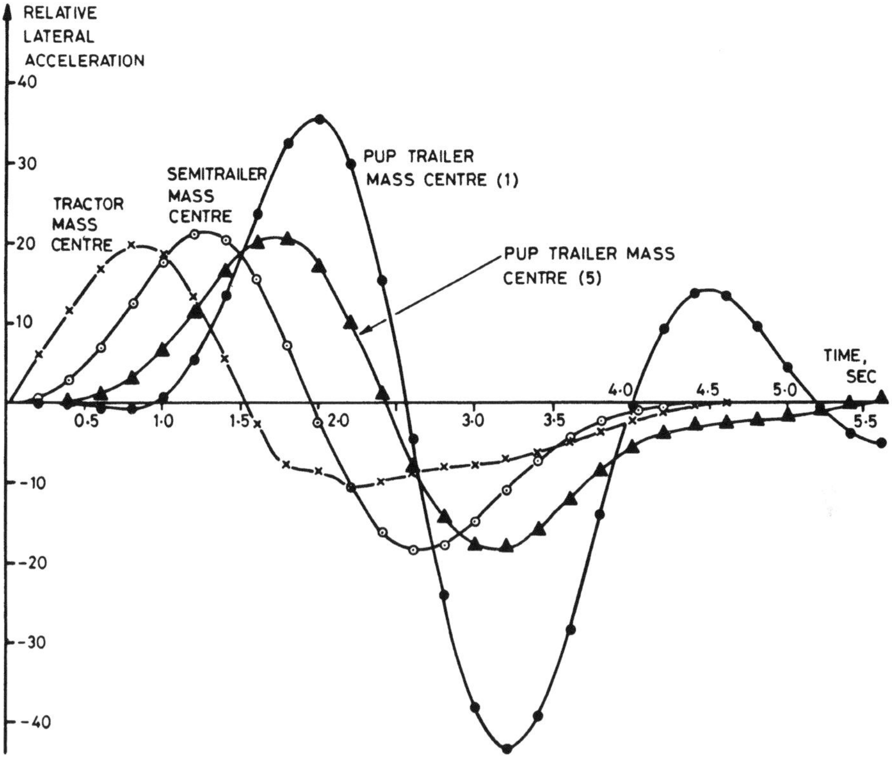


Fig. 7 Lateral acceleration responses of vehicle B at 88 km/hr to a single sine wave input of 2 second period

For steering input amplitudes sufficient to cause lateral accelerations of either semitrailer or pup trailer mass centres of 0.3g, the dolly to semitrailer articulation angle is always less than 5 degrees. It is shown in Appendix I that, for a two link hitch mechanism with links of a practicable length, the geometry of the modified hitch is virtually indistinguishable from that of the equivalent long tongue represented in the vehicle response theory, over this range of movement.

The low speed off-tracking behaviour, illustrated in Fig. 8, is poor with the modified hitch because the ratio of dolly steer angle to dolly lateral displacement relative to the semitrailer is low. For the dolly wheels to attain the necessary steer angle relative to those of the semitrailer, they must be displaced laterally through a relatively large distance. This necessary steer angle is determined only by the radius of the turn being negotiated and the longitudinal separation of the semitrailer and dolly axles. Since the low ratio of steer angle to lateral displacement is the main basis for the improvement of the frequency response and lane change behaviours, the

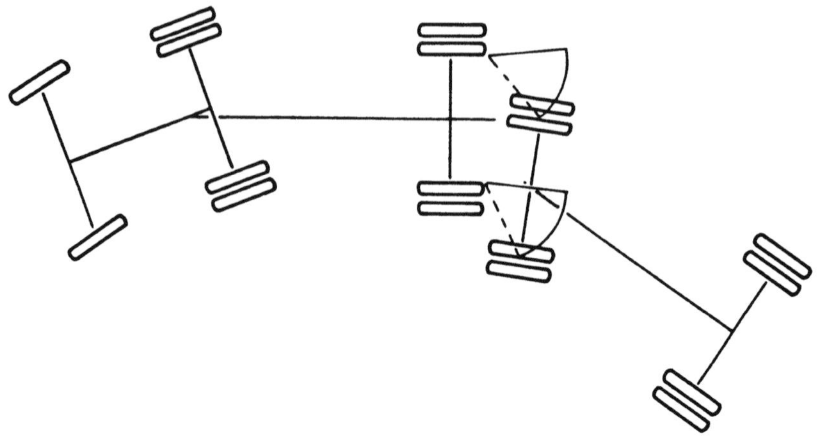

Fig. 8 Diagrammatic representation of low speed off-tracking behaviour of vehicle B in condition (5), turning a 15 m radius curve

fundamental nature of the conflict between good high speed responses and good low speed tracking, mentioned by Jindra [1], can be appreciated.

Adjustment of the hitch geometry so that for low speed running it is conventional, while for high speed running it effectively moves the pintle pin well forwards will allow achievement of both good low speed and good high speed behaviours. With the proposed hitch arrangement, this is mechanically simple. The two semitrailer ball joints can be mounted on a transverse drive mechanism so that they can be arranged in the "in" position for low speed operation, and in the "out" position for high speed operation, see Fig. 2. Operation of the mechanism could be automatic or manual, but safeguards would probably be advisable, particularly to avoid high speed running with the ball joints "in".

## 5. CONCLUSIONS

The hitch geometry proposed will improve the stability characteristics of dolly and pup type trailers somewhat, and will dramatically improve their frequency response and lane change response behaviours by reducing the amplification of tractor lateral accelerations at the pup trailer. Other measures considered to achieve comparable improvements do not appear capable of doing so.

The physical bases of the frequency response improvements are: (a) the pintle pin is effectively moved forwards in the semitrailer to a position which undergoes less lateral motion than before; and (b) the ratio of dolly steer to lateral displacement of the dolly relative to the semitrailer is reduced substantially. Of the two factors, the latter is, in general, the more powerful.

It is fundamental to the modified hitch geometry that the low speed off tracking behaviour is worsened by it, and adjustment of the geometry for low speed running is proposed as a necessary means to achieve good low speed tracking and also good high speed response behaviours.

## 6. ACKNOWLEDGEMENT

Much of the research reported in this paper was carried out while the author was a Visiting Associate Research Scientist at the University of Michigan's Highway Safety Research Institute. Assistance with computational matters received from C. Mallikarjunarao is gratefully acknowledged.

## 7. APPENDIX 1

*Hitch Geometry*

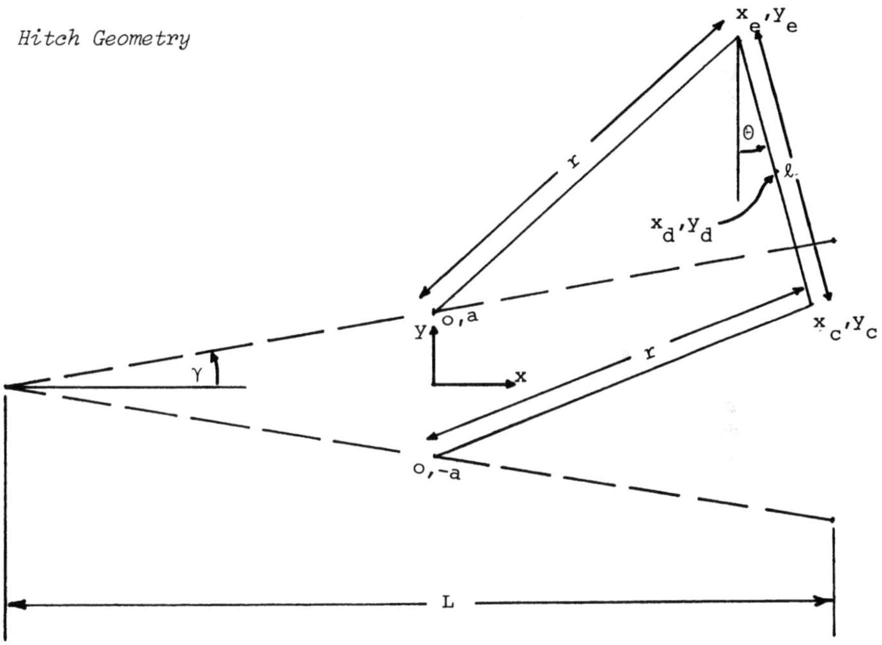


Fig. 9 Hitch in displaced condition

From the geometry of Fig. 9, the following relations apply:

$$L = \ell\{r^2-(0.5\ell-a)^2\}^{\frac{1}{2}}/(\ell-2a) \quad (1)$$

$$x_e^2 + (y_e-a)^2 = r^2 \quad (2)$$

$$x_c^2 + (y_c+a)^2 = r^2 \quad (3)$$

$$(x_e-x_c)^2 + (y_e-y_c)^2 = \ell^2 \quad (4)$$

$$y_d = (y_e + y_c)/2 \quad (5)$$
$$\Theta = \tan^{-1}\{(x_c - x_e)/(y_e - y_c)\} \quad (6)$$

ratio of steer angle to lateral displacement is $\Theta/y_d$

For a link of length L of equivalent geometry for small displacements, the ratio of steer angle to lateral displacement is $\Theta/(L \sin \Theta)$, where $\Theta$ is the angle through which the link articulates.

To compare the geometries of the long link treated in the vehicle response theory, and of the proposed two link arrangement, a computer program was written to solve the equations above iteratively to yield $(\Theta/y_d)$ and $\Theta/(L \sin\Theta)$ for various link lengths and angles. The shortest practicable links would be about 2 m long, and for this case $(\Theta/y_d)$ and $\Theta/(L \sin\Theta)$ are compared with each other for $0 \leq \Theta \leq 12^o$, in Table 3.

| $\Theta$(deg) | 2 | 4 | 6 | 8 | 10 | 12 |
|---|---|---|---|---|---|---|
| $\Theta/y_d$ | 0.10123 | 0.09553 | 0.10798 | 0.12632 | 0.14714 | 0.16900 |
| $\Theta/(L \sin \Theta)$ | 0.10258 | 0.10265 | 0.10275 | 0.10290 | 0.10309 | 0.10332 |

Table 3. Steering ratios of real and equivalent linkages.

It is clear from these results that for the articulation angles less than $5^o$ which were required in the lane change manoeuvres, the geometries of the proposed linkage and the one simulated are very closely similar.

## 8. APPENDIX 2

*Calculation of minimum amplitude on semitrailer centreline in frequency response test*

Suppose the semitrailer mass centre has a lateral acceleration $A_1 \sin wt$, and the pintle pin has acceleration $A_2 \sin(wt+\phi)$. The two points are a longitudinal distance X2B apart. It is easily shown that the acceleration amplitude of a point on the semitrailer centreline, a distance $\lambda$ in front of the pintle pin is given by

$$\frac{1}{X2B}\left[\lambda^2 A_1^2 + 2(X2B-\lambda)\lambda A_1 A_2 \cos \phi + (X2B-\lambda)^2 A_2^2\right]^{\frac{1}{2}}$$

The condition for this amplitude to be minimum is

$$\lambda = A_2 L(A_2 - A_1 \cos \phi)/(A_1^2 - 2A_1 A_2 \cos \phi + A_2^2)$$

Employing values of $A_1$, $A_2$, and $\phi$ obtained from the frequency response calculations, a short computer program was written to obtain the results of Table 4, which show where the point of minimum amplitude lies through the frequency range for the standard vehicles. From these results it is apparent that the pintle pin need not be placed in a position where it will provide an amplified input to the trailer, that for the higher input frequencies, an attenuated motion as compared with the tractor mass centre is possible, that the point of minimum amplitude moves rearwards in the semitrailer as the frequency increases from 1 rad/sec. and then forwards again as the frequency increases further, and that the point of minimum amplitude is always well forwards of the conventional pintle pin.

| ω, rad/sec | vehicle A | | vehicle B | | vehicle D | |
|---|---|---|---|---|---|---|
| | λ,m | MRA | λ,m | MRA | λ,m | MRA |
| 1 | 9.09 | 1.007 | 12.32 | 0.979 | 12.63 | 1.036 |
| 2 | 7.33 | 1.049 | 7.20 | 1.137 | 8.59 | 1.045 |
| 3 | 5.76 | 1.091 | 4.75 | 1.032 | 6.48 | 0.655 |
| 4 | 4.67 | 0.860 | 3.59 | 0.585 | 5.42 | 0.424 |
| 5 | 3.65 | 0.406 | 3.33 | 0.446 | 6.58 | 0.446 |
| 6 | 2.96 | 0.296 | 3.99 | 0.311 | 8.00 | 0.360 |

Table 4. Frequency response minimum relative amplitude results (MRA is the minimum amplitude relative to that of the tractor mass centre).

APPENDIX 3

*Parameter values - in the notation of* [4]

| parameter | vehicle A | vehicle B | vehicle C | vehicle D |
|---|---|---|---|---|
| $M_1$,kg | 6261 | 5620 | 5620 | 7800 |
| $M_2$,kg | 32249 | 13400 | 15250 | 24490 |
| $M_3$,kg | 2053 | 1400 | 1435 | 2000 |
| $M_4$,kg | 27212 | 13400 | 15250 | 24490 |
| $I_1$,kgm | 240.0 | 172.3 | 172.3 | 330.3 |
| $I_2$,kgm | 2007.2 | 680.9 | 667.7 | 3288.5 |
| $I_3$,kgm | 25.4 | 8.15 | 12.2 | 16.3 |
| $I_4$,kgm | 919.2 | 680.9 | 667.7 | 3288.5 |
| X11,m | 1.676 | 1.394 | 1.394 | 2.119 |
| X12,m | 1.092 | 1.856 | 1.856 | 1.881 |
| X13,m | 2.362 | - | - | - |
| X21,m | 1.295 | 2.474 | 2.342 | 4.529 |
| X22,m | 2.362 | - | - | - |
| X23,m | 3.429 | - | - | - |
| X31,m | -0.533 | 0.081 | 0.050 | 0.050 |
| X32,m | 0.533 | - | - | - |
| X41,m | 0.051 | 2.474 | 2.342 | 4.529 |
| X42,m | 1.118 | - | - | - |
| X43,m | 2.184 | - | - | - |
| X1A,m | 0.953 | 1.486 | 1.511 | 1.681 |
| X2A,m | 2.870 | 3.010 | 2.817 | 5.352 |
| X2B,m | 3.835 | 3.320 | 3.094 | 5.926 |
| X3B,m | 1.778 | 1.814 | 1.830 | 1.770 |
| X3C,m | 0.0 | 0.056 | 0.025 | 0.050 |
| X4C,m | 2.057 | 3.010 | 2.817 | 5.352 |

| parameter | vehicle | | | |
| --- | --- | --- | --- | --- |
| | A | B | C | D |
| y,m | 0.318 | 0.318 | 0.318 | 0.318 |
| C11,N/rad | 367020 | 155980 | 161570 | 168730 |
| C12,N/rad | 460810 | 314800 | 340010 | 618030 |
| C13,N/rad | 460810 | - | - | - |
| C21,N/rad | 426410 | 292240 | 331580 | 572900 |
| C22,N/rad | 426410 | - | - | - |
| C23,N/rad | 426410 | - | - | - |
| C31,N/rad | 426410 | 306370 | 334420 | 572900 |
| C32,N/rad | 426410 | - | - | - |
| C41,N/rad | 426410 | 292240 | 331580 | 572900 |
| C42,N/rad | 426410 | - | - | - |
| C43,N/rad | 426410 | - | - | - |
| N11,Nm/rad | 29906 | 7360 | 7625 | 7962 |
| N12,Nm/rad | 22682 | 14856 | 16047 | 29162 |
| N13,Nm/rad | 22682 | - | - | - |
| N21,Nm/rad | 19264 | 13788 | 15652 | 27032 |
| N22,Nm/rad | 19264 | - | - | - |
| N23,Nm/rad | 19264 | - | - | - |
| N31,Nm/rad | 19264 | 14461 | 15782 | 27032 |
| N32,Nm/rad | 19264 | - | - | - |
| N41,Nm/rad | 19264 | 13788 | 15652 | 27032 |
| N42,Nm/rad | 19264 | - | - | - |
| N43,Nm/rad | 19264 | - | - | - |
| CS12,N | 152970 | 143370 | 143370 | 143370 |
| CS13,N | 152970 | - | - | - |
| CS21,N | 143370 | 143370 | 143370 | 143370 |
| CS22,N | 143370 | - | - | - |
| CS23,N | 143370 | - | - | - |
| CS31,N | 143370 | 143370 | 143370 | 143370 |
| CS32,N | 143370 | - | - | - |
| CS41,N | 143370 | 143370 | 143370 | 143370 |
| CS42,N | 143370 | - | - | - |
| CS43,N | 143370 | - | - | - |

REFERENCES

1. Jindra, F., Lateral Oscillations of Trailer Trains. Ingenieur-Archiv., XXXIII (1964) p.194.
2. Hazemoto, T., Analysis of Lateral Stability for Doubles. SAE 730688, June 1973.
3. Hales, F.D., The Rigid Body Dynamics of Road Vehicle Trains. Proc.I.U.T.A.M. Symposium on The Dynamics of Vehicles on Roads and Tracks, Delft., August 1975, p.131.
4. Mallikarjunarao, C. and Fancher, P.S. Analysis of the Directional Response Characteristics of Double Tankers, SAE 781064, December 1978.
5. Nordmark, S. and Nordström, O., Lane Change Dynamics versus Geometric Design of Truck and Full Trailer Combinations. Proc.XVI Int.Auto.Tech.Congr. FISITA, Tokyo, p.1737.
6. Sharp, R.S., The Steering Responses of Doubles (Extensive summary of this paper) Veh. Syst.Dyn., 8, September 1979.

# SPECIAL RESONANT EFFECTS IN DYNAMIC RESPONSE OF VEHICLES

Jaroslav Šprinc[1] and Oldřich Kropáč[2]

[1] Institute of Theoretical and Applied Mechanics, Czechoslovak Academy of Sciences, Prague,
[2] Aeronautical Research and Test Institute, Prague – Letňany, CSSR

SUMMARY

The common need for reduction of vehicle vibration in operating conditions requires for all sources of vibration to be carefully analysed. In a many-mass model of a vehicle, an inverse effect of undercarriage damping may be observed under unfavourable combination of parameters, the elimination of which requires for an additional structural damping to be applied. Inacceptable increase of vibration may also appear on bridges and other transport structures with unsuitable elastic parameters and resonant properties. The undulation of the roadway pavement is considered to be the most important source of vehicle vibration. In more advanced analyses, two further important sources are to be considered, as well, namely the vibration of the driving unit and excitation due to wheel unbalance. For this combined loading case, a linear four-mass model has been designed and analysed. Examples of measurements of some unusual vibration phenomena obtained in real operating conditions of aircraft and cars are added. Desirable directions of further research are shortly outlined.

## 1. INTRODUCTION

The main dynamical phenomenon in vehicle behaviour in operating conditions consists in vibration of different structural elements, functional subsystems and aggregates, as well as of the vehicle as a whole. The prevailing part of vibration is due to unevennesses of roads. The problems of vibration and resonant effects in vehicles due to undulating pavement are being studied very extensively for several years in all technically advanced countries. The obtained results, both theoretical and experimental, enable the optimal solution of the complex system vehicle-roadway from the following points of view:

- operational capability as a global characteristic of the vehicle specified in Official Requirements,

- manoeuvrability of the vehicle and safety of its operation,

- riding comfort of passengers and the crew as well as admissible vibration of sensitive loads,

- fatigue, service life and reliability of critical elements of the vehicle.

In the paper, we shall be concerned in some special phenomena which may be caused partly as a result of an inverse effect of damping, partly owing to some external and internal sources of vibration of different origin.

For the analytical treatment of given problems, we shall use linear dynamical models with concentrated parameters. In order to preserve the lucidity in effects of individual sources, we shall limit our considerations to vertical vibration whereby the right-left symmetry of the vehicle will be considered. For the same reason of lucidity, analytical models will be assumed having only the minimum number of masses and elastic and damping elements necessary for evocation of the desired special phenomenon.

## 2. INVERSE EFFECT OF DAMPING ON THE MAGNITUDE OF RESONANT PEAKS

It is well known that with the one-mass vibrating system, the magnitude of the resonant peaks decreases with increasing damping. With the many-mass systems, however, the increasing damping may cause an increase of the response in some mass and in the frequency range where this is highly undesirable. As an example, a system shown in Fig. 1 may serve which represents an usual model of a vehicle moving on a undulating roadway where $M_1$ is the unsprung mass of the vehicle, $K_1$ is the stiffness parameter of the tyre, $K_2$ and $B_2$ are stiffness and damping parameters of the undercarriage, respectively, $M_2$ and $M_3$ are sprung masses whereby the mass $M_3$ may be interpreted to be the mass of passengers with the additional spring $K_3$ and damping $B_3$ with respect to $M_2$.

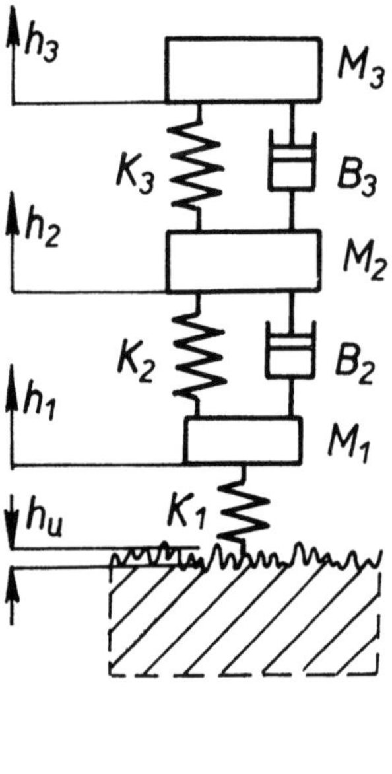


Fig. 1. Dynamic model of a vehicle

The dynamic behaviour of this system may be expressed by a matrix equation (1)

$$\begin{bmatrix} K_1+K_2-M_1\omega^2+i\omega B_2 & -K_2-i\omega B_2 & 0 \\ -K_2-i\omega B_2 & K_2+K_3-M_2\omega^2+i\omega B_2+i\omega B_3 & -K_3-i\omega B_3 \\ 0 & -K_3-i\omega B_3 & K_3-M_3\omega^2+i\omega B_3 \end{bmatrix} \begin{bmatrix} h_1 \\ h_2 \\ h_3 \end{bmatrix} = \begin{bmatrix} K_1 h_u \\ 0 \\ 0 \end{bmatrix}$$

from where the transfer functions for the displacement for masses $M_2$, $M_3$ may be easy derived in the form

$$F(i\omega)_{d2} = \frac{1-\left(\frac{M_3}{K_3}+\frac{B_2B_3}{K_2K_3}\right).\omega^2 + i\omega\left[\left(\frac{B_2}{K_2}+\frac{B_3}{K_3}\right)+\frac{B_2M_3}{K_2K_3}.\omega^2\right]}{(1-A\omega^2+B\omega^4-C\omega^6)+i\omega(D-E\omega^2+H\omega^4)} \qquad (2)$$

$$F(i\omega)_{d3} = \frac{1-\frac{B_2B_3}{K_2K_3}.\omega^2 + i\omega\left(\frac{B_2}{K_2}+\frac{B_3}{K_3}\right)}{(1-A\omega^2+B\omega^4-C\omega^6)+i\omega(D-E\omega^2+H\omega^4)} \qquad (3)$$

with expressions A, B, ... resulting from the evaluation of the determinant of the system (1). From the point of view of most performance characteristics, the force transfer functions are more informative, however, which may be obtained according to the relation

$$F(i\omega)_{fj} = -\frac{M_j\omega 2}{K_1}\,.\,F(i\omega)_{dj}\,, \quad j = 2,3. \qquad (4)$$

In Fig. 2. an example is given with parameters corresponding to a medium-size passenger car with $B_3 = 0$. When increasing the undercarriage damping, the resonance peak around the frequency 5 Hz increases, as well, while in the remaining resonance ranges, the gain remains practically unchanged. The discussed resonance peak may be essentially reduced by introducting the structural damping $B_3>0$ between the masses $M_2$ and $M_3$. The effect of the structural damping $B_3$ is shown in Fig. 3. It is evident that the positive effect of lowering the main second resonance peak has a negative side-effect in some increase of the response in the remaining resonance regions.

The inverse effect may appear also in the third resonance region. An example of such a case is given in Fig. 4., where, of course, the existance of this phenomenon probably would not be as dangerous as in the case shown in Fig. 2.

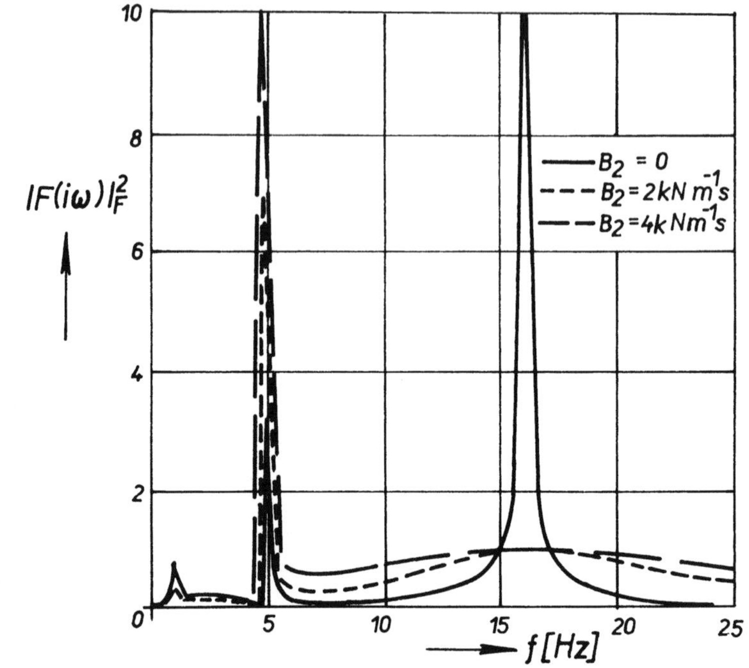


Fig. 2. Inverse effect of the undercarriage damping $B_2$ in a three-mass system with zero structural damping $B_3$

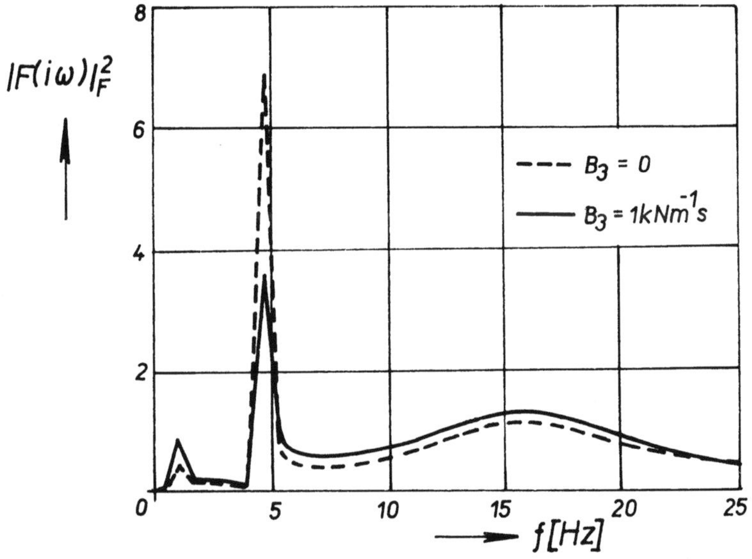


Fig. 3. Effect of structural damping $B_3$ ($B_2$ = 2 $kNm^{-1}s$ in both cases)

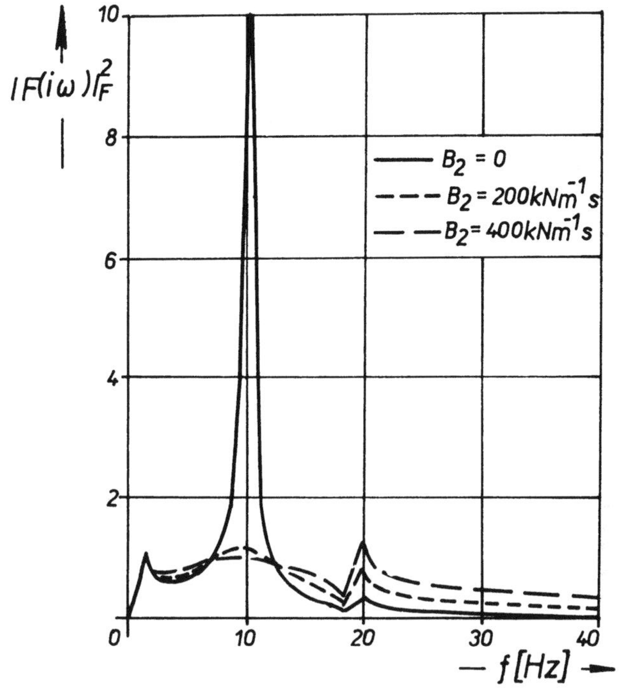


Fig. 4. Example of an inverse effect in the third resonance region

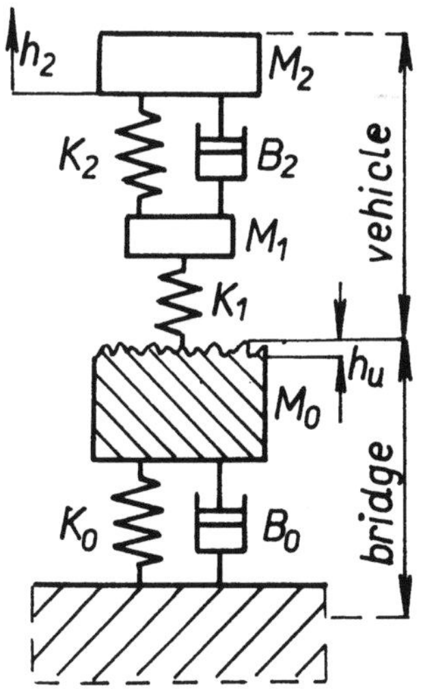


Fig. 5. Dynamic model of a vehicle on a bridge

## 3. RESONANT PHENOMENA ON BRIDGES

Special resonant phenomena may occur on bridges and similar transport structures which cannot be assumed to be rigidly connected with the earth but their mass and elastic and damping coupling with the earth is to be considered. This dynamical system has the same analytical model as heavy aircraft moving on an elastic runway. This problem has been discussed in an earlier paper of the authors [1] using a model similar to that shown in Fig. 5. where the interpretation of masses as a vehicle on a bridge has been used. It was shown in [1] that the effect of the elastic road surface results in an expressive resonance peak at the frequency $\omega_0 = \sqrt{K_o / M_o}$.

For completing the data enabling solution of particular problems, some common ranges for values of masses, stiffness and damping coëfficients and first natural frequencies for concrete and steel bridges for which the relevant data were available are given in Table 1. It should be noted, however, that the one-mass model for the bridge is applicable only for structures with relatively small span having quite high natural frequency of the first deformation mode.

In Fig. 6., an example of the dynamic response of a vehicle-bridge system is shown.

Table 1. Equivalent masses, eleastic and damping parameters and first natural frequencies of concrete and steel bridges

| Variables | Concrete bridges | Steel bridges |
|---|---|---|
| $M_o$ kg | 500 - 3000 | 300 - 2000 |
| $K_o$ $MNm^{-1}$ | 200 - 500 | 400 - 800 |
| $B_o$ $MNm^{-1}s$ | 1 - 2 | 0,5 - 1 |
| $f_o$ Hz | 0,5 - 2 | 1 - 5 |

## 4. INTERNAL SOURCES OF VIBRATION

On a moving vehicle, there are some internal sources of vibration which under unfavourable circumstances may reach inacceptably high values. There are two important internal sources of vibration, namely vibration of the driving unit and forces resulting from the wheel unbalance. In the first approximation, the later may be related to the total mass $M_1$ (which is equivalent to the most unfavourable case with all wheels having the same in-phase unbalance). In order to eliminate the vibrating effect of the driving unit, this unit is mounted on a flexible suspension provided with dampers. Thus, the vehicle model is to be adapted following Fig. 7 with the engine mass $M_4$ and suspension parameters $K_4$ and $B_4$. In this model, the damping of the tyre $B_1$ has been included for completeness.

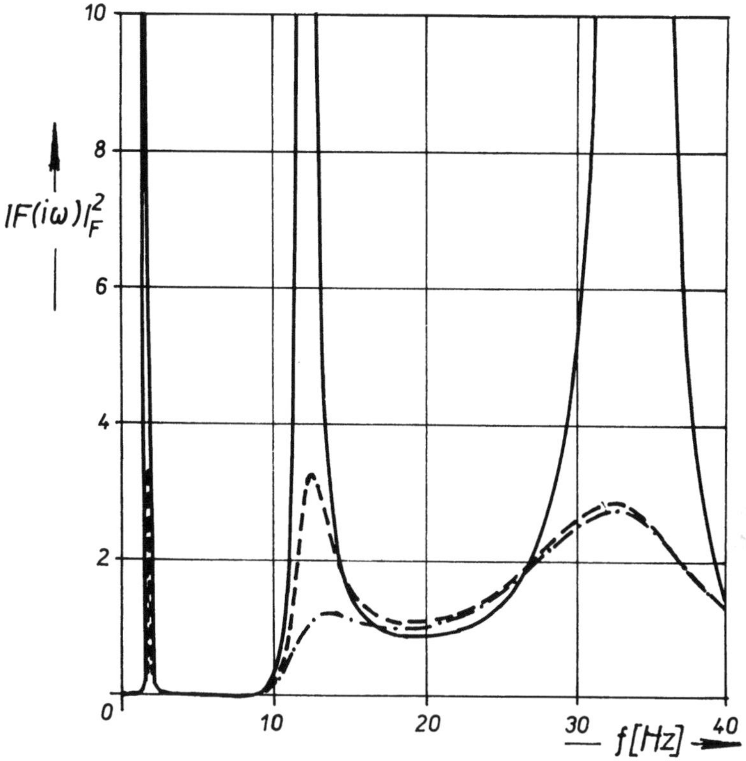


Fig. 6. Example of a dynamic response of a vehicle bridge system (B in $kNm^{-1}s$)

| | $B_o$ | $B_2$ |
|---|---|---|
| ———— (solid) | 200 | 0 |
| — — — — (dashed) | 200 | 2000 |
| —·—·— (dash-dot) | 1000 | 2000 |

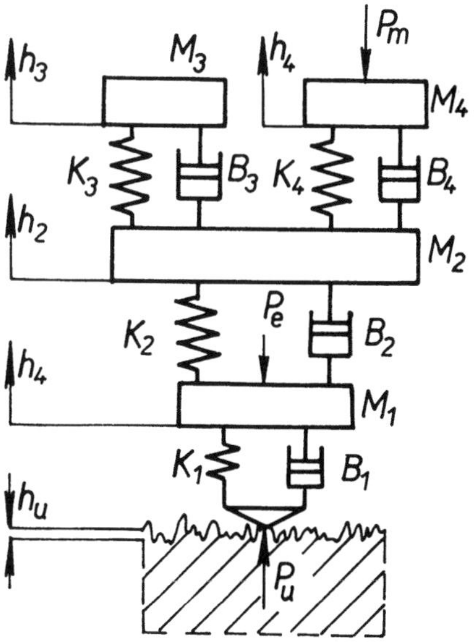


Fig. 7. Dynamic model of a vehicle with main components of external and internal excitation

When analysing the wheel unbalance cases in detail, one should consider first the unbalance of one wheel only producing force acting only on a corresponding part of mass $M_1$. In further steps, different unbalances with different phase shifts would be assumed for all wheels whereby the shifts may change during the drive due to nonequal tyre inflation, nonequal tyre wear, drive in a curve and nonuniform wheel slip. In the given formulation, the problem is much complicated so that solution is to be limited to most unfavourable modes of vehicle vibration.

## 5. COMBINATION OF EXTERNAL AND INTERNAL SOURCES OF VIBRATION

The model shown in Fig. 7. enables to express the response in the desired mass of the vehicle under simultaneous action of three basic sources, namely the pavement undulation, driving unit vibration, and wheel unbalance. Assuming for these inputs to be mutually uncorrelated, the total response may be expressed as a sum of partial responses to the partial inputs. The elasticity of the runway or of the bridge is not considered here. This would require for the model to be complemented by adding the mass $M_o$ with stiffness and damping parameters $K_o$ and $B_o$ respectively. In the realized analyses, the model shown in Fig. 7 has been exploited, the dynamic behaviour of which may be expressed by the equation

$$\underline{G}.\underline{h} = \underline{P}, \tag{5}$$

where $\underline{G}$ is the system matrix,
$\underline{h}$, is the displacement vector, and
$\underline{P}$ is the vector of applied forces.

The system matrix takes the form

$$\begin{bmatrix} K_1+K_2-M_1\omega^2+i\omega(B_1+B_2) & -K_2-i\omega B_2 & 0 & 0 \\ -K_2-i\omega B_2 & K_2+K_3+K_4-M_2\omega^2+i\omega(B_2+B_3+B_4) & -K_3-i\omega B_3 & -K_4-i\omega B_4 \\ 0 & -K_3-i\omega B_3 & K_3-M_3\omega^2+i\omega B_3 & 0 \\ 0 & -K_4-i\omega B_4 & 0 & K_4-M_4\omega^2+i\omega B_4 \end{bmatrix}$$

The input forces are as follows,

$P_u = h_u.(K_1+i\omega B_1)$ acting on mass $M_1$, whereby the power spectral density $S_{hu}(\Omega)$ of the pavement unevennesses is given which is transformed into $(1/v).S_{hu}(\omega/v)$, where $v[m.s^{-1}]$ is the forward speed of the vehicle,

$P_m = h_4.(-\omega^2.M_4)$ acting on mass $M_4$, whereby the power spectral density $S_{h4}.(\omega)$ of the driving unit vibration is given,

$P_e = M_1.e.v^2/r^2$ acting on mass $M_1$, where $v[m.s^{-1}]$ is the forward speed, $e[m]$ is the unbalance excentricity, $r[m]$ is the distance of the wheel axis from the road surface.

For these inputs, the responses in all masses in terms of displacements and forces have been evaluated. The resulting expressions are too extensive to be reproduced here. They are similar to Eqs. (2) and (3) but with polynomials in $\omega$ up to the eighth power and with parameters resulting form the solution of the sytem determinant $|G|$.

This model helps to explain some phenomena detected by measurements of vehicles in their operational conditions, some examples of which will be given in the next paragraph.

## 6. EXAMPLES OF MEASUREMENTS IN REAL OPERATING CONDITIONS

In Fig. 8., a comparison of power spectral densities of acceleration in the centre of gravity of an IL-14 aircraft for the take-off and landing operations on the same runway are shown. For the take-off regime, a pronounced vibration range around 40 Hz may be observed in addition to some amplification of all other resonant peaks.

In Fig. 9., a comparison of power spectral densities of acceleration in the centre of gravity of a LADA 1500 car is given for fully balanced wheels and for one wheel artificially unbalanced. The additional peak appearing in the dotted curve corresponds to the applied forward speed $v = 30\ m.s^{-1}$

In Fig. 10., the results of measurement made on a YAK-40 aircraft (acceleration on the passenger seat) when crossing a row of runway lights with considerably high camber are compared with running on the same runway but in a smooth trace.

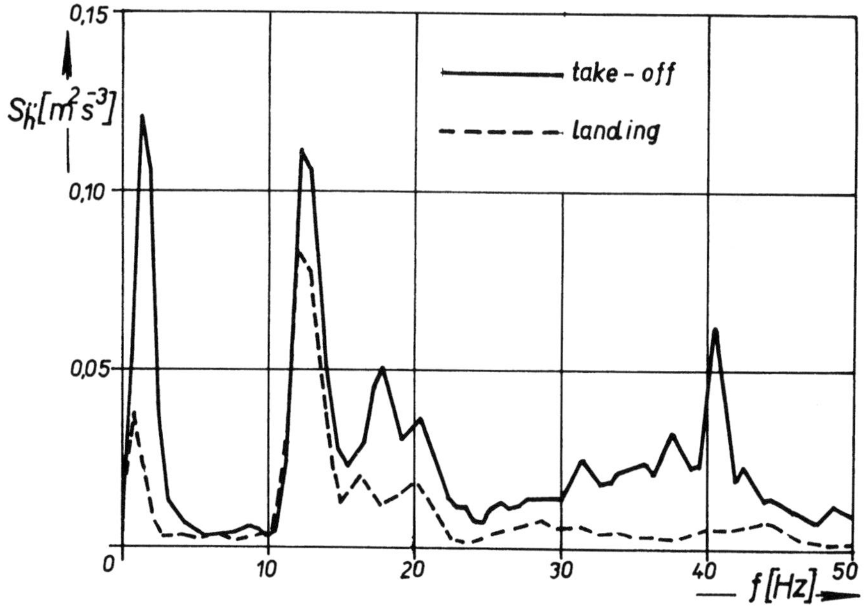


Fig. 8. Power spectral densities of acceleration in the centre of gravity of the IL-14 aircraft during take-off and landing operations

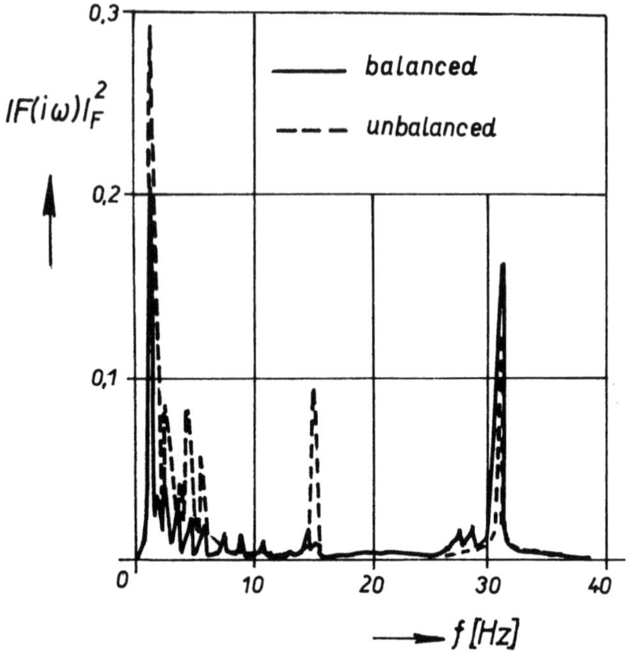


Fig. 9. Power spectral densities of acceleration in the centre of gravity of a LADA 1500 car with balanced and unbalanced wheel

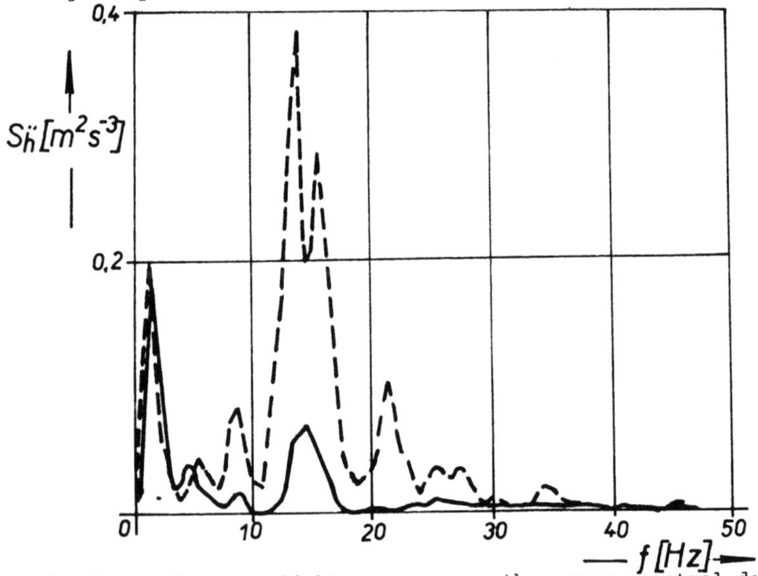


Fig. 10 Effect of runway lights overrun on the power spectral density of acceleration of a YAK-40 aircraft

——————— smooth trace — — — — row of runway lights

7. CONCLUSIONS

Extensive analysis of many-mass models of vehicles has shown that under some unfavourable combinations of system parameters a pronounced resonance regime in the response may occur which cannot be damped by means of a usual damping system of the vehicle undercarriage. It is also possible that an inverse effect of the damping parameter appear in a critical resonance region. In these cases, there is necessary to introduce structural damping - in some situation of relatively high magnitude - which suppresses this undesirable phenomenon but at the cost of some increase of the response in the remaining resonance ranges.

The effort in increasing the riding comfort of vehicles leads to necessity of a detailed analysis of all sources which produce vibration in a vehicle. In addition to the main external source resulting from the unevennesses and possible elasticity of roadways or bridges, there are some internal sources in vehicles the maximum elimination of which must be considered to be the primary interest of designers. In this paper, two internal sources have been considered in a rather short outline, namely the vibration of driving unit and forces resulting from the wheel unbalance. To obtain maximum lucidity of results, most simple analytical models of the cases in question have been considered.

For further improvement of obtained knowledge the following advanced problems may be formulated:

- take into account the nonhomogenety of the undercarriage wheel (tyre) i.e. the varibility of its stiffness and damping parameters along the tyre circumference,
- prove the existence of damping parameter dependence of decisive system elements on frequency and in the positive case, draw necessary conclusions,
- generalize the vehicle model as to be able to analyse not only the linear vertical vibration but also all components of a general motion of the vehicle as a whole and of its individual subsystems,
- study the problems of local energy transfer between individual vehicle subsystems under various operating conditions.

REFERENCES

1. Kropác O., Sprinc J.: Interaction Between Aircraft and Runway from the Point of View of Vehicle Dynamics. Proceed. IUTAM Symp. Dynamics of Vehicles on Roads and Railway Tracks, Delft, 1975, pp. 566-575.

# WHEELCLIMB DERAILMENT CRITERIA UNDER STEADY ROLLING AND DYNAMIC LOADING CONDITIONS

Larry M. Sweet, Amir Karmel and Peter K. Moy

Department of Mechanical and Aerospace Engineering,
Princeton University, Princeton, N.J., U.S.A.

ABSTRACT

Criteria for predicting wheelclimb derailment under steady rolling conditions are developed analytically and verified experimentally using a one-fifth scale model wheelset on tangent track. Criteria based on wheel and axle loads are compared, including the effect of applied roll moments. Derailment limits under dynamic wheelclimb are established using a nonlinear simulation model. Criteria for dynamic wheelclimb derailment are shown to be more complex than those for steady rolling conditions, being dependent on wheelset initial conditions, forward velocity, and duration of applied load. Limited experimental data provide a preliminary verification of the dynamic model.

## 1. INTRODUCTION

Analysis of derailment processes is critical to prediction of railroad vehicle safety. The mechanics of derailment are determined by the interaction of several nonlinear effects, including variation in wheel/rail contact point locations, contact angles, contact zone geometry, and forces due to creepage. Several theories have been published with experimental verification which characterize the wheelclimb derailment limit as a function of wheelset yaw angle relative to the track for steady rolling conditions on rigid rails [1-4].* Experimental results for dynamic conditions, however, indicate that the dynamic derailment limit may differ from that for steady rolling conditions, but no general theory has been developed that is consistent for both steady rolling and dynamic conditions. In this paper a unified theory is presented which provides the basis for synthesis of wheelclimb derailment criteria which may be applied for a variety of steady state and dynamic loading conditions for a single wheelset.

The results of previous quasisteady analyses of derailment under vertical and lateral axle loading are extended to include roll moment effects. Force ratio criteria based on loading of the axle and of the derailing wheel are compared. The derailment criteria for steady rolling conditions are validated with extensive data from experiments us-

* Numbers in brackets refer to List of References at end of paper

ing a dynamically scaled model wheelset on tangent track, which simulates realistically full scale derailment conditions. The theory is applied to dynamic conditions in which the incident lateral velocity of the wheelset at the initiation of the wheelclimb process is an important parameter in determining derailment limits. Computer simulation results show the effects of incident velocity, duration of lateral load, and forward velocity, and several candidate derailment criteria are explored. Limited experimental results from the wheelset scale model provide a preliminary verification of the dynamic analysis.

## 2. WHEELCLIMB DERAILMENT UNDER STEADY ROLLING CONDITIONS

### *Quasisteady Theory*

The steady rolling of a wheelset under load is a quasistatic process in which an equilibrium condition exists for all applied forces and moments due to vehicle and track loading. If for a given set of applied axle loads and wheelset yaw angle an equilibrium condition exists, characterized by a certain wheelset lateral position, wheelclimb derailment will not occur. If no stable equilibrium exists, wheelclimb derailment results. Yaw angle is defined to be positive when the derailing wheel is steered into the rail. For convenience, derailment is assumed to occur on the right rail, viewed in the direction of travel.

Since wheelset yaw angles are generally small, force and moment equilibrium conditions may be applied in a vertical plane passing through the axle. In this paper the effect of translation of the wheel/rail contact points out of this vertical plane is not considered. Analysis of the wheelset equilibrium conditions requires calculation of the forces at the contact points due to longitudinal, lateral and spin creep, each of which varies with wheelset lateral displacement, yaw angle, and axle angular velocity. The procedure for calculating the highly nonlinear wheel/rail contact forces in flange contact is presented in detail in [1], and is summarized schematically in Figure 1. The approach is similar to that presented in [3] and [4], each analysis including the Kalker creep force theory and iterative solutions for axle velocity and normal forces at the wheel/rail contact points. Numerical results are produced for specific wheel/rail contact geometries, computed for given wheel and rail profiles using algorithms developed in [5]. Creep forces at each contact point are determined through interpolation of numerical results from Kalker's Simplified Theory, tabulated for appropriate combinations of nondimensional creepages and contact ellipse geometry [6].

The results of the analysis may be presented in two forms. In the first, the lateral force (in track coordinates) applied to the wheelset at equilibrium for a given vertical load may be plotted as a function of lateral position and yaw angle (Figure 2). For each yaw angle the maximum lateral force represents the derailment limit. Dividing the maximum lateral force by the total axle load gives a wheelclimb derailment criteria in terms of an axle L/V ratio (also known as a derailment quotient). Using the coordinate systems and nomenclature in [1], the axle L/V ratio is derived as a function of dimensionless parameters $\Gamma_1$, $\Gamma_2$, $\ell_1$, $\ell_2$, and $\eta$.

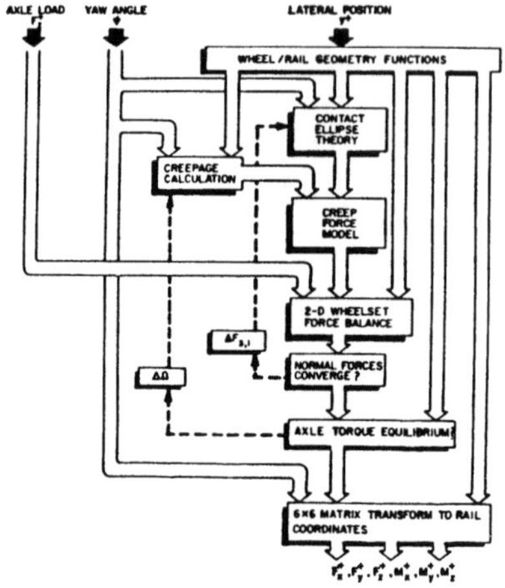


Fig. 1. Algorithm used to compute equilibrium forces acting on a wheelset under steady rolling conditions.

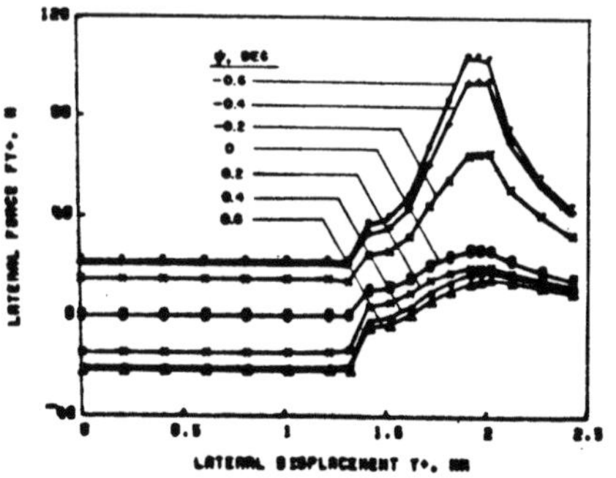


Fig. 2. Steady state wheelset lateral forces as a function of y and $\psi$ (44.5n axle load).

$$\left.\frac{L}{V}\right|_{axle} = [\ell_2\Gamma_1 + \ell_1\Gamma_2] + \frac{1}{2}\eta[-\Gamma_1 + \Gamma_2] \tag{1}$$

$$\text{where} \quad \Gamma_1 = \frac{\tan(\alpha_1+\phi) + \mu f_{2,1}}{1 - \mu f_{2,1}\tan(\alpha_1+\phi)} \qquad \ell_1 = \frac{L_1}{L_1 + L_2} \tag{2}$$

$$\Gamma_2 = \frac{\tan(\alpha_2+\phi) + \mu f_{2,2}}{1 - \mu f_{2,2}\tan(\alpha_2+\phi)} \qquad \ell_2 = \frac{L_2}{L_1 + L_2} \tag{3}$$

$$\eta = \frac{2M_x^+}{(L_1 + L_2)F_z^+} \tag{4}$$

Parameters $\Gamma_1$ and $\Gamma_2$ are recognized as the individual wheel L/V ratios for the left and right wheels, respectively. Parameter $\eta$ is the dimensionless roll moment, representing the degree of asymmetry in axle load between the extremes of wheel lift (with a practical range of $\pm 1$). In the above equations $f_{2,1}$ and $f_{2,2}$ are the dimensionless lateral plus spin creep coefficients on the left and right wheels, $\phi$ the wheelset roll angle, and $\alpha_1$ and $\alpha_2$ the associated contact angles, defined to be positive counterclockwise.

The L/V ratio for the wheel in flange contact is often used itself as a derailment criteria, and is defined using the above parameters as

$$\left.\frac{L}{V}\right|_{\text{flanged wheel}} = \Gamma_2 = \frac{\tan(\alpha_2+\phi) + \mu f_{2,2}}{1 - \mu f_{2,2} \tan(\alpha_2+\phi)} \tag{5}$$

For large yaw angles Eq.(5) approaches the classical limit of Nadal [2], but for the range of conditions $-3° < \psi < +3°$ the nondimensional creep $f_{2,2}$ may vary considerably with wheelset loading, wheel/rail profiles, and yaw angle, over the range $-1 < f_{2,2} < 1$. Because of the kinematic nature of the above relations for quasisteady conditions, Eqs.(1) through (5) are independent of forward velocity V.

*Comparison of Axle and Wheel L/V Ratio Criteria*

Due to the nonlinearity of the creep force phenomenon, wheelclimb derailment limits based on both axle and wheel forces vary with axle loading. Using numerical results for wheels and rails with new profiles extended from [1] to cases with non-zero roll moments, the degree to which each criterion may be applied universally is demonstrated. In Figure 3 wheel and axle L/V ratios are shown for varying axle vertical load. Both criteria are insensitive to vertical load, with minor variation evident only in the range $-1.4° < \psi < 0°$. The two criteria are related for all vertical axle loads by a single function. Figure 4 shows the effect of roll moment parameter $\eta$ on the two criteria. A positive roll moment, increasing the vertical force on the wheel in flange contact, produces a larger axle L/V ratio for wheelclimb derailment for all yaw angles. The scaling of the derailment limit with roll moment is reflected in the relative insensitivity in the wheel L/V ratio. Although the roll moment parameter $\eta$ does not appear explicitly in Eq.(5), it does affect the solution for wheelset force equilibrium, causing variations in $f_{2,2}$ over the range $-1.0° < \psi < 1.0°$.

Table 1 summarizes the relative advantages of use of criteria of each type. For situations in which axle loads including roll moments are known, from simulation or vehicle measurements, axle L/V ratios are better since the individual wheel forces are not required. If the roll moment is not known, as may be the case for field experiments on full-scale vehicles, wheel L/V ratios are better when applied to data from instrumented wheelsets.

Table 1. Comparison of Axle and Wheel L/V Ratio Criteria

| Advantage | Axle L/V | Wheel L/V |
|---|---|---|
| Insensitive to axle vertical load | Yes | Yes |
| Insensitive to axle roll moment | No | Yes |
| Shows variation with yaw angle | Yes | Yes |
| Does not require simulation or measurement of individual wheel contact forces | Yes | No |

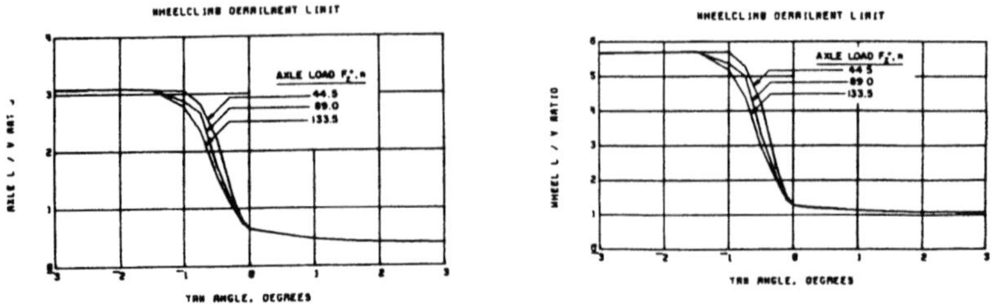


Fig. 3. Effect of axle load $F_z$ on axle (left) and wheel (right) L/V ratio derailment limits.

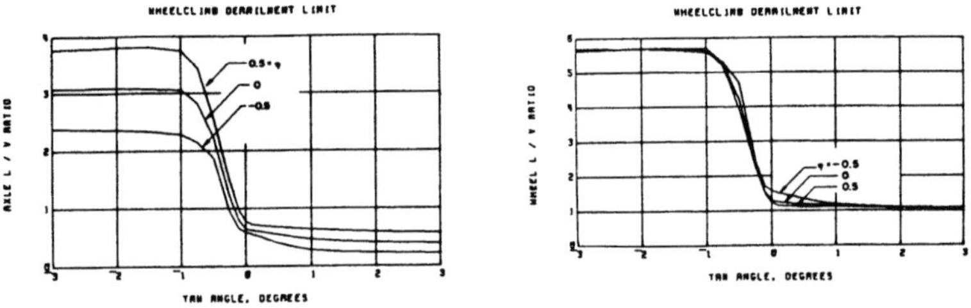


Fig. 4. Effect of nondimensional roll moment $\eta$ on axle (left) and wheel (right) L/V ratio derailment limits.

## *Experimental Verification of Quasisteady Theory*

An extensive series of experiments has been conducted to verify the wheelclimb derailment criteria presented. The apparatus employed is a dynamically scaled, instrumented single wheelset running on tangent track. The one-fifth scale model wheelset shown in Figure 5 is described in detail in [2]. Dynamic similitude of the creep forces is achieved through substitution of a polycarbonate resin of reduced elastic modulus for steel at the wheel/rail contact surfaces; in this manner the relationships among inertial, gravitational, suspension, and wheel/rail contact forces are maintained to reproduce full-scale phenomena under controlled laboratory conditions. Results from these scale model experiments are related to full scale through the following similitude laws, $\lambda_{length} = 0.2 \quad \lambda_{force} = 4.8 \times 10^{-4} \quad \lambda_{moment} = 9.6 \times 10^{-5}$

$$\lambda_{angle} = 1 \qquad \lambda_{L/V} = 1 \tag{6}$$

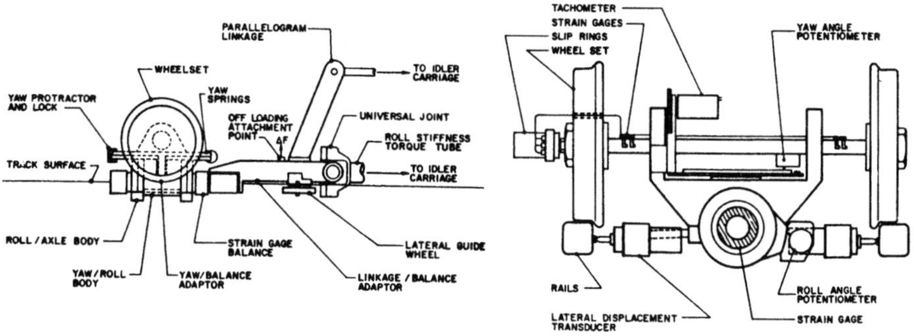


Fig. 5. Scale model wheelset experimental apparatus and instrumentation.

The wheelset model is supported by linkage and gimbal systems which provide vertical and lateral displacement, yaw, roll and pitch angular degrees of freedom. In the experiments reported in this paper the yaw angle is always locked to a specified value. The linkage system is attached to a separate carriage with independent suspension system running on the same rails. The carriage provides propulsion and serves as a platform for application of external forces. The instrumentation used in the experiments is shown in Figure 5. The principal load measurement device is a six-component, internally compensated strain gage balance, which measures all wheel/rail contact forces acting on the wheelset in the rail coordinate system. Cross-sensitivities among the six force and moment channels are known and used to resolve true forces and moments in post-experiment data processing.

The rails are mounted on an adjustable track structure 244m in length which is aligned to tolerances which are superior, in scale, to Class 6 track (U.S. DOT track specification). The track structure is sufficiently stiff to eliminate rail deflection under load. The wheels and rails are machined from the polycarbonate resin to the profiles of a new AAR wheel and new 135 lb CF and I rail, respectively, as documented in [5]. The theoretical and experimental results presented in this paper are specific to these profiles, and may differ for other wheel and rail geometries.

The quasisteady wheelclimb experiments consist of a lateral force ramp loading applied to the wheelset, with vertical axle load, yaw angle and forward velocity held fixed. The loading rate is held small to preserve quasisteady conditions. Typical time histories of axle L/V ratio and lateral displacement are shown in Figure 6.

The time histories indicate a very well-defined derailment limit. Fluctuations in the measured responses are due to variations in local

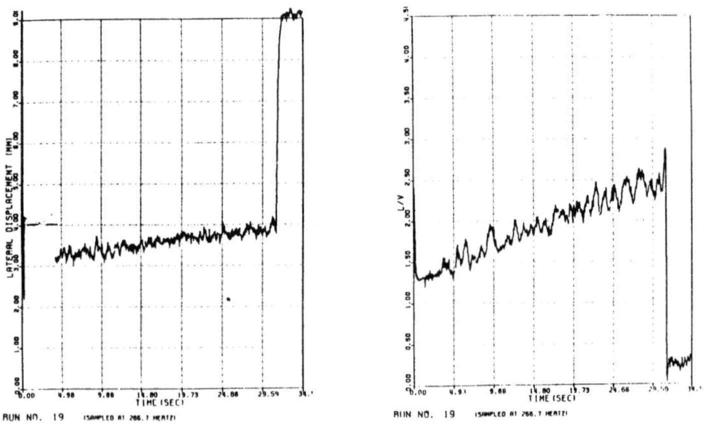


Fig. 6. Typical experimental time histories of lateral displacement and axle L/V ratio during quasisteady wheelclimb derailment.

track geometry; the "track signature" is quite repeatable when the experiment is duplicated at the same track location, even at different forward velocities. At other track locations, the fluctuations over the time history differ, but the L/V derailment limits are consistent within a range of ±5%. The temporal frequencies of these fluctuations are sufficiently low to maintain quasisteady conditions. Since the variations in track alignment and rail profile in the scale model are less than that expected in full scale, the fluctuations in measured responses shown are probably smaller than would be expected in field testing.

Experiments such as that illustrated in Figure 6 were conducted for the range of yaw angles $-3° \leq \psi \leq 3°$. The maximum axle L/V ratio immediately before derailment for each yaw angle is recorded in Figure 7 for the case of zero applied roll moment. Agreement between theory and experiment is generally good, although the theory overestimates the derailment limit for negative angles in the transition region between -0.5° and -2.0°.

Three possible sources of error may account for this discrepancy, two experimental and one theoretical. In the vicinity of the derailment location on the track, local variation in rail profile curvature from nominal values may produce significant changes in nondimensional lateral creep coefficient $f_{2,2}$. Error in $f_{2,2}$ or friction coefficient $\mu$ would produce an error in the predicted derailment limit. The third possible source of error results from the high ratio of the semi-axes of the contact ellipse in flange contact. For the wheel and rail profiles used in the experiment the ratio a/b is 16, which is beyond the numerical range computed by Kalker and the experimental range measured by Brickle [3]. Therefore numerical error in the calculated creep forces is not excluded as an explanation for the difference between theory and experiment.

The effect of roll moment on the measured axle L/V derailment limit is shown in Figure 8. Data points shown without lines are touching

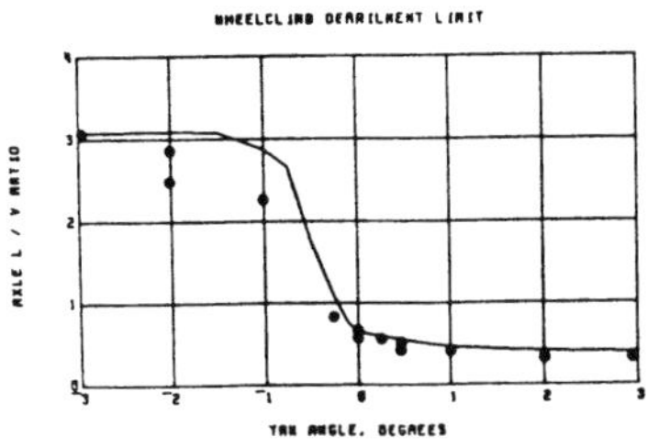

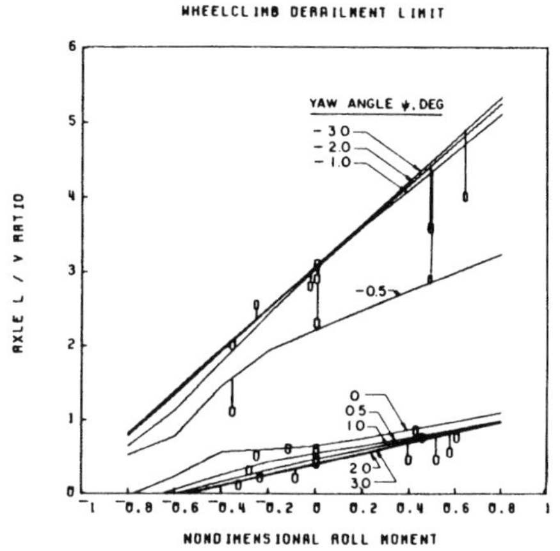


Fig. 7. Comparison of theory and experiment for wheelclimb derail- with no applied roll moment.

Fig. 8. Comparison of theory and experiment for wheelclimb derailment with roll moments applied.

the predicted derailment limit at the associated yaw angle. Data points with lines indicate the differences between theory and experiment. The application of constant roll moments under rolling conditions on the track was difficult experimentally, resulting in some scatter in the data. The agreement between theory and experiment is good, with the same differences as discussed above for negative yaw angles.

## 3. WHEELCLIMB DERAILMENT UNDER DYNAMIC LOADING CONDITIONS

### *Nonlinear Dynamic Theory*

The equations of motion for a wheelset become highly nonlinear at the onset of flange contact. Equivalent linearization techniques (statistical linearization, describing functions, etc.) are not applicable to derailment prediction since they do not represent accurately the dynamics at the extremes of wheelset excursion. The approach used here is a direct digital computer simulation, including nonlinear effects due to gravitational stiffness and creep force saturation. For the present analysis yaw angle is specified to be time invariant, so that the nonlinear variation in conicity is not required; this effect may be readily added for simulation of two degree-of-freedom (lateral displacement plus yaw angle) motions. The effects of collision dynamics at wheel/rail impact are neglected.

The single degree-of-freedom equation of motion for lateral displacement of the wheelset is,

$$M\ddot{y} + B\dot{y} + Ky + F_{cr}(y,\dot{y},\psi) + F_g(y) = F_o(t) \qquad (6)$$

The wheelset mass M has linear suspension elements B and K fixed to an inertial reference. The gravitational stiffness force $F_g(y)$ results from the changes in contact angles $\alpha_1$ and $\alpha_2$, roll angle $\phi$, and lengths $\ell_1$ and $\ell_2$, as the wheelset displaces laterally.

$$F_g(y) = F_z[\ell_2 \tan(\alpha_2+\phi) + \ell_1 \tan(\alpha_1+\phi)] \tag{7}$$

The creep force $F_{cr}$ in the lateral direction results from lateral creep due to $\dot{y}$ and $\psi$ and spin creep due to y, taking into account the variation in contact angles and contact geometry associated with wheelset lateral displacement y for specific wheel/rail profiles and the instantaneous contact forces. Digital simulation of the creep force requires repeated use of algorithms applying Kalker's theory of creep, which may be prohibitively expensive. In this paper the wheel/rail contact forces are approximated to produce a simplified model which is very useful for parametric studies. In this approximation, lateral creep forces due to lateral velocity $\dot{y}$ are separated from those due to y and $\psi$. The quasi-steady wheel/rail contact forces are

$$F_{QS}(y,\psi) = F_g(y) + F_{cr}(y,\psi) \tag{8}$$

which have been previously computed in [1] and stored for use during the simulations. These forces, shown in Figure 2, include the effects of gravitational stiffness, lateral creep force due to yaw angle $\psi$, and spin creep.

The lateral creep force on the wheel in tread contact (non-derailing wheel) is given by

$$F_1(\dot{y}) = \frac{f_L}{V}\dot{y} \quad ; \quad |F_1| \leq \frac{\mu F_z^+}{2} \tag{9}$$

where $f_L$ is the lateral creep coefficient and V the forward velocity. The magnitude of the lateral creep force on this wheel is constrained by the adhesion limit, as indicated in Eq.(9). For the flanging wheel, the lateral creep force due to $\dot{y}$ is approximated as

$$F_2(\dot{y}) = \frac{f_L}{V\cos^2\alpha_2} \cdot \dot{y} \tag{10}$$

The lateral creep force is defined in the contact plane at angle $\alpha_2$, resulting from a lateral velocity $V_2$ in the plane. Equating the mechanical power dissipated in the contact plane $[(f_L/V)V_2^2]$ to the mechanical power dissipated by creep forces due to lateral velocity in the horizontal direction ($\dot{y} = V_2 \cos \alpha_2$), yields the equivalent lateral creep force in Eq.(10). The normal force on the flange during derailment is high, so that $F_2$ will not exceed the adhesion limit.

The approximate equation of motion is now summarized as,

$$M\ddot{y} + B\dot{y} + F_1(\dot{y}) + F_2(\dot{y}) + Ky + F_{QS}(y,\psi) = F_o(t) \tag{11}$$

Dynamic wheelclimb simulations for the scale model wheelset (parameters summarized in Table 2) are illustrated in Figure 9. Shown are responses to step changes in applied lateral force, starting from $\dot{y}(0) = 0$ in flange contact at the opposite rail. Motion in tread contact is characterized by the response of an overdamped, second-order system. After flange contact, the gravitational stiffness force $F_g(y)$ results

Table 2. Scale Model Wheelset Parameters used in Dynamic Simulations

$M = 11.904$ kg  $B = 96.5$ n-sec/m  $K = 0$

$f_L = 1724$ n  $\mu = 0.3$  $F_z = 44.5$ n

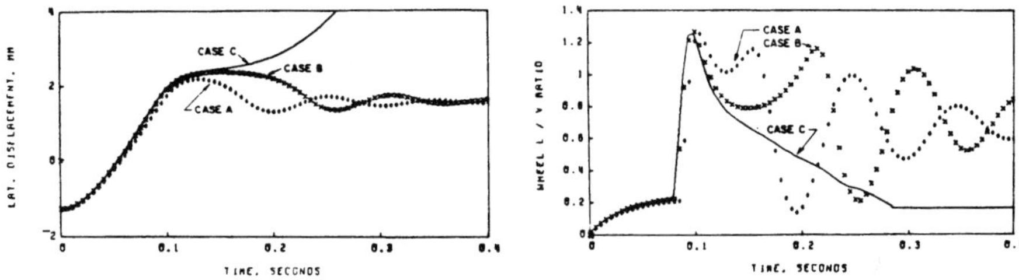


Fig. 9. Simulated dynamic responses to lateral loads of $F_o$=14.0n, (A), $F_o$=16.0n (B), and $F_o$=16.25n (C).

in a response which is typically underdamped, with lateral displacement exceeding the final equilibrium value in flange contact. As the magnitude of the applied lateral force is increased, both the peak and equilibrium lateral displacements increase, until the dynamic wheelclimb derailment limit; beyond the limit the restoring forces are insufficient to prevent derailment. The dynamic wheelclimb derailment limit exists primarily because the gravitational stiffness $F_g(y)$ decreases after the maximum in contact angle $\alpha_2$ is passed at large lateral displacements.

The conditions associated with the dynamic limit are defined by considering the point of maximum displacement where both lateral velocity and acceleration are zero. At the derailment limit all static and quasisteady forces are equal,

$$Ky + F_{QS}(y,\psi) = F_o(t) \tag{12}$$

The nonlinear nature of $F_{QS}(y,\psi)$ results in two solutions for displacement y that satisfy Eq.(12); the first solution is the equilibrium lateral displacement, or static limit $y_{SL}$; the second is the displacement $y_{DL}$ at the dynamic derailment limit. If $y_{DL}$ is exceeded and $F_o(t)$ continues to be applied, derailment will occur.

The maximum value for lateral displacement depends on the time history of wheelset lateral load $F_o(t)$, wheelset parameters, nonlinear functions $F_{QS}(y,\psi)$, $F_1$ and $F_2$, and the initial conditions. The functional interdependence of several of these factors is shown in Figs. 10 and 11. In Fig. 10 the relationship between constant applied lateral

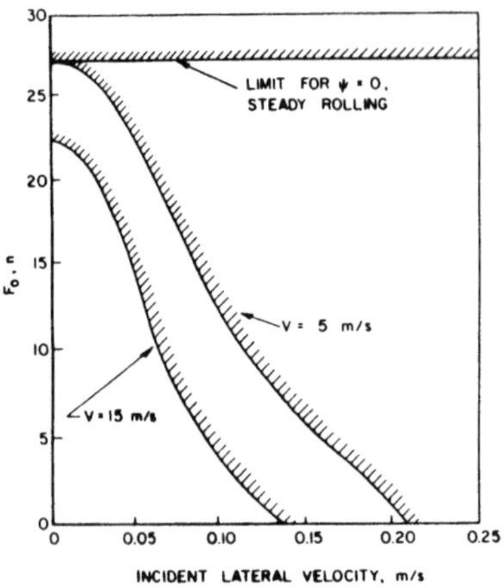


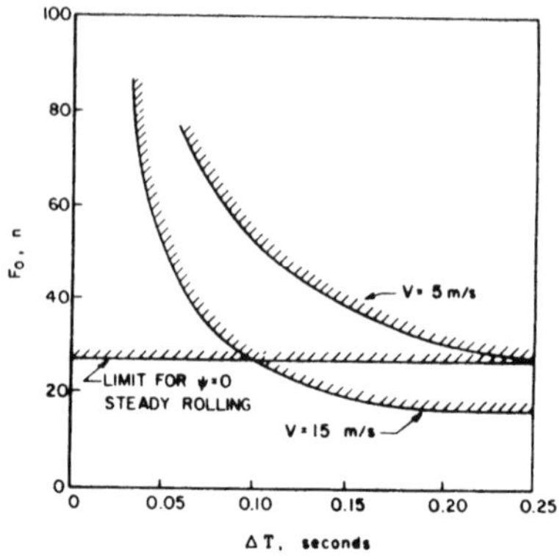


Fig. 10. Effect of incident lateral velocity on maximum applied lateral force (44.5n axle load).

Fig. 11 Effect of load duration on maximum applied lateral force (44.5n axle load).

force and lateral velocity at flange contact is shown. Dynamic derailment may occur even at L/V = 0 at large incident velocities. The derailment limit varies with forward velocity V. As velocity is increased lateral damping decreases, resulting in greater overshoot in lateral displacement and sensitivity to derailment. As the incident lateral and forward velocities approach zero, the dynamic derailment limit approaches the limit for steady rolling. The velocities shown are for the wheelset model, and correspond to 40.5 km/hr (24.7 mph) and 121.5 km/hr (74.1 mph) in full scale.

In. Fig. 11 the sensitivity of the derailment limit to duration of applied load $\Delta T$ is shown, with the initial condition being $\dot{y}(0) = 0$ with the wheelset in flange contact on the opposite rail. As $\Delta T$ is decreased, the wheelset may not have time to exceed the $y_{DL}$, or reversal of $F_o(t)$ may restore stable running if an ultimate displacement limit (taken to be 5mm in scale) has not been violated. Since the maximum displacement is reached in about 0.1 seconds, the ultimate limit exceeded in less than 0.05 seconds after $y_{DL}$, very large lateral forces may be applied for $\Delta T$ less than 0.15 seconds without derailment. As discussed previously, the dynamic derailment limit is lower at larger forward velocities, with the steady rolling limit approached for long $\Delta T$ and zero forward velocity.

*Dynamic Wheelclimb Derailment Criteria*

The derailment limits shown in Figures 10 and 11 are dependent on specific initial conditions or loading time histories. While these limits may be useful in analytical and simulation studies, dynamic wheelclimb derailment criteria that may be applied to variables that are readily measured in the field, such as wheel or axle loads, would be valuable. In the following paragraphs several candidate criteria are explored, using the simulated responses summarized in Figures 10 and 11 plus additional cases on both sides of the derailment limit as a data base. Of particular interest is the validity of the well known JNR criteria presented in [4].

For each simulated response the loads experienced by the wheelset are calculated. Such loads, as would be measured by an instrumented wheelset, are shown in Figure 9. A large number of load time histories (160) covering a wide range of incident lateral velocities, applied lateral force levels, and applied force durations were analyzed in terms of the following criteria:

a) Peak axle L/V
b) Peak wheel L/V
c) Time axle L/V exceeded specific levels
d) Time wheel L/V exceeded specific levels.

The peak L/V ratio limits for various forward velocities and load durations are shown in Figure 12, all for a common initial condition; different initial conditions for wheelset lateral displacement and velocity produce different peak ratios. Because wheelset dynamic derailments under realistic conditions involve ranges of initial conditions, forward velocities, and load durations, no single value for L/V ratio, either axle or wheel based, is a valid indicator of the derailment limit. At a minimum the dynamic limit must be a function of the above variables, but it is not apparent that the dependence on initial conditions may be removed as desired.

It has been reported that the length of time $t_1$ that a particular L/V ratio is exceeded is a valid indicator of dynamic derailment, with larger L/V ratios permissible at decreasing $t_1$. The JNR criteria is an example of this type of derailment limit [4]. The results of the 160 simulations are very much to the contrary. When exceedance times $t_1$ are plotted against L/V ratios, no distinct boundary separates derailments from non-derailments. The simulation results shown in Figure 9 are a strong counterexample to the JNR criteria; the duration of the peaks in L/V ratio actually <u>decreases</u> as the derailment limit is approached. It is noteworthy that the displacement derailment limit $y_{DL}$ occurs later in time than does the peak L/V ratio. Since the displacement limit $y_{DL}$ depends on both the loading and the past time history, it is not surprising that analysis of the measured L/V ratio alone does not determine uniquely the derailment limit.

The theory presented in this paper differs from that presented in [4] in that collision and wheelset roll inertia effects are not included in the former, while in the latter the contact angle and friction force are treated as constants. In a future study the authors of this paper will include collision and roll inertia effects so that the two theories may be compared directly. The existence of a relatively well-

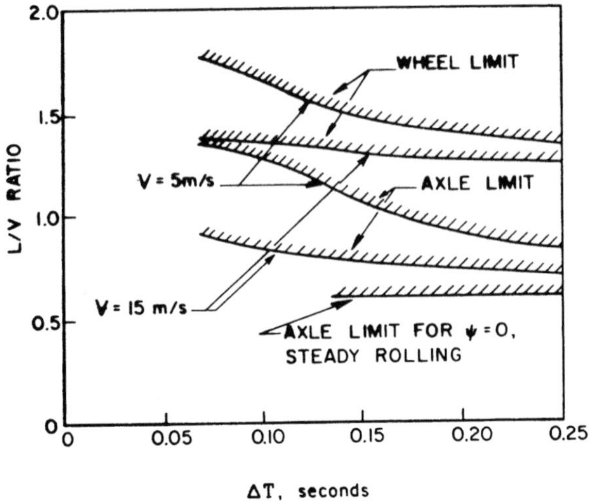


Fig. 12. Peak L/V ratios at the dynamic wheelclimb derailment limit.

defined derailment limit based on the Japanese experimental data may be due to a uniformity in parameters defining the derailment experiments (forward velocity, hunting motion only, no applied lateral force). The degree to which the Japanese results may be generalized to other conditions has not been established. Based on the simulation results in this paper, this questions clearly merits further serious investigation.

### *Dynamic Wheelclimb Experiments*

Dynamic wheelclimb experiments have been conducted using the dynamically scaled wheelset model described previously. The yaw angle is locked, limiting motion to single degree of freedom lateral displacement. Lateral forces of varying magnitudes and durations are applied using electronically controlled pneumatic actuators. Although the limited amount of data available at this writing is insufficient to validate the theory, the qualitative agreement of the predicted and measured time histories provides a preliminary positive verification of the analysis.

## 4. CONCLUSIONS

The forces acting on a single wheelset at equilibrium are analyzed using a quasisteady theory to determine wheelclimb derailment criteria under steady rolling conditions. Roll moments applied to the wheelset have an important effect on the derailment limit, when derailment criteria are based on axle loads. Criteria based on wheel loads are less sensitive to roll moment, but require knowledge of individual wheel forces from instrumented wheelsets or more complex calculations than for axle loads. The theoretical derailment limits are in general agreement with scale model wheelset experiments for the ranges of axle loads, roll moments, and yaw angles expected in full scale. The derailment criteria for steady rolling conditions are independent of forward velocity.

A dynamic wheelclimb derailment theory is developed which is consistent with that for steady rolling. The incident lateral velocity of the wheelset at flange contact, the forward velocity, and the duration of the applied load are shown by simulation to be critical parameters in determining the dynamic limit. Simulation results also indicate that analysis of wheelset forces alone, in the absence of information regarding the above parameters, does not provide a well-defined derailment limit. This conclusion is in contrast to the JNR criterion based on L/V ratio and force duration time. The dynamic analysis agrees qualitatively with preliminary experimental data from the scale model wheelset.


## ACKNOWLEDGMENTS

The authors acknowledge the assistance of J.M. Karohl and M. Tomeh in conducting the experiments, and E. Griffith in design and fabrication of the apparatus. The research is supported by the U.S. Department of Transportation under Contract No. DOT-TSC-1603.


## NOMENCLATURE

| | |
|---|---|
| $a, b$ | semi-axes of wheel/rail contact ellipse |
| $B$ | wheelset lateral suspension mechanical damping |
| $f_L$ | lateral creep coefficient |
| $f_{2,i}$ | nondimensional lateral creep force (contact plane coordinates, $i = 1$ for left wheel in tread contact, $i = 2$ for right wheel in flange contact) |
| $F_{cr}$ | total lateral forces due to creep |
| $F_g$ | gravitational stiffness force |
| $F_o$ | externally applied lateral force on wheelset |
| $F_{QS}$ | quasi-steady force due to lateral creep, spin creep, and gravitational stiffness |
| $F_z$ | axle load |
| $F_1, F_2$ | lateral creep force due to lateral velocity $\dot{y}$ |
| $K$ | wheelset lateral suspension mechanical stiffness |
| $L_1, L_2$ | horizontal distances from wheelset center to wheel/rail contact points |
| $\ell_1, \ell_2$ | nondimensional distances from wheelset center to wheel/rail contact points |
| L/V | ratio of lateral to vertical forces |

M wheelset mass

$t_1$ time specified L/V ratio is exceeded

$\Delta T$ duration of externally applied load $F_o(t)$

V forward velocity of wheelset

y wheelset lateral displacement from track centerline

$y_{SL}, y_{DL}$ wheelset lateral displacement at the static and dynamic derailment limits

$_{1}, _{2}$ contact angles, defined positive counterclockwise

$_{1}, _{2}$ wheel L/V ratios on left and right wheels

nondimensional roll moment applied to wheelset

friction coefficient

yaw angle

REFERENCES


1. Sweet, L.M. and Sivak, J.A., "Nonlinear Wheelset Forces in Flange Contact--Part I: Steady State Analysis and Numerical Results." To appear in ASME Trans., J. of Dyn. Systems, Meas. and Control, September 1979.
2. Sweet, L.M., Sivak, J.A. and Putman, W.F., "Nonlinear Wheelset Forc in Flange Contact--Part II: Measurements Using Dynamically Scal Models." To appear in ASME Trans., J. of Dyn. Systems, Meas. a Control, September 1979.
3. Arai, S. and Yokose, K., "Simulation of Lateral Motion of 2-Axle Railway Vehicles in Running." In The Dynamics of Vehicles on Roads and Railway Tracks, ed. H.B. Pacejka, Swets and Zeitlinge B.V., Amsterdam, 1976.
4. Gilchrist, A.O. and Brickle, B.V., "Re-examination of the Proneness to Derailment of a Railway Wheelset." J. of Mech. Engr. Sci., Vol. 13, pp. 131-141, June 1976.
5. Heller, R. and Cooperrider, N.K., "User's Manual for Asymmetric Wheel/Rail Contact Characterization Program." DOT Report FRA/ORD-78/05, December 1977.
6. Goree, J.G. and Law, E.H., "User's Manual for Kalker's Simplified Nonlinear Creep Theory." DOT Report FRA/ORD-78/06, December 197

# MODEL OF VIBRATION AND NOISE TRANSMISSION IN MOTOR VEHICLES

P. Urban

Motor Vehicle Research Institut (UVMV), Prague, CSSR

SUMMARY

The noise-control in vehicles is often limited by the possibility of reducing the components in the frequency band 80 - 160 Hz radiated by large parts of the vehicle structure. The sound transmission of the structure can be substituted principally by two multiports.

The calculation of the transmiddibility of both multi-ports leads either to a set of integral equations or after a Fourier transformation to komplex linear equations. The input signals of the multiports are correlated and so the set cannot be solved in a simple manner.

An alternative way of calculation with the model lies in the use of the principle of reciprocity in linear systems.

Theoretical solution is completed by experimental data.

INTRODUCTION

In the treatment of many problems in mechanical structures their dynamic properties must be known. Such a case is the problem of noise and vibration of the bodies and cabs of motor vehicles which we will further deal with. As far as we shall in this context discuss the vibration we shall understand it in direct connection with the problem of noise. When determining the dynamic properties of real structures we can measure either on structures excited by a natural signal or on structures excited by an artificial signal. In both cases direct processing of the structures transmission function is complex when the exciting signals are non-harmonic. The situation is simpler when the exciting signal is at least periodical, a favourable state exists when excitation is purely harmonic. On the other hand the artificial harmonic excitation demands bulky and costly equipment, particularly when the excited structure is of big size and mass and excited in a wide frequency band. Corresponding complications we can find even in excitation by non-harmonic periodic signals or by broad band signals (white or pink noise).

Lately - thanks to the development of A/D converters and digital processing - a method with very good prospects appears to be the method of pulse excitation.

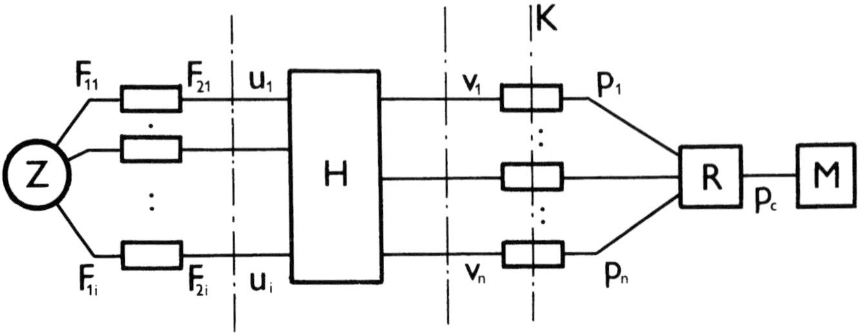


Fig. 1

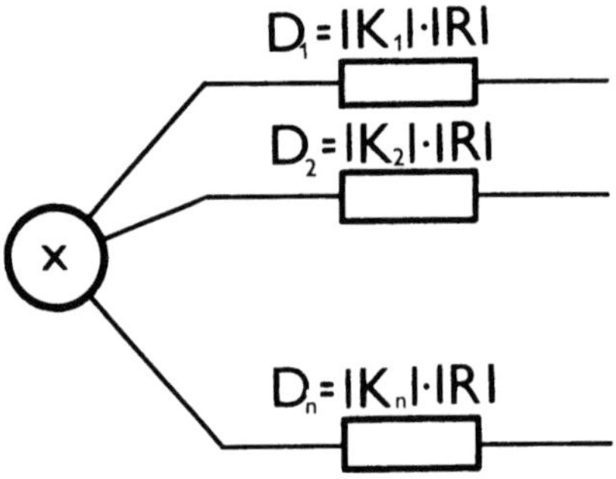


Fig. 2

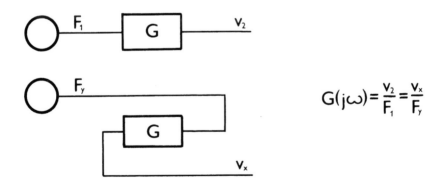

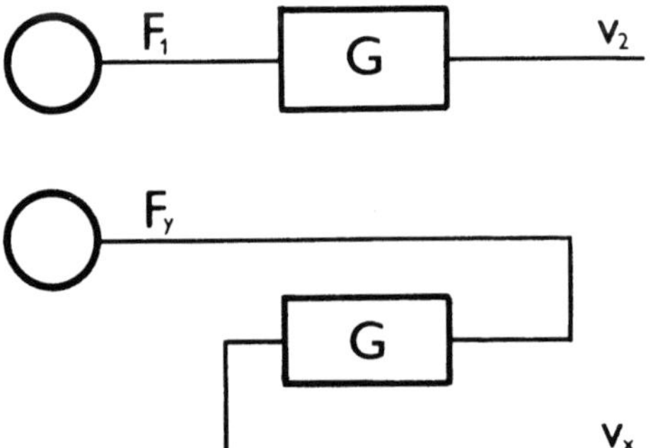


$$G(j\omega) = \frac{v_2}{F_1} = \frac{v_x}{F_y}$$

Fig. 3

The application of a pulse method in the investigation of the emission of noise from the vehicle bodies is the subject of the chapters which follow.

## MECHANISM OF PROPAGATION OF NOISE IN MOTOR VEHICLES

Let us consider only cases when the noise is transmitted inside a motor vehicle by structure-borne sound. The noise source (e.g. the engine, axle, etc.) is usually joined to the structure of a vehicle by a defined number of couplings through which the oscillating energy passes from the source to the structure. The vibrating energy propagates in the structure to the surface surrounding the passenger compartement from which it is emitted as a sonic energy.

The system of transmission of oscillating energy can be described by a substitute diagram according to Fig. 1. There Z stands for the source of energy acting at the individual coupling points of the structure of the vehicle by forces $F_{11}$ - $F_{1i}$. The oscillating energy is transmitted from these points to the structure of the vehicle usually by elastic suspension, represented by transmission members (two-ports) $G_1$ - $G_i$. In view of the properties of an elastic suspension the value of the energy transmitted and thus also the value of velocity of the oscillation at suspension points ($u_1$ - $u_i$) are a function not only of the transmissibility of members G but also of the position of the suspension points. The part of a vehicle structure further transmitting the material borne sound can be replaced by a multiport H with <u>i</u> inputs at the side of the source. The multiport has, in actual fact, an infinite number of outputs which are the individual oscillating points of the walls surrounding the passenger's compartement. In a real model we are forced to assume, on the output side of the multipole, a finite number <u>n</u> of outputs and we assume, at every individual output, a synchronous oscillation of the belonging part of the surface.

Another step in our substitute diagram are n transmission members K replacing the transmission of the oscillating energy from the solid body (the surface bounding the interior compartement of the vehicle) to air (the interior space for passengers).

Element R characterizes the properties for the internal space for passengers (absorption, reflections) and finally M represents the output member of the whole chain, i.e. either the ear of the disturbed person or, in the case of measurements, the measuring microphone.

When we want to control the noise in the vehicle under investigation, we have the possibility of interventions in any of the members of the whole transmission chain, i.e. at the source, in its suspension to the structure, in the actual structure of the vehicle as well as in the interior space of the vehicle.

## SOLUTION

Let us consider elementary area dS of the surface of an engine which oscillates with the same phase at a rate v (perpendicularly to area dS) When continuity is maintained also the neighbouring particles of mass of the gas in the adjacent gaseous atmosphere must oscillate at the identical rate v.

The value of power output dP emitted by the area under consideration is

$$dP = p.dS.v \tag{1}$$

where p is the acoustic pressure in the vicinity of areas dS.

In the close proximity of the area dS we may consider the wave produced to be plane and may calculate with the expression

$$\frac{p}{v} = \rho\, c_o \tag{2}$$

where $\rho\, c_o$ is the characteristic impedance of air for a plane wave ($\rho$ density of air, $c_o$ speed of sound).

The acoustic output emitted inside the vehicle by area dS may then be expressed by the relation

$$dP = v^2.\rho\, c_o.dS \tag{3}$$

As long as the oscillations of the individual areas dS (in Fig. 1 the transmission members K) would be uncorrelated it would be possible to determine, in principle, the total emitted output power by an integral of the type

$$P = \rho\, c_o \iint v^2.dS \tag{4}$$

Since, however, the oscillations at the individual points of the surface are correlated, particularly in the region of the natural frequencies, the whole system must be solved with the help of the velocity potential.

The precision of such a purely theoretical solution is nevertheless not high. Therefore the following relation for the expression of the acoustic output emitted by the surface is usually used

$$P = \rho\, c_o \cdot v_m^2 \cdot S \cdot \sigma \tag{5}$$

where $v_m$ is the average of the vibration-velocity (RMS value) of the walls surrounding the space under consideration

S the total area of the surrounding walls

$\sigma$ coefficient combining the directional and phase properties of emission of correlated sources.

Another problem is the determination of the relation between the acoustic output emitted inside the compartement and the acoustic pressure at the point of judgement. The type of the acoustic field in the small enclosed space of passenger cars and lorries can be determined and described only with difficulty (small volume, boundary conditions of walls with finite rigidity, absorption). The sound field can be considered, at least approximately to be a diffuse field. Then, the following relation can be written

$$P = \frac{p^2 . A}{4 \rho c_o} \quad (6)$$

where A is the absorption of the compartement.

The ratio between the oscillation of the surface of the cab or body and the noise at the point under consideration can then be expressed by the relation

$$p^2 = \frac{4 v_m^2 . (\rho c_o)^2 S.\sigma}{A} \quad (7)$$

Let us now express the transmission between v and $P_c$ in Fig. 1 as a function of time

$$p_c (t) = \sum_i \int_o^\infty \int_o^\infty k_i (\theta).r (\lambda). v_i(t-\theta-\lambda).d\theta.d\lambda \quad (8)$$

Functions $k(\theta)$ and $r(\lambda)$ respectively are Fourier transformations of functions $K(j\omega)$ and $R(j\omega)$ respectively.

The double integrals can be eliminated by a Fourier transformation for a multiport with n inputs and a single output. The input signals are correlated.

Such a solution leads to a system of n linear equations of the type (see Bibliography 1)

$$\begin{aligned} W_{1p} &= K_1 (j\omega) R(j\omega) .W_{11}+K_2 (j\omega) R(j\omega) W_{12}..+K_n (j\omega) R(j\omega) W_{1n} \\ W_{2p} &= K_1 (j\omega) R(j\omega) .W_{21}+K_2 (j\omega) R(j\omega) W_{22}..+K_n (j\omega) R(j\omega) W_{2n} \qquad (9) \\ &\cdot \\ &\cdot \\ &\cdot \\ W_{np} &= K_1(j\omega)R(j\omega).W_{n1}+K_2(j\omega)R(j\omega)W_{n2}..+K_n(j\omega)R(j\omega)W_{nn} \end{aligned}$$

where functions W are cross spectral densities of signals (according to indices).

$P_c$ and $v_1 \ldots v_n$
$W$ are complex functions of frequency $\omega$.

A measurement of transmissions by this method would be possible, but rather complicated ($\frac{n^2-n}{2}$ cross spectra, n spectra, solution of system 9. Practically it is possible to measure transmissions much more simply by utilizing the procedure outlined below.

With linear passive systems the principle of reciprocity applies. When we divide the transmission system in Fig. 1 by section v-n into a left hand and a right hand part we may understand both halves as independent multigates with a single input and n outputs or as n twoports with identical input signals, see Fig. 2 for the righthand side.

The left hand part of the substitute diagram in Fig. 1 can be measured, on the one hand, by natural signals (e.g. from the known driving force and its frequency at a four-cylinder engine), on the other hand by artificial signals.

The right hand side of the substitute diagram from Fig. 1 can be measured in a reciprocal cycle (see Fig. 2 and 3) only by the introduction of artificial signals.

For this measurement principle let us revert to relation 7 which can be rewritten, from the point of view of transmission of signals, in the form

$$p^2(\omega) = D^2 \cdot v^2(\omega) \qquad (10)$$

where function $p^2(\omega)$ is the square of the root-mean-square value of the acoustic pressure and it holds true that $p^2(\omega) \sim W_{pp}$ it holds true analogously also for $v^2(\omega)$

D is also a function of frequency

$$D^2 = \frac{4(\rho\, c_o)^2\, \sigma S}{A} \qquad (11)$$

$\rho\, c_o$ as well as S are independent of the frequency. On the other hand A and $\sigma$ are functions of frequency. The functions $p^2(\omega)$ as well as $v^2(\omega)$ are real functions of $\omega$. It must therefore also hold true that

$$D^2 = D^2(\omega)$$

In relation to equation (9) it must then hold true that

$$D_i^2(\omega) = K_i(j\omega).R(j\omega).K_i^*(j\omega).R^*(j\omega)$$

where the functions marked with an asterisk* are complex conjugate functions to the original function.

In such a case - by using the reciprocity principle - we need to solve a multiport with one input and n outputs.

Pulse Excitation, Appraisal of Responses

With artificial excitation of both the mechanical and the acoustic systems of a vehicle an exciting signal must be obtained which is sufficiently intensive in the desired frequency range for its separation from interfering signals (background) to allow processing undisturbed by random deviations. From that point of view it is expedient to process the time behaviour of the exciting pulse as well as the responses by calculation of the cross spectra eliminating non-correlated noises of the measuring apparatus and sometimes also spurious signals in the structure (microphone effect on large surfaces).

In a practical application of the working procedure outlined above the

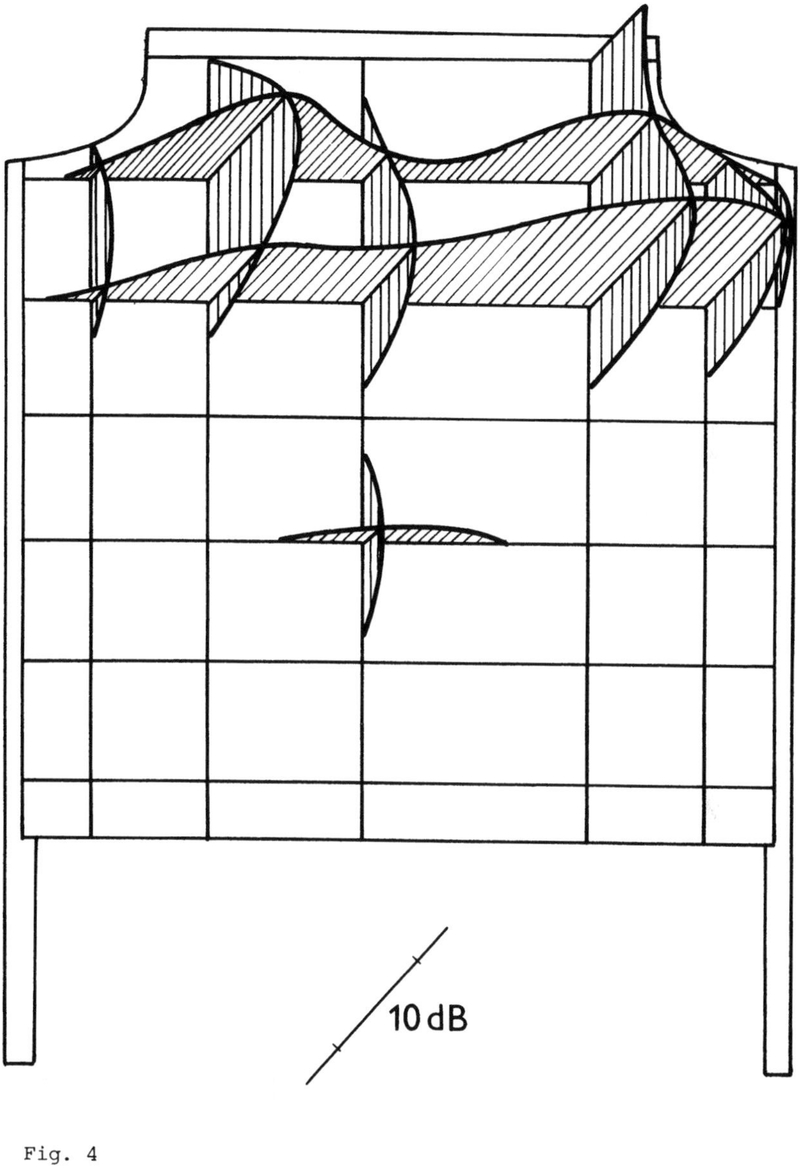


Fig. 4

following two types of exciting pulses were used:

a) a pulse of force excited by the elastic impact of a material body (striker) at the point and in the direction of action of forces really existing during the operation of the vehicle. The frequency properties of the pulse can be set by the elasticity of the point of the striker.

b) a pulse of sonic pressure, excited by an electric discharge. The shape and properties of the exciting pulse can be determined, with certain difficulties, indirectly from the calibrating measurement of the relation between the emitted acoustic output and the behaviour of the discharge current. The frequency properties of the pulse can be set by the inductance of the line to the spark gap.

For the actual measurement the surfaces surrounding the interior space of the body have to be divided by a net structure of measuring points. The density of the net must be determined by the mode of the oscillation (frequency, mass and ridigity of surfaces). In this structure transmissions have to be measured with excitation from both sides.

An example of such a measurement on the floor part of a passenger car is given in Fig. 4. First transmission measurement: force of 2nd harmonic of engine - vibrations on floor). Second transmission measurement: acoustic presure - vibrations on floor. Third measurement - point impedances on floor. From these measurements the influence of sound radiation from individual parts of the floor can be combined and assessed.

## CONCLUSION

The pulse method has fully proved its worth in the determination of parts of the surfaces emitting noise, inside the passenger compartement.

The use of reciprocity method enables transmission measurement in simple manner.

## BIBLIOGRAPHY


(1) P. Urban - Theory and methods of measurement of vibration transmission by methoanical parts (in Czech), Academia Praha 1973

(2) D. Mayer - Theory of electrical networks (in Czech), SNTL Praha 1979

# List of Participants

## A

**Apetaur, M.**, Doc. ing., Praha 3, U vodàrny 12, CSSR–130 00
**Appel, H.**, Prof. Dr.-Ing., Technische Universität Berlin, 100 Berlin 12
**Aylward, R. W.**, Dr., London Transport, Acton works, London W3 8B2, GB
**Auer, C.**, Bayerische Motoren Werke AG, 8000 München 40

## B

**van Baardwijk, P.**, ir., Volvo Car B. V., 5700 MC Helmond, The Netherlands
**Bane, O.**, Volvo Car Corporation, S-405 08 Gothenburg, Sweden
**Beermann, H.-J.**, Prof. Dr.-Ing., Technische Universität Braunschweig, 3300 Braunschweig
**Bisimis, E.**, Dr.-Ing., Fa. Alfred Teves, 6000 Frankfurt
**Böhm, F.**, Prof. Dr. techn., Institut für Mechanik, 1000 Berlin 12

## C

**Chatelle, P.**, Construction Ferroriaires et Metalliques, B-1040 Bruxelles, Belgium
**Chenchanna, P.**, Dr.-Ing., Adam Opel AG, Technische Berechnung, 6090 Rüsselsheim a.M.
**Christ, H.**, Prof. Dr.-Ing., Daimler-Benz AG, 7000 Stuttgart 60,
**Clark, R. A.**, British Rail, London Road, GB-Derby

## D

**Dahlberg, T.**, Chalmers University of Technology, S-41296 Gothenburg, Sweden
**Dalsenius, R.**, Volvo Car Corporation, S-40508 Gothenburg, Sweden
**Dödlbacher, G.**, Dipl.-Ing., Ford-Werke AG, 5000 Köln 60
**Dokainish, M. A.**, Mc Master University Hamilton, Ontario L8S–4L7, Canada
**Dorgham, M. A.**, Dr., The Open University, Walten Hall, Milton Keynes MK7 6AA, GB

## F

**Falk, E.**, Volvo Car Corporation, S 40508 Gothenburg, Sweden

## G

**Garavy, K.**, Dipl.-Ing., Technische Universität Berlin, 1000 Berlin 12
**Gasch, R.**, Prof. Dr.-Ing., Technische Universität Berlin, 1000 Berlin 10
**v. Glasner, E. C.**, Dr., Daimler-Benz AG, 7000 Stuttgart 60
**Gnadler, R.**, Prof. Dr.-Ing., Universität Karlsruhe, 7500 Karlsruhe 1
**Good, M. C.**, University of Melbourne, Victoria 3052, Australia

## H

**Hauschild, W.,** Dipl.-Ing., Technische Universität Berlin, D-1000 Berlin 12
**Hedrick, J. K.,** Prof., Massachussetts Institute of Technology, Cambridge, Mass., 02139, USA
**Heller, R.,** Clemson University, Clemson, S.C., 29631, USA
**Harada, H.,** Toyota Motor Co., Shizuoka-Ken, Susono-Shi, Mishuku 1200, J-410-11, Japan
**Havlićek, V.,** Dipl.-Ing., Research Institute of Locomotives, 19002 Praha 9, CSSR

## J

**Jaksch, F. O.,** Volvo opt 56200, Car Division, S-40508 Göteborg, Sweden
**Jaschinski, A.,** Deutsche Forschungs- und Versuchsanstalt für Luft- und Raumfahrt e. V., Institut für Dynamik der Flugsysteme, D-8031 Oberpfaffenhofen
**Johnson, D. A.,** Université Catholique de Louvain, B-1348 Louvain-la-Neuve, Belgique

## K

**Kalker, J. J.,** Dr., University of Technology, Juliana aan 132, 2628 BL Delft, Netherlands
**Keizer, C. P.,** ir., Delft University of Technology, 2600 GA Delft, Netherlands
**Knothe, K.,** Prof. Dr.-Ing., Technische Universität Berlin. 1000 Berlin 12
**Koch, J.,** Dipl.-Ing., 2411 Hollenbeck
**Koenen, C.,** Delft University of Technology, Delft, Netherlands
**Kuhla, E.,** Forschungsgemeinschaft Rad/Schiene, 6800 Mannheim 1
**Krettek, O.,** Prof. RWTH Aachen, 5100 Aachen

## L

**Law, E. H.,** Dr., Clemson University, Clemson S.C., 29631, USA
**Lemke, M.,** Dipl.-Ing., Technische Universität Berlin, 1000 Berlin 12
**Lukowski, S.,** Dr., Technical University of Wroclaw, 53-312 Wroclaw, Poland
**Lyon, D.,** British Rail, Wilmorton Derby DE2 8UP, GB

## M

**MacAdam, C. C.,** University of Michigan, Ann Arbor, Michigan 48109, USA
**Matsui,** Tokyu Car Corporation, Yokohoma, 236 Japan
**Meinke,** Dr.-Ing., MAN, Neue Technologie, 8000 München 50
**Michels, W.,** Dipl.-Ing., Fried. Krupp GmbH, 4300 Essen
**Mitschke, M.,** Prof. Dr.-Ing., Technische Universität Braunschweig, 3300 Braunschweig

**Movelle, D.,** Dipl.-Ing., Technische Universität Berlin, 1000 Berlin 12
**Müller, P. C.,** Prof. Dr., Technische Universität München, 8000 München 2

## N

**Nagai, M.,** Dr. of Eng., Tokyo University of Agriculture and Technology, Tokyo 184, Japan
**Nordström, O.,** National Swedish Road and Traffic, S-58101 Linköping, Sweden

## P

**Pacejka, H. B.,** Dr. ir., Delft University of Technology, NL-2600 GA Delft, Netherlands
**Panik, F.,** Dr.-Ing., Daimler-Benz AG, 7000 Stuttgart 60
**Pauly, A.,** Dr., Bayerische Motoren Werke AG, 8000 München 40
**de Pater, A. D.,** Prof. Dr. Ir., Delft University of Technology, NL-2628 Delft, Netherlands
**Popp, K.,** Dr.-Ing., habil., Technische Universität München, 8000 München 2

## R

**Richter, B.,** Dr.-Ing., Volkswagenwerk AG, 3180 Wolfsburg
**Richter, R.,** Deutsche Forschungs- und Versuchsanstalt für Luft- und Raumfahrt e. V. – DFVLR, Oberpaffenhofen, 8031 Weßling
**Rönitz, R.,** Dipl.-Ing., Adam Opel AG, 6090 Rüsselsheim
**Rompe, K.,** Dr., Technischer Überwachungs-Verein, Rheinland e.V., 5000 Köln 1
**Ruf, G.,** Dipl.-Ing., Dr.-Ing. h.c. F. Porsche AG, 7251 Weissach 1
**Rus, L.,** Dr.-Ing., CKD Praha, Závod Lokomotivka, 19002 Praha 9, CSSR

## S

**Sachs, H. K.,** Prof. Dr.-Ing., Wayne State University, Detroit, Michigan 48202, USA
**Saito, Y.,** Prof. Dr., Tokyo University of Agriculture and Technology, Koganei-shi, Tokyo, Japan
**Samin, J. C.,** Universite Catholique de Lauvain, B-1348 Lauvain, Belgium
**Sayers, M.,** University of Michigan, Ann Arbor, Michigan 48109, USA
**Segel, L.,** Prof., The University of Michigan, Ann Arbor, Michigan 48109, USA

**Sharp, R. S.,** Dr., University of Leeds, Leeds LS2 9JT, GB
**Simic, D.,** Prof. Dr., University of Belgrade, 34000 Kragujevac, Jugoslavia
**Slibar, A.,** Prof. Dipl.-Ing. Dr. techn., Technische Universität Wien A-1040 Wien, Austria
**Sprinc, J.,** Czechoslovak Academy of Sciences, 12849 Praha 2, CSSR
**Sweet, L. M.,** Prof., Princton University, Department of Mechanical and Aerospace Engineering, The Engineering Quadrangle, Princeton, N. J. 08544, USA
**Schiehlen, W. O.,** Prof. Dr.-Ing., Universität Stuttgart, 7000 Stuttgart 80
**Schutzbach, H.,** GST Gesellschaft für Systemtechnik, 4300 Essen

## W

**Wache, D.,** Dipl.-Ing., Bayerische Motoren Werke AG, 8000 München 40
**Wickens, A. H.,** Dr. Dir., British Rail, Wilmorton Derby DE2 8UP, Derby 49203 Extn. 4203, GB
**Wiese, J.,** Dipl.-Ing., Deutsche Eisenbahn Consulting GmbH, 6000 Frankfurt/Main 70
**Willumeit, H.-P.,** Prof. Dr.-Ing., Technische Universität Berlin, 1000 Berlin 12

## Z

**Zabadneh, M. K.,** Dr.-Ing., Daimler-Benz AG, Gutekantstr. 6, 7251 Heimsheim

Printed and bound by CPI Group (UK) Ltd, Croydon, CR0 4YY
23/10/2024
01777667-0018